Physical Supramolecular Chemistry

NATO ASI Series

Advanced Science Institutes Series

A Series presenting the results of activities sponsored by the NATO Science Committee, which aims at the dissemination of advanced scientific and technological knowledge, with a view to strengthening links between scientific communities.

The Series is published by an international board of publishers in conjunction with the NATO Scientific Affairs Division

A Life Sciences **B Physics**	Plenum Publishing Corporation London and New York
C Mathematical and Physical Sciences **D Behavioural and Social Sciences** **E Applied Sciences**	Kluwer Academic Publishers Dordrecht, Boston and London
F Computer and Systems Sciences **G Ecological Sciences** **H Cell Biology** **I Global Environmental Change**	Springer-Verlag Berlin, Heidelberg, New York, London, Paris and Tokyo

PARTNERSHIP SUB-SERIES

1. Disarmament Technologies	Kluwer Academic Publishers
2. Environment	Springer-Verlag / Kluwer Academic Publishers
3. High Technology	Kluwer Academic Publishers
4. Science and Technology Policy	Kluwer Academic Publishers
5. Computer Networking	Kluwer Academic Publishers

The Partnership Sub-Series incorporates activities undertaken in collaboration with NATO's Cooperation Partners, the countries of the CIS and Central and Eastern Europe, in Priority Areas of concern to those countries.

NATO-PCO-DATA BASE

The electronic index to the NATO ASI Series provides full bibliographical references (with keywords and/or abstracts) to more than 50000 contributions from international scientists published in all sections of the NATO ASI Series.
Access to the NATO-PCO-DATA BASE is possible in two ways:

– via online FILE 128 (NATO-PCO-DATA BASE) hosted by ESRIN, Via Galileo Galilei, I-00044 Frascati, Italy.

– via CD-ROM "NATO-PCO-DATA BASE" with user-friendly retrieval software in English, French and German (© WTV GmbH and DATAWARE Technologies Inc. 1989).

The CD-ROM can be ordered through any member of the Board of Publishers or through NATO-PCO, Overijse, Belgium.

Series C: Mathematical and Physical Sciences – Vol. 485

Physical Supramolecular Chemistry

edited by

Luis Echegoyen

and

Angel E. Kaifer

Department of Chemistry,
University of Miami
Coral Gables, Florida, U.S.A.

Kluwer Academic Publishers

Dordrecht / Boston / London

Published in cooperation with NATO Scientific Affairs Division

Proceedings of the NATO Advanced Research Workshop on
Physical Supramolecular Chemistry
Miami, U.S.A.
January 7–10, 1996

A C.I.P. Catalogue record for this book is available from the Library of Congress.

ISBN 978-94-010-6628-0 e-ISBN-13:978-94-009-0317-3
DOI: 10.1007/978-94-009-0317-3

Published by Kluwer Academic Publishers,
P.O. Box 17, 3300 AA Dordrecht, The Netherlands.

Kluwer Academic Publishers incorporates the publishing programmes of
D. Reidel, Martinus Nijhoff, Dr W. Junk and MTP Press.

Sold and distributed in the U.S.A. and Canada
by Kluwer Academic Publishers,
101 Philip Drive, Norwell, MA 02061, U.S.A.

In all other countries, sold and distributed
by Kluwer Academic Publishers Group,
P.O. Box 322, 3300 AH Dordrecht, The Netherlands.

This book contains the proceedings of a NATO Advanced Research Workshop held within the programme of activities of the NATO Special Programme on Supramolecular Chemistry as part of the activities of the NATO Science Committee.

Other books previously published as a result of the activities of the Special Programme are:

WIPFF, G. (Ed.), *Computational Approaches in Supramolecular Chemistry.* (ASIC 426) 1994. ISBN 0-7923-2767-5

FLEISCHAKER, G.R., COLONNA, S. and LUISI, P.L. (Eds.), *Self-Production of Supramolecular Structures.* From Synthetic Structures to Models of Minimal Living Systems. (ASIC 446) 1994. ISBN 0-7923-3163-X

FABBRIZZI, L., POGGI, A. (Eds.), *Transition Metals in Supramolecular Chemistry.* (ASIC 448) 1994. ISBN 0-7923-3196-6

BECHER, J. and SCHAUMBURG, K. (Eds.), *Molecular Engineering for Advanced Materials.* (ASIC 456) 1995. ISBN 0-7923-3347-0

LA MAR, G.N. (Ed.), *Nuclear Magnetic Resonance of Paramagnetic Macromolecules.* (ASIC 457) 1995. ISBN 0-7923-3348-9

SIEGEL, JAY S. (Ed.), *Supramolecular Stereochemistry.* (ASIC 473) 1995. ISBN 0-7923-3702-6

WILCOX, C.S. and HAMILTON A.D. (Eds.), *Molecular Design and Bioorganic Catalysis.* (ASIC 478) 1996. ISBN 0-7923-4024-8

MEUNIER, B. (Ed.), *DNA and RNA Cleavers and Chemotherapy of Cancer and Viral Diseases.* (ASIC 479) 1996. ISBN 0-7923-4025-6

KAHN, O. (Ed.), *Magnetism: A Supramolecular Function.* (ASIC 484) 1996. ISBN 0-7923-4153-8

Table of Contents

INTRODUCTION

The field of Supramolecular chemistry has reached a high level of sophistication and maturity, especially in the last 5-10 years. The sophistication is primarily evident in the highly complex chemical architectures that have been synthesized to accomplish specific functions such as selective binding and transport of substrates, in catalysis, and for the formation of assemblies, to name just a few of the applications. Many other molecular structures have been synthesized to form unusual materials with interesting and/or useful properties. From this point of view, the field is mature, as would have been anticipated from its historical development, starting in 1967 with Pedersen and the crown ethers.

Since that time, the field has evolved mainly from a synthetic organic perspective. There are notable exceptions to this statement, but in general it holds true that the major advances were made in the laboratories of organic chemists with considerable topological insight. The design and synthesis of the molecules was also partly guided by physical insight provided by measurements and calculations done by physical chemists. As a consequence, many chemical systems were synthesized in relatively small amounts to afford their full characterization but not in large enough quantities to fully explore their intended properties or structures. Thus in the decades of the 70's and 80's physical chemists played an important role in guiding the future directions in the field, providing needed structural and energetic information and providing improved experimental designs.

As the supramolecular systems have progressed in structural complexity and their functional properties have become more esoteric, the routine physical methods of characterization have quickly become outdated. It is thus necessary to have a more direct input from physical chemists who are capable of generating new methods or modifying old ones to make the necessary measurements to establish structural and functional characteristics of the materials being synthesized. As an example, the sizes of supramolecular assemblies have grown at an almost exponential rate in recent years. Standard mass spectral and NMR techniques no longer suffice to fully characterize these structures in detail. New and sensitive techniques, many borrowed from biophysical laboratories, are constantly being applied in the field of supramolecular chemistry. Probing the structures, dynamics, and energetics of these complex molecular architectures is an important frontier that continues to change as the level of complexity of the systems studied also continues to increase. More recent techniques, such as STM and AFM are offering detailed structural information of supramolecular systems and, perhaps more importantly, affording new ways to actually construct novel architectures on surfaces. The latter are rapidly becoming powerful techniques to construct functional assemblies on selected surfaces with atomic resolution.

It thus seemed appropriate to hold an ARW in the area of Physical Supramolecular chemistry. Many other such international workshops and symposia are held every year which concentrate on the synthetic aspects of the field. Novel structures are presented at these along with some physical characterization data. However, the central theme is invariably a synthetic one, based on what are mainly synthetic approaches. The idea behind the present symposium was to provide a forum for those practitioners in the field to show what is possible today in their respective specialties and to show the way for future developments. Areas such as calorimetry, mass spectroscopy, and X-ray crystallography have been and continue to be of fundamental importance in the characterization of supramolecular assemblies. But even these well established techniques need to meet the new challenges imposed by the growing complexity of the structures.

Perhaps more importantly, the workshop was designed to provide the perspective of the physical chemist's point of view in the field of supramolecular chemistry. While many problems can be solved synthetically by an appropriate structural modification, the right physical measuring technique may preclude the need for what sometimes is tedious and laborious synthetic work. It is therefore expected that the workshop will develop this physical approach, as a way of doing supramolecular chemistry.

Since the number of physical chemists interested in supramolecular topics seems somewhat small, certainly less than the corresponding number of organic chemists in the field, the need for this workshop seems evident. More competent practitioners in the field are needed. The field is sufficiently mature that the time to explore its physical chemistry in detail is here and now.

From the list of speakers, it should be evident that most areas of physical chemistry that explore supramolecular chemistry were represented. These include: theory, calorimetry, photochemistry, electrochemistry, NMR spectroscopy, X-ray crystallography, mass spectrometry, monolayers and membranes, STM and AFM, and magnetic, optical, and electrical properties. At least two people in each sub-area were selected in order to provide differing views and to complement each other's presentation.

As this is written, it is possible to say that the workshop was a resounding success. The novel approach to supramolecular topics, the mix of the speakers and the friendly and stimulating atmosphere were praised uniformly by the participants and speakers.

We wish to thank Dr. Alain Jubier of the NATO Scientific Affairs Division for his valuable support. We are grateful to the NATO Special Program in Supramolecular Chemistry which provided most of the funding for the workshop. We also acknowledge the important financial contributions from the University of Miami.

It would be unfair to close the acknowledgements without thanking the members of our research groups and staff whose effort and dedication contributed in many important ways to the success of the workshop. Their names are: Pierre L. Boulas, René Castro, Luis Godínez, Marielle Gómez-Kaifer, Donna Johnston, Sandra Mendoza, Eduardo Pérez-Cordero, Matthew Repasky, Miriam Richardson, Richard Sachleben, Ruth Signorile and Betty Vilaboy.

Luis Echegoyen and Angel E. Kaifer
Miami, U.S.A.
April 1996

Organizing Committee

Luis **Echegoyen**
Department of Chemistry
University of Miami
Coral Gables, FL 33124-0431, USA

Angel **Kaifer**
Department of Chemistry
University of Miami
Coral Gables, FL 33124-0431, USA

Vicenzo **Balzani**
Dipartamento of Chimica
Universita deli Studi di Bologna
40125 Bologna, ITALY

David **Reinhoudt**
Department of Chemistry
University of Twente
7500 AE Enschede, THE NETHERLANDS

List of Speakers and Participants

Héctor **Abruña** (S)

Department of Chemistry
Baker Laboratory
Cornell University
Ithaca, N.Y. 14853, USA

D. B. **Amabilino** (P)

Laboratoire de Chimie Organo-Minerale
6 eme Sud - Institut Le Bel
Universite Louis Pasteur
4, rue Blaise Pascal
Strasbourg, 67070, FRANCE

Claudia **Arana** (P)

Dade International, Inc.
P.O. Box 520672
Miami, FL 33152-0672, USA

Jerry **Atwood** (S)

Department of Chemistry
University of Missouri-Columbia
601 S. College Avenue
Columbia, Missouri 65211, USA

Vincenzo **Balzani** (S)

Universita deli Studi di Bologna
Dipartamento di Chimica
40125, Bologna, ITALY

Paul **Baxter** (P)

Laboratoire de Chimie Supramoleculaire
Institut Le Bel
Universite Louis Pasteur
Strasbourg, FRANCE

Andrew C. **Benniston** (P)

Department of Chemistry
The University of Glasgow
Glasgow, G12 8QQ
Scotland, (UNITED KINGDOM)

Richard **Bissell** (P)

SOMS Centre
Chemistry Building
University of Leeds
Leeds, LS2 9JT, UNITED KINGDOM

Cornelia **Bohne** (P)

Department of Chemistry
University of Victoria
P.O. Box 3055
Victoria, B.C., V8W 3P6, CANADA

Pierre Louis **Boulas** (P)
Department of Chemistry
University of Miami
P.O. Box 249118
Miami, FL 33124-4571, USA

Jennifer **Brodbelt** (S)
Department of Chemistry & Biochemistry
University of Texas
Austin, Texas 78712, USA

Sebastiano **Campagna** (P)
Dipartimento di Chimica Inorganica
Universita di Messina
via Sperone 31
Vill. S. Agata
Messina, 98166, ITALY

Carmen **Casado** (P)
Department of Chemistry
Baker Laboratory
Cornell University
Ithaca, NY 14853-1301, USA

Alessandro **Casnati** (P)
Dipartimento di Chimica Organica e Industriale
Universita degli Studi
Viale delle Scienze
Parma, I-43100, ITALY

René **Castro** (P)
Department of Chemistry
University of Miami
1301 Memorial Dr.
Coral Gables, FL 33124, USA

Dana **Caulder** (P)
Department of Chemistry
UC Berkeley
c/o Ken Raymond
Berkeley, CA 94720, USA

Luisa **De Cola** (P)
Dipartimento di Chimica "G. Ciamician"
via Selmi, 2
Bologna, 40126, ITALY

David **Dearden** (S)
Department of Chemistry
Brigham Young University
225 Eyring Science Center
Provo, Utah 84602, USA

Pierre **Delhaes** (S)
Centre de Recherche Paul Pascal
Chateau Brivazae
Abenue A. Schweitzer
33600, Pessac Cedex, FRANCE

Christian **Detellier** (S)
Ottawa-Carleton Chemistry Institute
Ottawa University Campus
Ottawa
Ontario, KIN 6N5, CANADA

James L. **Dye** (S)
Department of Chemistry
Michigan State University
320 Chemistry Building
East Lansing, Michigan 48824, USA

Kenneth B. **Eisenthal** (S)
Department of Chemistry
Room 344 Havenmeyer Hall
Columbia University
New York, NY 10027, USA

Jeffrey D. **Evanseck** (S)
Department of Chemistry
University of Miami
315 Cox Science Building
Coral Gables, Florida 33124-0431, USA

Luigi **Fabbrizzi** (S)
Dipartimento di Chimica Generale
Universita di Pavia
via Taramelll 12
Pavia, 27100, ITALY

Mary Anne **Fox** (S)
Department of Chemistry
The University of Texas at Austin
Austin, TX 78712-1167, USA

Thomas M. **Fyles** (S)
Department of Chemistry
University of Victoria
P.O. Box 3055, Victoria,
British Columbia, V8W 3P6, CANADA

N. **García** (S)
Departmento de Física Fundamental
Universidad Autónoma de Madrid
E-28049
Madrid, SPAIN

Miguel A. **García-Garibay** (S)
Department of Chemistry
University of California at Los Angeles
405 Hilgard Ave.
Los Angeles, CA 90024-1569, USA

Giovanna **Ghirlanda** (P)
Centro Meccanismi Reazioni Organiche
Via Marzolo NR.1
Padova, 35100, ITALY

Luis **Godínez** (P)
Department of Chemistry
University of Miami
Coral Gables, FL 33124, USA

Marielle **Gómez-Kaifer** (P)
Department of Chemistry
University of Miami
Coral Gables, FL 33124, USA

Anthony **Harriman** (S)
Laboratoire de Photochimie
E.H.I.C.S.
1 rue Blaise Pascal
Strasbourg, 67008, FRANCE

Bernold **Hasenknopf** (P)
Laboratiore de Chimie Supramoleculaire
Universite Louis Pasteur
Strasbourg, 67070, FRANCE

Valerie **Heitz** (P)
Laboratoire de Chimie Organo-Minerale
Institut Le Bel
4, rue Blaise Pascal
Strasbourg, 67070, FRANCE

Thomas **Hofler** (P)
Polymerinstitut
Uni Karlsruhe
Hertzstr 16
Karlsruhe, 76187, GERMANY

Reed **Izatt** (S)
Brigham Young University
225 ESC
Provo, Utah 84602-4672, USA

William L. **Jorgensen** (S)
Department of Chemistry
Yale University
New Haven, Conneticut 06511, USA

Roger **Leblanc** (S)

Department of Chemistry
University of Miami
315 Cox Science Building
Coral Gables, Florida 33124-0431, USA

Yi **Li** (P)

Coulter Corporation
P.O. Box 169015
MC 22-A03
Miami, FL 33116-9015, USA

Maurizio **Licchelli** (P)

Dipartimento di Chimica Generale
Via Taramelli 12
Pavia, I-27100, ITALY

Mathai **Mammen** (P)

Department of Chemistry
Harvard University
12 Oxford St.
Cambridge, MA 02138, USA

Ilse **Manet** (P)

Laboratorio di Fotochimica
Dipartimento di Chimica "G. Ciamician"
Bologna, 40126, ITALY

Michael **Maskus** (P)

Department of Chemistry
Baker Laboratory
Cornell University
Ithaca, NY 14853-1301, USA

Sandra **Mendoza** (P)

Department of Chemistry
University of Miami
P.O. Box 249118
Coral Gables, FL 33124, USA

Eduardo E. **Pérez-Cordero** (P)

Department of Chemistry
University of Miami
P.O. Box 249118
Coral Gables, FL 33124, USA

Maria Luisa **Pérez-Garcia** (P)

Laboratorio de Química Orgánica
Facultad de Farmacia
Universidad de Barcelona
Av. de Joan XXIII, s/n
Barcelona, E-08028, SPAIN

Kaliappa G. **Ragunathan** (P)
FR Organische Chemie
Universitat Des Saarlandes
Saarbrucken, D-66123, GERMANY

David N. **Reinhoudt** (S)
Department of Chemistry
University of Twente
P.O. Box 217
7500 AE
Enschede, THE NETHERLANDS

Matthew **Repasky** (P)
Department of Chemistry
University of Miami
1301 Memorial Dr.
Coral Gables, FL 33124, USA

Ferenc **Reti** (P)
Department of Atomic Physics
Technical University of Budapest
H-1111 Budafoki ut 8.
Budapest, HUNGARY

Miriam **Richardson** (P)
Department of Chemistry
University of Miami
1301 Memorial Dr.
Coral Gables, FL 33124, USA

Richard **Sachleben** (P)
Department of Chemistry
University of Miami
1301 Memorial Dr.
Coral Gables, FL 33124, USA

Wolfram **Saenger** (S)
Institute fur Kristallographie
Freie Universitat Berlin
Takustrasse 6, D-14195
Berlin, GERMANY

Joachim **Sartorius** (P)
Universitat Des Saarlandes
FR Organische Chemie
Saarbrucken, D-66041, GERMANY

Martin **Saunders** (S)
Department of Chemistry
Yale University
225 Prospect Street
P.O.Box 6666
New Haven, Connecticut 06511, USA

Hans-Jorg **Schneider** (S)	Universitat des Saarlandes FR Organische Chemie D-6600 Saarbrucken, 11, GERMANY
D. **Schoemaker** (P)	Physics Department Universitaire Instelling Antwerpen Universiteitsplein 1 Antwerpen (Wilrijk), B-2610, BELGIUM
Daniel **Talham** (S)	Department of Chemistry University of Florida Gainsville, FL 32611-7200, USA
Nongjian **Tao** (P)	Physics Department Florida International. University Miami, FL 33119, USA
Rodica Paula **Turcu** (P)	Institute of Isotopic & Molecular Technology P.O. Box 700 Cluj-Napoca, R-3400, ROMANIA
Valeria **Van Axel Castelli** (P)	Dipartimento di Chimica Universita "La Sapienza" P.le Aldo Moro, 5 Roma, 185, ITALY
Carol A. **Venanzi** (S)	Department of Chemical Engineering Chemistry and Environmental Sciences New Jersey Institute of Technology University Heights Newark, NJ 07102-1982, USA
Diana **Watkins** (P)	Department of Chemistry Columbia University P.O. Box 861 Havemeyer Hall New York, NY 10027, USA
Karl **Wieghardt** (S)	Lehrstuhl fur Anorganische Chemie I der Universitat D-44780 Bochum, GERMANY

Craig **Wilcox** (S)
Department of Chemistry
University of Pittsburgh
Pittsburgh, Pennsylvania 15260, USA

Alan F. **Williams** (S)
Department of Inorganic,
Analytical & Applied Chemistry
University of Geneva
CH-1211
Geneva, 4, SWITZERLAND

George **Wipff** (S)
URA 422 CNRS
Institute de Chimie
4, rue Balise Pascal
67000
Strasbourg, FRANCE

Qingshan **Xie** (P)
Department of Chemistry
Michigan State University
East Lansing, MI 48824, USA

Renliang **Xu** (P)
Scientific Instruments
Coulter Corporation
P.O. Box 169015
MC 19500
Miami, FL 33116-9015, USA

Nelsi **Zaccheroni** (P)
Universita di Bologna
Dipartimento di Chimica "G. Ciamician"
Via Selmi 2
Bologna, 40126, ITALY

SUPRAMOLECULAR STRUCTURES FOR SWITCHING AND SENSING

David N. Reinhoudt*, André M. A. van Wageningen, Bart-Hendrik Huisman
Laboratory of Organic Chemistry, University of Twente
P.O. Box 217, 7500 AE Enschede, The Netherlands

Abstract

The synthesis of calix[4]arene-based carceplexes via two different methods is described, as well as a method for modifying the behavior of guests after incarceration. Furthermore, the sensing abilities of self-assembled monolayers of receptor molecules on gold is discussed. The expertise of the two different fields is combined for the development of a molecular switch.

1. Introduction

In our group new receptor molecules have been synthesized via the combination of known building blocks, calix[4]arenes and resorcin[4]arenes. Furthermore, self-assembled monolayers have been prepared using cavitands and calix[4]arenes. In this review it is shown that combining the knowledge concerning the synthesis of receptor molecules

L. Echegoyen and A.E. Kaifer (eds.), Physical Supramolecular Chemistry, 1–9.

with a well defined cavity and the preparation and characterization of self-assembled monolayers can be used for the development of new molecular switches.

2. Calix[4]arene-based Carcerands

Cram *et al.* [1] have shown that resorcin[4]arene-based carcerands can permanently incarcerate guest molecules. Although incarcerated guests can adopt different orientations this does not lead to different stereoisomers due to the symmetry of the carcerand. Combination of calix[4]arenes with resorcin[4]arenes leads to calix[4]arene-based carcerands **2** which possess a non-symmetric cavity [2]. Therefore different orientations of incarcerated guests lead to different diastereoisomers. This makes these molecules of interest because of their potential use as molecular switches. Due to the flexibility of the calix[4]arene skeleton direct coupling between a tetra-functionalized calix[4]arene and a tetrol cavitand does not result in the formation of a calix[4]arene-based carcerand. Therefore, a new method for the introduction of amino groups from iodo-substituted calix[4]arenes was developed [3]. Reaction of 1,2-bis(chloroacetamido)-3,4-

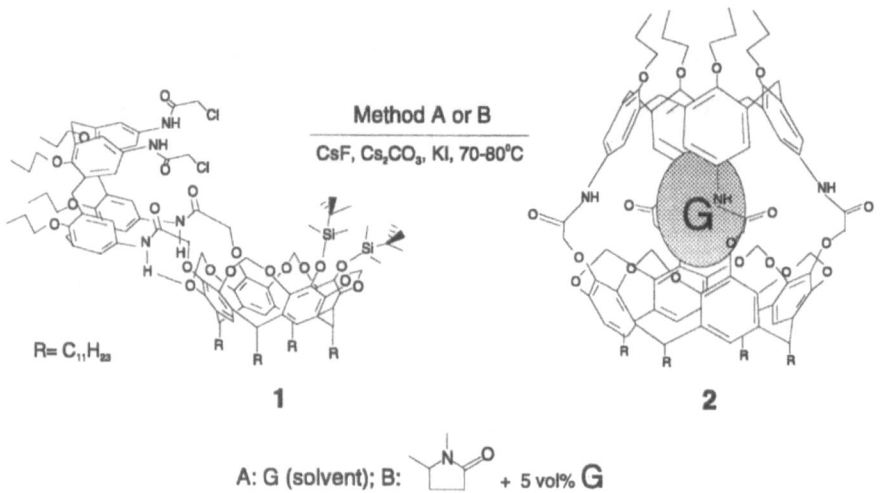

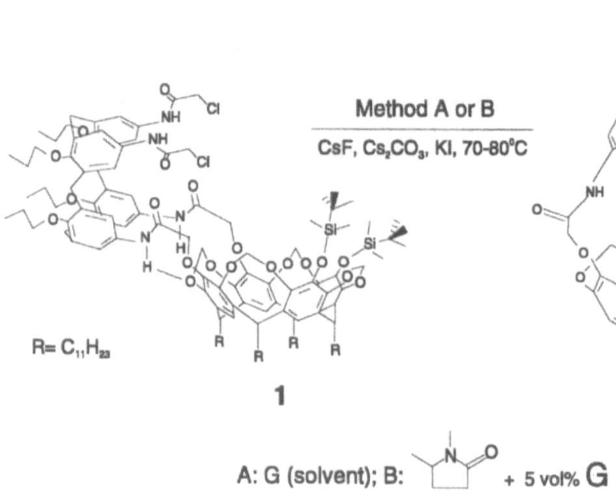

Scheme 1

dinitrocalix[4]arene with tetrol-resorcin[4]arene predominantly leads to an 1:1 *endo* coupled product. This preference for the *endo* orientation is probably a result of electrostatic interactions between the nitro groups on the calix[4]arene and the hydroxyl groups on the resorcin[4]arene. The 1:1 coupled product is converted into **1** via reduction of the remaining nitro groups and reaction with chloroacetyl chloride. This compound can be used for the synthesis of calix[4]arene-based carceplexes either via *solvent* or *doped* inclusion.

2.1. SYNTHESIS VIA SOLVENT INCLUSION

One method for the synthesis of calix[4]arene-based carceplexes comprises the formation of the final two bridges in an appropriate solvent (method A). During this reaction one solvent molecule is permanently incarcerated. Only highly polar solvents such as amides and sulphoxides can be used, *e.g.* DMF and ethyl methyl sulphoxide, respectively. Increasing the size of the solvent going from DMF to 1,5-dimethyl-2-pyrrolidinone leads to a dramatic decrease of the yield from quantitative to only 5%, respectively [4].

2.2. SYNTHESIS VIA DOPED INCLUSION

The possibility to use only highly polar solvents for the synthesis of calix[4]arene-based carceplexes limits the number of potential guests. Therefore we developed a synthesis method called *doped* inclusion (method B) [4]. By using 1,5-dimethyl-2-pyrrolidone as a solvent, which itself is a bad template for the closure reaction, and adding potential guests, *e.g.* 2-butanone or 3-sulfolene (5-10 vol%) the carceplex with the added guest is exclusively formed. The ability to synthesize carceplexes via doped inclusion enlarges the scope of potential guests to solids, *e.g.* 3-sulfolene, and deuterated guests, *e.g.* DMF-d_7 and DMSO-d_6.

2.3. MODIFICATION OF THE BRIDGES AFTER INCARCERATION OF GUESTS

The orientation of the guests inside the carcerand was studied by 2D NOESY and 2D ROESY NMR spectroscopy. With exception of *N,N*-dimethyl acetamide and *N*-methyl-2-pyrrolidone the guests either rotate fast on the NMR chemical shift timescale (DMF, DMSO) or preferably adopt one orientation in a temperature range of -40° to 100° C (ethyl methyl sulfoxide, 2-butanone). Therefore it is interesting to have a method for changing the behaviour of guests after incarceration. An apparent position for modification are the amide bridges between the resorcin[4]- and calix[4]arene moiety. The amide bridges in carceplexes **2** could be converted into thiamides in essentially quantitative yields using Lawessons reagent in refluxing xylene giving thiacarceplexes **3** [4]. The incarcerated guests *do not react* which means that they are not reactive under the reaction conditions. This indicates that the guests are very well shielded.

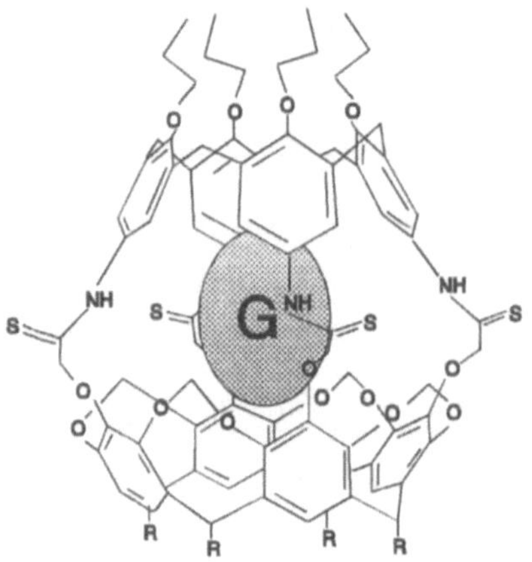

3

Chart 1

The energy barriers ($\Delta G^{\neq}$) for the interconversion between the different diastereoisomers were determined by 2D EXSY NMR spectroscopy. As shown in TABLE 1 the conversion of the amide bridges into thioamides results in an increase of the energy barriers for interconversion by 1.8-2.5 kcal mol^{-1}. An explanation for this behaviour may be the stronger hydrogen bond donating character of thioamides compared to amides. Furthermore, molecular modelling studies revealed that in the calix[4]arene-based carceplexes with thioamide bridges the distance between the diametrical NH protons (across the cavity) is significantly smaller [5].

TABLE 1. Energy barriers ($\Delta G^{\neq}_{273}$) for interconversion between different diastereoisomers of guests inside calix[4]arene-based carcerands determined by 2D EXSY NMR spectroscopy in $CDCl_3$ at 273 K and distance between diametrical NH protons determined using molecular mechanics calculations.

Guest	Bridge	$\Delta G^{\neq}_{273}$ (kcal/mol)	d (NH..HN) (Å)
N,N-Dimethylacetamide	Amide	12.7 ± 0.5	9.29 ± 073
	Thioamide	15.2 ± 0.5	8.56 ± 0.24
N-Methyl-2-pyrrolidinone	Amide	15.7 ± 0.5	9.60 ± 0.69
	Thioamide	17.5 ± 0.5	8.85 ± 0.51

3. Self-Assembled Monolayers on Gold

Self-assembled monolayers have been investigated extensively during the past decade [6], in particular because of the potential applications, like in piezoelectrical devices, nonlinear optics, and microsensors [7]. Highly ordered monolayers were obtained of resorcin[4]arene [8] and calix[4]arene [9] derivatives, substituted with four dialkylsulfides for the adsorption to gold as depicted in Chart 2. The driving force for the formation of a monolayer is the Au-S interaction, while van der Waals forces between the alkyl chains provide a high degree of order. Essential in the design of the adsorbates is that the alkyl chains cover a space which is comparable in size to the area covered by the 'arene' unit [10].

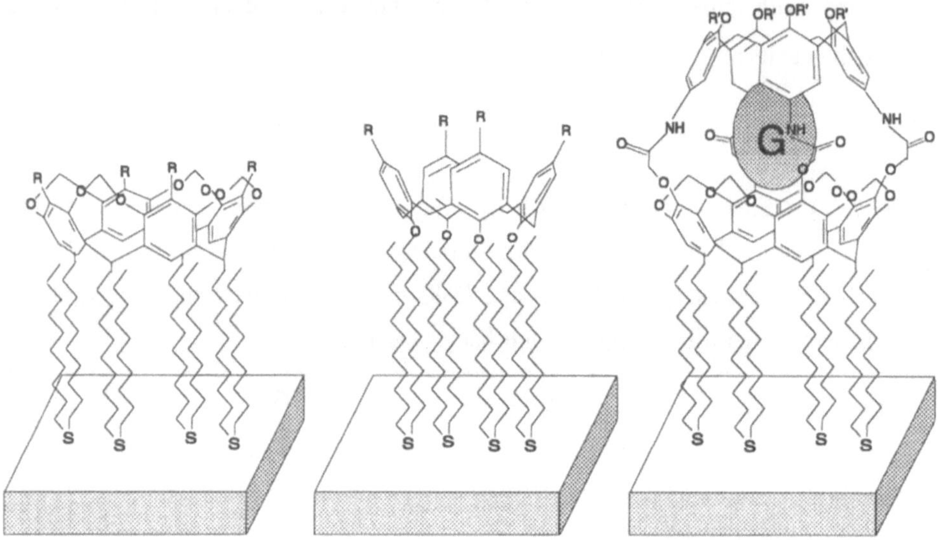

4 5 6

Chart 2

3.1. MONOLAYERS AS SENSING INTERFACE

In the highly ordered monolayers of cavitand **4** the cavity is exposed to the outer interface. It was shown in sensing experiments that in contact with various (halogenated) hydrocarbons the cavity acts as a molecular recognition site. The specific interactions were detected both by the use of a quartz crystal microbalance (QCM) [11] and surface plasmon resonance (SPR) measurements [12]. The cavity effect was further evidenced by comparison with an inert disulfide, for which a ten times smaller response was obtained compared to the monolayer with the cavity. Due to the short diffusion lengths, very fast response times could be obtained (t_{90} < 1 sec). Both the QCM and SPR are very sensitive techniques, of which SPR appears to be more stable in time. Therefore,

SPR enabled us to study the adsoption behavior more closely, to find that an increasing reponse occurs upon adsorption. As the sensing results with these layers is very promising, new recognition sites are under development. The calix[4]arene self-assembled monolayers (**5**, R = H, *t*-butyl) are important in this regard, as the modification of these compounds is more advanced than of resorcin[4]arenes.

3.2. DEVELOPMENT OF SWITCHING MONOLAYER

The experience with self-assembled monolayers was recently combined with the calix[4]arene-based carcerands. The carcerand can be oriented in space in a controlled manner by assembling this compound in a monolayer on gold. Thus the orientation of the carcerand is fixed, while the guest remains free to be switched independently between two different orientations. As appeared from the systematic study of guests described before, different energy barriers between the two states can be obtained, depending on the size of the cavity and of the incarcerated guest. Self-assembled monolayers of carceplex **6** (synthesized via method A with DMF as guest as shown in Scheme 1, R = $C_{10}H_{20}$-S-$C_{10}H_{21}$) were prepared from a 0.1 mM solution in ethanol on gold [13]. The monolayers were characterized by a number of surface sensitive techniques (like FTIR, ellipsometry, contact angle measurement, XPS, and SIMS). Well-ordered monolayers were obtained, for which a structure is proposed as depicted in Chart 2.

The monolayer infrared spectrum provides clear information about the order and orientation of the carceplex **6.DMF** on the monolayer (Table 2). The methylene C-H stretches have the same absorption maxima in the monolayer as in the bulk, which indicates a high degree of order. The amides in the carceplex appear in the bulk infrared spectrum as amide I (predominantly C=O stretch) and amide II. In the monolayer spectrum, however, only amide II appears at 1537 cm^{-1} as a clear absorption, while amide I is very weak or absent. The orientation of the transition dipole of amide I is in the proposed monolayer structure parallel to the gold, and therefore is not absorbed by the (p-polarized) infrared beam. This confirms that the orientation of the carceplex **6.DMF** is 'upright', similar to the previously reported adsorbates.

TABLE 2. Selected infrared peak positions for carceplex 2 in the bulk and the monolayer.

vibration	bulk [cm^{-1}]	monolayer [cm^{-1}]
$-CH_2-$, symmetrical	2853	2853
$-CH_2-$, asymmetrical	2924	2926
Amide I	1696	n.o.
Amide II	1526	1537

4. Summary

It has been shown that calix[4]arene-based carceplexes can be obtained *via* solvent or doped inclusion. Furthermore, the behavior of incarcerated guests can be altered after incarceration by modification of the bridges between the resorcin[4]- and calix[4]arene moiety. Self-assembled monolayers are an elegant tool to control molecular orientation. Thus resorcin[4]arene monolayers were formed, for which sensing applications based on molecular recognition were shown. The calix[4]arene-based carceplex was confined in space by means of self-assembly to gold. This approach is a step toward a new type of molecular switch.

References and notes

1. D. J. Cram and J. M. Cram, *Container Molecules and their Guests* in *Monographs in Supramolecular Chemistry Vol. 4*, ed. J. F. Stoddart, Royal Society of Chemistry, Cambridge (1994).

2. Timmerman, P.; Verboom, W.; van Veggel, F. C. J. M.; van Duynhoven, J. P. M.; Reinhoudt, D.N. *Angew. Chem. Int. Ed. Engl.* **1994**, *33*, 2345.

3. Timmerman, P.; Verboom, W.; Reinhoudt, D. N.; Arduini, A.; Grandi, S.; Sicuri, A. R.; Pochini, A.; Ungaro, R. *Synthesis* **1994**, 185.

4. van Wageningen, A. M. A.; van Duynhoven, J. P. M.; Verboom, W.; Reinhoudt, D. N. *J. Chem. Soc., Chem. Commun.* **1995**, 1941.

5. Molecular mechanics calculations were carried out using the CHARMm force field incorporated in Quanta.

6. For a recent review see: Dubois, L. H. ; Nuzzo, R. G. *Annu. Rev. Phys. Chem.* **1992**, *43*, 437-463.

7. Ulman, A. *An Introduction to Ultrathin Organic Films;* Academic Press: San Diego, CA, 1991.

8. Thoden van Velzen, E. U.; Engbersen, J. F. J.; Reinhoudt, D. N. *J. Am. Chem. Soc.* **1994**, *116*, 3597-3598.

9. Huisman, B.-H.; Thoden van Velzen, E. U.; van Veggel, F. C. J. M.; Engbersen, J. F. J.; Reinhoudt, D. N. *Tetrahedron Lett.* **1995**, *36*, 3273-3276.

10. Thoden van Velzen, E. U.; Engbersen, J. F. J.; de Lange, P. J.; Mahy, J. W. G.; Reinhoudt, D. N. *J. Am. Chem. Soc.* **1995**, *117*, 6853-6862.

11. Schierbaum, K.-D; Weiß, T.; Thoden van Velzen, E. U.; Engbersen, J. F. J., Reinhoudt, D. N.; Göpel, W. *Science* **1994**, *265*, 1413-1415.

12. Huisman, B.-H.; van Veggel, F. C. J. M.; Reinhoudt, D. N., submitted.

13. Huisman, B.-H.; Rudkevich, D. M.; van Veggel, F. C. J. M.; Reinhoudt, D. N., in preparation.

AN INCREMENTAL EMPIRICAL APPROACH TO NON-COVALENT INTERACTIONS IN AND WITH DNA

H.-J. SCHNEIDER*, J. SARTORIUS
FR Organische Chemie der Universität des Saarlandes
D 66041 Saarbrücken, Germany

Abstract. Non-covalent binding forces such as salt bridges, hydrogen bonds and intercalations in and with DNA are compared with similar interactions in synthetic supramolecular complexes. Based on many association free energies ΔG complemented in some cases by structural analyses with NMR methods it is shown that polyamine associations are predictable with an increment of (5+/- 1) kJ per mol and per salt bridge, with little contribution of hydrogen bonding. Azoniacyclophanes with 4 permanent charges also fall into this rule; differences in affinities can be explained by different geometric matching to larger or more narrow grooves in double stranded B- or A-DNA. A large cyclophane shows by NMR clear evidence of a rotaxane-like inclusion of single stranded DNA within the cavity. The association energies and modes of nucleoside derivatives as well as of over 50 related compounds in chloroform can be predicted by only one increment for the primary hydrogen bond of ΔG_p =7.9 kJ/mol, and one increment of ΔG_s=2.9 kJ/mol for secondary interactions, which can be either repulsive or attractive. NMR titrations are shown to provide unequivocal evidence for intercalation modes; they also allow determination of affinities based on the evaluation of several independent signals, in favourable cases even with the observed band width changes. Indole derivatives including tryptophane containing peptides, gramine etc, show intercalation with a stacking contribution of 1-2 kJ/mol only if ammonium centers in side chains provide a primary binding force with the known salt bridge increments. Benzene derivatives even with several charges never intercalate. Antimalarials like chloroquine and related quinoline or naphthalene derivatives again need help from positively charges side chains, yielding then rather constant stacking contributions of 8-10 kJ/mol. It is shown that contrary to literature assumptions the introduction of positive charges within the stacking aromatic unit does not enhance affinities, and that there is no evidence for only partial intercalation even with so-called weak intercalators. The effect of added salts or of polyamines on the affinity of these intercalators to ds-DNA are in line with the prediction of 5 kJ/mol and salt bridge.

L. Echegoyen and A.E. Kaifer (eds.), Physical Supramolecular Chemistry, 11–25.

Keywords DNA, Nucleic Aids, Nucleobases, Polyamines, Hydrogen Bonds, Salt Bridges, Intercalation, DNA Unfolding, Rotaxanes, LFER, Binding Increments, NMR-spectroscopy

Several non-covalent binding mechanisms are essential in chemistry and biological functions of nucleic acids. Their study will not only further our understanding of the molecular basis of life and corresponding endangering factors by biocides. The predictability of non-covalent interaction modes and strength can also greatly enhance the chances for a rational design of drugs directed against a variety of diseases. In the present chapter we describe preliminary attempts to quantify intermolecular forces in and with DNA on the basis of additive increments describing free associations energies, in line with strategies we developpped before for the analysis of synthetic supramolecular complexes [1]. We also discuss new insight into structural variations of DNA with artificial ligands such as cyclophanes, mostly on the basis of NMR spectroscopic methods.

1. Ion pairs/Salt Bridges with Polyamines

The condensation of metal cations on ds-DNA and its conformational consequences has been aptly demonstrated experimentally, and put on firm theoretical basis particular by Manning [2]. Polyamines play an important biological role [3], also as they can compete with binding of for instance groove binding antibiotics, or, if conjugated to other binders, enhance their affinity to DNA. Polyamines are also known to have a stabilizing effect on the native right-handed B-helix of double stranded DNA at lower concentrations, whereas concentrations above 2 μM (in the case of spermine) can lead to interconversion to left-handed Z-conformations [4].

What are the possibilities to predict quantitatively the affinity of biogenic and artificial polyamines to nucleic acids ? Stewart et al [5] have already shown 1988 a fairly linear correlation between the number of basic nitrogen atoms in the polyamine and DNA affinities, as far one can use the convenient assay based on the 50% release of the fluorescent dye ethidium bromide from double stranded DNA as a affinity measure. We found, that indeed many other amines, including cyclic ones (see below) also fit into the correlation (Figure 1), and that in fact the known true association energies of biogenic polyamines [6] are the same as predicted on the basis of a single empirical increment ΔG=(5+/- 1) kJ per mol and salt bridge [7], which we derived before on the basis of now over 70 ion pair complexes in water [1,8], if necessary corrected for negligible ionic strength [9]. One should emphasize, that the identification of the number of possible salt bridges with the DNA phosphate groups requires suitable computer aided molecular modelling of such complexes [10]. The close agreement between experimental and ion pair-derived ΔG values, which is found also for poly-lysine [7], already suggests, that ion pairing is the dominant binding mechanism for such polyamines. This was made further evident by replacing the acidic hydrogens at the ammonium centers by alkyl groups: the corresponding per-methylated polyamines showed the *same* affinities as the native polyamines, which have been generally believed to bind mostly by hydrogen bonding to the phosphate groups [7].

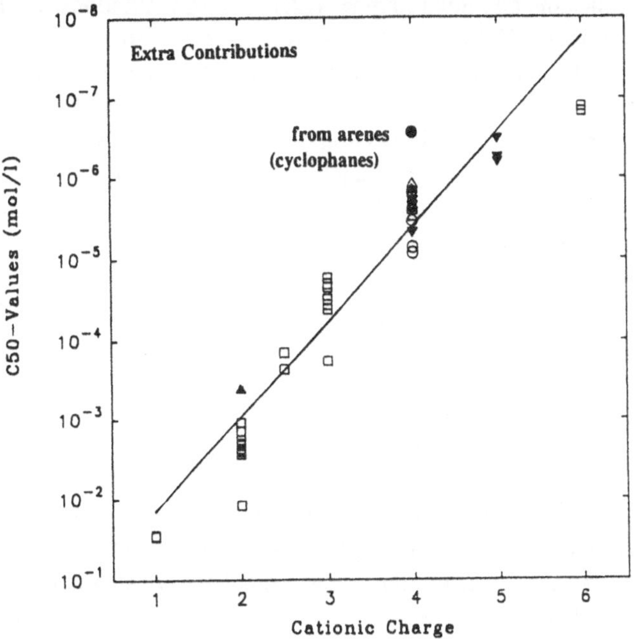

Figure 1. Affinities of polyamines to CT-DNA as evaluated by the C50 assay with ethidiumbromide vs. the numbers of charges in the polyamine; the numbers for the azoniacaclophanes CP44 is marked with a filled circle

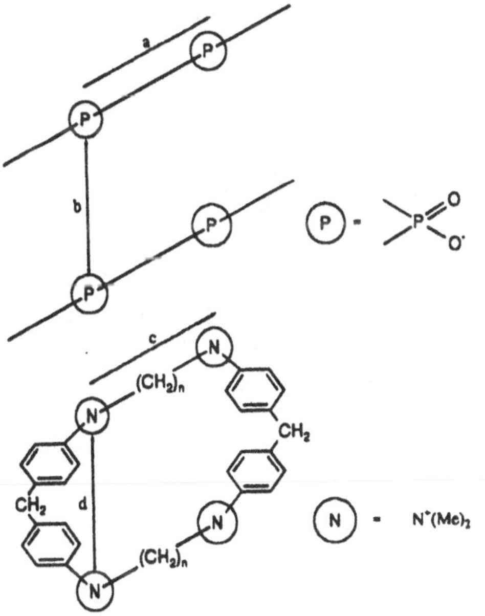

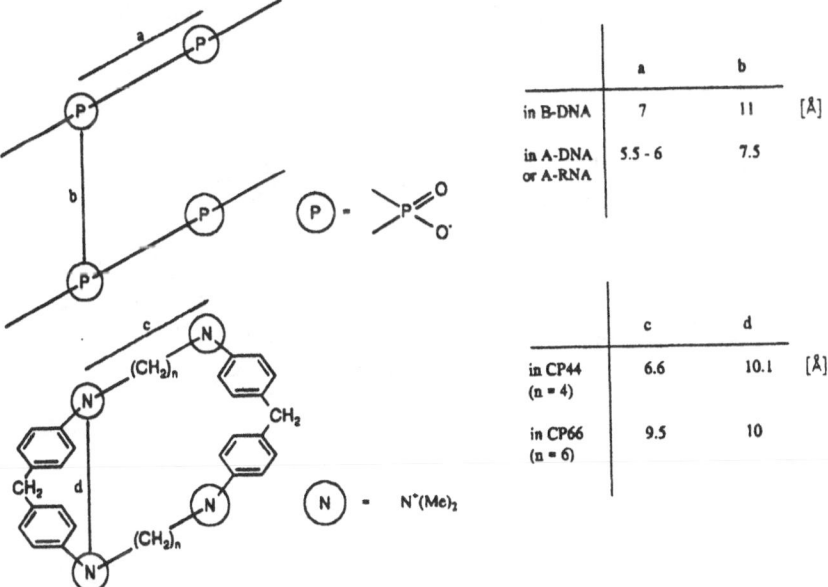

	a	b	
in B-DNA	7	11	[Å]
in A-DNA or A-RNA	5.5 - 6	7.5	

	c	d	
in CP44 (n = 4)	6.6	10.1	[Å]
in CP66 (n = 6)	9.5	10	

Figure 2. Azoniacyclophane CPnn interactions with ds-nucleic acids in the B- and in the A conformation; distances from computer aided molecular modelling; explanations see text.

Azoniacyclophanes [11] with 4 positive charges are found to have affinities to DNA which are close to polyamines bearing 4 charges like spermine [7]. Whereas our earlier analysis [7] was based on less direct affinity measurements by competition with the fluorescence indicator ethidiumbromide (EB) , NMR titrations of a selected cyclophane with sonicated calf thymus - DNA now shows a association free energy of ΔG= - 18 kJ/mol (+- 2 kJ/mol). Although the affinities of all cyclophanes CPnn fall into the expected range , there is a significant increase for CP44 in which the charges are separated by 4 methylene groups [7]. Inspection of molecular models of ds-DNA reveal, that indeed this cyclophane geometrically matches the phosphate anion sites particularly well (Figure 2). Inspired by recent results of Wilson et al [12], who found that our smaller cyclophanes like CP44 stabilize also RNA, but that CP66 unfolds RNA dramatically, we also checked the geometries with the smaller RNA groove. And indeed we found that again CP44, but not the larger CP66 matches the geometry of the smaller A-RNA groove (Figure 2).

2. Rotaxane-like Inclusion of Single-stranded Nucleic Acids within a Cyclophane

With *single* stranded nucleic acids CP66 is found to react in an unprecedented way: above melting which is increased with DNA, but very much decreased with RNA [12] - the signals of the central methylene protons of the cyclophane are shifted upfield by up to 0.25 ppm. These and other shifts correspond very closely to the shifts we found for CP66 by encapsulation of adenosinphosphate (Table 1). The only explanation for this is a rotaxane-like structure shown in Figure 3. Computer aided molecular modelling supports, that the cavity of CP66 indeed can take up a polynucletotide chain without distortion even of the stacking preserved also in the single stranded nucleic acid.

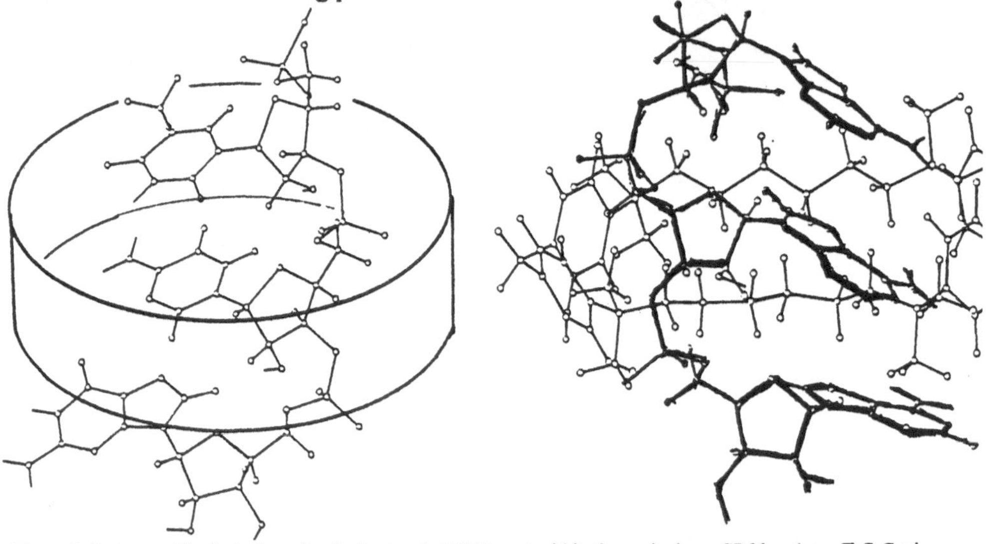

Figure 3. Rotaxane-like inclusion of a single stranded DNA part within the cyclophane CP66 (a: a T-C-G trimer in energy-minimized stacked conformation; b: a A-A-A trimer; simulations with the CHARMm force field).

Table 1. Comparison between complexation induced NMR shifts (CIS , in [ppm]) between insertion of AMP, GMP and a single stranded DNA or RNA inside the CP66 cavity.

	γ-CH_2	β-CH_2	α-CH_2
by AMP^{2-}	-0.25	-0.20	-0.10
by GMP^{2-}	-0.15	-0.15	-0.10
by ss-DNA	-0.25	-0.19	-0.15

3. Hydrogen Bonds

Hydrogen bonds are as Watson-Crick base pairs the heart of the double helix; they also play an essential role in groove binding of many natural and artificial ligands, which may exhibit sequence specificity by Hoogsteen-type of bonds. The energies involved in base pairing have been investigated earlier with several nucleobases which were alkyl-substituted for measurements in lipophilic solvents [4a] like chloroform. In view of the uncertain accuracy of older measurements [4a,13] we have re-investigated such equilibria with improved NMR methods, and with nucleoside derivatives like **1** (scheme 1). These have the advantage to be closer to the natural system, and to restrict to some degree the multitude of possible association modes of the natural nucleobases themselves [14]. Still, the non-linear least square fit of a NMR titration with G and C derivatives **1c** and **1d** shows already substantial deviations from a simple 1:1 model. Evaluation taking into account also self-association of G and C removes the systematic deviation from fit ; the ΔG's obtained for the G and C dimer are also close to the ones obtained from dilution measurements of G and C alone. Similar results were obtained from calorimetric titrations in CCl_4. The ΔG values in this solvent are due to the absence of an acidic hydrogen as in $CHCl_3$ somewhat larger. The dramatic effects of hydrogen bonding by the medium is illustrated by the fact, that addition of only 1% methanol to CCl_4 lowers the association constants by a factor of 25.

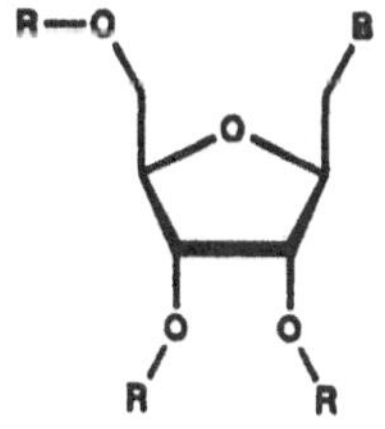

R = $SiMe_2tBu$ (TBDMS)

B =		
	A	1a
	T	1b
	G	1c
	C	1d

1

Scheme 1 Structures of the used silylated ribonucleosides

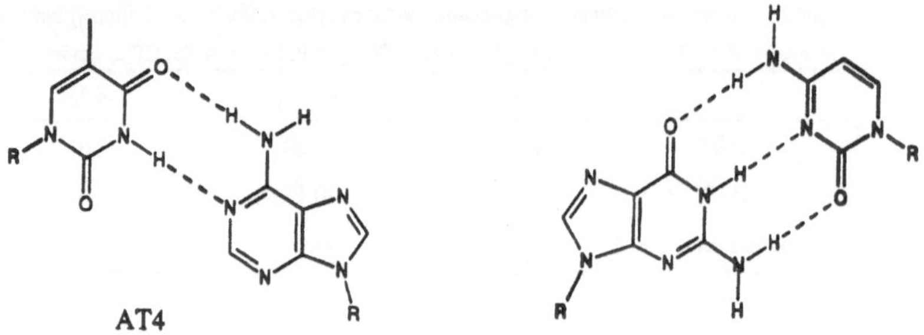

AT4

GC1

AT4	GC1
- 8.7	- 24.0
- 10.0	*- 23.8*

Scheme 2 Observed and calculated ΔG_t values for the Watson-Crick pairs A-T and G-C

Preliminary NOE measurements are in line with the association modes shown in scheme 2, which also contains the free energies ΔG of association. Measurements with A and T nucleosides [15] (Scheme 1, 2) showed less interference with self-association and general agreement with literature data [4a]. After having secured the relevant binding affinities and modes we set out to compare them with other associations and to derive , if possible, a simple and general set of increments describing such complexes.

Jorgensen et al [13] have on the basis of charge density calculations demonstrated how the hydrogen bond in base pairs will invariably be accompanied by either repulsive or attractive *secondary* interactions, depending on whether the primary bond site is flanked by negatively charged functions acceptor units **A**, or by positively donor functions **D** (see scheme 2 for an illustration with the Watson-Crick base pairs). Several workers [16,17] have shown this concept to hold for many synthetic complexes, which indeed show increasing stabilities if -unlike the natural bases- all **D** sites are on one , and the **A** sites on the other partner. The artificial complexes are designed to give in contrast to the natural systems essentially only one type of complex; they lend themselves therefore much easier for mechanistic interpretation.

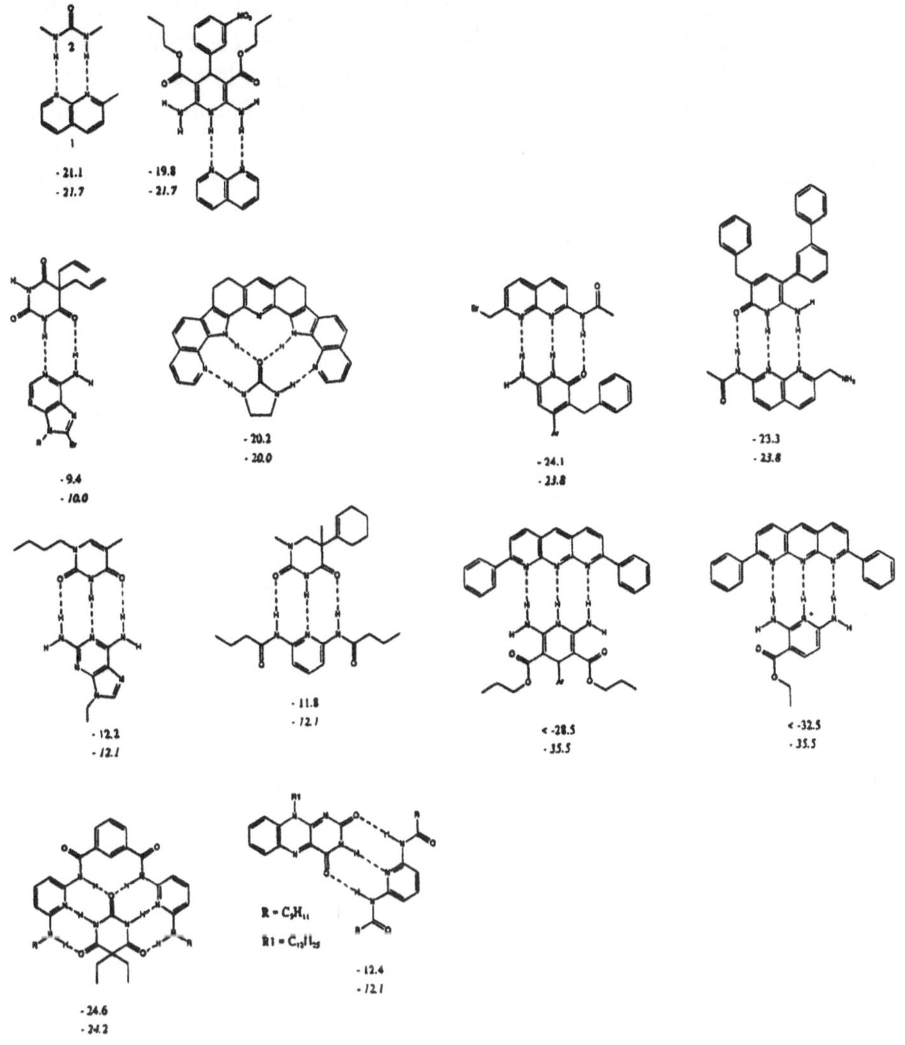

Scheme 3 Overview of representative synthetic host-guest complexes used for the analyses, with exp. and *calc.* ΔG_t values in chloroform

Although in particular Zimmerman et al [16] have given already many data, we synthesized and measured some additional **AA/DD** combinations which were until now scarcely represented (Scheme 3). Quite in contrast to mechanistic reasoning [16e] we find that one describe the association free energies of more than 50 complexes , including some peptide-like structures, with single increments of *ΔGp=7.9 kJ/mol for the primary hydrogen bond, and ΔG_S=2.9 kJ/mol for the secondary interaction,* irrespective of ΔG_S being negative or positive (all values with $CHCl_3$ as solvent).

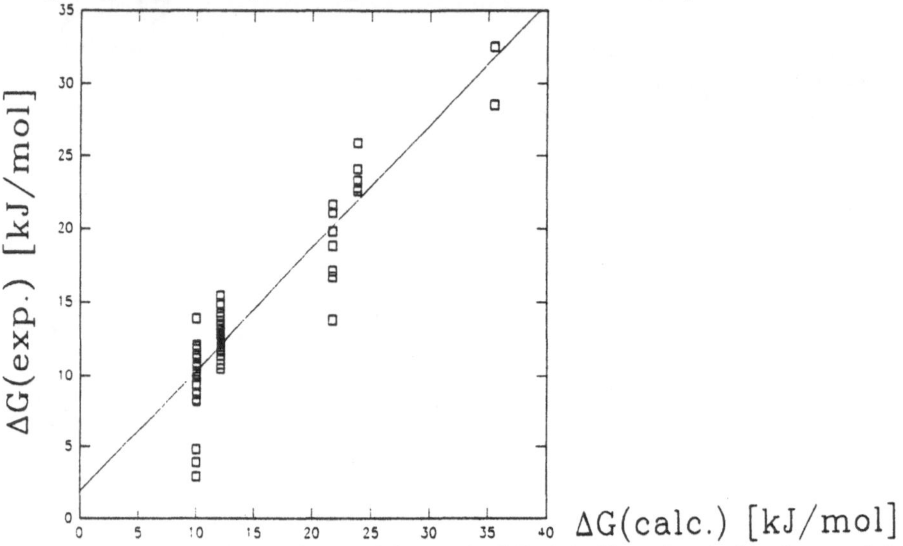

Figure 4. Experimental *vs.* calculated free association energies [kJ/mol] for 58 complexes in chloroform;correlation coefficient r=0.913, slope m=0.84

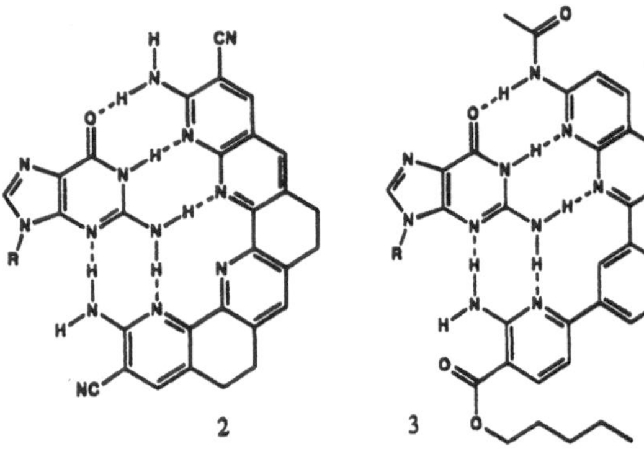

- 30.1 - 13.0

- 39.5 *- 23.7*

Scheme 4 Ligands 2 and 3 and their complexes [18] with a guanine derivative (exp. ΔG_t values in [kJ/mol] in DMSO-d_6/$CDCl_3$ mixture (v/v=1:4) and *calc.* ΔG_t in [kJ/mol] in chloroform)

The surprising agreement between calculated and experimental ΔG values is illustrated with representative examples in Schemes 2 and 3, and with Figure 4. It is in line with electrostatic forces as main factors; an increased charge separation will strengthen the primary hydrogen bond, but at the same time may increase in most complexes of the data set then repulsive interactions, *and vice versa* . The reader is reminded that almost all linear free energy correlations are based on substantial compensation effects, as they are obviously present here . We note, that even compounds such as **2** and **3** (scheme 4) which have different flexibilities fit into the correlation, if solvent effects by added DMSO are taken into account [18].

The practical value of the derived increments is not only, that they allow to design suitable artificial ligands, but also to predict easily stability constants. These otherwise would require extensive free energy pertubation calculations, which also must rely upon properly parametrized force field potentials. Moreover, the possible *structures* of base pairs become again easy to predict. Thus, the nucleobase complexes calculated by Kollman et al [14] are in line with interaction modes obtained from the simple increments. Also, the known tendencies [19] of the bases to form triples can be explained by the presence or absence of additional stabilizing secondary terms in the Hoogsteen pattern of the purines.

4. Intercalation

Stacking between the nucleobases provides another essential factor stabilizing the DNA double helix; it is of paramount importance also for the interaction of for instance many antibiotics or antitumor agents. In spite of numerous theoretical [20] and experimental [21] studies there is to the best of our knowledge until now no general way to predict even semiquantitatively for which compounds intercalation will occur, and what affinities are expected. It has been postulated for instance that benzene derivatives with aminoalkyl sidechains intercalate [22] , or that socalled weak intercalators destack the base pairs only partially, or that introduction of heteroatoms or charge separations in the aromatic intercalator unit will enhance association energies [23]. These claims are challenged by our studies, which makes heavy use of modern NMR techniques for both the distinction of binding modes, and for measuring association constants with DNA [24]. These methods also allow to remove uncertainity about binding modes with simple peptides [25] and with some antibiotics [26] and antimalarials [27]. Other methods generally recommended for the study of intercalators [28] like melting point, hypochromicity, or viscosity changes, were also used . However, they were found to be much less conclusive with respect to securing binding mode and affinity, and will therefore not be discussed here.

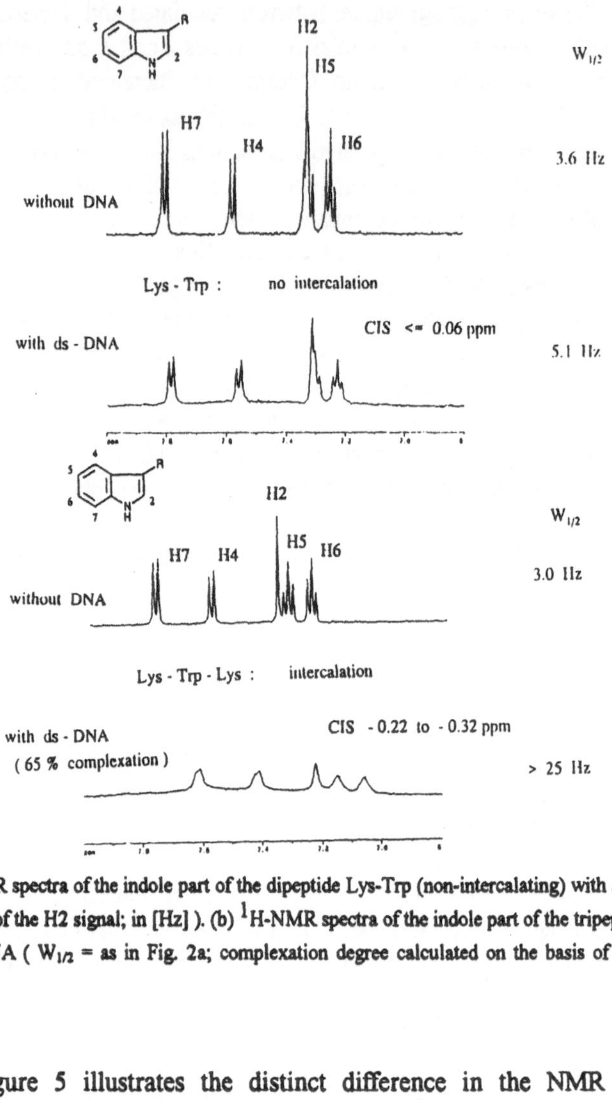

Figure 5. (a) ^{1}H-NMR spectra of the indole part of the dipeptide Lys-Trp (non-intercalating) with and without DNA ($W_{1/2}$ = half height of the H2 signal; in [Hz]). (b) ^{1}H-NMR spectra of the indole part of the tripeptide Lys-Trp-Lys with and without DNA ($W_{1/2}$ = as in Fig. 2a; complexation degree calculated on the basis of a next neighbour exclusion model)

Figure 5 illustrates the distinct difference in the NMR spectra with intercalating and only groove-binding peptides. The shielding effect $\Delta\delta$ on intercalating units by the nucleobases has been studied earlier theoretically [29] and experimentally [22,30] , as well as the line broadenening $W_{1/2}$ resulting from the slow tumbling of the biopolymer. We found that the NMR shift changes can be used for complex titrations even with native sonicated calf thymus (CT) DNA, yielding with non-linear least square fit consistent equilibrium constants from *several* signals, as well as corresponding intrinsic shifts at 100% comlexation (CIS values). The calculated K values depend on the chosen calculational model, which is to a lesser degree so for the CIS values.(Figure 6) In some cases one can even use the line width $W_{1/2}$ change for titration; the intrinsic (100% complexation) $W_{1/2}$ value can be obtained at practically more useful low degrees of complexation by simple correction on the basis of known association constants . As there is no dependence of $W_{1/2}$ on the shift differences or CIS values it is clear that exchange of the ligands studied here is still fast, and that the broadening does indeed result from the slow movement of the polymer.

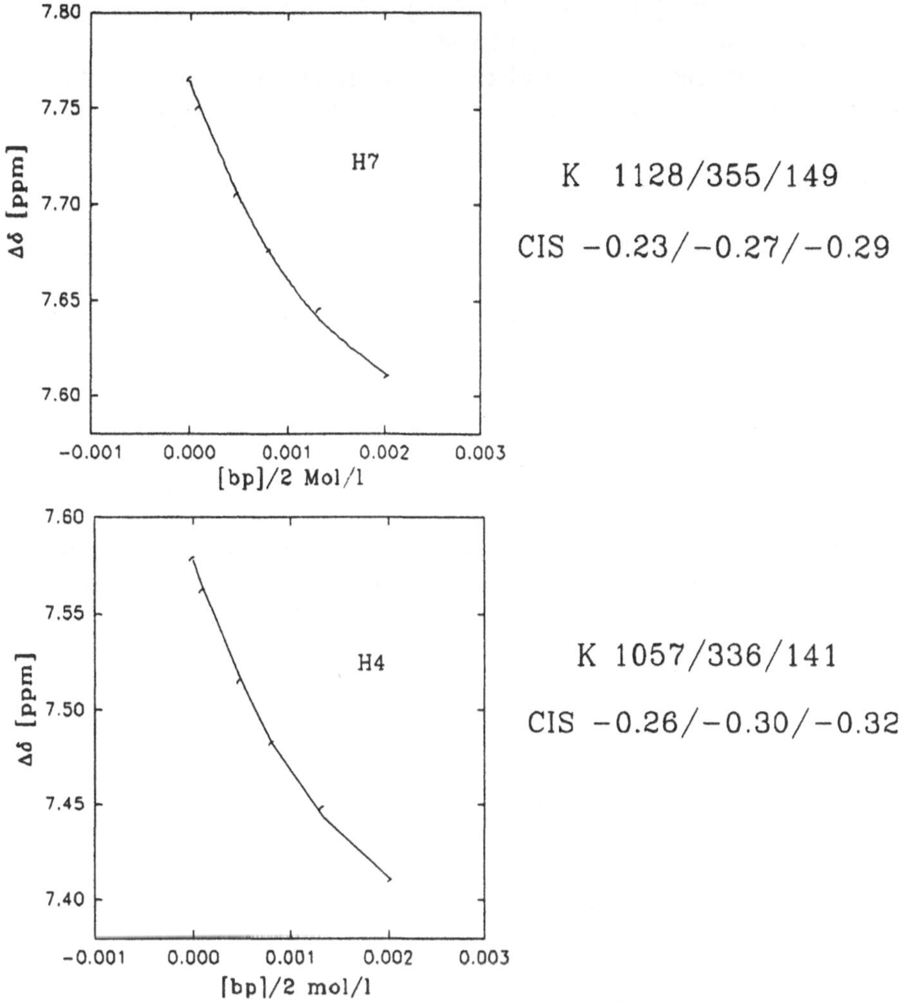

Figure 6. Plots for the determination of the formal association constant in $[M^{-1}]$ between tripeptide Lys-Trp-Lys and sonicated CT-DNA for two different protons of the indole part on the basis of next neigh-bour exclusion/without neighbour exclusion/with independant behavior of each phosphate (numbering scheme see Fig. 2 ; CIS = complexation induced shift (for 100% complexation, from line fit), in [ppm]; in D2O at pD=7.4, T = 303 +-0.1K, buffer 20 mM Na-phosphate)

Based on analyses of CT-DNA complexes with 48 different ligands, including classical intercalators like quinacrine , one can can generalize that *intercalators must show CIS > 0.1 ppm;* $W_{1/2} > 10$ *Hz* at signals in the intercalating unit, whereas groove binding leads to CIS < 0.05 ppm and to $W_{1/2} < 3\text{-}5$ Hz. Signals of side chain atoms are intermediate between these values as long they are not more than two carbon atoms away from the intercalating unit.

Most of the compounds studied contained in variable amount and position positively charged nitrogen atoms in the side chain, providing salt bridge to the groove phosphate groups. In consequence,association constants were either measured at low ionic strenght (I(NaCl)=0 M, [Na-phosphate] = 20 mM) or at known higher salt concentrations; the latter titrations allowed to extract by competition the Coulomb contribution to binding with values which were gratifying close to the value of 5 kJ per mol and bridge derived before (see section I). Similar to the analysis of nucleotide complexes with cyclophanes [7] we could factorize lipophilic and ionic binding contributions (ΔG_l, ΔG_i, respectively) by assigning the independently secured value of ΔG_i= 5 kJ per mol and salt bridge to each contact ion pair. It must be stressed, however, that the prediction of intercalation forces becomes difficult if size and substitution pattern particularly of larger aromatic units requires specific orientations. Further, the *in vivo* activity of lipophilic stacking ligands may depend on other factors such as transport; there is some recent evidence of simple correlations with hydrophobicity [20b].

The results of these NMR studies can be summarized as follows:

1. benzene derivatives never intercalate (ΔG_l << 0.5 kJ/mol);

2. indole derivatives intercalate only if binding to DNA is assisted by simultaneous ion pairing to phosphate by an attached aminogroup, which must be at least be separated by one carbon atom from the indole unit (examples: gramine, tryptamine); ΔG_l= 1-2 kJ/mol

3. tryptophane-containing peptides only intercalate if the indole-bearing aminoacid is flanked by at least *two* basic aminoacids like lysine;

4. naphthalene derivatives intercalate only if supporting aminoalkyl substituents are present (positional conditions see (2); ΔG_l= 8-10 kJ/mol

5. heteroatoms or even positive charges in the aromatic unit do not lead to increased association constants (examples: quinoline, N-methylquinolinium salts : no intercalation without alkylamino sidechain).

6. antimalarials like chloroquine, quinine and primaquine therefore intercalate with similar affinities as corresponding naphthalene derivatives depending on the size and flexibility of the sidechain;

7. all these intercalators switch to groove binding as function of ionic strength or of added polyamine (spermine); the salt effect can again described on the basis of the 5 kJ increment for ion pairing

8. anthracene-shaped compounds like acridins intercalate also without supporting positive charges. (ΔG_l= 26 to 27 kJ/mol for unsubstituted acridine).

9. very large aromatic units like a pyrene derivative cannot intercalate for steric reasons, however shows evidence for strong lipophilic contributions in groove binding.

10. "Partial" intercalation mode: neither NMR shifts nor preliminary NOEs (NOESY measurements with a ds-octamer) give any evidence for an incomplete intercalation or destacking with ligands such as indole or napthalene derivatives.

5. Conclusion

There is promise that even the complex binding phenomena of ligand binding to DNA can be tackled by the empirical approaches of physical organic chemistry, mostly based on NMR measurements with suitable selected ligand structures. The results can be explained by additive increments for salt bridges with side chain charges around invariably 5 kJ/mol , and lipophilic interactions which seem to be a function of the intercalating surface as long as the geometry between the then destacked Watson-Crick - bound base pairs tolerates penetration. That charge variations within the intercalator by heteroatoms does not lead to changed association constants is in line with our recent finding with porphyrin complexes in water [31]. There we found the association energies to be only a function of the size of the bound aromatic system, independent of the presence of heteroatoms in the ligand.

6. Acknowledgement

Our work is supported by the Deutsche Forschungsgemeinschaft, Bonn, and the Fonds der Chemischen Industrie, Frankfurt. We acknowledge valuable discussions with Drs. Abraham and Raevsky supported by a NATO CRG, and with Drs Wilson and Fernandez-Saiz.

7. References and Notes

1 Schneider, H.-J. *Chem. Soc. Rev.* **1994**, 22, 227

2 Manning, G.S. *Quart. Rev. Biophys.* **1978**, *11*, 179

3 Tabor, C.W.; Tabor, H. *Annu. Rev. Biochem.* **1984**, *53*, 749

4 a) Saenger, W. in *Principles of Nucleic Acid Structure,* Chap. 12 Springer New York 1988 and references cited therein b) Rich, A.; Nordheim, A.; Wang, A. H.-J. *Ann.Rev.Biochem.* **1984**, *53*, 791. c) Wang, A. H.-J.; Quigley, G.J.; Kolpak, F.J.; van der Marel, G.; van Boom, J.H.; Rich, A. *Science* **1980**, *211*, 171.

5 Stewart, K.D. *Biochem. Biophys. Res. Commun.* **1988**, *152*, 1441; Stewart, K.D.; Gray, T.A. *J.Phys.Org.Chem.* **1992**, *5*, 461.

6 Braunlin, W.H.; Strick, T.J.; Record, M.T. *Biopolymers* **1982**, *21*, 1301.

7 Schneider, H.-J.; Blatter, T. *Angew. Chem.* **1992**, *104*, 1244; *Angew. Chem.,Int.Ed.Engl.* **1992**, *31*, 1207

8 Schneider,H.-J.; Theis, I. *Angew.Chem.* **1989**, *101*, 757 ; *Angew.Chem.,Int.Ed.Engl.* **1989**, *28*, 753; Schneider, H.-J.; Schiestel, T.; Zimmermann, P. *J.Am.Chem.Soc.* **1992**, *114*, 7698.

9 Schneider,H.-J.; Kramer,R.; Simova,S; Schneider,U. *J.Am.Chem.Soc.* **1988**, *110*, 6442;

Schneider, H.-J.; Theis, I. *J.Org.Chem.* **1992**, *57*, 3066.

10 Stekowski, L.; Harden, D.B., Wydra, R.L.; Stewart, K.D.; Wilson, W.D. *J. Mol. Recognit.* **1989**; *2*, 158.

11 Odashima, K.; Koga, K in : *Cyclophanes,* Vol2 ; Academic Press, New York, **1983**, 629; Diederich, F. Cyclophanes (Monographs in Supramolecular Chemistry) J.F. Stoddart (Ed.) **1991**, Royal Soc.Ch.Cambridge; Schneider, H.-J.; Philippi, K.; Pöhlmann, J. *Angew.Chem.* **1984**, *96*, 907; *Angew.Chem., Int.Ed.Engl.* **1984**, *23*, 908; Schneider, H.-J.; Busch, R. *Chem.Ber.* **1986**, *119*, 747.

12 Fernandez-Saiz, M.; Wilson,D.; Sartorius, J.; Schneider, H.-J., manuscript submitted for publication.

13 a) Jorgensen, W.L.; Pranata, J. *J.Am.Chem.Soc.* **1990**, *112*, 2008. b) Pranata, J.; Wierschke, S.G.; Jorgensen, W.L. *J.Am.Chem.Soc.* **1991**, *113*, 2810

14 a) Leach, A.R.; Kollman, P.A. *J.AmChem.Soc.* **1992**, *114*, 3675. b) Cieplak, P.; Kollman, P.A. *J.Am.Chem.Soc.* **1988**, *110*, 3734. c) Gould, I.R.; Kollman, P.A. *J.Am.Chem.Soc.* **1994**, *116*, 2493. d) Dang, L.X.; Kollman, P.A. *J.Am.Chem.Soc.* **1990**, *112*, 503

15 Junker H.-D. Diplomarbeit Saarbrücken 1995

16 a) Zimmerman, S.C.; Murray, T.J. *Phil.Trans.R.Soc.Lond.A* **1993**, *345*, 49. b) Murray, T.J.; Zimmerman, S.C.; Kolotuchin, S.V. *Tetrahedron* **1995**, *51*, 635. c) Zimmerman, S.C.; Wu, W.; Zeng, Z. *J.Am.Chem.Soc.* **1991**, *113*, 196. d) Zimmerman, S.C.; Murray, T.J. *Tetrahedr.Lett.* **1994**, *34*, 4077. e) Zimmerman, S.C.; Murray, T.J. In: *Computational Approaches in Supramolecular Chemistry* ; Wipff, G., Ed.; NATO ASI Series C Vol. 426 Kluwer Academic Publishers, Dordrecht 1994

17 Bell, D.A.; Anslyn, E.V. *Tetrahedron* **1995**, *51*, 7161

18 Bell, T.W.; Hou, Z.; Zimmerman, S.C.; Thiessen P.A. *Angew.Chem.* **1995**, *107*, 2331

19 a) Cheng, Y.-K.; Pettitt B.M. *Prog.Biophys.Molec.Biol.* **1992**, *58*, 225-257. b) Thuong, N.T.; Helene, C. *Angew.Chem.* **1993**, *105*, 697

20 a) Pullman, B.; Jortner, J. in: *Molecular Basis of Specificity in Nucleic Acid-Drug Interactions* ; The Jerusalem Symposia on Quantum Chemistry and Biochemistry, Vol. 23, Kluwer Academic Publishers, Dordrecht 1990 and references cited therein b) Gupta, S.P. *Chem.Rev.* **1994**, *94*, 1507

21 a) Hashimoto, Y.; Shudo, K. *Life Chemistry Rep.* **1988**, *6*, 231. b) Wilson, W.D. in : *Nucleic Acids in Chemistry and Biology* ; Blackburn, G.M.; Gait, M.J. , eds. ; Oxford University Press Oxford 1990 c) Kumar, S.; Jaseja, M.; Zimmermann, J.; Yadagiri, B.; Pou, R.T.; Sapse, A.-M.; Lown, J.W. *J.Biomol.Struct.Dyn.* **1990**, *1*, 99. d) Pindur, U.; Haber, M.; Sattler, K. *J.Chem.Educ.* **1993**, *70*, 263.

22 a) Kapicak, L.; Gabbay, E.J. *J.Am.Chem.Soc.* **1975**, *97*, 403. b) Gabbay, E.J.; Scofield, R.; Baxter, C.S. *J.Am.Chem.Soc.* **1973**, *95*, 7850

23 a) Strekowski, L.; Mokrosz, J.L.; Wilson, W.D.; Mokrosz, M.J.; Strekowski, A. *Biochemistry* **1992**, *31*, 10802. b) Davidson, M.W.; Griggs, B.G.; Boykin, D.W.; Wilson, W.D. *J.Med.Chem.* **1977**, *20*, 1117

24 Sartorius, J.; Schneider H.-J. *FEBS Lett.* **1995**, *374*, 387

25 Rajeswari, M.R.; Montenay-Garestier, T.; Helene, C. *Biochemistry* **1987**, *26*, 6825 and references cited therein

26 Addess, K.J.; Feigon, J. *Biochemistry* **1994**, *33*, 12397 and references cited therein

27 a) Meshnick, S.R. *Parasitology Today* **1990**, *6*, 87. b) Kwakye-Berko, F.; Meshnick, S.R.

Mol.Biochem.Parasitol. **1989**, *35*, 51

28 Long, E.C.; Barton, J.K. *Acc.Chem.Res.* **1990**, *23*, 271

29 Lavery, R.; Pullman, B. *J.Biomol.Struct.Dyn.* **1985**, *2*, 1021

30 a) Chandrasekaran, S.; Kusuma, S.; Boykin, D.W.; Wilson, W.D. *Magn.Res.Chem.* **1986**, *24*, 630. b) Feigon, J.; Denny, W.A.; Leupin, W.; Kearns, D.R. *J.Med.Chem.* **1984**, *27*, 450. c) Patel, D.J.; Canuel, L.L. *Biopolymers* **1977**, *16*, 857. d) Patel, D.J.; Shen, C. *Proc.Natl.Acad.Sci. U.S.A.* **1978**, *75*, 2553. e) Wilson, W.D.; Krishnamoorthy, C.R.; Wang, Y.H.; Smith, J.C. *Biopolymers* **1985**, *24*, 1941. f) Chandrasekaran, S.; Jones, R.L.; Wilson, W.D. *Biopolymers* **1985**, *24*, 1963. g) Delbarre, A.; Gourevitch, M.I.; Gaugain, B.; LePecq, J.B.; Roques, B.P. *Nucleic Acids Res.* **1983**, *11*, 4467. h) Hudson, B.P.; Dupureur, C.M.; Barton, J.K. *J.Am.Chem.Soc.* **1995**, *117*, 9379

31 Schneider, H.-J.; Wang, M. *J.Org.Chem.* **1994**, *59*, 7464

LANGMUIR-BLODGETT SUPRAMOLECULAR ASSEMBLIES: INCORPORATING INORGANIC EXTENDED LATTICE STRUCTURES

CANDACE T. SEIP, HOUSTON BYRD, JOHN K. PIKE,
SCOTT WHIPPS AND DANIEL R. TALHAM
Department of Chemistry, University of Florida
Gainesville, Florida 32611-7200

Abstract

Inorganic extended lattice structures are incorporated into Langmuir-Blodgett films to form mixed organic/inorganic layered assemblies. Langmuir-Blodgett films of divalent and tetravalent transition metal phosphonates are described and shown to be single-layer analogs of the known solid-state layered structures. Using this approach, it is shown that traditional solid-state properties, such as magnetic exchange and high lattice energies, can be incorporated into Langmuir-Blodgett assemblies.

1. Introduction

A promising application of the concepts of supramolecular chemistry is achieving synthetic control over thin film and surface structures. Catalysis, sensing, separations, information storage, energy storage and non-linear optics are just some examples of areas where efficiencies can be greatly enhanced by optimizing intermolecular interactions at interfaces. Two approaches currently used to achieve organized assemblies of organic molecules at solid surfaces are Langmuir-Blodgett film methods and organic self-assembly methods. There is potential benefit to incorporating inorganic entities into these primarily organic assemblies as in many areas of application, such as magnetism and catalysis, inorganic structures currently offer superior performance. In this article, we will review some of our efforts to incorporate extended-lattice inorganic structures into Langmuir-Blodgett films to achieve "mixed organic/inorganic" LB films.

The description "mixed organic/inorganic" has been applied to materials that contain within the same solid-state structure separate components featuring organic-organic interactions and inorganic-inorganic interactions[1, 2]. Several classes of mixed organic/inorganic assemblies are currently receiving a great deal of attention as researchers have realized that by combining organic and inorganic networks a vast array of new solid-state structures is possible often with tunable bulk and interfacial properties. Some examples of recent interest are layered inorganic solids with organic intercalates[3, 4], organic/inorganic layered solids[1, 2, 5, 6], zeolites and other open framework inorganic

L. Echegoyen and A.E. Kaifer (eds.), Physical Supramolecular Chemistry, 27–37.

host materials with organic guests[7, 8], and metal oxide mesostructures formed with the aid of organic surfactants[9-11]. The ability to fabricate a solid or surface with control over features such as particle size, the shape and size of pores, and the chemical nature of cavities and interfaces has enormous potential for many applications.

Langmuir-Blodgett (LB) films[12] with their separate hydrophilic and hydrophobic components should be ideal materials for investigating the controlled formation of mixed organic/inorganic layered structures. Historically, however, little attention has been paid to the variety of possible inorganic structures that might be incorporated in the hydrophilic portion of LB films. An approach that we are investigating is to utilize known solid-state layered structures to organize LB films[13, 14]. There are several examples in the solid-state literature of mixed organic/inorganic layered compounds where polar ionic networks are separated by nonpolar organic networks[2]. Two examples are shown in Figure 1. Transition metal organophosphonates frequently form layered structures where the binding within the metal phosphonate layer is ionic/covalent and the interaction between layers is van der Waals in nature[5, 15, 16]. The structure[17] of $Ca(O_3PC_6H_{13})_2$ is shown in Figure 1, but layered structures are known for many other organophosphonates with a variety of divalent, trivalent and tetravalent metal ions[5, 15-17]. The alkylammonium layered perovskites[2], also pictured in Figure 1, are another family of mixed organic/inorganic layered structures based on organic amphiphiles. When comparing structures of this type it becomes clear that it is the inorganic lattice energy that determines these structures as the same ionic lattices are found in examples without organic substituents. For example, the metal halide lattice in the alkylammonium layered perovskites is essentially isostructural with that in the purely inorganic K_2MnF_4. Since we can identify inorganic interactions that clearly favor layered structures, it is worthwhile to ask whether or not these inorganic structure types can be used to control in a predictable way the structures of

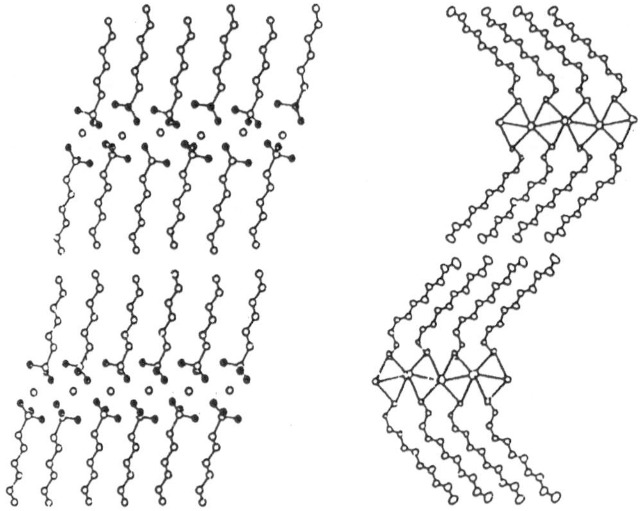

Figure 1. Examples of mixed organic/inorganic layered solids: left, the layered metal phosphonate $Ca(HO_3PC_6H_{13})_2$ (reference 17) and, right, the layered perovskite $(C_{10}H_{21}NH_3)_2CdCl_4$ (reference 2).

mixed organic/inorganic LB films. We review here our attempts to employ this strategy for formulating LB films and describe the preparation of LB films based on transition metal organophosphonate layered structures using octadecylphosphonic acid with divalent and tetravalent metal ions. We also discuss assemblies formed by self-assembling organophosphonate molecules at metalated Langmuir-Blodgett template layers.

2. Divalent Metal Phosphonate LB Films

The divalent metal phosphonate Langmuir-Blodgett films are fabricated using normal vertical deposition methods[13]. We normally begin with OTS-covered substrates so that the first organic layer is deposited on the down-stroke. Upon withdrawing the substrate through the film to complete a conventional Y-type bilayer, the film crystallizes forming the extended lattice metal phosphonate structure. A single "bilayer" consists of one metal phosphonate layer sandwiched between two organic layers. Important variables when depositing the films are the deposition speeds and subphase pH. A slow deposition speed on the upstroke allows the water to drain from the film, facilitating crystallization of the metal phosphonate extended lattice. The optimum pH of the subphase depends on the identity of the metal ion. The metal phosphonates are soluble in acid so that if the pH is too low, metal ions are not incorporated into the film. On the other hand, if the pH is too high, the phosphonate monolayer becomes too rigid to transfer. In general, the pH is raised to the highest value where the films are still processible. For example, the manganese phosphonate films are transferred in a pH range of 5.2-5.5. The cadmium films are transferred at lower pH and the calcium and magnesium films are transferred at higher pH. Elemental ratios in the films are determined by XPS, and in each case the metal to phosphorus ratio is 1:1, consistent with the solid-state metal phosphonate structures. The deposited films are clearly layered and several orders of the *(001)* Bragg peak are seen in X-ray diffraction from the films. The interlayer spacings derived from X-ray diffraction are 48.5 Å for the manganese octadecylphosphonate film, 48.2 Å for the cadmium film, 47.6 Å for the magnesium film, and 46.7 Å for the calcium film.

Infrared analyses on the films show that the film structure is consistent with the solid-state materials. Figure 2 compares mid-IR spectra of manganese octadecylphosphonate LB film and solid-state manganese ethylphosphonate. Bands associated with the long alkyl chain are prominant in the LB film spectrum with the asymmetric methyl stretch ($\nu_a(CH_3)$), at 2958 cm^{-1}, the asymmetric methylene stretch ($\nu_a(CH_2)$) at 2917 cm^{-1}, the symmetric methylene stretch ($\nu_s(CH_2)$) at 2850 cm^{-1}, and the methylene bending mode at 1467 cm^{-1}. The positions and shapes of the bands are consistent with an organized array of all-trans alkyl chains[18, 19]. The similarity between the manganese octadecylphosphonate film and the solid-state manganese ethylphosphonate is clearly demonstrated in the P-O region. The asymmetric phosphonate stretch ($\nu_a(PO_3^{2-})$) at 1088 cm^{-1} and the symmetric phosphonate stretch ($\nu_s(PO_3^{2-})$) at 978 cm^{-1} are observed in the film, and the analogous bands appear at 1088 cm^{-1} and 968 cm^{-1} in the solids. The absence of the strong P=O stretch in the 1350-1250 cm^{-1} region or the 1250-1110 cm^{-1} region for free and hydrogen bonded modes, respectively, suggests that the all of the phosphonate groups in the film are ionized. The appearance of the H-O-H bend at 1608 cm^{-1} also indicates that Mn^{2+} ions are present in the deposited films. In the bulk manganese phosphonates, each Mn^{2+} ion is bound by five oxygen atoms from the phosphonate anion and one H_2O molecule fills out the

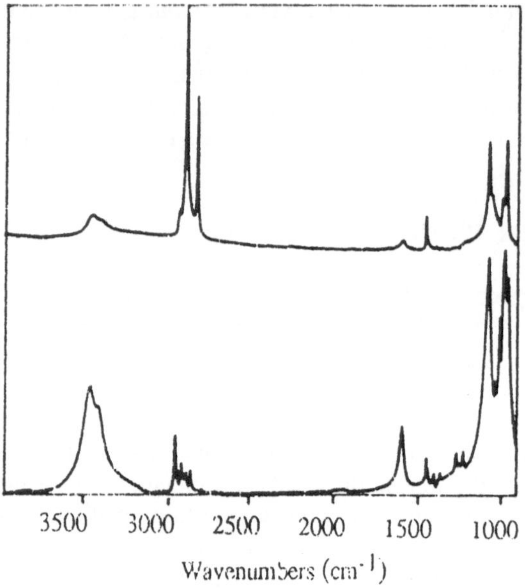

Figure 2. Mid-IR spectra of the manganese octadecylphosphonate LB film (top) and the solid-state manganese ethylphosphonate (bottom).

coordination sphere. The H_2O bend (1608 cm^{-1}) is also observed in the bulk manganese phosphonate materials.

The Mn^{2+} ion is paramagnetic allowing us to use magnetic properties to further compare the LB films to the solid-state analogs[13]. The manganese ions in the solid-state manganese phosphonates experience antiferromagnetic exchange, and several of the solids have been investigated as examples of layered antiferromagnets[20, 21]. For example, we previously prepared a powder sample of $Mn(O_3PC_6H_5)\cdot H_2O$ and observed an antiferromagnetic ordering transition at 12K by SQUID magnetometry[22]. Meanwhile, Carling *et al.*[20, 21] have shown that the series $MnC_nH_{2n+1}PO_3\cdot H_2O$ (n = 1-4) are all canted antiferromagnets with ordering temperatures in the range 14.8-15.1 K. They are termed "canted antiferromagnets" because below the ordering temperature the antiferromagnetically coupled moments do not exactly cancel resulting in a small net magnetization in the material. Our initial magnetic studies of the LB films have been with EPR. EPR is sensitive enough to allow observation of the films, and our studies have been on films of 50 bilayers in thickness. The temperature dependence of the EPR signal shows evidence for antiferromagnetic exchange and short-range antiferromagnetic order in the manganese phosphonate LB films. The integrated area of the EPR signal is proportional to the spin susceptibility, and a plot of 1/area vs. temperature yields a classic Curie-Weiss plot for the film, Figure 3. Extrapolation of the high-temperature data to intercept the temperature axis gives a negative intercept, consistent with antiferromagnetic exchange. The deviation from Curie-Weiss behavior at low temperature is characteristic of antiferromagnetic exchange in a low-dimensional lattice. The temperature dependent intensity data have been fit with a numerical expression for the susceptibility of a quadratic-layer Heisenberg antiferromagnet[23] and the fit yields a value for antiferro-

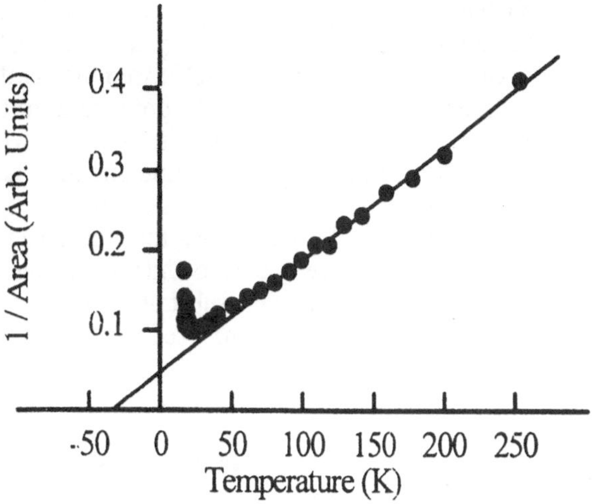

Figure 3. The integrated area of the EPR signal of the manganese octadecylphosphonate LB film is plotted as 1/area *vs*. temperature to show Curie-Weiss behavior. Extrapolation of the high-temperature data to intercept the temperature axis yields a negative Weiss constant incicating antiferromagnetic exchange.

magnetic exchange of J/k = -2.8 K, which is within the range of values observed for powder samples of the solid-state manganese phosphonates. The observed magnetic exchange arises from the crystalline extended-lattice structure of the LB layers, and the magnetic behavior of the manganese film is further evidence that the metal phosphonate extended lattice structures are formed during the Langmuir-Blodgett deposition procedure.

3. Zirconium Phosphonate LB Films

Perhaps the most extensively studied metal phosphonates are zirconium salts[5, 15], and we have attempted to prepare zirconium phosphonate LB film analogs[14, 24]. Zirconium (4+) alkylphosphonates are less soluble than their divalent metal analogs, and this difference manifests itself when attempting to prepare LB films. Octadecylphosphonic acid, when spread at the air-water interface of a subphase containing Zr^{4+} ions, binds the metal ions so tightly that the Langmuir monolayer becomes too rigid to transfer using conventional LB film deposition methods. While the strong Zr^{4+}/phosphonate binding interaction works against forming a processible Langmuir monolayer, this same feature allowed us to develop a novel stepwise procedure for depositing monolayer and multilayer films of zirconium octadecylphosphonate[14] that involves a combination of LB and "inorganic" self-assembly methods. This procedure is outlined in Figure 4. The first step creates an LB template of octadecylphosphonic acid suitable for binding Zr^{4+} ions by transferring a single LB layer of octadecylphosphonic acid from a pure water subphase onto an OTS-covered substrate that is dipped through the film and into a vial sitting in the subphase. The vial containing the substrate immersed in subphase is then removed. In step 2, Zr^{4+} ions "self-assemble" at the newly formed organic template upon adding $ZrOCl_2$ to the vial to produce a 5mM zirconium ion solution. After 30 minutes in the zirconium solution, the substrate with the zirconated octadecylphosphonic acid layer is removed and placed into another vial containing pure water. Step 3 of the deposition procedure is to cap the zirconated layer with another LB layer to complete the bilayer assembly. To accomplish this, the substrate in pure water is placed back into the LB trough where a new octadecylphosphonic acid film is compressed over the vial and then transferred to the substrate creating a Y-type zirconium octadecylphosphonate bilayer. Multilayers can be produced by repeating this three-step

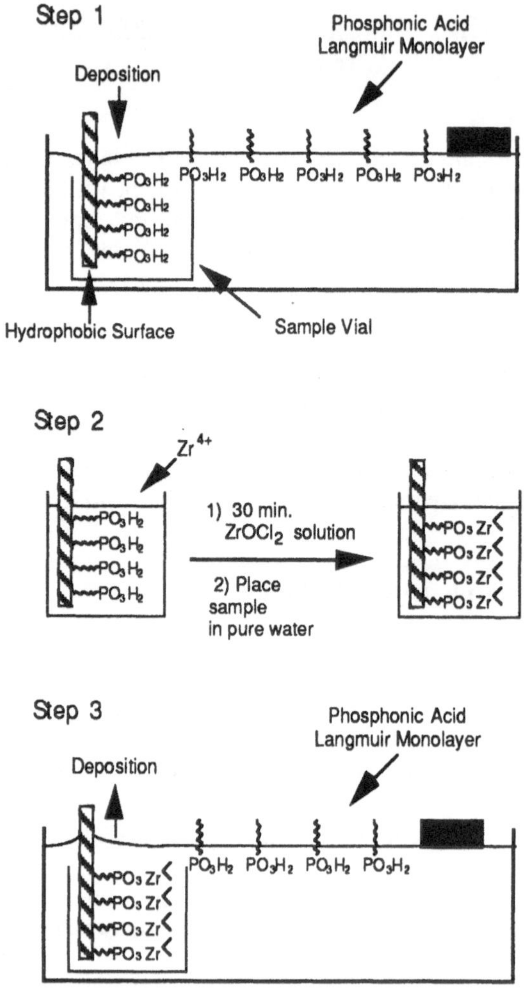

Figure 4. Three step procedure for depositing zirconium phosphonate LB films.

deposition procedure on the resulting hydrophobic surface. Each step of the deposition process has been carefully analyzed using XPS, FTIR, ellipsometry, contact angle measurements and X-ray diffraction. The zirconium phosphonate deposition process clearly forms layered LB films, although the in-plane structure has not been easy to discern due to the poor crystallinity of the layers. This is similar to most solid-state zirconium phosphonates which normally are also poorly crystalline and is a consequence of the strong zirconium-oxygen binding in the phosphonates which does not allow the structure to aneal on the time scale of the deposition process. Another consequence of the strong zirconium-phosphonate binding is that solid-state zirconium phosphonates are highly insoluble in both water and organic solvents, and the zirconium phosphonate LB films are similarly insoluble. For example, we have observed a 10-bilayer film of zirconium octadecylphosphonate to remain intact after soaking for several hours immersed in chloroform[14]. Although the zirconium phosphonate headgroup binding does not appear to afford much control over LB film structure, these films are extremely stable once deposited.

4. Self-Assembled Multilayers

The zirconium phopshonate binding interaction has also been used in layer-by-layer self-assembly schemes. In the original procedure, developed by Mallouk *et al.*[25, 26] a molecule bearing the phosphonic acid functionality is adsorbed to a surface, and layers are built-up by alternately adsorbing Zr^{4+} ions and α,ω-diphosphonic acid molecules from solution, Figure 5a. Thin films formed in this way have been used as the dielectric in metal-insulator-semiconductor and metal-insulator-metal devices[27], and these procedures have also been adapted to prepare films containing oriented assemblies of a wide variety of functionalized molecules[5, 6, 24-26, 28, 29].

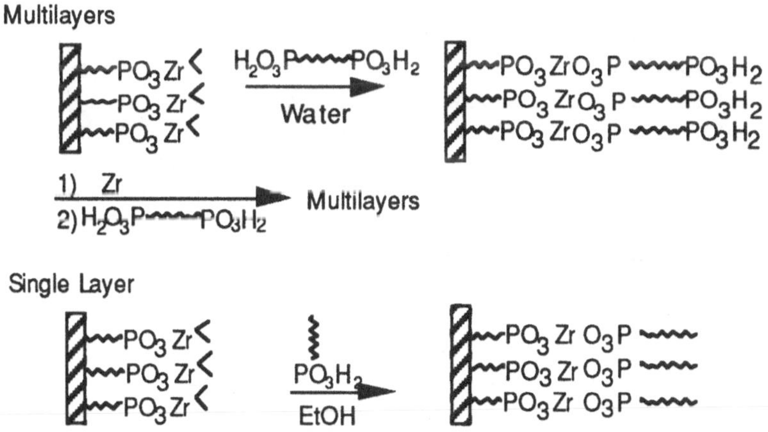

Figure 5. **a**. Self-assembly scheme for depositing zirconium phosphonate multilayers (after Lee *et al*, reference 26). **b**. Scheme for self-assembly of a monolayer of octadecylphosphonic acid at the zirconated LB film template layer.

The zirconated octadecylphosphonate LB film that results in Step 2 of Figure 4 should be an ideal starting point for attempting the self-assembled multilayer deposition. In any layer-by-layer deposition process the organization of the first layer will affect how molecules pack in subsequent layers. XPS demonstrates[14] that the Zr:P ratio of the zirconated LB film in Step 2 is 1:1. The Langmuir-Blodgett process produces a close-packed array of octadecylphosphonic acid sites that when exposed to the solution of zirconium ions bind one zirconium ion per site producing a monolayer of close-packed zirconium sites.

We have investigated the self-assembly of phosphonic acid molecules at this surface[24]. Octadecylphosphonic acid can be self-assembled at the zirconated surface, according to Figure 5b, to form a monolayer with a level of organization indistinguishable from the LB film used as the template. We have also investigated the layer-by-layer deposition of α,ω-diphosphonic acid molecules at the zirconated LB template using the procedure described in Figure 5a[24]. The most highly organized assemblies were achieved with rigid molecules that cannot fold over and bind both phosphonic acid groups to the same layer. An example is quaterthiophenediphosphonic acid (QDP)[28], shown below.

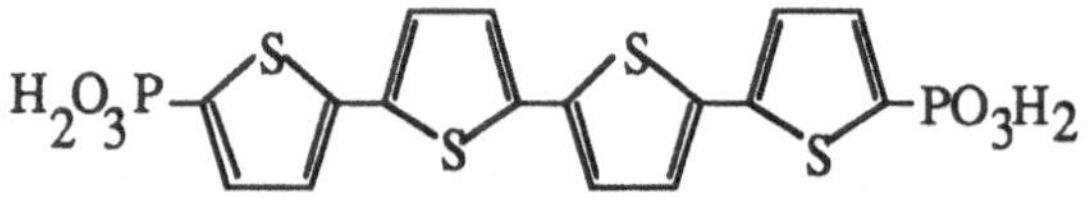

The QDP sample for our experiments was provided by Dr. Howard Katz at AT&T. The layer-by-layer assembly of QDP was monitored by UV-vis spectroscopy with the incident beam perpendicular to the substrate. There is no change in the UV-vis spectrum when the time of deposition is increased beyond 1 hr. suggesting that the deposition is complete. This is confirmed by XPS which shows that for deposition times of 1 hr and 7 hrs the Zr:P:S ratios are the same within experimental error[24]. A plot of the UV-vis absorbance at 390nm versus the number of self-assembled QDP layers is shown in Figure 6. The linear increase in absorbance demonstrates that the same amount of QDP is deposited during each cycle. X-ray diffraction from a ten layer film of QDP assembled at a zirconated octadecylphosphonic acid LB film yielded first and second order reflections corresponding to a (001) d-spacing of 20.19 Å. The layer thickness agrees well with the size of the QDP molecule, and the diffraction proved that the multilayer film has a layered structure. It is clear that the well organized LB template layer played an important role in organizing the subsequent self-asseembled layers, as diffraction was not observed from QDP multilayers assembled at less organized template layers.

Conclusions

The work described here illustrates that inorganic extended lattice structures can be incorporated as part of Langmuir-Blodgett films to form mixed organic/inorganic layered supramolecular assemblies. Furthermore, results on the manganese phosphonate films demonstrate that physical properties normally associated with inorganic materials, such as solid-state magnetism, can be introduced into LB films through the inorganic lattice. It should not be overlooked, however, that perhaps the most important feature

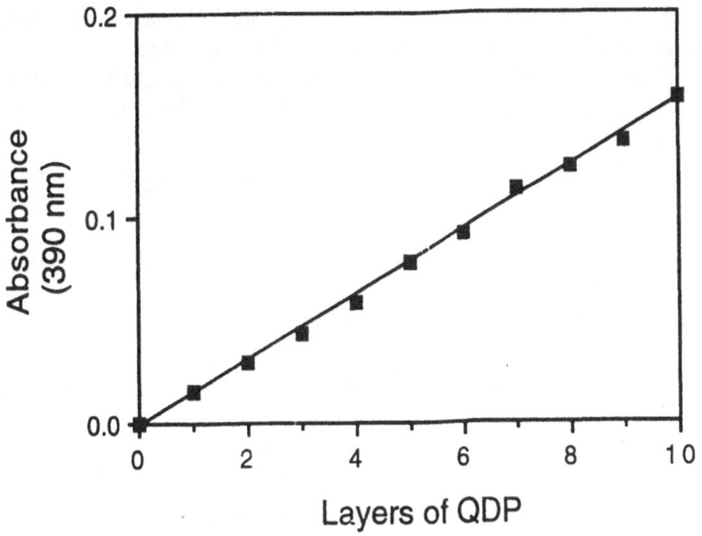

Figure 6. The absorbance of QDP at 390 nm *vs.* the number of QDP layers assembled a the zirconated LB template.

that the inorganic component adds to these films is increased lattice energy. The metal phosphonate LB films have crystalline order dictated by the metal phosphonate lattice, giving the LB films increased crystallinity and enhanced stability relative to a purely organic LB film. There are many applications where LB assemblies have failed due to the metastable nature of the layered structure. By incorporating the metal-phosphonate inorganic lattice energy, the layered structure becomes favored. Future work will be aimed at forming LB films of functional organic molecules designed to complement the extended lattice inorganic systems.

5. Acknowledgments

We thank the National Science Foundation for support of this research.

6. References

1. Day, P. and Ledsham, R. D. (1982) Organic-Inorganic Molecular Composites as Possible Low-Dimensional Conductors: Photo-Polymerization of Organic Moities Intercalated in Inorganic Layer Compounds *Mol. Cryst. Liq. Cryst.* **86**: 163-174.

2. Day, P. (1985) Organic-Inorganic Layer Compounds: Physical Properties and Chemical Reactions *Phil. Trans. R. Soc. Lond. A* **314**: 145-158.
3. Bringley, J. F. and Averill, B. A. (1990) Aromatic Hydrocarbon Intercalates: Synthesis and Characterization of Iron Oxychloride Intercalated with Perylene and Tetracene. *Chem. Mater.* **2**(2): 180-186.
4. Kanatzidis, M. C., Wu, C.-G., Marcy, H. O. and Kannewurf, C. R. (1989) Conductive Polymer Bronzes. Intercalated Polyaniline in V2O5 Xerogels *JACS* **111**(11): 4139-4141.
5. Cao, G., Hong, H.-G. and Mallouk, T. E. (1992) Layered Metal Phosphates and Phosphonates: From Crystals to Monolayers *Acc. Chem. Res.* **25**: 420-427.
6. Thompson, M. E. (1994) Use of Layered Metal Phosphonates for the Design and Construction of Molecular Materials *Chem. Mater.* **6**(8): 1168-1175.
7. Ozin, G. A. (1992) Nanochemistry: Synthesis in Diminishing Dimensions. *Adv. Mater.* **4**(10): 612-649.
8. Bein, T., Ed. (1992) *Supramolecular Architecture. Synthetic Control in Thin Films and Solids.* ACS Symposium Series. Washington DC, American Chemical Society.
9. Beck, J. S., Vartuli, J. C., Roth, W. J., Leonowicz, M. E., Kresge, C. T., Schmitt, K. D., Chu, C. T.-W., Olson, D. H., Sheppard, E. W., McCullen, S. B., Higgins, J. B. and Schlenker, J. L. (1992) A New Family of Mesoporous Molecular Sieves Prepared with Liquid Crystal Templates *J. Am. Chem. Soc.* **114**: 10834-10843.
10. Monnier, A., Schuth, F., Huo, Q., Kumar, D., Margolese, D., Maxwell, R. S., Stucky, G. D., Krishnamurty, M., Petroff, P., Firouzi, A., Janicke, M. and Chmelka, B. F. (1993) Cooperative Formation of Inorganic-Organic Interfaces in the Synthesis of Silicate Mesostructures *Science* **261**: 1299-1303.
11. Kresge, C. T., Leonowicz, M. E., Roth, W. J., Vartuli, J. C. and Beck, J. S. (1992) Ordered Mesoporous Molecular Sieves Synthesized by a Liquid-Crystal Template Mechanism *Nature* **359**: 710-713.
12. Roberts, G. G., Ed. (1990) *Langmuir-Blodgett Films.* New York, Plenum Press.
13. Byrd, H., Pike, J. K. and Talham, D. R. (1994) Extended-Lattice Langmuir-Blodgett Films: Manganese Octadecylphosphonate LB Films are Structural and Magnetic Analogs of Solid-State Manganese Phosphonates *J. Am. Chem. Soc.* **116**: 7903-7904.
14. Byrd, H., Pike, J. K. and Talham, D. R. (1993) Inorganic Monolayers Formed at an Organic Template: A Langmuir-Blodgett Route to Monlayer and Multilayer Films of Zirconium Octadecylphosphonate *Chem. Mater.* **5**: 709-715.
15. Clearfield, A. (1990) Layered Phosphates, Phosphites, and Phosphonates of Group 4 and 14 Metals *Comm. Inorg. Chem.* **10**: 89-128.
16. Dines, M. B. and DiGiacomo, P. M. (1981) Derivatized Lamellar Phosphates and Phosphonates of M(IV) Ions *Inorg. Chem.* **20**: 92-97.
17. Cao, G., Lynch, V. M., Swinnea, J. S. and Mallouk, T. E. (1990) Synthesis and Structural Characterization of Layered Calcium and Lanthanide Phosphonate Salts *Inorg. Chem.* **29**: 2112-2117.
18. Maoz, R. and Sagiv, J. (1984) On the Formation and Structure of Self-Assembling Monolayers *J. Colloid Interface Sci.* **100**: 465-496.
19. Porter, M. D., Bright, T. B., Allara, D. L. and Chidsey, C. E. D. (1987) Spontaneous Organized Molecular Assemblies. 4. Structural Characterization of n-

Alkyl Thiol Monolayers on Gold by Optical Ellipsometry, Infrared Spectroscopy, and Electrochemistry *J. Am. Chem. Soc.* **109**: 3559-3568.

20. Carling, S. G., Day, P. and Visser, D. (1995) Dimensionality crossovers in the magnetization of the weakly ferromagnetic two-dimensional manganese alkylphosphonate hydrates MnCnH2n+1, n = 2-4 *J. Phys.: Condens. Matter.* **7**: L109-L113.
21. Carling, S. G., Day, P., Visser, D. and Kremer, R. K. (1993) Weak Ferromagnetic Behavior of the Manganese Alkylphosphonate Hydrates *J. Solid State Chem.* **106**: 111-119.
22. Crews, M. L. (1992). Master's Thesis, University of Florida.
23. Rushbrooke, G. S. and Wood, P. J. (1958) On the Curie Points and High Temperature Susceptibilities of Heisenberg Model Ferromagnetics *Molec. Phys.* **1**: 257.
24. Byrd, H., Whipps, S., Pike, J. K., Ma, J., Nagler, S. E. and Talham, D. R. (1994) Role of the Template Layer in Organizing Self-Assembled Films: Zirconium Phosphonate Monolayers and Multilayers at a Langmuir-Blodgett Template *J. Am. Chem. Soc.* **116**: 295-301.
25. Lee, H., Kepley, L. J., Hong, H.-G., Akhter, S. and Mallouk, T. E. (1988) Adsorption of Ordered Zirconium Phosphonate Multilayer Films on Silicon and Gold Surfaces *J. Phys. Chem.* **92**: 2597-2601.
26. Lee, H., Kepley, L. J., Hong, H.-G. and Mallouk, T. E. (1988) Inorganic Analogues of Langmuir-Blodgett Films: Adsorption of 1,10 Decanebisphosphonate Multilayers on Silicon Surfaces *J. Am. Chem. Soc.* **110**: 618-620.
27. Kepley, L. J., Sackett, D. D., Bell, C. M. and Mallouk, T. E. (1992) Metal-Insulator-Semiconductor and Metal-Insulator-Metal Devices Derived From Zirconium Phosphonate Thin Films *Thin Solid Films* **208**: 132-136.
28. Katz, H. E., Schilling, M. L., Chidsey, C. E. D., Putvinski, T. M. and Hutton, R. S. (1991) Quaterthiophenediphosphonic acid (QDP): A Rigid, Electron-Rich Building Block for Zirconium-Based Multilayers *Chem. Mater.* **3**: 699-703.
29. Katz, H. E., Scheller, G., Putvinski, T. M., Schilling, M. L., Wilson, W. L. and Chidsey, C. E. D. (1991) Polar Orientation of Dyes in Robust Multilayers by Zirconium Phosphate-Phosphonate Interlayers *Science* **254**: 1485-1487.

CATION TRANSPORT ACROSS BILAYER MEMBRANES
Techniques and mechanisms

T.M. FYLES
Department of Chemistry, University of Victoria
Box 3065, Victoria, B.C. Canada V8W 3V6

1. Introduction

Membrane transport of cations has played a role in the development of supramolecular chemistry from the outset. Three-phase transport measurements, and associated two-phase extraction experiments are frequently one of the first tools used to characterize a new supramolecular host and in many cases they provide the only physical characterization of the supramolecular properties of a new host. Moreover, transport systems appear to have significant potential for eventual applications in separations, drug delivery, or analytical sensors, hence their enduring popularity.[1-3]

Membrane transport is dominated by diffusion - an inherently sluggish process. The quest for faster or more efficient systems inevitably leads to very thin membranes - to bilayers. Synthetic ion transporters which are active in bilayer membranes have been reported by a number of groups over the past decade.[3, 4] At the same time, the structural and mechanistic characteristics of naturally occurring oligopeptide channels, such as gramicidin and alamethicin [5], have been explored in detail. Combining information on both natural and synthetic ion transporters leads to three general criteria for the design of membrane-active transporters. The first is a requirement for compatibility of the transporter with the lipid environment. The transporter should be amphiphilic and should have a roughly columnar shape to insert into a bilayer leaflet in place of one or more lipid molecules. Secondly, the transporter must stabilize an ion as it crosses the hydrophobic region of the bilayer. This might involve direct transporter-ion interactions, or might simply involve transporter-stabilized water deep within the bilayer. A third practical requirement is a feasible synthesis. The active structures required are large, capable of spanning a 4 nm bilayer thickness and surrounding an ion of at least 0.3 nm diameter. Consequently, an efficient synthesis or an efficient aggregation process is required.

Our effort initially focused on the construction of large unimolecular transporters.[6] The most active of these materials had molecular weights of 3500-5000 g/mol, and apparently acted as channel-type transporters in vesicle bilayers.[7] The synthesis of these structures used a modular set of components that permitted rapid and efficient

L. Echegoyen and A.E. Kaifer (eds.), Physical Supramolecular Chemistry, 39–46.

construction of the targets. Nonetheless, the syntheses were exacting, and the characterization of the target compounds pressed the limits of conventional proof-of-structure techniques (NMR, MS). At the same time, we explored the behaviours of much simpler bola-amphiphiles made from the same construction set.[8] Although not as active in vesicles as the unimolecular transporters, these structures were much lower molecular weight (<1500 g/mol) and therefore much simpler to prepare and characterize. During the same period Regen[9], Menger[10], and Kobuke[11], all reported very simple compounds which apparently induce transport across vesicle bilayers.

Among the many unanswered questions from this preliminary work are two of overriding practical importance. The first is: what is(are) the mechanism(s) that allow these simple compounds to provoke ion transport? And secondly: how can those mechanisms be identified by experiment? Our current emphasis is directed to resolving these issues for the simplest of the active transporters, and our progress is summarized here.

2. Transport mechanisms

The process of describing a transport "mechanism" differs from elucidating a conventional chemical mechanism in both the spatial and temporal dimensions. In a conventional chemical reaction each elementary step occurs extremely rapidly - within a bond vibration - and involves molecular scale dimensions of 0.2 - 0.3 nm. Intermediates in a conventional chemical process could have long lifetimes, and during their lifetimes might diffuse over large distances, but this is usually ignored in the description of a mechanism. In contrast, the rapid elementary steps of a transport mechanism (the ligand and solvent exchange reactions) are usually ignored, and the emphasis is placed on the diffusional events which occur over distances much greater than the molecular scale. Transport "mechanisms" also involve a description of the coupling between transmembrane flux and the transmembrane gradients which drive the process.[12] At its core, a transport "mechanism" is a description of the behaviour of a collection of molecules. Transport is inherently supramolecular.

A cartoon of the main types of mechanisms which have been described for ion transport across bilayer membranes is given in Figure 1. The carrier mechanism (Figure 1, A) is well known from studies involving liquid membranes.[2] The cation forms an inclusion complex with the host-carrier, and the cation complex diffuses across the membrane. Typically this diffusional step is rate-limiting. Cation release to the aqueous phase, followed by return diffusion of the carrier completes the transport cycle. The ion channel mechanism (Figure 1, B) is derived from the gramicidin paradigm.[5] The transporter is held in a transmembrane orientation and transport occurs by diffusion of the cation within the transporter. The distinction between a carrier and a channel is obvious: in the former the cation moves *with* the transporter, while in the latter the

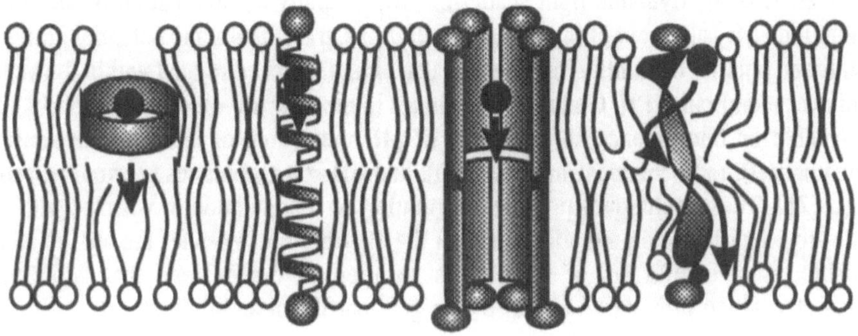

Figure 1. Schematic mechanisms for ion transport across bilayer membranes: A, a carrier mechanism; B, an ion channel; C, an aggregate pore; D, a membrane-disrupting event.

cation moves *past* the transporter. Despite the obvious physical differences between these two mechanisms they are kinetically indistinguishable and are simply the extremes of a continuum.[13]

A variety of ways to organize a transmembrane channel can be envisaged. The simple dimer illustrated (Figure 1, B) might be replaced by a unimolecular structure (not illustrated), or could involve an aggregate of transporter molecules as illustrated in Figure 1, C. This aggregate pore is akin to the active structure proposed for the antibiotic amphotericin [14], or the transmembrane clusters formed by amphiphilic peptide helices.[3] Cation diffusion would occur within the aggregate, presumably involving a deep fjord of water, or a continuous water channel. Channel-like activity can also be envisaged as a result of defect-formation or "membrane-disruption" [9] as illustrated in Figure 1, D. The role of the transporter in this case is to disrupt the bilayer organization sufficiently to reduce or remove the barrier to free diffusion across the membrane. This is short of complete disruption of the membrane as a detergent might provoke. The membrane structure and morphology is preserved on a large scale, but the barrier has been breached at defined defect sites.

3. Experimental probes of transport mechanism.

There are two principle experimental systems for the investigation of bilayer transport phenomena: vesicles (or liposomes), and planar bilayers (black lipid membranes). The former are closed spherical structures formed by dispersion of lipid in water. There are numerous ways to prepare vesicles, ranging from simple swirling of the lipid in a suitable buffer, to formation of equilibrium emulsions or suspensions followed by removal of the suspending agent.[14] Vesicle size and morphology (uni- *versus* multi-bilayer) can be controlled by suitable choice of dispersion conditions, and by subsequent size separation. Electron microscopy can distinguish uni- and multi-lamellar structures, and can be used to construct a size distribution histogram.[15] Particle size is more easily assessed by dynamic light scattering.[16] Usually a combination of techniques is required. Measurement of internal concentrations, transmembrane potential gradients, membrane polarity or fluidity, requires the use of indicators entrapped within the vesicle or bilayer membrane.[17] Common techniques to monitor transport include NMR [18], UV-visible or fluorescence spectrometry of pH, metal ion, or environment sensitive indicators[19], radiochemical methods [20], and pH-stat techniques to monitor proton efflux.[21] These techniques provide information on a population of vesicles, although some are selective for a particular subset of the distribution. Vesicles are also relatively unstable, so a careful set of controls is required to demonstrate that addition of transporter which provokes a collapse of a gradient is not an artifact.

Direct analysis of the behaviour of single transporter molecules is possible using planar bilayer techniques. Both solvent derived, and solvent-free bilayers can be formed across a small hole in a hydrophobic barrier.[22] This experiment provides independent access to both sides of the membrane, and the transport of ions can be directly monitored by potentiometric, conductimetric, and voltage-clamped techniques.[23] Both the conductivity (specific conductance, ion transport rate) and the lifetime of the active states can be measured, leading directly to the determination of mechanism.

The experimental demonstration of a particular transport mechanism is a complex process involving simultaneous testing of multiple hypotheses. Some general guidelines are summarized in Table 1. Since a carrier mechanism requires that the ion diffuse with the transporter, the transport rate will be strongly dependent on the membrane fluidity. Activation energy has also be used as a key criterion.[18] An ion channel mechanism is most readily demonstrated by the observation of step-conductance changes in a voltage-clamp experiment using planar bilayers. Aggregate pores require more than one molecule in the active unit, therefore a high kinetic order is a good indicator of this type of mechanism. Non-specific membrane disruption is most easily demonstrated by the release of large entrapped molecules under the same conditions as the ion transport.

TABLE 1. Summary of transport mechanisms and key experimental tests.

General mechanism	Key mechanistic feature	Experimental signature
Carrier	coupled diffusion of transporter and ion	controlled by vesicle fluidity
Ion channel	diffusion within the transporter	step-conductance changes conductivity appropriate to small pore
Aggregate pore	aggregation within bilayer, diffusion within aggregate	high kinetic order
Membrane-disruption	non-specific permeability increase	release of large molecular indicators vesicle morphology intact

4. Transport behaviours of bolaamphiphiles.

This section discusses some specific examples of transport mechanism using the transporters illustrated in Figure 2 as examples. The bis-macrocyclic tetraester A8TrgAP8TrgA is an example of the type of structure which can be assembled from our modular construction set.[7] Bolaamphiphiles of this general type are quite active transporters in vesicles [8], and this particular compound is among the most active discovered to date. In vesicles it has an apparent kinetic order of 1.7, and is cation selective with an Eisenmann III sequence (K > Rb > Na > Cs > Li).[24] In planar bilayers it shows multiple step-conductance changes of equal conductance per step. We conclude from the body of available data that this compound is an example of an aggregate pore involving a small number of molecule in the active structure.[24]

The bis-crown ether illustrated in Figure 2 is conceptually derived from the Regen diester, by substitution of the linear glycol sections by crown ethers. The synthesis follows directly from work reported in the early 80's which gave amide-acid crown ethers such as 18NH18C6A.[25] The latter are known to act as carriers in liquid membranes and are potent surfactants.[26] We anticipated that the addition of crown ethers would impart cation selectivity to the Regen diester, while preserving the membrane-disrupting activity of the latter.[9] This expectation was fulfilled: the bis-crown ether is quite inactive relative to the Regen diester, but shows significantly higher activity with potassium than sodium.

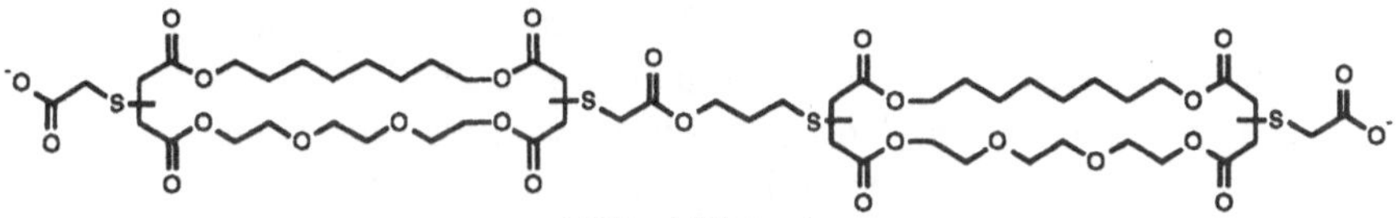

A8TrgAP8TrgA

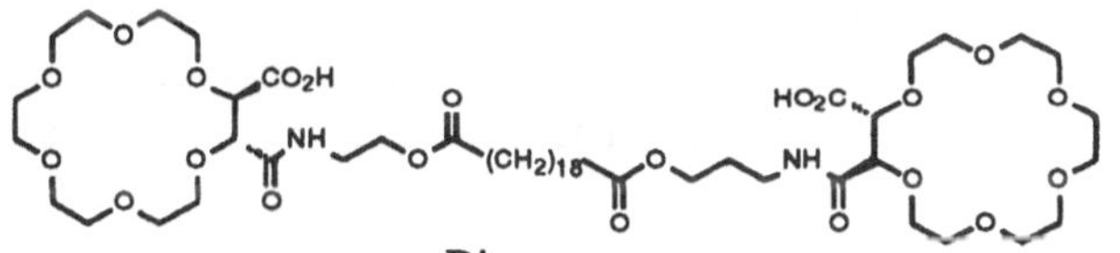

Bis-crown

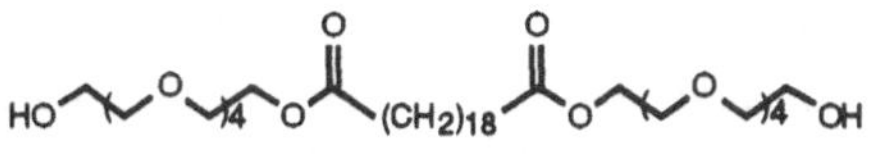

Regen diester

18NH18C6A

Figure 2 Structures of the compounds discussed in the text.

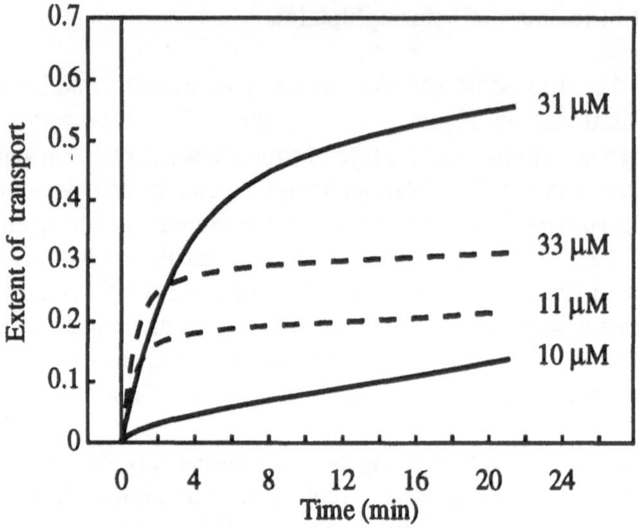

Figure 3 Experimental pH-stat data for A8TrgPA8TrgA (solid lines) and bis-crown ether (dashed lines) at two concentrations.

The behaviours of the bis-crown ether and A8TrgPA8TrgA as assessed by a pH-stat experiment are compared in Figure 3. The aggregate pore shows a regular, near-first-order increase in the extent of transport as a function of time and a significant increase in rate with increase in concentration. The bis-crown ether shows a rapid burst of proton efflux followed by a slow transport process with a rate which is nearly independent of concentration. This is consistent with the expectation of a membrane disruption event - the rapid burst is the main disruption of the most susceptible sub-population of vesicles.

The membrane disruption is more clearly identified by the experiments summarized in Figure 4. The assay is based on the release of 5[6]-carboxyfluorescein from vesicles hence a relative large pore is required. The Regen diester, 18NH18C6 and the detergent Triton X-100 are all active membrane disrupting agents. The activity of the bis-crown ether depends on the presence of a divalent cation such as Ba^{2+}, but even with sufficient Ba^{2+} only 80% of the vesicle contents are released. This corresponds to the fraction of vesicles which are uni-lamellar. The bis-crown ether is quite a selective disrupting agent - it will only attack large unilamellar vesicles, and requires a divalent ion to do so. Regen has demonstrated that U-shaped compounds are the most active membrane disrupting agents.[9] It is tempting to speculate that the bis-crown is held in a U-shape by formation of a 1:1 sandwich complex. Consistent with this proposal is the observation of a saturation of disruption at an apparent stoichiometry of 1:1, and an apparent association constant about 10^3 M^{-1}.

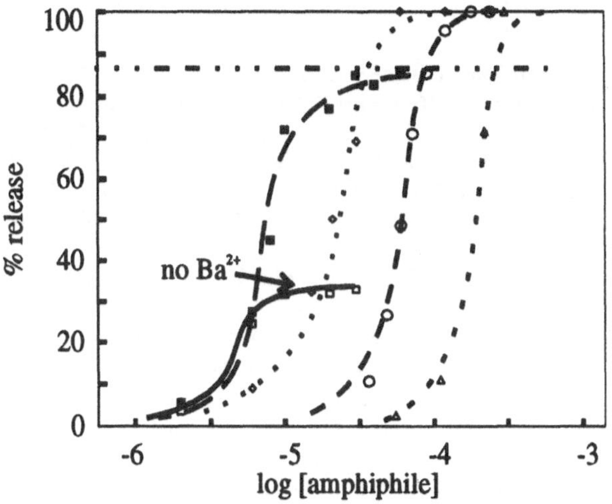

Figure 4 Extent of 5[6]-carboxyfluorescein release as a function of amphiphile concentration. Squares = bis-crown ether; diamonds = Regen diester; circles = 18NH18C6A; triangles = Triton X-100.

5. Summary

The behaviour of a large suite of bolaamphiphile transporters can be rationalized by the mechanistic sketch of Figure 5. The starting point is surface adsorption of the bolaamphiphile on the vesicle (A). This state progresses to a U-shaped state (B) which in some systems does not proceed further. The head group region is strongly perturbed by the U-shaped state and this leads to membrane disruption. When the hydrophilic-lipophilic balance is correct, the U-shaped state can progress to an isolated transmembrane state (C). This state must have relatively little impact on the bilayer as it is invisible to the planar bilayer technique. Clustering of monomers from state C to D leads to aggregate pores. This sequence of events is consistent with a large body of

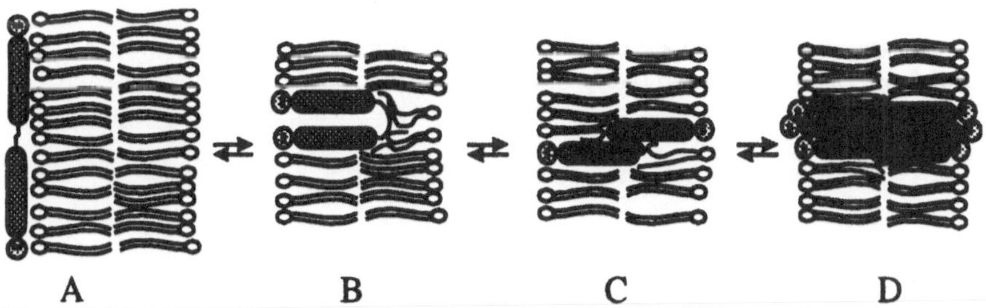

Figure 5 Schematic mechanism for bolaamphiphile interactions with bilayer membranes.

data but the structural factors which control the individual rates, partition equilibria, and overall behaviours are only slowly coming under experimental scrutiny.

6. References

1. Fyles, T.M. (1990) *Biorg. Chem. Frontiers* **1**, 72.
2. Tsukube, H. (1993) *Liquid Membranes*, C.R.C. Press, Boca Raton.
3. Fyles, T.M. and van Straaten-Nijenhuis, W.F. (1996) Ion Channel Models, in Reinhoudt, D. (ed.), *Comprehensive Supramolecular Chemistry*, Elsevier, Amsterdam, Volume 10, chapter 3.
4. Previous systems are cited in recent reports: Tanaka, Y., Kobuke, Y, and Sokabe, M. (1995) *Angew. Chem. Int. Ed. Engl.* **34**, 69; Stadler, E. Dedek, P. Yamashita, Y, and Regen S.L. (1994) *J. Am. Chem. Soc.* **116**, 6677.
5. Sansom, M.S.P. (1991) *Prog. Biophys. Mol. Biol.* **55**, 140; Wooley, G.A. and Wallace, B.A. (1992) *J. Membr. Biol.* **129**, 109.
6. Fyles, T.M., James, T.D. and Kaye, K.C. (1993) *J. Am. Chem. Soc.* **115**, 12315.
7. Fyles, T.M., James, T.D., Pryhitka, A. and Zojaji, M. (1993) *J. Org. Chem.* **58**, 7456.
8. Fyles, T.M., Kaye, K.C., Pryhitka, A., Tweddell, J. and Zojaji, M. (1994) *Supramol. Chem.* **3**, 197.
9. Jayasuriya, N., Bosak, S. and Regen, S.L. (1990) *J. Am. Chem. Soc.* **112**, 5844.
10. Menger, F.M., Davis, D.S., Perischetti, R.A. and Lee, J.-J. (1990) *J. Am. Chem. Soc.* **112**, 2451.
11. Kobuke, Y., Ueda, K. and Sokabe, M. (1992) *J. Am. Chem. Soc.* **114**, 7618.
12. Goddard, J.D. (1985) *J. Phys. Chem.* **89**, 1825.
13. Lauger, P. (1972) *Science*, **172**, 24.
14. New, R.R.C. (ed.) (1990) *Liposomes: a practical approach*, IRL Press, Oxford.
15. Warren, R.C. (1987) *Physics and the Architecture of Cell Membranes*, Adam Hilger, Bristol.
16. Menger, F.M., Lee, J.-J., Aitkens, P. and Davis, S. (1989) *J. Colloid Interfac. Sci.* **129**, 185.
17. Starzak, M.E. (1984) *The Physical Chemistry of Membranes*, Academic Press, Orlando.
18. Grandjean, J. and Laszlo, P. (1984) *J. Am. Chem. Soc.* **106**, 1472; Riddell, F.G., Arumugan, S. and Patel, A. (1990) *Inorg. Chem.* **29**, 2397.
19. Clement, N.R. and Gould, J.M. (1981) *Biochemistry* **20**, 1534; Blau, L. and Weissmann, G. (1988) *Biochemistry* **27**, 5661.
20. Brunner, J., Graham, D.E., Hauser, H. and Semenza, G. (1980) *J. Membr. Biol.* **57**, 133.
21. Castaing, M., Morel, F., and Lehn, J.-M. (1986) *J. Membr. Biol.* **89**, 251.
22. Alvarez, O. (1986) in Miller, C. (ed.) *Ion Channel Reconstitution*, Plenum Press, New York; Nikolelis, D.P. and Krull, U.J. (1992) *Talanta* **39**, 1045.
23. Hille, B. (1984) *Ion Channels of Excitable Membranes* Sinauer Assoc., Sunderland.
24. Fyles, T.M., Heberle, D., van Straaten-Nijenhuis, W.F., and Zhou, X. (in press) *Supramol. Chem.*
25. Fyles, T.M., Malik-Diemer, V.A. and Whitfield, D.M. (1981) *Can. J. Chem.* **59**, 1734.
26. Fyles, T.M., Malik-Diemer, V.A., McGavin, C.A. and Whitfield, D.M. (1982) *Can. J. Chem.* **60**, 2259; Fyles, T.M. (1985) *J. Chem. Soc. Farad. I* **82**, 617.

Acknowledgements The enthusiastic contributions of co-workers indicated by the references is gratefully acknowledged. This ongoing project is supported by the Natural Sciences and Engineering Research Council of Canada.

LASER STUDIES OF MOLECULES AT LIQUID INTERFACES BY SECOND HARMONIC AND SUM-FREQUENCY GENERATION

E. BORGUET, D. ZHANG AND K. B. EISENTHAL
Department of Chemistry,
Columbia University,
New York, NY 10027, USA

1. Introduction.

Interfaces are of considerable scientific and technological interest. However, the investigation of chemical and physical phenomena at interfaces has not been as straightforward as in the bulk. The interface region, defined as the atomic or molecular layers that constitute the junction between two dissimilar media, contains far fewer species than are typically probed in bulk experiments. This difficulty is compounded when the species of interest are present both in the bulk and the interface. Most experimental techniques are unable either to probe the interface exclusively or to discriminate between the bulk and interface response. In the past, surface tension or surface potential methods have been applied to investigate molecules adsorbed at the air/water interface, and quantities such as the adsorption isotherm have been determined [1-3]. However, these methods are limited by their inability to distinguish between distinct chemical species and to provide fundamental insight at a molecular level. Spectroscopic techniques have the potential to overcome these limitations. One obstacle to the use, in interface investigations, of the powerful spectroscopic methods developed for the study of bulk processes is the necessity to differentiate between the response of the overwhelming number of bulk molecules and that of the much smaller number of surface species.

Recently, the investigation of interfaces using second order nonlinear spectroscopic techniques, such as Second Harmonic Generation (SHG) and Sum Frequency Generation (SFG) has been the subject of intense scientific activity [4-9]. SHG and SFG are coherent second order nonlinear optical processes which are forbidden, in the electric-dipole approximation, in centrosymmetric media. SFG results from the action of two incident light fields, E_{ω_1} and E_{ω_2}, centered at optical frequencies ω_1 and ω_2, on a material characterized by a second order nonlinear susceptibility, $\chi^{(2)}_{\omega_1+\omega_2}$. SHG is the special case when $\omega_1 = \omega_2$. These fields induce a second order polarization, $P^{(2)}_{\omega_1+\omega_2}$, in the medium which is related to the incident fields by equation (1).

L. Echegoyen and A.E. Kaifer (eds.), Physical Supramolecular Chemistry, 47–64.

$$P^{(2)}_{\omega_1+\omega_2}=\chi^{(2)}_{\omega_1+\omega_2}\, E_{\omega_1}\, E_{\omega_2} \quad (1)$$

The inversion symmetry is broken at the interface as a consequence of its asymmetric environment, exemplified by the observation that molecules at liquid surfaces are aligned in a preferential direction rather than randomly oriented as they are in the bulk [10, 11]. Hence, the induced coherent second order polarization can radiate from the interface. SHG and SFG enable the concentration, orientation and identity of interfacial species to be determined, with ultrafast time resolution if necessary.

The asymmetry of the forces acting at interfaces is a universal characteristic. This has important consequences for the chemical and physical properties of the interfacial region. It has been demonstrated experimentally, using SHG, that fundamental chemical behavior such as acidity (pH) [12] and equilibrium constants [13] are dramatically different at the interface. SFG has enabled new molecular behavior such as phase transitions involving small mutually soluble molecules at liquid interfaces to be observed [14].

2. Barrierless and Activated Relaxation Dynamics at Aqueous Interfaces

2.1 EXPERIMENTAL TECHNIQUE: SHG

The nonlinear response of the interface is intrinsically weak. Even with the use of pulsed lasers, which provide the high peak powers (>10 MW/cm^2) necessary to drive the second order nonlinear response, one typically detects on the order of one second harmonic (SH) photon for every 10^{12} incident fundamental photons. For the experiments on Malachite Green reported here a 10 Hz amplified colliding-pulse mode-locked dye laser provides 130 fs pulses at 625 nm. The pump/probe approach employed is illustrated in Figure 1. The probe laser beam is focused at the interface and the second harmonic radiation (312.5 nm) is separated from the reflected fundamental

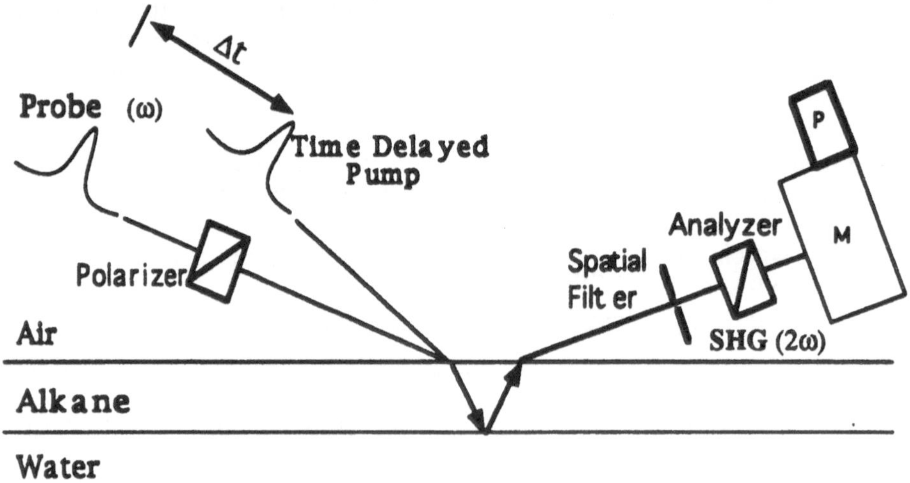

Figure 1

using filters and a monochromator (M), and detected using a gated photomultiplier (P). The pump and probe beams, both incident at 70 degrees to the surface normal, propagate in slightly different directions to enable spatial separation of the SH radiation generated by each beam. The liquid-air and liquid-liquid interface experiments were performed on samples contained in Teflon beakers. Experiments at the fused silica/aqueous interface used a Teflon sample cell tightly covered with a fused silica equilateral prism. Polarizers enable distinct components of the surface nonlinear susceptibility to be probed. Aqueous solutions of Malachite Green Chloride in doubly distilled water were prepared. The bulk acidity was controlled using hydrochloric acid solutions. Gibbs surface excess at the air/aqueous interface was determined from surface tension measurements using the Wilhelmy plate method [1].

2.2 DYNAMICS OF BARRIERLESS PROCESSES AT INTERFACES

The dynamics of chemical and physical processes depend on the properties of the medium in which they occur. Time-resolved Second Harmonic Generation (TRSHG), by avoiding the complications of bulk contributions, enables surface species to be probed directly and on an ultrafast timescale [4, 7, 9]. Using this technique, progress has been made in our understanding of elementary molecular dynamics at interfaces [4, 7]. The rotational motion of Rhodamine 6G, a large organic molecule, at the air/aqueous interface has been studied using picosecond lasers and the dynamics were observed to be slower than in the bulk, and dominated by out of plane motions [12]. TRSHG has also been used to investigate the dynamics of activated processes, and in the case of the photoisomerization of DODCI (3,3'-diethyloxadicarbocyanine iodide) at the air/aqueous interface the rate was measured to be considerably faster than in the bulk [15]. TRSHG has been employed to probe the photophysics of a number of molecules, including Rhodamine 6G and Malachite Green, at silica/liquid and silica/air interfaces [16, 17]. Studies of intermolecular electronic energy transfer at an air/liquid interface have been performed, and behavior consistent with our understanding of the process in the bulk was observed [18].

Barrierless processes constitute an important class of chemical isomerization dynamics, of relevance to primary electron transfer in the photosynthetic reaction center, oxygen transport by hemoglobin and fundamental mechanisms of vision [19]. In addition, they provide a means of probing the solvent structure in the interfacial region in particular how it affects processes occurring on an ultrafast timescale [20, 21]. The dynamics of barrierless chemical transformations are determined not by the presence of barriers that separate different configurational regions but rather chiefly by interactions with the surrounding molecules (frictional effects) and the shape and intersections of the potential energy surfaces of different electronic states. As a prototype system we have explored the excited state isomerization dynamics of a triphenylmethane (TMP) dye, Malachite Green, a molecule extensively studied in bulk media [22]. Malachite Green adsorbed on solid surfaces has also been the object of a number of investigations [16, 17, 23-

27]. In the bulk the ultrafast relaxation dynamics of TMP dyes, such as Malachite Green, is sensitive to the nature of the molecular environment [22, 28]. The fluorescence quantum yield and excited state relaxation rate show a strong dependence on solvent viscosity. The ultrafast dynamics and apparent lack of a temperature dependence, under conditions of constant viscosity, in low viscosity solvents indicate that the relaxation is barrierless. The rate limiting step is believed to involve the rotation of the aromatic groups about the bond between each one and the central carbon atom [29, 30]. The solvent drag forces opposing these rotations are manifested by the slowing down of the isomerization as the macroscopic bulk solvent viscosity increases. A number of theories have been developed to account for the experimental observations. Most assume that the dynamics involve diffusional motion on excited state potentials of differing shape, according to the model, until a sink, again of varying functional form, is encountered at which point internal

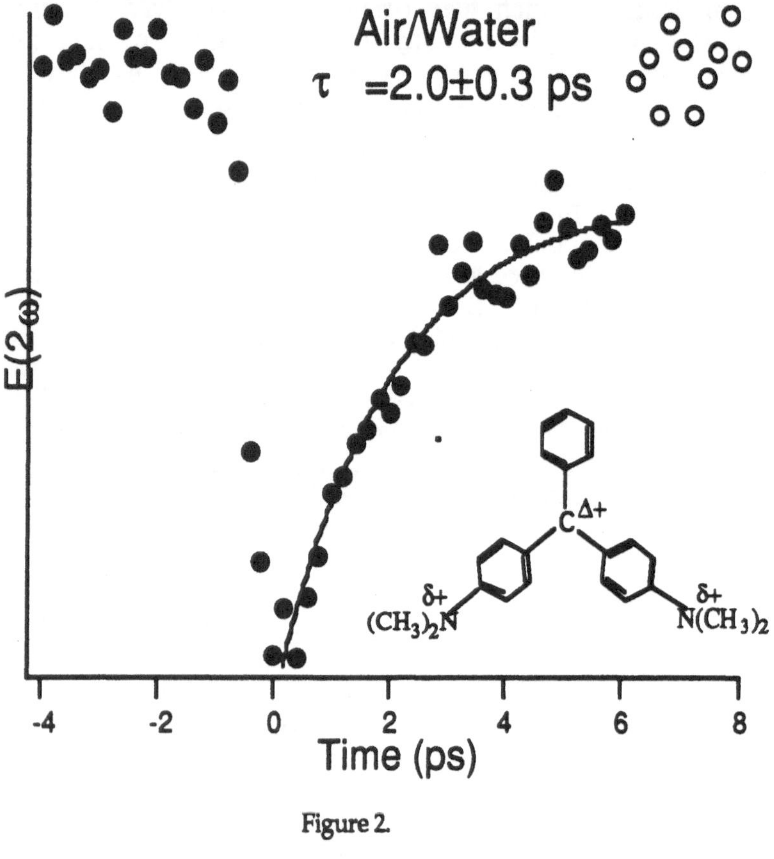

Figure 2.

conversion to the ground state occurs [31-35]. However, the relative contributions of the three aromatic groups, the necessity of concerted rotation, and indeed a complete understanding of bulk dynamics of TMP dyes are unclear at present. It is clear, nevertheless, that when the aromatic rings are not free to rotate, as in the structurally related xanthene dyes such as Rhodamine B, where the anilino rings are connected by an oxygen atom, there is no rapid excited state relaxation. Rather one observes strong fluorescence and nanosecond excited state lifetimes, ostensibly related to the fact that rotation of the terminal groups is precluded by a chemical bond.

The TRSHG dynamics observed at the air/aqueous interface, following promotion of Malachite Green to its first excited electronic state (S_1), are shown in Figure 2. We estimate, from surface tension measurements that the surface coverage is approximately 10^{13} molecules/cm^2 for the 100 μM solution used in the experiments reported in Figure 2. The TRSHG dynamics were independent of concentration in the range investigated (100-200 μM) and independent of energy density in the range studied.

The relaxation dynamics are best characterized by a single exponential decay (2.0±0.3 ps at the air/aqueous interface). This is significantly slower than reported for Malachite Green in bulk water (≈ 0.7ps) [36-38]. We found that molecular rotation at the interface is not significant on the timescale of the isomerization kinetics at the interface from the observation that the polarization of the second harmonic light did not change with delay between the pump and probe light pulses. This observation is consistent with previous TRSHG measurements which revealed that diffusional molecular motion at the air/aqueous interface for molecules of comparable size and charge is only significant on timescales greater than one hundred picoseconds [12].

The frictional resistance exerted by the solvent against the twisting of the aromatic moieties in the bulk isomerization dynamics is seen in the viscosity dependence of the kinetics. Solvent polarity is not observed to play a significant role in TMP relaxation dynamics as similar dynamics are observed in solvents of comparable viscosity yet greatly different dielectric constants and values of the solvent polarity parameter ET(30) [39, 40]. We expect that the isomerization at the interface will also be sensitive to the friction experienced along the twisting coordinates.

In order to investigate this hypothesis TRSHG experiments were performed at the alkane/aqueous interface. It is known from various studies, and consistent with our expectations, that the hydrophobic part of a molecule, in this case the phenyl group, would be pointed away from the aqueous phase and the more hydrophilic part, in this case the partially charged dimethylaniline groups, would be oriented into the aqueous phase [10, 41]. The polarization of the second harmonic fields radiated from the alkane/water, corrected for effects of refraction and angle of incidence, suggests a molecular axis tilt angle of $42^{o}\pm1^{o}$, very similar to that found for the air/aqueous interface. Also, the SH signal from the alkane/aqueous interface is observed to be an order of magnitude greater than for the air/aqueous interface, which indicates that the population of Malachite Green is about three times greater at the alkane/aqueous interface for similar bulk concentrations. This result points to a favorable interaction of Malachite Green with the alkane phase and an interface geometry in which the hydrophobic phenyl group projects into the alkane phase.

The inference that the phenyl group projects into the air and alkane phases, at the air/aqueous and alkane/aqueous interfaces, respectively, with about the same orientation enables us to examine the contribution of the different aromatic groups to the relaxation dynamics of Malachite Green. With respect to the phenyl group we expect the twisting motions about the carbon-phenyl group bond to be very sensitive to the frictional properties of the phase into which the phenyl group is projecting. We were surprised to find that the kinetics changed by only ≈20% in going from an octane/H_2O interface (3.0 ps) to a pentadecane/water interface (3.6 ps). The viscosities of octane and pentadecane differ by a factor of six and were expected to markedly change the twisting relaxation dynamics about the central carbon-phenyl bond because the phenyl group projects into the alkane phase. This result brings into question the presently accepted model of isomerization which assumes that the synchronous rotations of all three aromatic rings determines the rate of isomerization [22, 24, 39]. Our results suggest that it is the nitrogen containing dimethylaniline moieties that determine the isomerization rate. Because these more hydrophilic moieties project into the aqueous phase, at approximately the same orientation at the various interfaces, the twisting motions experience the same aqueous friction, and therefore yield the same isomerization dynamics at these various interfaces.

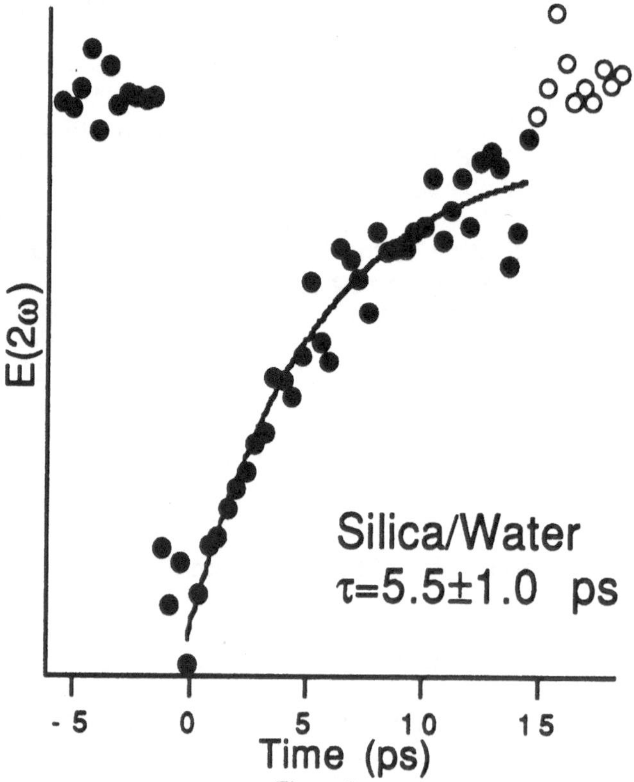

Figure 3

These observations support theoretical studies, based on molecular dynamics simulations, which concluded that the structure of water at the air/aqueous and alkane/aqueous interfaces is similar [42]. We cannot establish on the basis of these experiments whether the different relaxation times are due to the structure of water being different at the alkane/water and air/water interfaces or the structure of water being the same at these interfaces and the different relaxation times arise from the different positions of Malachite Green at these interfaces. In any event the results suggest that there is not a major difference in the water structure at the air/aqueous and alkane/aqueous interfaces.

At the silica/aqueous interface, unlike the air/aqueous and alkane/aqueous interfaces, the phenyl group as well as the dimethylaniline groups must be immersed in the water phase . The isomerization dynamics (Figure 3) is slower (5.5±1.0 ps) than at the air/aqueous and alkane/aqueous interfaces, indicating either a specific interaction of Malachite Green with charged silanol groups or that the structure of water at the silica interface in the vicinity of Malachite Green is different from that at the alkane and air interfaces. This latter result would have a bearing on experimentally based inferences, that water structure and therefore viscosity are greater at the silica/aqueous interface than in the bulk aqueous solution [43-45].

2.3 COMPARISON OF ACTIVATED AND BARRIERLESS ISOMERIZATION RATES AT THE AIR/AQUEOUS INTERFACE

Our study of photoisomerization of a cyanine dye, DODCI, where both a barrier and frictional effects determine the dynamics, indicate a decrease in excited state lifetime at the air/aqueous interface [15]. This is in contrast to the observed slowing of Malachite Green relaxation dynamics discussed above. We believe that the molecular environment, orientation and solvent structure in the vicinity of the different chromophores of Malachite Green and DODCI dictate the observed interfacial behavior. The expected orientation of DODCI leads to one end of the molecule pointing away from the bulk water phase towards the air, analogous to what is observed for the phenyl group of Malachite Green. For DODCI, photoisomerization can be achieved by the motion of one end of the cyanine dye about its methylene bonds, through the air side of the air/aqueous interface, a region of low solvent coupling. However, for Malachite Green the motion that effects the isomerization is that of the anilino groups, which are immersed in the aqueous phase, a region of strong coupling to the interfacial water molecules. Thus the friction experienced by DODCI and Malachite Green is different though they are both at air/aqueous interfaces. The geometry of DODCI at the interface suggests that the polarity experienced would be less at the air aqueous interface than in the bulk aqueous phase because solvation should be greater in the bulk solution. Based on bulk studies we expect the barrier for DODCI isomerization would be higher at the less polar interface and therefore the kinetics would be slower. Our experimental results are opposite to this, suggesting that it is the decreased friction and not the barrier that yields the faster rate for DODCI isomerization at the interface, in agreement with molecular dynamics simulations [46].

3. INTERFACE PHASE TRANSITIONS USING INFRARED+VISIBLE SUM FREQUENCY GENERATION

Vibrational spectroscopy is a powerful means of identifying chemical species and their interactions with the surrounding environment. An interface-specific analog, infrared+visible sum frequency generation (SFG), has been used in this group to investigate the behavior of soluble [14, 47] and insoluble [48] organic molecules at aqueous interfaces. By taking advantage of vibrational resonant enhancement, SFG is capable of acquiring vibrational spectra of molecules at the interface. This is achieved by tuning the infrared light through the characteristic vibrational frequencies of molecules at the interface. A vibrational mode appears in the SFG spectrum, if it is both infrared and Raman active. Distinct parts of a molecule can probed by tuning the IR laser to the vibrational frequencies of distinct chromophores within the molecule.

3.1 EXPERIMENTAL TECHNIQUE: SFG

The SFG experimental setup is illustrated in Figure 4. Briefly, a picosecond visible laser beam (λ=532 nm) and a picosecond tunable infrared (IR) beam ($\lambda\approx$4.5 mm) are overlapped on a common spot on the surface of interest. The two beams are in a counter-propagating geometry and incident at an angle of

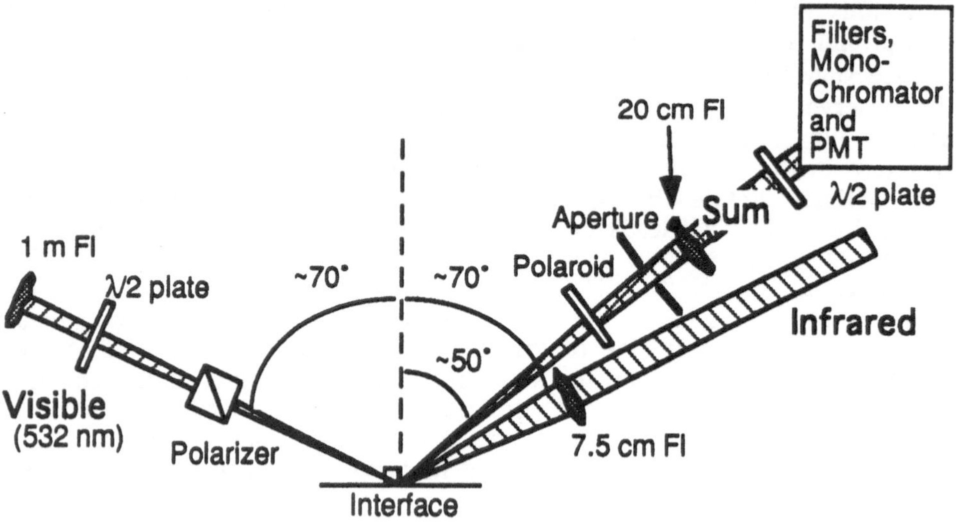

Figure 4

70° from the surface normal. Both incoming beams are polarized at 45° to the surface normal and parallel to each other. The SF signal is then collected in the wavelength range of 470-480 nm at an angle of approximately 50° from the surface normal as dictated by the wavevector matching conditions at the interface [49]. The signal is separated from the background through the use of spatial apertures, optical filters and a monochromator. The SF spectra were obtained by scanning the infrared laser in 2 cm^{-1} increments. Interfacial density of acetonitrile (CH_3CN) is varied using different bulk concentrations, indicated by mole fraction of acetonitrile (X_{ACN}) in water. Samples of long chain nitrile were prepared by spreading a measured amount of hexane solutions of $CD_3(CH_2)_{19}CN$ on to a thoroughly cleaned water surface, and allowing the hexane to evaporated. All measurements were performed by placing the sample in a sealed housing during the experiment to prevent the evaporation of the aqueous solution and eliminate the possibility of thermal convection. Further experimental details of the SFG technique have been reported elsewhere [14].

3.2 STRUCTURAL PHASE TRANSITIONS OF ACETONITRILE AT THE AIR/AQUEOUS INTERFACE

IR+visible Sum-Frequency spectra of acetonitrile (CH_3CN) at two different bulk concentrations and hence surface concentrations, are shown in Figure 5. The SF spectra in the CN stretching region appear very different for these two concentrations. At the lower bulk concentration (X_{ACN}=0.03), the CN stretching vibrational frequency of interfacial acetonitrile is higher than that of neat bulk acetonitrile. This shift is characteristic of nitriles hydrogen

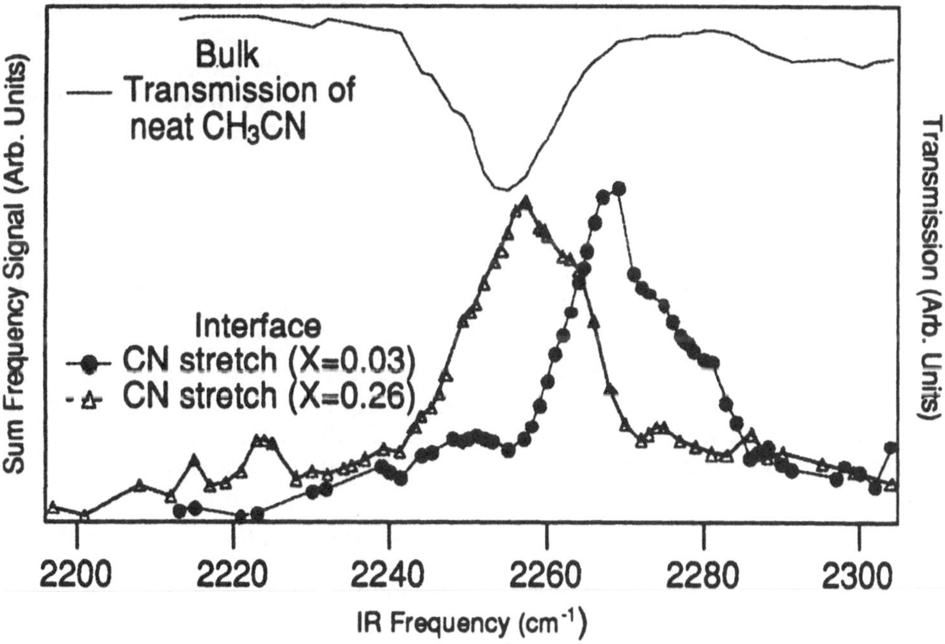

Figure 5

bonded to proton donors such as water in bulk solution [50, 51]. At a higher bulk concentration (X_{ACN}=0.26), the CN vibrational frequency for interfacial acetonitrile is red shifted to a value similar to that of neat bulk acetonitrile (Figure 5). This dependence of vibrational frequency indicates that the environment of acetonitrile molecules at the surface changes with the bulk composition.

In order to obtain a more complete physical picture of the behavior of acetonitrile absorbed at the air/water interface, we have also performed SFG experiments at many other bulk concentrations. The frequency shifts for the CN stretching vibration at the air/aqueous interface versus X_{ACN} are shown in Figure 6. The interface CN frequency shifts ~14 cm^{-1} when the bulk concentration of acetonitrile is varied in a very narrow range around 0.07 mole fraction. It is known that vibrational frequencies of a molecule are sensitive to its surrounding owing to the change of intermolecular interactions. The frequency shift in the CN vibration observed in our experiments suggests the existence of two different molecular environments for the acetonitrile molecules at the air/water interface as the bulk composition varies. Manifested by the CN vibrational frequency shifting in the blue direction ~14 cm^{-1} relative to that of the bulk value, our data suggest that at a low bulk concentration ($X_{ACN} < 0.07$), the interface acetonitrile molecules are solvated and hydrogen bonded with water molecules. The dominant intermolecular interaction at the interface for low bulk concentration is hydrogen bonding between surface acetonitrile molecules and water molecules. At high bulk concentration ($X_{ACN} > 0.07$), hydrogen bonding is no longer important as indicated by the shift in the surface CN vibrational frequency back to that of the neat liquid, and acetonitrile molecules are in an environment where the dominant interactions are between acetonitrile molecules. The abruptness of the shift in the CN frequency suggests the existence of a structural discontinuity for acetonitrile adsorbed at the air/water interface. As the bulk composition of the mixtures varies, surface acetonitrile molecules go through a transition from an environment in which they are hydrogen bonded to water to one where such hydrogen bonds are broken.

Structural information about acetonitrile at the air/water interface is revealed by the polarization of the SF signal which is related to the orientation of the transition dipole at the interface. The SF polarization for the CN stretching vibration of CH_3CN on water is shown versus acetonitrile bulk concentration in Figure 6. The CN sum frequency polarization changes abruptly at a bulk acetonitrile concentration of 0.07 mole fraction, the same concentration at which the CN vibrational frequency undergoes a sudden shift. Using a phenomenological model [6] we can estimate that, at a concentration below 0.07 mole fraction, the molecular symmetry axis is tilted on average about 40° from the surface normal. Above this concentration the molecular tilt of ACN is about 70°, indicating that molecules lie more horizontally on the surface at higher concentrations ($X_{ACN} > 0.07$). Surface potential measurements indicate that acetonitrile lies nearly flat at the neat acetonitrile/air interface [52]which is consistent with our SFG results. Surface tension measurements of acetonitrile-water mixtures, found to be consistent with those in the literature [2, 53] indicate that the maximum in the Gibbs excess occurs in the vicinity of the bulk concentration (X_{ACN}=0.07) at which we observe the sudden spectral and structural changes. The monolayer is likely to be formed near this maximum.

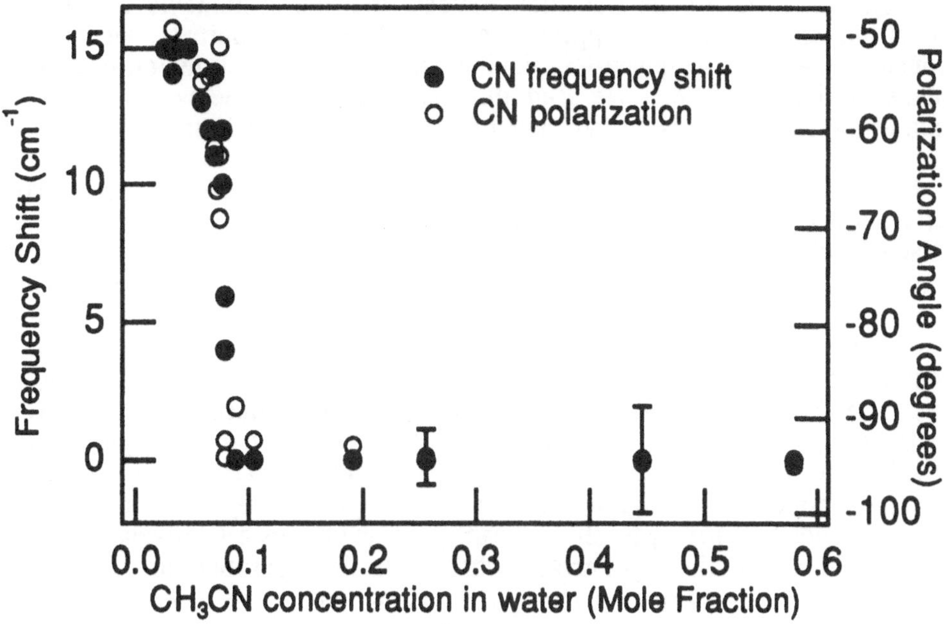

Figure 6

Combining the spectroscopic and orientational data obtained from the SFG experiments, we have constructed a physical picture of acetonitrile molecules adsorbed at the air/water interface. When acetonitrile molecules are initially introduced to the interface, they probably first enter cavities in the hydrogen-bonded water network. The intermolecular interactions are mainly hydrogen bonds between acetonitrile molecules and water molecules. This gives rise to the blue shift in the CN vibrational frequency at low bulk concentrations (X_{ACN} < 0.07). At these low concentrations interface acetonitrile is oriented ~ 40° from surface normal, probably the favored geometry of acetonitrile hydrogen bonded with water. As the surface density of aligned CN dipoles grows, the dipole-dipole repulsive interactions between acetonitrile molecules greatly increase. At a critical interface population, corresponding to X_{ACN}=0.07, a cooperative action of surface acetonitrile molecules occurs to reduce these repulsive interactions by reorienting their electric dipoles closer to the surface plane. This sudden reorientation of acetonitrile molecules at the interface probably triggered the rupturing the hydrogen bonds between acetonitrile and water. Thus the CN vibrational frequency shifts back abruptly to that of the neat bulk value. The abruptness of the changes in orientation and vibrational frequency as a function of concentration of acetonitrile at the interface leads us to attribute these findings to a phase transition rather than a continuous change in structure and intermolecular interactions. Whether this phase transition involves a phase separation into an acetonitrile rich region and a water rich region at the interface, analogous to what occurs in bulk acetonitrile-water mixtures at a lower temperature (272 K) and at a higher bulk acetonitrile mole fraction, is not known at this time. Although the sum frequency results on the high concentration side of the transition indicate that the acetonitrile vibrational spectrum and orientation approaches that of neat acetonitrile, we cannot conclude that a phase separation yielding an acetonitrile rich region has taken place at the interface. We find this possibility nonetheless to be quite interesting since there is no evidence of a phase separation in the bulk region.

3.3 STRUCTURAL PHASE TRANSITIONS OF LANGNUIR MONOLAYERS AT THE AIR/AQUEOUS INTERFACE

Water insoluble long chain amphiphilic compounds form monolayer films at the air/water interface known as Langmuir monolayers [54]. These provide ideal two dimensional systems to study intermolecular interactions, pattern formation and phase transitions. Recently, structural information at a molecular level has been provided by a number of techniques, including x-ray diffraction, neutron scattering, fluorescence microscopy, electron diffraction, reflection IR spectroscopy, surface second harmonic generation (SHG) and IR+visible sum-frequency generation (SFG). The currently accepted model for Langmuir monolayers, supported by data from these studies, holds that the hydrophylic head group maintains a relatively constant orientation at the liquid interface while it is the hydrophobic tail whose conformation and orientation changes as the monolayer is compressed [54, 55].

In this laboratory, the spectral and orientational behavior of both the head (-CN) and terminal (CD_3) groups of a long chain amphiphile (1-cyano-eicosane, $CD_3(CH_2)_{19}CN$) as a function of density have been investigated for the first time [48]. This has provided a more complete picture and detailed physical picture of the behavior of Langmuir monolayers, calling into question some of the basic tenets of the currently accepted model of phase transitions in these systems.

The pressure-area (P-A) isotherm for the long chain nitrile, $CD_3(CH_2)_{19}CN$, spread on water indicates two distinct regions separated by a turning point at A=27 Å^2/molecule [48, 56]; a gas-condensed coexistence (GCC) region for A>27 Å^2/molecule and a condensed phase (CP) region for A<27 Å^2/molecule. In the GCC region, the surface is viewed as consisting of high density (condensed phase) islands dispersed in regions of low density (gas phase) molecules. SFG spectra and polarization were recorded in the IR region of 2000 - 2300 cm^{-1} on samples spread at between 100 and 22 Å^2/molecule, from the GCC region, through the turning point and into the CP region. Our results indicate that the orientations of the head and tail group of the amphiphile vary with surface density, but with markedly different dependence on density.

The spectral and polarization behavior of the CN (head group) is contrary to what might be predicted from the consensus model. The polarization of SFG from the CN group remains constant throughout the GCC region, as shown in Figure 7, as the intermolecular separation of molecules in the high density islands, the origin of most of the SFG signal, remains the same (27 $Å^2$/molecule). The orientation of the CN changes abruptly when the transition from the GCC region to the CP occurs, as molecules are squeezed to a density less than 27 $Å^2$/molecule.

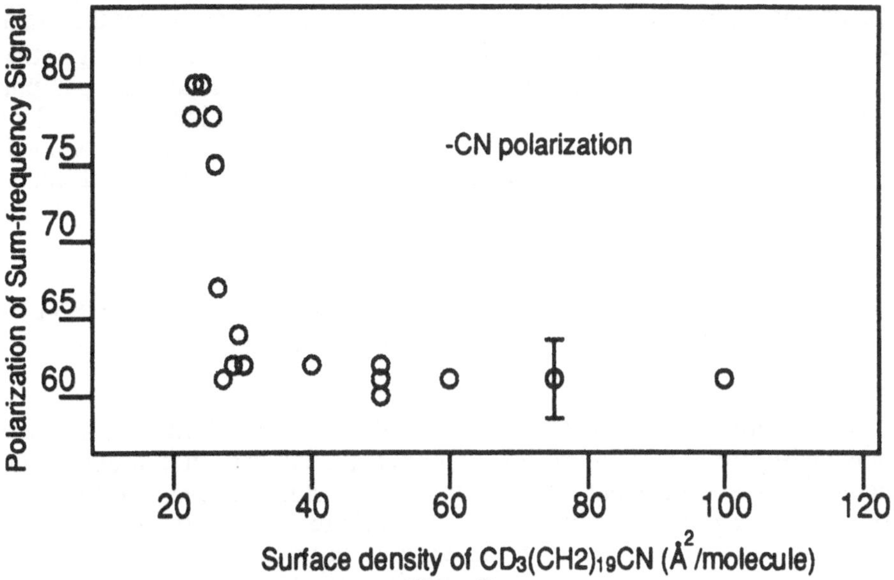

Figure 7

Similar behavior is observed for the peak of the CN stretching band; constant value, blue shifted by 10 cm^{-1} from the bulk value, for A>27 $Å^2$/molecule and a sudden reduction in the CN frequency, back to the bulk value, for A<27 $Å^2$/molecule. In addition, the FWHM of the CN sum frequency feature at low density is ~24 cm^{-1} and at high density it is ~18 cm^{-1}.

As with acetonitrile we attribute the blue shift in the CN frequency to hydrogen bonding of nitrile head groups with water molecules in the GCC region. The shifting back of the CN frequency of the monolayer to that of the bulk nitrile value, associated with the GCC-CP phase transition triggered by monolayer compression to A<27 $Å^2$/molecule, indicates that hydrogen bonds to water molecules are broken and water molecules between head groups are squeezed out of the CP monolayer. The variation in spectral width of the CN vibration of the $CD_3(CH_2)_{19}CN$ as the surface density is changed was not observed in the case of acetonitrile, though the CN vibrational frequencies behave similarly. We believe that the enlarged spectral width in the GCC region reflects a greater variety of local environments for the CN head group in this region than in the more compressed CP region. In the GCC region, the nitrile group with a long alkyl chain attached to it, is likely to experience more environments than that of acetonitrile, since the long chain may be arranged in more conformations than the single methyl group of acetonitrile. For A<27 $Å^2$/molecule the more densely packed and thereby more ordered chains are constrained to a narrower distribution of geometries yielding a narrower CN vibration spectrum.

The terminal methyl group behaves differently, as probed by polarization of SFG enhanced by resonance with the CH_3 symmetric stretch. While no abrupt change in polarization is detected, the transition to the homogeneous CP at the 27 $Å^2$/molecule is still observable as a change in the slope of polarization versus density curve for CD_3. We infer that the gradual polarization change of SFG in the GCC region is due to the varying ratio of the number of molecules close to the perimeter of the islands and gas phase regions to the number of molecules in the center of the high density islands. Molecules close to the perimeter are more likely to have their tails extended in a more horizontal orientation towards the gas phase region, than molecules in the center of the islands, which should exhibit the more upright orientation of the CP.

This work in combination with other studies [57] suggests that the consensus model of Langmuir monolayer behavior needs to be revised, with special consideration given to the role of the head group.

4. Conclusion

The use of nonlinear optical techniques (SHG and SFG) has furthered our understanding of liquid interfaces. Our results indicate that water is more structured at the air/aqueous, alkane/aqueous and silica/aqueous interfaces than in the bulk, in agreement with theory. The orienting property of the interface has enabled the contribution of the various aromatic groups of malachite green to the isomerization process to be determined, indicating that rotation of the dimethylanilino groups about their bond to the central carbon atom is the dominant relaxation pathway. The contribution of the phenyl group rotation about its bond to the central carbon atom is determined to be slight, contrary to the accepted model. The importance of the orientation and position of isomerizing groups in determining the rates of activated and barrierless isomerization has been highlighted. A previously unknown interfacial phase transition involving mutually miscible molecules has been discovered and investigated. The behavior of head and tail groups of amphiphilic molecules in Langmuir monolayers have been determined separately as a function of density for the first time. These studies offer new insight into the previously neglected role of the head group in two-dimensional phase transitions.

Acknowledgment

The authors thank the Division of Chemical Science of the Department of Energy for their support and the National Science Foundation for their equipment support.

References

1. A.W. Adamson (1982) *Physical Chemistry of Surfaces*, 4 ed.; John Wiley & Sons: New York.

2. W.J. Cheong and P.W. Carr (1987) The surface tension of mixtures of methanol, acetonitrile, tetrahydrofuran, isopropanol, tertiary butanol and dimethylsulfoxide with water at 25° C, *J. Liq. Chrom.* **10** 561.
3. Z. Koczorowsk and S. Kurowski (1992) A "macroscopic" approach to the adsorption potential at the air-solution interface, *J. Electroanal. Chem.* **329** 25.
4. Y.R. Shen (1989) Optical second harmonic generation at interfaces, *Ann. Rev. Phys. Chem.* **40** 327.
5. G.L. Richmond, J.M. Robinson and V.L. Shannon (1988) Second harmonic generation studies of interfacial structure and dynamics, *Prog. Surf. Sci.* **28** 1.
6. T.F. Heinz in *Nonlinear Surface Electromagnetic Phenomena*, H.-E. Ponath and G. I. Stegeman, Eds.; Elsevier: Amsterdam, 1991
7. K.B. Eisenthal (1992) Equilibrium and dynamic processes at interfaces by second harmonic and sum frequency generation, *Ann. Rev. Phys. Chem.* **43** 627.
8. K.B. Eisenthal (1993) Liquid interfaces, *Acct. Chem. Res.* **26** 636.
9. R.M. Corn and D.A. Higgins (1994) Optical second harmonic generation as a probe of surface chemistry, *Chem. Rev.* **94** 107.
10. J.M. Hicks, K. Kemnitz, K.B. Eisenthal and T.F. Heinz (1986) Studies of liquid surfaces by second harmonic generation, *J. Phys. Chem.* **90** 560.
11. T. Rasing, Y.R. Shen, M.W. Kim, J. P. Valint and J. Bock (1985) Orientation of surfactant molecules at a liquid-air interface measured by optical second-harmonic generation, *Phys. Rev. A* **31** 537.
12. A. Castro, E.V. Sitzmann, D. Zhang and K.B. Eisenthal (1991) Rotational Relaxation at the air/water interface by time-resolved second harmonic generation, *J. Phys. Chem.* **95** 6752.
13. X. Zhao, S. Subrahmanyan and K.B. Eisenthal (1990) Determination of pKa at the air/water interface by second harmonic generation., *Chem. Phys. Lett.* **171** 558.
14. D. Zhang, J.H. Gutow, K.B. Eisenthal and T.F. Heinz (1993) Sudden structural change at an air/binary liquid interface: Sum frequency study of the air/acetonitrile-water interface, *J. Chem. Phys.* **98** 5099.
15. E.V. Sitzmann and K.B. Eisenthal (1988) Picosecond dynamics of chemical reaction at an air-water interface studied by second harmonic generation, *J. Phys. Chem.* **92** 4579.
16. S.R. Meech and K. Yoshihara (1990) Time-resolved surface second harmonic generation: a test of the method and its application to picosecond isomerization in adsorbates, *J. Phys. Chem.* **94** 4913.
17. S.R. Meech and K. Yoshihara (1990) Picosecond dynamics at the solid-liquid interface: a total internal reflection time-resolved surface second harmonic generation study, *Chem. Phys. Lett.* **174** 423.
18. E.V. Sitzmann and K.B. Eisenthal (1989) Dynamics of intermolecular electronic energy transfer at an air/liquid interface, *J. Chem. Phys.* **90** 2831.
19. V. Sundstrom and U. Aberg (1993) Dynamics of barrierless reactions in condensed phases, *J. Mol. Liq.* **57** 149.

20. E. Borguet, X. Shi and K.B. Eisenthal (1994) Ultrafast isomerization dynamics at interfaces by time-resolved second harmonic generation, in *Ultrafast Phenomena IX*, P. F. Barbara, W. H. Knox, G. Mourou and A. H. Zewail, Eds.; Springer, Berlin.
21. X. Shi, E. Borguet, A.N. Tarnovsky and K.B. Eisenthal (in press) Ultrafast dynamics and structure at interfaces by second harmonic generation, *Chemical Physics*
22. D.F. Duxbury (1993) The photochemistry and photophysics of triphenylmethane dyes in solid and liquid media, *Chem. Rev.* **93** 381.
23. K. Kemnitz and K. Yoshihara (1990) Free volume effect of organic dye monolayer in the adsorbed state, *J. Phys. Chem.* **94** 8805.
24. K. Kemnitz and K. Yoshihara (1990) Malachite green as a sensitive free volume probe, *Chem. Lett.* 1789.
25. M.J.E. Morganthaler and S.R. Meech (1993) Picosecond dynamics of torsional motion of malachite green adsorbed on silica. A Time-resolved surface second harmonic generation: study:, *Chem. Phys. Lett.* **202** 57.
26. M.J.E. Morganthaler and S.R. Meech (1993) Ultrafast torsional motion dynamics in adsorbates: An SSHG study in *Ultrafast Phenomena VIII*, J. L. Martin, A. Migus, G. A. Mourou and A. H. Zewail, Eds.; Springer. Berlin.
27. M.A. Bell, B. Crystall, G. Rumbles, G. Porter and D.R. Klug (1994) The influence of a solid/liquid interface on the fluorescence kinetics of the triphenylmethane dye malachite green, *Chem. Phys. Lett.* **221** 15.
28. D. Ben-Amotz and C.B. Harris (1987) Torsional dynamics of molecules on barrierless potentials in liquids. II. Test of theoretical models, *J. Chem. Phys* **86** 5433.
29. M. Vogel and W. Rettig (1985) Efficient intramolecular fluorescence quenching in triphenylmethane-dyes involving excited states with charge separation and twisted conformations, *Ber. Bunsenges. Phys. Chem.* **89** 962.
30. W. Rettig (1988) Photophysical and photochemical switches based on Twisted Intramolecular Charge Transfer (TICT) States, *Appl. Phys. B* **45** 145.
31. G. Oster and Y. Nishijima (1956) Fluorescence and internal rotation: their dependence on viscosity of the medium, *J. Am. Chem. Soc.* **78** 1581.
32. T. Förster and G. Hoffmann (1971) Die Viskositätsabhängigkeit der Fluoreszenzquantenausbeuten einiger Farbstoffsysteme, *Z. Phys. Chem.* **75** 63.
33. B. Bagchi, G.R. Fleming and D.W. Oxtoby (1983) Theory of electronic relaxation in solution in the absence of an activation barrier, *J. Chem. Phys.* **78** 7375.
34. D. Ben-Amotz and C.B. Harris (1987) Torsional dynamics of molecules on barrierless potentials in liquids. 1. Temperature and wavelength dependent picosecond studies of triphenylmethane dyes, *J. Chem. Phys* **86** 4856.

35. G.V. Raviprasad and A.M. Jayannavar (1994) Friction dependence of reaction rates in simple barrierless reactions, *Chem. Phys. Lett.* **220** 353.
36. S. Saikan and J. Sei (1983) Investigation of the conformational change in triphenylmethane dyes via polarization spectroscopy, *J. Chem. Phys.* **79** 4154.
37. A. Migus, A. Antonetti, J. Etchepare, D. Hulin and A. Orszag (1985) Femtosecond spectroscopy with high-power tunable optical pulses, *J. Opt. Soc. Am. B* **2** 584.
38. A. Mokhtari, L. Fini and J. Chesnoy (1987) Ultrafast conformational equilibration in triphenylmethane dyes analyzed by time resolved induced photoabsorption, *J. Chem. Phys.* **87** 3429.
39. D.A. Cremers and M.W. Windsor (1980) A study of the viscosity-dependent electronic relaxation of some triphenylmethane dyes using picosecond flash photolysis, *Chem. Phys. Lett.* **71** 27.
40. M.M. Martin, E. Breheret, F. Nesa and Y.H. Meyer (1989) Picosecond relaxation path of ethyl violet, *Chem. Phys.* **130** 279.
41. A. Pohorille and I. Benjamin (1991) Molecular dynamics of phenol at the liquid-vapor interface of water, *J. Chem. Phys.* **94** 5599.
42. A. Pohorille and M.A. Wilson, Isomerization Reactions at Interfaces, ed.; Kluwer Academic: Dordrecht, 1994; Vol. p 207.
43. T.G. Fillingim, S.-B. Zhu, S. Yao, J. Lee and G.W. Robinson (1989) Chemically stiff water: ions, surfaces, pores, bubbles and biology, *Chem. Phys. Lett.* **161** 444.
44. M. Yanagimachi, N. Tamai and H. Masuhara (1993) Excited-state proton transfer of 1-naphtol in liquid-solid interface: Picosecond total internal reflection fluorescence study, *Chem. Phys. Lett.* **201** 115.
45. Q. Du, E. Freysz and Y.R. Shen (1994) Vibrational spectra of water molecules at quartz/water interfaces, *Phys. Rev. Lett.* **72** 238.
46. I. Benjamin and A. Pohorille (1993) Isomerization reaction and equilibrium at liquid-vapor interface of water. A molecular-dynamics study, *J. Chem. Phys.* **98** 236.
47. D. Zhang, J.H. Gutow and K.B. Eisenthal (1995) Structural phase transitions of small molecules at the air/water interface, *Submitted*
48. D. Zhang, J.H. Gutow and K.B. Eisenthal (1994) Vibrational spectra, orientations and phase transitions at the air/water interface: probing head and tail groups by sum frequency generation, *J. Phys. Chem.* **98** 13729.
49. J.H. Hunt, P. Guyot-Sionnest and Y.R. Shen (1987) Observation of C-H stretch vibrations of monolayers of molecules by optical sum-frequency generation, *Chem. Phys. Lett.* **133** 189.
50. H. Abramczyk and W. Reimschüssel (1985) Vibrational relaxation and frequency shifts of proton acceptors in hydrogen-bonded systems. benzonitrile in solution, *Chem. Phys.* **100** 243.
51. D. Ben-Amotz, M.-R. Lee, S.Y. Cho and D.J. List (1992) Solvent and pressure-induced perturbations of the vibrational potential surface of acetonitrile, *J. Chem. Phys.* **96** 8781.
52. W.C. Duncan-Hewitt (1991) Oriented dipoles at interfaces: Calculation of surface potential and surface tension, *Lang.* 7 1229.

53. A.-L. Vierk (1950) *Z. Anorg. Chem.* **261** 283.
54. G.L. Gaines Jr. (1966) *Insoluble Monolayers at Liquid-Gas Interfaces*, Interscience Publishers: New York.
55. M.A. Moller, D.J. Tildesley, K.S. Kim and N. Quirke (1990) Molecular dynamics simulation of a Langmuir-Blodgett film, *J. Chem. Phys.* **94** 8390.
56. L.E. Copeland and W.D. Harkins (1942) The pressure-area-temperature and energy relations of monolayers of octadecanenitrile, *J. Am. Chem. Soc.* **64** 1600.
57. X. Zhao and K.B. Eisenthal (1995) Monolayer orientational fluctuations and new phase transition at the air/water interface detected by second harmonic generation., *J. Phys. Chem.* **102** 5818.

LIQUID-LIQUID EXTRACTION OF ALKALI CATIONS BY CALIX[4]CROWN IONOPHORES: CONFORMATION AND SOLVENT DEPENDENT Na^+ / Cs^+ BINDING SELECTIVITY. A MD FEP STUDY IN PURE CHLOROFORM AND MD SIMULATIONS AT THE WATER / CHLOROFORM INTERFACE

M. Lauterbach, G. Wipff *

Laboratoire de Modélisation et Simulations Moléculaires, URA 422 CNRS, Institut de Chimie, 4 rue Blaise Pascal, 67000 Strasbourg (France)

1. Summary

We present a theoretical study of the stereochemical dependence of alkali metal ion complexation and liquid-liquid extraction by 1,3-dimethoxy-calix[4]arene-crown6 (**L**). Molecular dynamics and free energy calculations are reported for the *cone*, *1,3-alternate* and *partial cone* conformers in chloroform. The conformation of **L** free and complexed is determined by two opposed components: the internal energy of the solute (which favours the *1,3-alternate* form) and the solute / chloroform interaction energy (which favours the cone form). In dry chloroform, the binding sequence is close to the gas phase, *i.e.* Na^+ preferred by all three conformers. In the presence of the picrate counterion, the Na^+ - Cs^+ free energy difference is reduced, due to the differences in the uncomplexed states of the cations. For the 1,3-alternate form, it drops to almost zero. Modelling the cation *extraction* from water to chloroform via a thermodynamic cycle leads to a *conformation dependent binding sequence*, *i.e.* Cs^+ is better extracted than Na^+ by the 1,3-alternate form, but Na^+ preferred by the cone form, with or without explicit account of counterions. This is in agreement with experimental extraction data. Finally, we report MD simulations on the free and Cs^+ complexed **L** at a chloroform/water interface, with and without counterions, with several starting configurations. Contrary to expectations, no spontaneous migration to the organic phase takes place. Instead, these solutes remain "adsorbed at the interface" on the chloroform side and behave therefore as surfactants.

L. Echegoyen and A.E. Kaifer (eds.), Physical Supramolecular Chemistry, 65–102.

2. Introduction

Calix[4]arenes bridged by crown ethers in the 1,3 position, and O-alkylated in the 2,4 position display interesting stereochemical dependent ionophoric properties[1-3]. Schematically, the calixarene platform may adopt three conformations: cone, partial cone, and 1,3-alternate [4-6], which show different cation binding characteristics. In $CDCl_3$ saturated with water, the 1,3-dimethoxy-*p*-*tert*-butylcalix[4]arene-crown5, which is conformationally mobile, complexes alkali picrates and displays a higher K^+ / Na^+ selectivity and K^+ affinity in its partial cone, than in the cone conformation [7]. Under the same conditions, its -crown6 homologue extracts preferentially Cs^+, in the sequence $Na^+ < K^+ < Rb^+ < Cs$ [7]. NMR data in CD_3CN or CD_3OD indicate that the dimethoxy-calix[4]arene-crown6 (**L** with X = H; tb-**L** with X = *tert*-butyl, see Chart 1) binds Na^+ in the cone conformation, but binds Cs^+ in the 1,3-alternate conformation. [7] Substitution of O-Me by bulkier O-Alkyl groups prevents interconversions of the calixarene. The conformationally locked 1,3-alternate calix[4]-crown-6 derivatives also display a clear Cs^+ /Na^+ affinity, while the cone form is a poor cation binder[8]. There is thus an interesting conformation dependent Cs^+ / Na^+ binding selectivity which may be of particular interest in the context of Cs^+ separation from radioactive wastes[9].

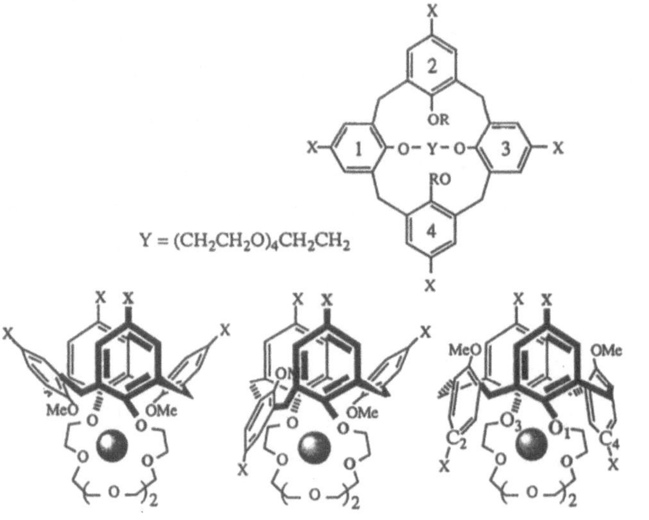

Cone Partial Cone 1,3-Alternate

L: X = H, tb-**L**: X = $C(CH_3)_3$

Chart 1 : Scheme of calix[4]crown6 derivatives. Cone, partial cone ("paco") and 1,3 alternate ("1,3-alt") conformers of 1,3-dimethoxy-calix[4]crown6.

In a recent paper we presented a computational study on 1,3-dimethoxy-calix[4]-crown6 and -crown5 hosts in their free state, and complexed with Na^+ to Cs^+ alkali cations[10]. They were first considered in the gas phase in order to determine their intrinsic conformational preference and binding affinities. Then, we modelled the solution state with water as solvent, first as a mimic of the methanol solutions where the calixarenes are soluble and second as the source phase in liquid-liquid extraction experiments. Based on free energy perturbation ("FEP") calculations on X-calix[4]crown6 complexes (X = H or *tert*-butyl) we found a clear *conformation dependent binding selectivity*. The cone forms preferred Na^+ over Cs^+, while the 1,3-alternate form preferred Cs^+ and the partial cone form did not display clear Na^+ / Cs^+ preference[10]. The conformation dependent lifetime of inclusive complexes was in agreement with this conclusion. In water, Na^+ initially encapsulated inside the crown of the 1,3-alternate host decomplexed in less than 50 ps while Cs^+ stayed inside. Conversely, for the cone conformer, Na^+ remained inside and Cs^+ rapidly decomplexed. This difference was analyzed in terms of conformation dependent host-guest interactions and shielding of the cation. We speculated that in non-aqueous solvents like acetonitrile or methanol, the binding sequence should be the same as in water. After the calculations were completed, thermodynamic data for complexation have been reported in methanol[8]. They fully support our predictions on the conformation dependent Cs^+ / Na^+ binding selectivity. In relation with the extraction experiments we modelled the extraction from water to an organic phase by a thermodynamic cycle, assuming that the structures and relative stability of the complexes in the organic phase can be represented by the gas phase situation.

In this paper, we report subsequent MD and FEP studies of calix[4]crown6 complexes in chloroform solution, with an explicit representation of the solvent. Counterions which were neglected in the aqueous phase simulations are now considered. Conformational equilibria in apolar solvents such as chloroform are generally believed to be close to the gas phase equilibria. In this study we show that the solute / chloroform interaction energies depend on the conformation of the solute. Finally, in relation with the cation extraction experiments by ionophores, we report new investigations at the chloroform/water interface, using an explicit representation of the solvents.

3. Methods

The molecular dynamics and free energy perturbation calculation have been carried out with the AMBER4.0 software[11] using the following representation of the potential energy:

$$E_T = \sum_{bonds} K_r (r - r_{eq})^2 + \sum_{angles} K_q (\theta - \theta_{eq})^2 + \sum_{dihedrals} \sum_{\eta} \frac{V_\eta}{2} (1 + cos n\Phi) + \sum_{j=1}^{atoms} \sum_{i>j}^{atoms} \left(e_{ij} \left(\frac{R_{ij}^*}{R_{ij}} \right)^{12} - 2e_{ij} \left(\frac{R_{ij}^*}{R_{ij}} \right)^{6} + \left(\frac{q_i q_j}{e R_{ij}} \right) \right) \quad (1)$$

The bonds and bond angles are treated as harmonic springs, and a torsional term is associated to the dihedral angles. The interactions between atoms separated by at least three bonds are described within a pairwise additive scheme by a 1-6-12 potential. The parameters for the calixarene and for the crown moieties (with explicit CH_2 groups) are taken from AMBER force field, using the atom types and charges described in reference[10].

The starting structures of the three conformers were modelbuilt. The cation was set at the center of the polyether ring and surrounded by the ether oxygens. For the simulations with a counterion the last structure of *in vacuo* simulations of the charged complex was taken and the picrate counterion placed at the average $M^+...O_{ether}$ distance to the cation.

The chloroform molecules are represented in the united atom approximation with OPLS van der Waals and electrostatic parameters, fitted by Jorgensen *et al.* on the pure liquid by Monte Carlo simulations[12,13]. The solvent box was "cubic", represented with periodic boundary conditions. The solute was placed at the center of the box and all solvent molecules within 3 Å and beyond a distance (ranging from 10 to 15 Å) from the solute were deleted. For the free host, we compared several protocols of simulations, depending on the size of the box, the cut-off for non-bonded interactions and on the choice of (N, P, T) versus (N, V, T) ensembles. Since the OPLS parameters were derived with a rigid chloroform molecule, we also investigated three representations of its internal freedom, keeping the same reference geometry as in the Monte Carlo calculations (3 C-Cl bonds at 1.758 Å and 3 Cl-Cl pseudo-bonds at 2.903 Å): the first model is "rigid" with large k_{C-Cl} and k_{Cl-Cl} force constants (500 kcal/mol/Å^2) to prevent deformations, without SHAKE. The second and third model use weaker force constants (k_{C-Cl} = 216 kcal/mol and k_{Cl-Cl} =100 kcal/mol), with respectively "rigid" (SHAKE on C-Cl and Cl-Cl "bonds") and "flexible" bonds (no SHAKE). Unless otherwise specified, the OPLS charges ($q_{CH} = 0.42$ and $q_{Cl} = -0.14$) are used on chloroform molecules. However, for comparison, several calculations were repeated with a "polar" model ($q_{CH} = 0.84$, $q_{Cl} = -0.28$) where the dipole moment of the OPLS molecule is scaled by 2.0.

For the MD and FEP calculations in chloroform on calix[4]crown6 M^+ complexes (with or without counterion) 100 ps in the (N, P, T) ensemble were performed without SHAKE, using a cutoff distance of 15Å and the "flexible" chloroform (model 3) in "box15" (see next). After 1000 - 4000 steps of conjugate gradient energy minimization, the MD simulations were run at 1 atm and 300K using the Verlet algorithm, starting with random velocities. A residue based cut-off of 15 Å was used for non-bonded interactions, taking the solute(s) as a single residue. The temperature was maintained to 300 K by velocity scaling in the gas phase and coupling to a thermal bath in solution.

The chloroform / water interface is represented by the contact region of two adjacent boxes of pure solvents (OPLS [12,13] and TIP3P [14]) containing 200 - 300 chloroform and 800 - 1000 water molecules respectively. As for the pure liquid systems, periodic boundary conditions were applied in the three directions. Different starting positions of the solutes with respect to the interface were compared during 350 ps of MD at constant volume using a time step of 1 fs. For other details see footnote of Table 8.

The free energy perturbation calculation (FEP) were performed with the windowing technique[15,16], changing the ε, R* parameters of M^+ linearly with λ: $V_\lambda = \lambda V_{M_1^+} + (1-\lambda) V_{M_2^+}$. Mutations were performed in the sequence Na^+ -> K^+ -> Rb^+ -> Cs^+. The mutation from one cation (free or complexed) into the next one was achieved in 11 (K^+ -> Rb^+, Rb^+ -> Cs^+) or 101 (Na^+ -> K^+ when indicated) windows. At each window, 1 ps of equilibration and 4 ps of data collection were performed applying the same protocol as for the corresponding MD calculations. The free energy change ΔG was averaged from forward and backward cumulated values.

To analyze the energy we consider the molecular mechanics minimized energy $E_{opt\text{-}Solute}$ of the solute extracted from the last set of MD, the average MD potential energy <E> and its components, recalculated from the trajectories which were saved every 0.5 ps.

4. Results

We first present the results of the free hosts **L** and tb-**L** in chloroform, with a comparison of several protocols of simulation. Then, based on MD and FEP calculations using a same protocol, we report structural and energy features of the LM^+ complexes, with and without picrate counterion, and relative free energies of complexation and extraction. Finally, the free and complexed **L** ionophore is described at the chloroform/water interface.

4.1. CALIX[4]CROWN6 UNCOMPLEXED IN CHLOROFORM: COMPARISON OF SEVERAL PROTOCOLS.

This section is devoted to methodological investigations on the way to simulate chloroform solutions. Given the low polarity of this solvent, we wanted to determine if we could save computer time by restraining the solvent box to a rather "small" number of solvent molecules. We first investigate the influence of the size of the box and the cutoff distance on the results of simulations for **L** uncomplexed in the 1,3-alternate conformation (Table 1). The ligand is considered in a small box of chloroform ("box10") and in larger boxes ("box12", "box15"), with cut-off distances ("cut10" to "cut15") increasing accordingly from 10 to 15 Å. The edge of the box ranges from 25 to 44 Å (Table 1). The V_x, V_y, V_z dimensions of "box*n*" are obtained by increasing the X_{max}, Y_{max}, Z_{max} coordinates of the solute by a distance of *n* Å (in the EDIT section of AMBER). For the (N,V,T) calculations, we scaled the coordinates of the carbon atom of $CHCl_3$ molecules iteratively to "fill the holes" between adjacent elementary boxes. For the (N,P,T) calculations, this solvent "compression" was achieved during the equilibration step. This is why, for a given "Box*n*" value, the number of solvent molecules differs in the (N,P,T) / (N,V,T) simulations (Tables 1 and 2).

Then, using the large box and cut-off ("box15" and "cut15"), we compare the three conformers of tb-**L** in different simulation conditions. This is followed by a brief discussion of the structural and energy features of tb-**L** and **L**.

4.1.1. *The 1,3-alternate form of tert-butylcalix[4]-crown6 tb-**L** uncomplexed.*

The simulation conditions have a clear incidence on the density of the solution. With the largest box at constant pressure the density (1.45) is close to the experimental density of pure chloroform (1.47). With smaller boxes, the density is lower (1.36-1.42).

Simulations at constant volume should in principle fit the experimental density of chloroform. It is however lower because of edge effects of the elementary neighbouring box units.

	nb chloro	box size	$\langle E_{solute/chloro} \rangle$	$\langle E_{solute} \rangle$	density
Protocol 1: const V, without SHAKE, rigid $CHCl_3$, 50 ps [a]					
"box10"	223	31,32,35	-62.7 ± 4	252 ± 12	1.31
"box15"	516	41,42,44	-66.6 ± 4	255 ± 10	1.40
"box15"	516	41,42,44	-67.8 ± 5	255 ± 7	1.40
Protocol 2: const P, SHAKE, 100 ps [b]					
"box10"	145	25,28,32	-69.5 ± 3	218 ± 6	1.36
"box15"	388	37,38,39	-70.4 ± 4	216 ± 6	1.45
"box12"	228	31,32,34	-74.3 ± 4	217 ± 5	1.42
"box15"	388	37,38,39	-77.0 ± 3	215 ± 6	1.45

Table 1: The free ligand tb-**L**, 1,3-alternate form, in chloroform. Comparison of different box sizes (Å) and cutoffs (Å). Number of solvent molecules, average ligand-solvent interaction energies $\langle E_{solute/chloro} \rangle$, average energy of the solute (kcal/mol) and average chloroform density. a) $k_{C\text{-}Cl} = k_{Cl\text{-}Cl}$ =500 kcal/mol; b) $k_{C\text{-}Cl}$ = 216 kcal/mol and $k_{Cl\text{-}Cl}$ =100 kcal/mol.

Common to all simulations is an attractive solute/solvent interaction energy (-63 to -77 kcal/mol). With a given cutoff, it increases somewhat with the box size (from "box10" to "box15") by 4 kcal/mol for the (N,V,T) calculations and by 0.9 kcal/mol in the (N,P,T) calculations. For a given box size ("box 15") increasing the cutoff has a negligible effect for the (N,V,T) simulations but increases this energy by 7 kcal/mol in the (N,P,T) run. With the same "box15" and cut-off, differences are noted between results from Protocol 1 and Protocol 2. We did not investigate whether they are due to SHAKE/no-SHAKE or to the choice of constant V/ P.

The average solute energy E_{solute} displays only weak dependence on the cutoff and the size of the box, compared to statistical fluctuations. It is about 30 kcal/mol lower when SHAKE is used on the solvent molecules and on the C-H bonds of the solute.

After these preliminary investigations in order to represent more adequately the solution we decided to use "box15" with a cutoff distance of 15 Å for our following calculations in pure chloroform.

4.1.2. *p-tert-Butylcalix[4]crown6* (tb-**L**) *uncomplexed: comparison of the 1,3-alternate, partial cone and cone forms.*

As in this section a same box size ("box 15") is used, the density is fairly constant and close to that of bulk chloroform for (N,P,T) simulations (Table 2).

Structural features For one given conformer of tb-**L** the average structural parameters are nearly independent of the simulation conditions. Its overall structure is also very close in chloroform solution to the one in the gas phase[10]. For instance, for the cone form, the average distances between opposite phenolic oxygens $O_2...O_4$ and $O_1...O_3$ are 3.4 and 5.0 Å in both cases. The average angles formed between each aromatic system and the plane of the four methylene-bridge carbon atoms (ω_1, ω_3 / ω_2, ω_4) illustrate that the cone is "flattened" and of average C_{2v} symmetry (85°, 82° / 40°, 41° respectively).

In the cone form of **L**, the cone is more open than in the tb-**L** derivative, because it complexes one solvent molecule, with one C-Cl bond pointing into the cavity, which becomes more "squared" (ω_1, ω_3 decrease to 77° while ω_2, ω_4 increase to 43°). The crown ether cavity, too small to accommodate one $CHCl_3$ molecule, remains empty for all conformers (Figure 1).

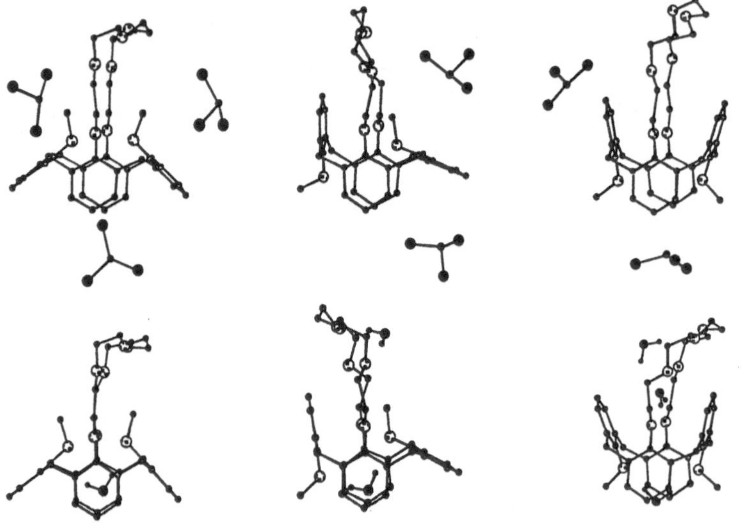

Figure 1: Snapshots of the free ligand **L** in the cone, partial cone and 1,3-alternate form in chloroform (top) and in water (bottom) including selected solvent molecules: last set after 100 ps of MD.

Energy analysis. We focus on the potential energy of the solute, and on the solute/chloroform interaction energy $E_{solute/chloro}$, as a function of the protocol and conformation of the solute.

In all cases, the difference ΔE_{opt} in energy minimized structures of tb-**L** (extracted from the last set of dynamics) follows the order: 1,3-alt < paco < cone as in the gas phase[10] (Table 2). Depending on the protocol, the preference for the 1,3-alternate over the cone form is 5 - 10 kcal/mol. For **L**, the 1,3-alternate and paco forms have similar energies, about 5 kcal/mol lower than the cone form.

		nb chloro	box size	$<E_{solute/chloro}>$	ΔE_{opt_solute} a)	density
	Protocol 1: const V, without SHAKE, rigid $CHCl_3$, 50 ps					
tb-**L**	1,3-alt	516	41,42,44	-69.9 ± 5	0.0 (119.9)	1.40
	paco	522	41,42,44	-71.9 ± 4	3.3	1.40
	cone	608	41,45,47	-73.1 ± 4	10.2	1.41
	Protocol 2: const P, SHAKE, 100 ps b)					
tb-**L**	1,3-alt	388	37,38,39	-77.0 ± 3	0.0 (119.9)	1.45
	paco	516	40,40,44	-79.2 ± 4	1.3	1.46
	cone	437	37,40,42	-79.9 ± 5	9.9	1.45
	Protocol 3: const P, without SHAKE, "flexible $CHCl_3$", 100 ps					
tb-**L**	1,3-alt	345	34,36,38	-77.5 ± 5	0.0 (122.7)	1.47
	paco	317	33,34,38	-79.3 ± 4	2.7	1.46
	cone	339	34,36,38	-80.1 ± 4	6.3	1.45
L	1,3-alt	356	35,36,39	-68.1 ± 4	0.0 (112.8)	1.46
	paco	365	35,36,39	-68.7 ± 4	0.5	1.45
	cone	361	35,36,39	-74.4 ± 4	5.3	1.46

Table 2 : The free ligands tb-**L** and **L** in chloroform. Comparison of three simulation protocols for "box15" and a cutoff of 15 Å. Number of solvent molecules, box size (Å), average solute/solvent interaction energy, energy of the solute (kcal/mol) and average chloroform density. a) relative energy of the solute, after molecular mechanics minimisation of the structure extracted from the last set of MD. b) SHAKE applied on the C-H bonds of the solute and on the C-Cl and Cl...Cl "bonds" of $CHCl_3$.

For a given conformer of tb-**L**, the solute/solvent interaction energy depends on the simulation conditions and varies by up to 7 kcal/mol. However the *differences* $\Delta E_{solute/chloro}$ from one conformer to the other are weakly dependent of the simulation conditions and indicate that the cone is best solvated, in the sequence cone > paco > 1,3-alt. In the unsubstituted **L** calix-crown the conformation dependence of $E_{solute/chloro}$ is still larger. It favours the cone over the 1,3-alternate form by 6.3 kcal/mol because extra stabilisation results from a chloroform molecule complexed "on the top" of the cone

(Figure 1). Such complexation was prevented with bulky *tert*-butyl substituents at the upper rim of *tb*-**L**.

In water, we calculated a different sequence of solute/solvent interaction energies due to specific interactions and difference in polarity and granularity of the solvent (for more details see reference[10]). The 1,3-alternate **L** coordinates three water molecules directly, while the cone and partial cone conformers coordinate one water molecule only, resulting in the sequence of **L**/water interaction energies: 1,3-alt (-85) > paco (-76) > cone (-73.6 kcal/mol). In *tb*-**L**, the *tert*-butyl substituents prevent water to coordinate the polar sites of the solute, except in the paco form where one water molecule is hydrogen bonded to the two phenolic oxygens after 70 ps of MD. As a consequence the partial cone form is best hydrated: paco > cone > 1,3-alt. Thus, formation of "supermolecule" between the solute and solvent molecules is clearly dependent of the conformation of the solute and on the solvent.

NMR data in chloroform indicate that **L** displays an equilibrium between the three forms: cone (60%) > paco (25%) > 1,3-alt (15%). The order is the same in CD_3CN and in CD_3OD where the cone is also dominant[8]. Intrinsically (in the gas phase), using different electrostatic and force field representations of the phenolic groups, we calculated a reversed sequence of stability[10]. The calculations therefore strongly suggest that, even in chloroform, *solvation contributes to the conformational state of these ligands and that the cone is dominant because it is best solvated.*

4.2. M^+ CALIX[4]CROWN6 COMPLEXES IN CHLOROFORM. INFLUENCE OF THE PICRATE COUNTERION.

In this section we focus on the alkali cation binding by **L** first without and with Pic^- counterion. This is followed by energy features of the complexes. We chose the unsubstituted **L** calix-crown instead of its tb-**L** derivative, in order to allow for cation - anion interactions in the complex. Such interaction might be prevented by the bulky alkyl groups of tb-**L**.

Schematically, **L** possess a "calix cavity" delineated by the four aromatic rings and the "crown cavity" surrounded by the six crown ether oxygens. We define the position of M^+ with respect to the center of these cavities, *i.e.* the center of mass cmO_{ph} of the four phenolic oxygens (cone) and the two opposite phenolic oxygens respectively (1,3-alternate) and the center of mass cmO_{cr} of the six crown oxygens, respectively.

			cone			1,3-alt		
			Na^+	K^+	Cs^+	Na^+	K^+	Cs^+
in vacuo	LM^+	$\langle d_{M^+..cmOph}\rangle$ a)	2.5	2.9	3.4	2.6/4.0	3.0/4.7	3.3/5.0
		$\langle d_{M^+..cmOcr}\rangle$ b)	1.5	1.3	1.2	1.6	1.0	0.7
	LM^+Pic^-	$\langle d_{M^+..cmOph}\rangle$ a)	2.5	2.8	3.3	4.3/6.4	3.8/5.8	3.5/5.4
		$\langle d_{M^+..cmOcr}\rangle$ b)	2.0	1.6	1.3	1.2	1.0	0.8
in chloro	LM^+	$\langle d_{M^+..cmOph}\rangle$ a)	2.4	2.7	3.2	2.4/3.3	2.8/4.3	3.2/5.0
		$\langle d_{M^+..cmOcr}\rangle$ b)	1.6	1.3	1.2	2.6	1.3	0.8
	LM^+Pic^-	$\langle d_{M^+..cmOph}\rangle$ a)	2.4	2.8	3.3	4.2/6.4	3.1/5.1	3.4/5.4
		$\langle d_{M^+..cmOcr}\rangle$ b)	2.0	1.6	1.4	1.1	1.1	1.0

Table 3 : $\mathbf{LM^+}$ and $\mathbf{LM^+Pic^-}$ complexes in chloroform. a) Average distances between M^+ and the four phenolic oxygens (cone) and the two opposite phenolic oxygens respectively (1,3-alternate $O_{1,3}/O_{2,4}$) ($\langle d_{M^+...cmOph}\rangle$) and b) between M^+ and the center of mass of the six crown ether oxygens ($\langle d_{M^+...cmOcr}\rangle$).

Structural features of the complexes. During the dynamics, all $\mathbf{LM^+}$ complexes remain of inclusive type as in the gas phase[10]. Snapshots are presented in Figure 2. The position of M^+ with respect to the calix- and crown-cavities is characterized by the $M^+...cmO_{ph}$ and $M^+...cmO_{cr}$ distances reported Table 3. It depends on M^+ and on the counterion.

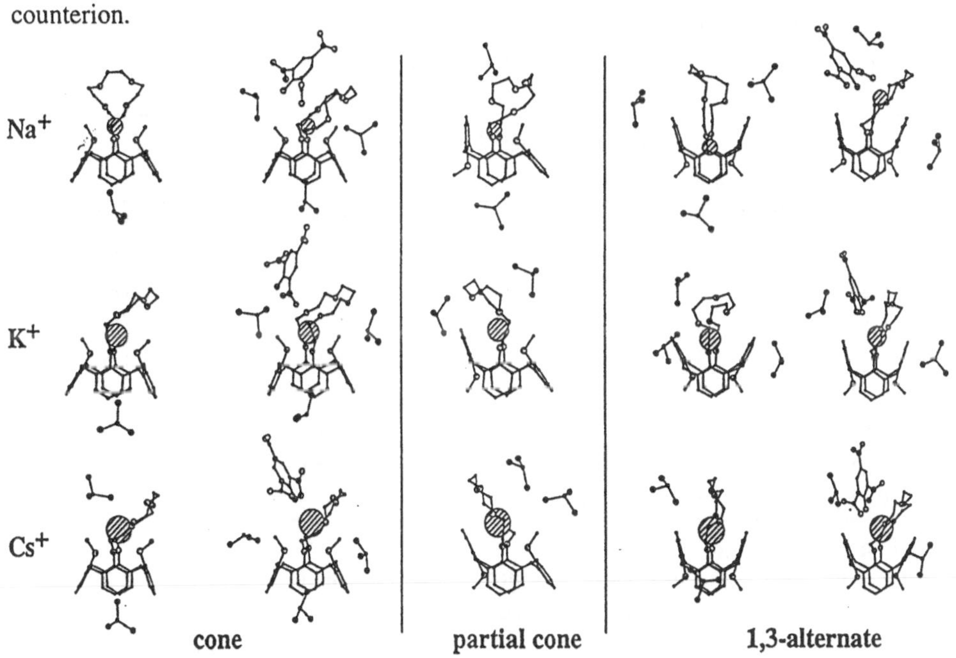

Figure 2 : Snapshots of the $\mathbf{LNa^+}$, $\mathbf{LK^+}$ and $\mathbf{LCs^+}$ complexes in chloroform without picrate counterion in the cone, partial cone and 1,3-alternate form and with counterion in the cone and 1,3-alternate form: last set after 100 ps of MD.

In the absence of Pic^-, the smallest cation Na^+ is closer to the O_{ph} phenolic oxygens, than to the O_{ether} oxygens. As M^+ gets bigger, it moves towards the crown cavity. The structure of a given complex in chloroform is close to the one simulated *in vacuo*[10]. The only exception is the 1,3-alternate Na^+ complex where Na^+ sits deeper inside the calix cavity in chloroform than *in vacuo*: the distance $M^+...cmO_{crown}$ decreases from 2.6 Å in chloroform to 1.6 Å *in vacuo* (Table 3). This is indicative of the poor fit between Na^+ and the 1,3-alternate **L**.

When the Pic^- counterion is included in the simulation, it stays in intimate contact with the cation and pulls it toward the crown cavity. As a result, the crown ether chain bends away from Pic^- (Figure 2) and the $M^+...cmO_{ph}$ distance increases. For a given cation, the anion pulling effect is larger in the 1,3-alternate ($<d_{M^+..cmOph}>$: between 0.2 Å for Cs^+ and 3.1 Å for Na^+) than in the cone (0.1 Å) complex, in relation with the different modes of Pic^- coordination. In the 1,3-alternate complexes, Pic^- coordinates M^+ via its O^- and O_{NO_2} oxygens and oscillates in a plane between the crown ether chain and the nearest aromatic ring. In the cone complexes, it mainly coordinates via one O_{NO_2} oxygen only and rotates during the simulation around its C_2 symmetry axis. This versatility of picrate binding to Cs^+ is consistent with the two modes observed in the solid state: via its O^- oxygen and one nitro group in the $\mathbf{L}Cs^+Pic^-$ complex [2], but via only one nitro oxygen in the 1,3-diisopropoxy-calix[4]crown6 Cs^+Pic^- complex[8].

	MD simulations a)			X-ray structure b)		
	in vacuo	in chloro	in water	**1a**	**1b**	**2**
$<Cs^+..O_1>$ c)	3.7	3.4	3.5	3.319(5)	3.468(5)	3.189(5)
$<Cs^+..C_2>$ c)	3.4	3.5	3.5	3.43(1)	3.354(7)	3.486(8)
$<Cs^+..O_3>$ c)	3.3	3.4	3.5	3.300(5)	3.269(4)	3.188(5)
$<Cs^+..C_4>$ c)	3.7	3.6	3.6	3.568(7)	3.422(7)	3.58(1)
$<Cs^+..O_{cr1}>$ d)	3.1	3.2	3.1	3.265(9)	3.196(7)	3.100(5)
$<Cs^+..O_{cr2}>$ d)	3.4	3.4	3.3	3.67(1)	3.401(7)	3.276(9)
$<Cs^+..O_{cr3}>$ d)	3.3	3.3	3.4	3.333(6)	3.398(6)	3.475(7)
$<Cs^+..O_{cr4}>$ d)	3.3	3.4	3.3	3.212(5)	3.294(5)	3.245(6)
$<Cs^+..OPic^->$	3.0	3.0		3.094(6)	3.143(6)	

Table 4 : Interatomic distances in the 1,3-alternate $\mathbf{L}Cs^+Pic^-$ complex. a) Average values from simulations *in vacuo*, in chloroform and in water; statistical fluctuations are about 0.3Å; b) solid state structures of $\mathbf{L}Cs^+Pic^-$ (two independent molecules in the unit cell [2,8]: **1a**, **1b**) and of 1,3-diisopropoxy-calix[4]crown6 Cs^+Pic^- [8] **2**); c) see atom labels in Chart 1; d) O_{cr1} to O_{cr4} are the four crown ether oxygens.

As for **L** free, specific interactions with the solvent can be observed. In the cone complexes (with and without Pic^-) one $CHCl_3$ molecule sits over the cone, with one C-Cl bond pointing inside. Such pattern is not observed for the 1,3-alternate form (Figure 2).

The calculated features of **L**Cs^+ can be compared with those observed in the X-ray structures of the 1,3-alternate Cs^+Pic^- complexes of **L** [2], with two independent molecules present in the unit cell, and of its diisopropoxy-calix[4]-crown6 analogue [8] (Table 4). Their $Cs^+...C_2$, $Cs^+...C_4$ (see Chart 1) contact distances range from 3.354 to 3.58 Å. Our calculated average values in **L**Cs^+Pic^- are quite close (from 3.4 to 3.7 Å). This compares well with other systems where Cs^+...arene interactions have been documented [17,18] and may be taken as evidence for "cation-π interactions"[19,20]. This does not mean, as pointed out previously[10], that such interactions alone are the source of the high affinity of the 1,3-alternate forms of calixarenes for Cs^+, compared to other alkali cations. Intrinsically, aromatics display *larger* attractions with Na^+ than with Cs^+[21]. The Cs^+ selectivity results from solvation effects and from the fit of Cs^+ inside the 1,3-alternate calix-crown host.

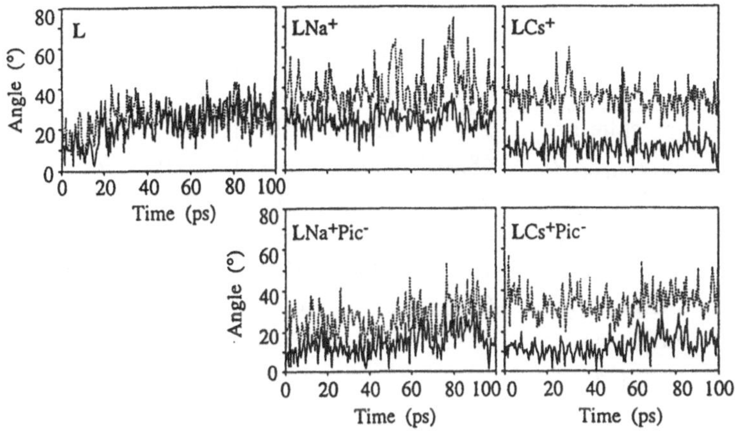

Figure 3 : Free ligand **L**, **L**Na^+, **L**Cs^+, **L**Na^+Pic^- and **L**Cs^+Pic^- complexes (1,3-alternate conformer) in chloroform. Angles ω_{13} (—) and ω_{24} (-----) between opposite aromatic rings as a function of time.

Finally, concerning the calculated overall shape of the calix-cavity, it is found to depend on the complex and time. This is illustrated by the time evolution of the ω_{13} and ω_{24} angles between opposed phenolic rings of **L** 1,3-alternate (Figure 3). For the **L** ligand free, both angles are very close (26°, 25° respectively). In all M^+ complexes, ω_{13} becomes smaller than ω_{24}. The asymmetry is largest in **L**Cs^+ ($\omega_{13} = 11° \pm 5$, $\omega_{24} = 34° \pm 6$). The Pic^- counterion coordination to M^+ brings minor perturbation to these angles in

LCs^+ (ω_{13} = 16° ± 6, ω_{24} = 35° ± 7), indicative of the good fit between Cs^+ and the crown cavity of the ligand. In the Na^+ complex, ω_{13} and ω_{24} decrease, as a result of the cation pulling by Pic^- (from ω_{13} = 24°± 5, ω_{24} = 42° ± 12 to ω_{13} = 18°± 7, ω_{24} = 30° ± 8). The other conformers display similar cation and time dependent modulation of the shape of the calixarene core.

Energy analysis. We reported in Table 5 average energy components concerning the internal energy of the **LM**$^+$ and **LM**$^+$Pic$^-$ solutes and the interactions of M^+, Pic^- and **L** with chloroform. First, with or without counterion, the potential energies E_{opt_solute} of the complexes are lowest with Na^+ as guest. The sequence ($Na^+ > K^+ > Rb^+ > Cs^+$) follows the M^+/**L** interaction energies which are larger in the cone than in the 1,3-alternate form.

Second, each conformer of a **LM**$^+$ complex interacts significantly with the solvent. The M^+ contribution (about -13 kcal/mol) is nearly independent on M^+ for the three conformers. The **L** contribution (from -63 to -69 kcal/mol) is, as for **L** free, larger for the cone than for the 1,3-alternate form (Table 5).

		LM$^+$			LM$^+$Pic$^-$		
		Na^+	K^+	Cs^+	Na^+	K^+	Cs^+
cone	nb chloro a)	299	297	296	493	496	488
	E_{opt_solute} b)	48.1	67.9	84.2	-116.6	-95.8	-83.5
	$<E_{M+/L}>$	-91.8	-72.6	-56.6	-79.3	-68.2	-52.6
	$<E_{solute/chloro}>$	-81.7	-82.4	-79.4	-108.7	-105.6	-104.6
	$<E_{M+/chloro}>$	-13.7	-13.4	-13.3	0.8	0.2	0.8
	$<E_{L/chloro}>$	-65.9	-69.0	-66.1	-69.3	-64.5	-64.1
	$<E_{Pic-/chloro}>$				-40.2	-41.3	-41.3
	$<E_{M+/Pic-}>$				-79.7	-67.7	-75.2
1,3-alt	nb chloro a)	283	293	296	404	419	423
	E_{opt_solute} b)	55.7	60.2	69.4	-120.4	-108.9	-96.6
	$<E_{M+/L}>$	-66.6	-59.7	-52.5	-54.8	-53.3	-48.2
	$<E_{solute/chloro}>$	-75.6	-77.1	-76.6	-98.9	-93.8	-100.3
	$<E_{M+/chloro}>$	-12.6	-13.5	-13.7	1.9	2.7	1.4
	$<E_{L/chloro}>$	-62.9	-63.6	-63.0	-62.0	-60.6	-60.3
	$<E_{Pic-/chloro}>$				-38.8	-35.9	-41.4
	$<E_{M+/Pic-}>$				-92.0	-81.4	-75.8

Table 5 : LM$^+$ and **LM**$^+$Pic$^-$ complexes in chloroform. Potential energy of the solute minimized structures (E_{opt_solute}). Average $E_{M+/L}$, $E_{M+/CHCl_3}$, $E_{L/CHCl_3}$, E_{M+/Pic^-}, $E_{Pic^-/L}$ and $E_{Pic^-/CHCl_3}$ interaction energies (kcal/mol). Statistical fluctuations are about 2 to 5 kcal/mol. a) Number of chloroform molecules in the simulation box; b) see Table 2.

Since the counterion remains in contact with **LM**$^+$, it is most relevant to analyze the **LM**$^+$Pic$^-$ complexes. Their total interaction with chloroform is larger for the cone than for the 1,3-alternate complexes (by about 10 kcal/mol with Na^+, 12 kcal/mol with K^+ and 4 kcal/mole for Cs^+) due to the contributions of Pic$^-$ and **L** (-36 to -41 kcal/mol and -60 to -69 kcal/mol, respectively). As M^+ is shielded from the solvent, its contribution is nearly zero.

NMR data in chloroform solution indicate that **LNa**$^+$ is present only in the cone conformation, while for **LCs**$^+$ only the 1,3-alternate form is detected[8]. Based on the simulations, the relative stability of two conformers cannot simply be compared because the systems contain different numbers of solvent molecules and of the large statistical fluctuations (about 50 kcal/mol). However, as for the free ligand **L**, the energy analysis strongly suggests that the conformational preferences in chloroform result not only from intrinsic energies, but also from solvation energies which favour the cone over the 1,3-alternate complexes. Although quantitative agreement may be fortuitous, it is noticeable that for **LNa**$^+$Pic$^-$, the intrinsic preference for the 1,3-alternate/cone (ΔE_{opt} = 3.8 kcal/mol) is overcome by the difference in "solvation energies" ($\Delta_{solvation}$ = -9.8 kcal/mol) and the cone would be preferred by 6 kcal/mol. For **LCs**$^+$Pic$^-$, ΔE_{opt} = 13.1 kcal/mol and $\Delta_{solvation}$ = -4.3, *i.e.* the 1,3-alternate form would be favoured by 8.8 kcal/mol, in qualitative agreement with NMR data.

Some other points, involving the counterion, may be noted. The stabilization energy $\Delta E_{opt\_solute} = E_{opt\_LM^+Pic^-} - E_{opt\_LM^+}$ of the solute brought about by the counterion (from 164 to 176 kcal/mol) depends on M^+ and on the conformation of **L**. For a given cation, the M^+/Pic$^-$ attractions are larger in the 1,3-alternate than in the cone complexes. They decrease as expected with the size of M^+ in the 1,3-alternate complexes, but not in the cone complexes where the sequence is changed ($Na^+ > Rb^+ > Cs^+ > K^+$) as K^+ interacts the least with the anion. The **L** /M^+ attractions are somewhat reduced in the presence of Pic$^-$ (by about 12 kcal/mol both forms), because the contacts between M^+ and **L** become less tight. These results suggests that, in addition to its lipophilicity, the size and electrostatic features of the counterion modulates its interactions with the complex and the related affinity for a given cation.

4.3. RELATIVE CATION BINDING AFFINITIES OF THE THREE CONFORMERS OF CALIX[4]CROWN6 *IN PURE CHLOROFORM*. INFLUENCE OF THE COUNTERION.

$$\begin{array}{ccc} & \Delta G_1 & \\ M_1^+(Pic^-) + \underline{L} & \text{------------>} & \underline{L}M_1^+(Pic^-) \\ \Delta G_3 \downarrow & & \downarrow \Delta G_4 \\ M_2^+(Pic^-) + \underline{L} & \text{------------>} & \underline{L}M_2^+(Pic^-) \\ & \Delta G_2 & \end{array} \quad (2)$$

In solution, the binding selectivity of cation M_1^+ compared to M_2^+ is measured experimentally by $\Delta G_c = \Delta G_1 - \Delta G_2$. The computer approach does not simulate the complexation process, but follows the alchemical route [22-24] by mutating M_1^+ into M_2^+ in their free (ΔG_3) and complexed states (ΔG_4). As a result of the thermodynamic cycle, $\Delta G_c = \Delta G_3 - \Delta G_4$. The results are reported in Table 6, first without counterion, then with Pic^- forming an intimate ion pair with the free or complexed cation.

In the absence of counterion, for the three conformers, ΔG_4 increases from **L**Na^+ to **L**Cs^+ ($Na^+ > K^+ > Rb^+ > Cs^+$), due mostly to the M^+ interactions with **L** (Table 5). ΔG_4 depends markedly on the conformer: 17.9, 26.9 and 35.4 kcal/mol respectively for the 1,3-alternate, paco and cone complexes. These numbers are larger than the calculated Na^+ / Cs^+ difference in solvation free energies (ΔG_3 = 6.0 kcal/mole). As a result, $\Delta G_c = \Delta G_3 - \Delta G_4$ is negative which means that, *in the absence of counterion, the three conformers prefer Na^+ over Cs^+*, in the sequence $Na^+ > K^+ > Rb^+ > Cs^+$ (Table 5 and Figure 4).

We investigated the counterion effect on binding selectivity in the cone and 1,3-alternate complexes. The counterion increases the ΔG_3 energies for the free Na^+Pic^- / Cs^+Pic^- pairs (from 6.0 to 17.6 kcal/mol). It has less effect on ΔG_4, since the complexed cation interacts less with Pic^- than in its free state. The **L**Na^+ / **L**Cs^+ ΔG_4 free energy increases by less than 2 kcal/mol in the presence of Pic^- for the two conformers. The $\Delta G_c = \Delta G_3 - \Delta G_4$ difference is thus reduced in the presence of Pic^-, but the preference for Na^+ is retained, at least for the cone form (ΔG_c = -19 kcal/mol). For 1,3-alternate form, ΔG_c becomes too small (-1.6 kcal/mol) to conclude safely.

				Na+ -> K+	K+ -> Rb+	Rb+ -> Cs+	Na+ -> Cs+
M^+		in chloro	ΔG_3	3.2 a)	1.1 a)	1.7 a)	6.0
			ΔG_{3polar}	5.5 a)	1.9 a)	3.1 a)	10.5
LM^+	cone	*in vacuo*	ΔG_4	18.3	7.0	9.8	35.1
		in chloro	ΔG_4	18.5 a)	7.8	9.1	35.4
			ΔG_3-ΔG_4	-15.3	-6.7	-7.4	-29.4
			ΔG_{3polar}-ΔG_4	-13.0	-5.9	-6.0	-24.9
	paco	*in vacuo*	ΔG_4	16.2	4.4	6.3	26.9
		in chloro	ΔG_4	17.9 a)	4.4	7.0	29.3
			ΔG_3-ΔG_4	-14.7	-3.3	-5.3	-23.3
			ΔG_{3polar}-ΔG_4	-12.4	-2.5	-3.9	-18.8
	1,3-alt	*in vacuo*	ΔG_4	10.0	3.2	4.7	17.9
		in chloro	ΔG_4	9.7 a)	3.0	5.0	17.7
			ΔG_3-ΔG_4	-6.5	-1.9	-3.3	-11.7
			ΔG_{3polar}-ΔG_4	-4.2	-1.1	-1.9	-7.2
M^+Pic^-		in chloro	ΔG_3	9.8	3.2	4.6	17.6
			$\Delta G_{3\text{-}polar}$	10.4	3.5	5.0	18.9
LM^+Pic^-	cone	in chloro	ΔG_4	18.9 a)	7.9	10.1	36.9
			ΔG_3-ΔG_4	-9.1	-4.7	-5.5	-19.3
			ΔG_{3polar}-ΔG_4	-8.5	-4.4	-5.1	-18.0
	1,3-alt	in chloro	ΔG_4	9.4	3.4	6.4	19.2
			ΔG_3-ΔG_4	0.4	-0.2	-1.8	-1.6
			ΔG_{3polar}-ΔG_4	1.0	0.1	-1.4	-0.3

Table 6: LM^+ and LM^+Pic^- complexes *in vacuo* and in chloroform: Free energy differences ΔG_4 and relative binding affinities $\Delta G_c = \Delta G_3 - \Delta G_4$ (kcal/mol). Unless otherwise specified calculations are performed in OPLS chloroform. a) Calculated for a total simulation time of 105 ps (21 windows).

At this stage, it is important to address the question of the electrostatic representation of chloroform, since the OPLS parameters, like other models fitted on the pure liquid[25], may not be appropriate to describe interactions with polar or charged solutes. This is why we repeated the ΔG_3 calculations on M^+ and M^+Pic^- (free states) in the "polar" chloroform model, as a crude mimics of solvent polarisation by the ions. This increases ΔG_3 from 6.0 to 10.5 kcal/mol for Na^+ / Cs^+ and from 17.6 to 18.9 kcal/mol only for Na^+Pic^- / Cs^+Pic^-. Since the complexed cations are shielded from chloroform, the ΔG_4 energies should be similar with the "polar" as with the OPLS model. Thus combining

ΔG_3 ("polar" solvent) - ΔG_4 ("OPLS" solvent) retains the preference for Na^+ over Cs^+ for the three conformers of **L** in the absence of counterion (Table 6 and Figure 4). In the presence of Pic^- the cone still prefers clearly Na^+ over Cs^+ (ΔG_C = -18 kcal/mol), but the 1,3-alternate form display no clear preference (ΔG_C = -0.3 kcal/mol).

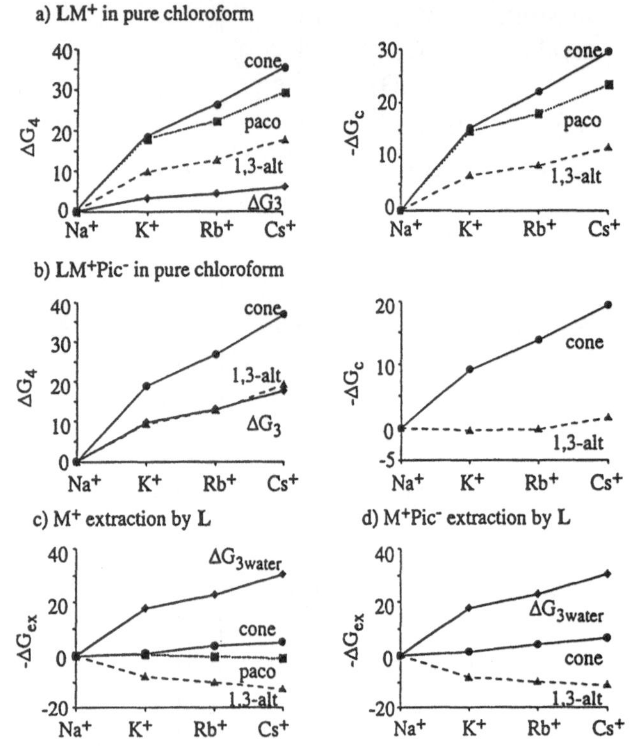

Figure 4: Graphical representation of FEP results for $\mathbf{LM^+}$ and $\mathbf{LM^+Pic^-}$ complexes in chloroform in kcal/mol. Top and middle left: relative free energy for mutating the free cation (ΔG_3) and the complexed cation (ΔG_4). Top and middle right: relative binding affinities $-\Delta G_c = -(\Delta G_3 - \Delta G_4)$ of $\mathbf{LM^+}$ (a) and of $\mathbf{LM^+Pic^-}$ (b) complexes. Bottom: relative free energies of cation extraction from water to chloroform $-\Delta G_{ex} = -(\Delta G_{3aq} - \Delta G_{4org})$ of M^+ (c) and of M^+Pic^- (d) by **L**.

This binding selectivity in chloroform differs from the one calculated previously in water[10], where **L** 1,3 alternate binds Cs^+ better Na^+, but the cone form binds Na^+ better than Cs^+. In both solvents, the ΔG_4 free energy differences are close. However, the difference in solvation free energies ΔG_3 is larger (about 30 kcal/mol for Na^+ / Cs^+) than in chloroform.

Based on *extraction* experiments, the free energies of binding alkali picrates by **L**, tb-**L** and the 1,3-alternate 1,3-diisopropoxy-calix[4]-crown6 have been determined in $CHCl_3$ [2,7,8], using the Cram's extraction method [26,27]. They display a clear preference for Cs^+ over Na^+, in the order $Na^+ < K^+ < Rb^+ < Cs^+$. From the computational side, the model used for chloroform has to be questioned as the neglect of polarisation effects may underestimate the Na^+ / Cs^+ ΔG_3 energy, and therefore exaggerate the binding affinity for Na^+, compared to Cs^+. On the other side, experimental data correspond to a chloroform phase saturated with water. We have shown previously that the assumption for the organic phase is more water-like than gas phase-like, led to the prediction of a large Cs^+ / Na^+ extraction selectivity[10]. Specific interactions of the complex with water play an important role in determining the binding selectivity. Theoretical studies, including mixed chloroform - water solvents should contribute to elucidate this question.

4.4. EXTRACTION OF M^+ CATIONS BY CALIX[4]CROWN6 L FROM WATER TO *CHLOROFORM*. INFLUENCE OF THE COUNTERION.

The extraction of M_1^+ and M_2^+ cations by **L** from an aqueous to an organic phase can be modelled using the following thermodynamic cycle:

$$\begin{array}{ccc} & \Delta G_{ex1} & \\ M_1^+(Pic^-)_{aq} + \underline{\mathbf{L}}_{org} & \text{------------>} & \underline{\mathbf{L}}M_1^+(Pic^-)_{org} \\ \Delta G_{3\_aq} \downarrow & & \downarrow \Delta G_{4\_org} \\ M_2^+(Pic^-)_{aq} + \underline{\mathbf{L}}_{org} & \text{------------>} & \underline{\mathbf{L}}M_2^+(Pic^-)_{org} \\ & \Delta G_{ex2} & \end{array} \quad (3)$$

The difference $\Delta G_{ex} = \Delta G_{ex1} - \Delta G_{ex2} = \Delta G_{3\_aq} - \Delta G_{4\_org}$ defines the extraction selectivity. Such a scheme assumes that the free ions are present in water only and that their complexes are present in the organic phase only. If the organic phase is dry, the free energy difference ΔG_{4_org} is equal to ΔG_{4_chloro}. In fact, in extraction experiments the organic phase may contain significant amounts of water, up to saturation. As a consequence the "real" ΔG_{4_org} value is expected to be intermediate between ΔG_{4_aq} (pure water) and ΔG_{4_chloro} (dry chloroform). In this scheme the Pic^- anion is indicated in the two liquid phases. However, in water the ions are dissociated and ΔG_{3_aq} corresponds to the cations only. In the organic phase, the counterion is bound to the complex, and ΔG_{4_org} should be calculated accordingly. In Table 7 we report the results with and without counterion for purpose of comparison.

			Na^+ -> K^+	K^+ -> Rb^+	Rb^+ -> Cs^+	Na^+ -> Cs^+
	ΔG_{3_aq}	M^+	17.6	5.1	7.7	30.4
cone	ΔG_{ex}	LM^+	-0.9	-2.7	-1.4	-5.0
		LM^+Pic^-	-1.3	-2.8	-2.4	-6.5
paco	ΔG_{ex}	LM^+	-0.3	0.7	0.7	1.1
1,3-alt	ΔG_{ex}	LM^+	7.9	2.1	2.7	12.7
		LM^+Pic^-	8.2	1.7	1.3	11.2

Table 7 : Extraction of M^+ from water to "dry" chloroform by **L** without and with picrate counterion. Calculated relative free energies (kcal/mol). $\Delta G_{ex} = \Delta G_{3\_aq} - \Delta G_{4\_chloro}$ (see Table 6).

These results and those displayed in Figure 5 show a clear *conformation dependent extraction selectivity: the 1,3-alternate form of* ***L*** *extracts* Cs^+ *better than* Na^+ (ΔG_{ex} = 11 kcal/mol with counterion and ΔG_{ex} = 13 kcal/mol without counterion). The cone ligand would extract Na^+ preferentially (ΔG_{ex} = -7 kcal/mol with counterion and ΔG_{ex} = -5 kcal/mol without counterion). These results are in agreement with experimental data on extraction selectivity[7,8]. For the partial cone form, the calculations are reported only without counterion and a very small preference is found for Cs^+ (ΔG_{ex} = 1 kcal/mol).

4.5. THE CALIX[4]-CROWN6 L IONOPHORE FREE AND COMPLEXED AT THE WATER / CHLOROFORM INTERFACE.

In reference [10] we reported the first MD investigations on the tb-**L** ionophore free and complexed at the chloroform / water interface. The cone Na^+ and 1,3-alternate Cs^+ complexes, simulated for 100 ps were found to remain "adsorbed" at the interface, sitting mostly in the chloroform phase. Repeating the simulation of the Cs^+ complex with a Pic^- counterion did not lead to migration to the organic phase either. These simulations were run at 300K in the (N, V, T) ensemble, using a "rigid" model of chloroform (k = 500 kcal/mol; no SHAKE). In this paper, we report longer simulations (350 ps) on the unsubstituted **L** ionophore, performed under the same conditions. We compare **L** free, its Cs^+ complex with two different starting positions and its Cs^+Pic^- complex with two different starting positions. We repeated these two 350 ps simulations of the LCs^+Pic^- complex, using the "polar" chloroform model.

Figure 5 depicts the position and orientation of the solutes at the beginning of the simulation. The time evolution of the distance between the center of mass of **L**, Cs^+ and

Pic⁻ with respect to the interface can be followed in Figure 6. Snapshots at the end of the dynamics, including selected solvent molecules are represented in Figure 7.

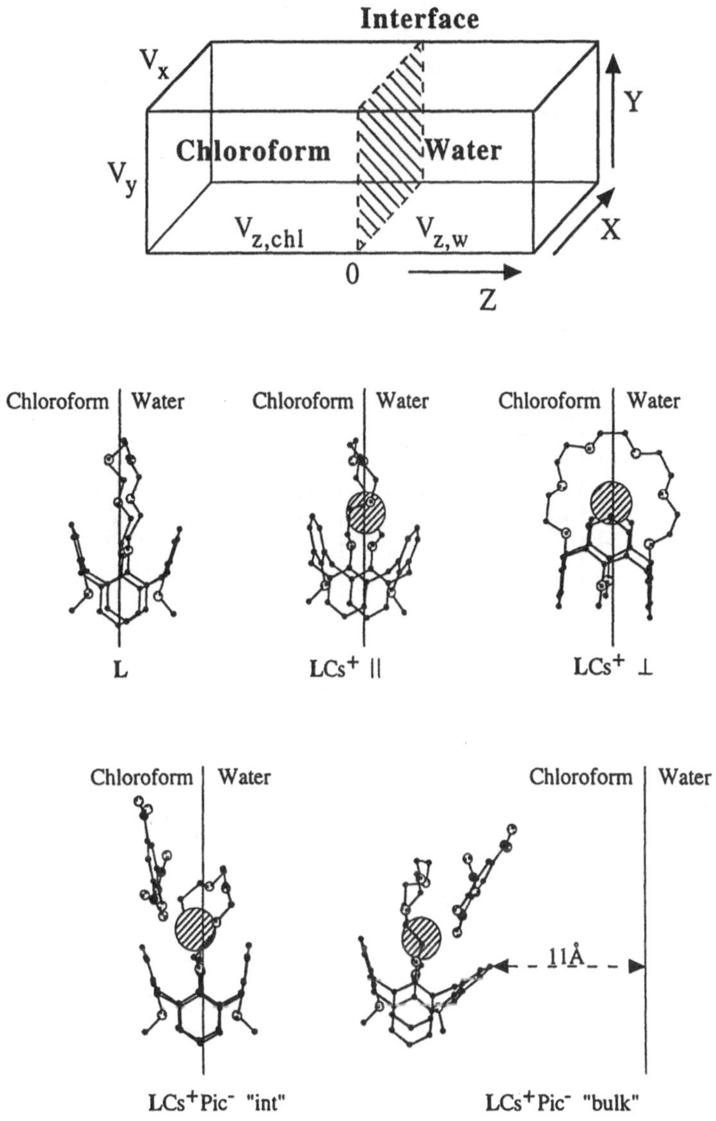

Figure 5: Simulation box and starting positions of the solutes at the chloroform/water interface: the free ligand **L**, the **LCs^+** complex (starting orientations with the crown ether parallel || and perpendicular to the interface ⊥). **LCs^+Pic^-** "int": complex starting at the interface. **LCs^+Pic^-** "bulk": complex starting in chloroform.

The 1,3-alternate free ligand **L**. The simulation starts with the ligand equally shared between the two liquid phases, oriented with its pseudo C_2 symmetry axis and the ether crown in the plane of the interface (Figure 5). During the first 160 ps **L** moves to the chloroform phase. The phenolic groups are entirely surrounded by chloroform molecules while the crown moiety remains close to the water phase. Then the ligand returns to the interface and captures two water molecules hydrogen bonded to its crown-ether oxygens. One of them bridges over the two phenolic oxygens. The second one makes a hydrogen-bond relay with the interface.

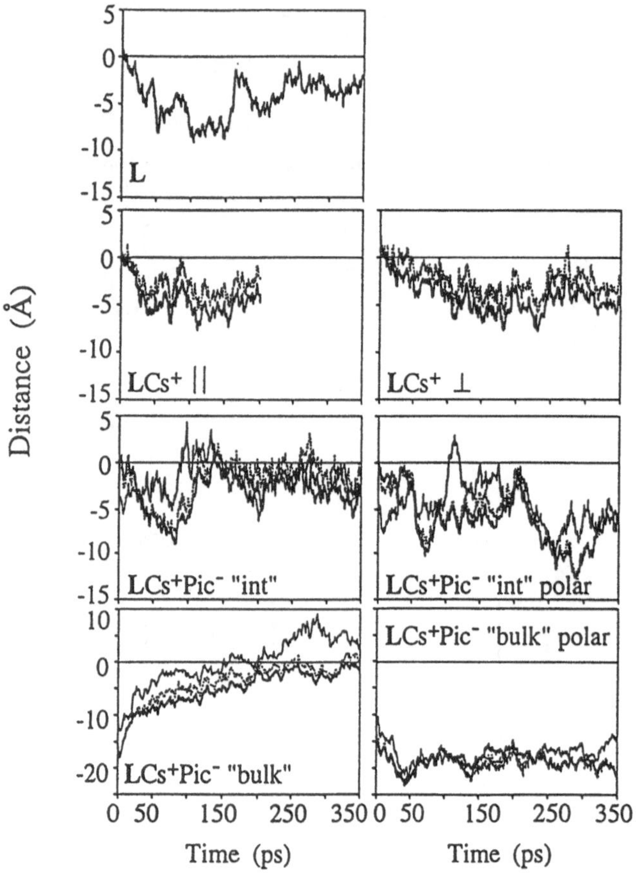

Figure 6 : L, LCs⁺ and LCs⁺Pic⁻ at the chloroform/water interface: Time evolution of the distances (Å) between the interface and the center of mass of the ligand **L** ($d_{L..Int}$, bold line —), of the cation ($d_{Cs^+..Int}$, dotted line ----) and of the anion ($d_{Pic^-..Int}$, full line —). The **LCs⁺** complex (starting orientations || and ⊥; see Figure 5). **LCs⁺Pic⁻** "int": starting at the interface. **LCs⁺Pic⁻** "bulk": starting in chloroform. Simulated with two chloroform models (see text).

The remote face of **L** pointing to chloroform is not hydrated. This location of **L** at the interface remains unchanged until the end of the simulation (Figure 7). Thus, although **L** is more soluble in chloroform than in water, it does not migrate fully to the chloroform phase.

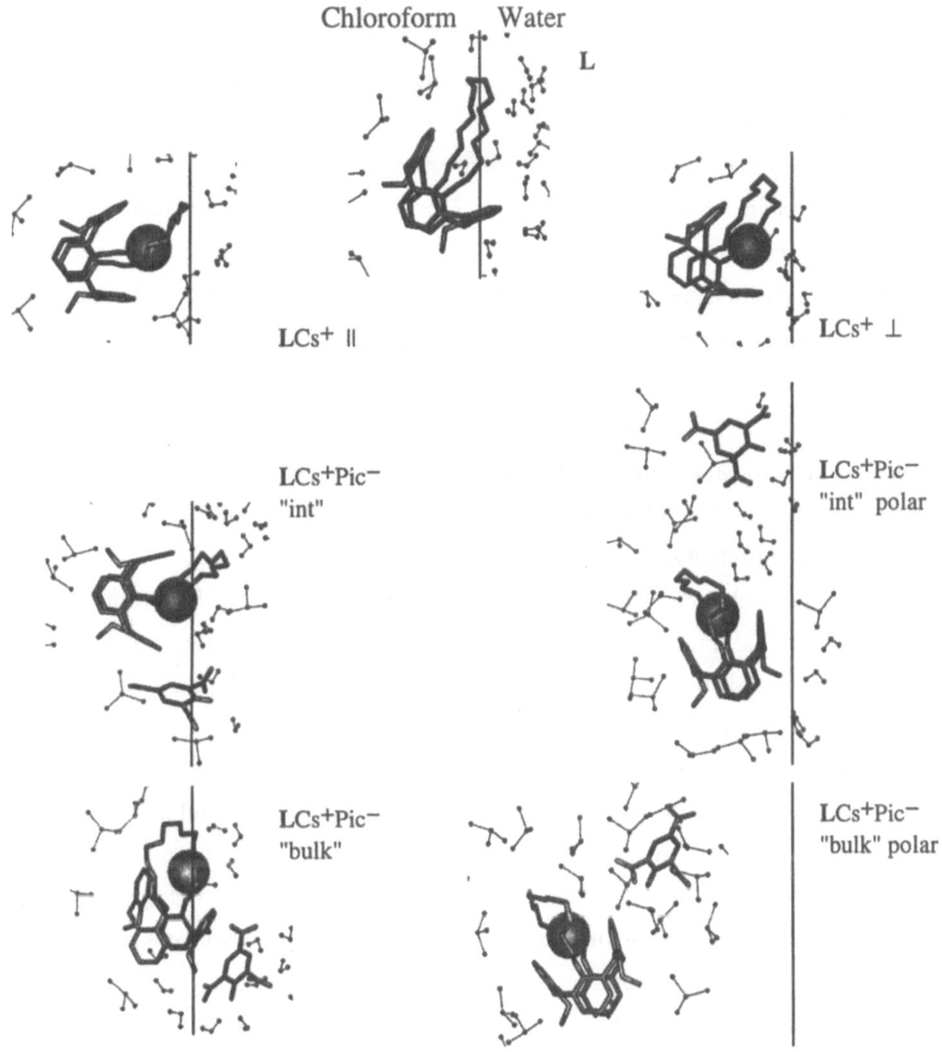

Figure 7: **L**, **LCs**$^+$ and **LCs**$^+$**Pic**$^-$ at the chloroform/water interface: Snapshots after 350 ps of simulation (except **LCs**$^+$ || : 200 ps). **LCs**$^+$ (starting orientations || and ⊥); **LCs**$^+$Pic$^-$ "int" and **LCs**$^+$Pic$^-$ "bulk"; see Figure 5. Simulated with two chloroform models (see text).

The LCs^+ *complex.* It has been simulated with two starting orientations at the interface. In the first simulation, it is oriented as the free ligand described above (Figure 5) with the crown moiety and Cs^+ in the plane of the interface. During the first 10 ps, the aromatic moiety moves towards chloroform whereas the crown-Cs^+ part stays at the water surface. One water molecule coordinates Cs^+ at 3.1-3.5 Å until the end of the simulation. After another 40 ps the complex has almost completely migrated into chloroform (Figure 6) but remains hydrogen bonded with water molecules via the ether crown, which bends to allow for these interactions (Figure 7). It remains so until the end of the simulation, with its C_2 symmetry axis "perpendicular" to the interface.

At the beginning of the second simulation **LCs**$^+$ is also equally shared between the two liquid phases, but the ether crown is perpendicular to the interface (Figure 5). In about 40 ps it adopts a position similar to the one described above (Figures 6 and 7). Thus, the two different starting orientations of the solute converge to a similar situation.

The 1,3-alternate LCs^+Pic^- *complex.* Four simulations have been performed with this complex: the two first used the OPLS chloroform model, and start respectively at the interface and in the bulk chloroform. The two others use the "polar" chloroform model with these two different starting positions.

In the first simulation, **LCs**$^+$**Pic**$^-$ is oriented initially as above with the crown ether ring in the plane of the interface. The anion, nearly parallel to the interface, at 3 Å on the chloroform side, is coordinated to Cs^+ (Figure 5). During the 350 ps of simulation the anion and the complex stay at the interface, but dissociate. The distance between Pic^- and Cs^+ increases to 7 Å. Such relationship differs from the one in pure chloroform (where Pic^- is bound to Cs^+) or in pure water (Pic^- migrates in the "equatorial plane" of **L**; see Figure 10). The **LCs**$^+$ complex, more mobile than with counterion oscillates between orientations parallel or (more frequently) perpendicular to the interface (Figures 6 and 7). Thus, the solute remains again more or less anchored at the water phase instead of migrating to the bulk chloroform.

In order to examine whether this is not biased by the choice of starting state, we performed a second simulation starting with the solute completely immersed in chloroform (Figure 5). The smallest distance between **L** and the interface is 11 Å, *i.e.* 1 Å less than the cutoff distance. After 200 ps, **LCs**$^+$**Pic**$^-$ migrates to the interface with its crown in contact with the water phase, as found with the simulation which started at the interface (Figure 7). The counterion first diffused into water, but after 350 ps has returned close to the interface. These computer experiments thus confirm the larger affinity of the solute for the interface, compared to the bulk organic phase.

At this stage, the force field representation of the system, and of chloroform in particular has to be questioned. This is why we reran these two simulations on LCs^+Pic^-, but using the "polar" chloroform model. The behaviour becomes quite different. Now, the complex initially set at the interface moves deeply into chloroform, accompanied by one water molecule coordinated to Cs^+ (Figures 6 and 7). Conversely, in the simulation which starts with LCs^+Pic^- in "polar" bulk chloroform, the solute remains in the bulk instead to migrate to the interface. In both simulations with the "polar" chloroform model some water molecules diffuse into the chloroform phase while some chloroform molecules diffuse into the water phase.

Such solvent diffusion does not occur with "standard" OPLS chloroform. A superposition of the solvent density profiles of simulations with "standard" and "polar" chloroform quantifies the related increase of interfacial width *IW* (Figure 8). We define *IW* as the Z distance where the density of water decreases from 90% to 10% of the bulk value. If the density is averaged over the 50-350 ps of dynamics, *IW* increases from 5.7 (OPLS chloroform) to 7.4 Å ("polar" chloroform). However, solvent mixing increases with time and during the last 50 ps simulated, *IW* increases from 7.6 Å (OPLS chloroform) to 13.6 Å ("polar" chloroform). Beyond 5 Å from the interface, there is no solvent mixing with the OPLS chloroform, whereas with the "polar" chloroform, these density curves are non zero, indicating enhanced solvent diffusion and mixing (Figure 8).

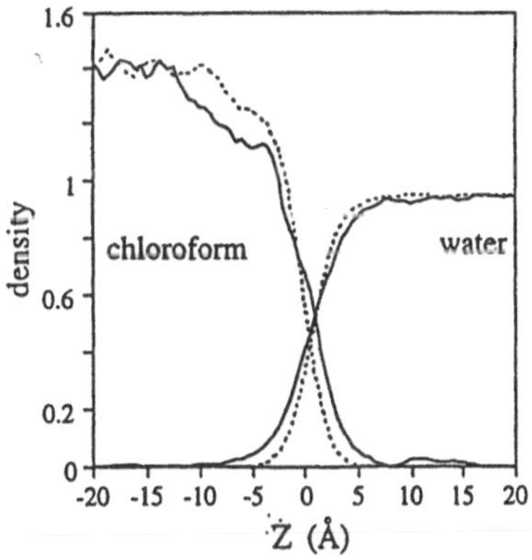

Figure 8 : LCs^+Pic^- complex at the interface: density profiles of chloroform and water. Simulation with "standard" OPLS chloroform (dotted lines -----) and "polar" chloroform (full lines ——).

Energy analysis at the interface: Why do the solutes remain adsorbed at the interface on the chloroform side? In this section, we analyze the affinity of the different solutes for each of the liquid phases, to investigate why they do not migrate to the bulk organic phase when chloroform is described by the OPLS model. The interaction energy between each solute (**L**, **LCs$^+$** and **LCs$^+$Pic$^-$**) and the two solvents is followed as a function of time (Figure 9). Average values are reported in Table 8. In all cases, the solutes have attractive interactions with *the two* solvents. Although they sit almost entirely in the chloroform phase, they enjoy significant interactions with the water phase. The time evolution analysis illustrates why **L** prefers to remain at the interface, and the crucial role played by a few water molecules.

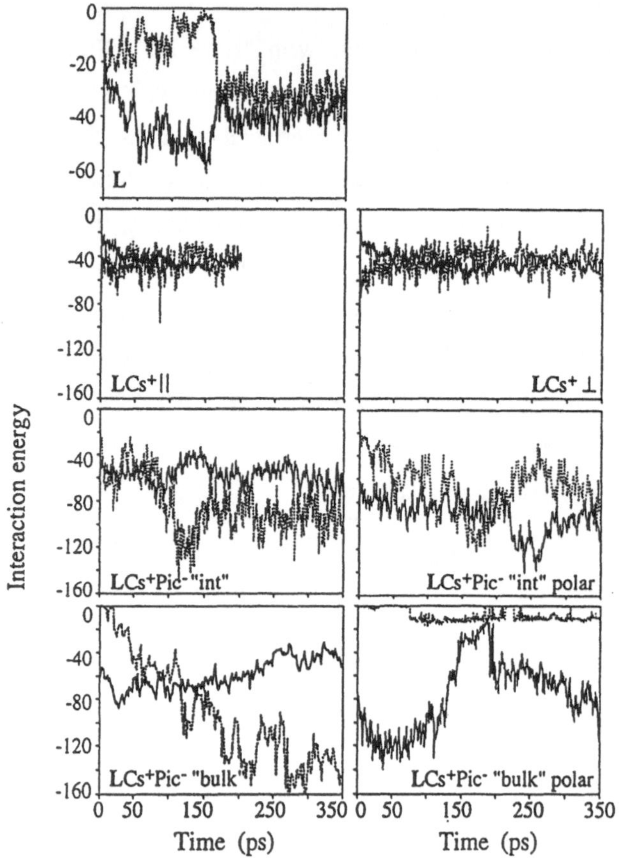

Figure 9 : L, **LCs$^+$** and **LCs$^+$Pic$^-$** at the chloroform/water interface: Interaction energies (kcal/mol) of the solute with water (dotted lines -----), and with chloroform (full lines ——) as a function of time. The **LCs$^+$** complex (starting orientations || and ⊥; see Figure 5). **LCs$^+$Pic$^-$** "int": starting at the interface. **LCs$^+$Pic$^-$** "bulk": starting in chloroform. Simulated with two chloroform models (see text).

	Starting at the interface					Starting in bulk chloroform	
Solute	**L**	**LCs+ || d)**	**LCs+ ⊥**	**LCs+Pic-**	**LCs+Pic-**	**LCs+Pic- e)**	**LCs+Pic-**
Chloroform model	OPLS	OPLS	OPLS	OPLS	"polar"	OPLS	"polar"
nb_{water}, $nb_{chloroform}$ a)	772, 221	850, 225	849, 209	1123, 365	1123, 365	1007, 322	947, 387
V_x, V_y, $V_{z\text{-wat}}$, $V_{z\text{-chl}}$ b)	33,30,28,34	33,30,29,34	33,29,30,34	39,35,27,39	39,35,27,39	34,39,24,44	34,39,24,43
Average Interaction Energies (kcal/mol) c)							
$\langle E_{chloroform/water} \rangle$	-101 ± 15	-124 ± 17	-111 ± 18	-176 ± 20	-501 ± 37	-159 ± 20	-479 ± 33
$\langle E_{L/water} \rangle$	-33 ± 6	-5 ± 3	-1 ± 3	-5 ± 6	-2 ± 4	-4 ± 4	-12 ± 2
$\langle E_{Cs^+/water} \rangle$		-38 ± 9	-43 ± 9	-33 ± 13	-29 ± 12	-43 ± 9	+3 ± 1
$\langle E_{Pic^-/water} \rangle$				-54 ± 12	-53 ± 16	-85 ± 7	-1 ± 1
$\langle E_{LCs^+Pic^-/water} \rangle$	-33 ± 6	-43 ± 9	-44 ± 9	-92 ± 13	-84 ± 16	-132 ± 9	-10 ± 2
$\langle E_{L/chloroform} \rangle$	-37 ± 4	-42 ± 3	-41 ± 4	-35 ± 7	-44 ± 5	-37 ± 4	-40 ± 5
$\langle E_{Cs^+/chloroform} \rangle$		-3 ± 2	-3 ± 3	-1 ± 2	-12 ± 5	-3 ± 2	-2 ± 4
$\langle E_{Pic^-/chloroform} \rangle$				-24 ± 6	-37 ± 15	-3 ± 3	-39 ± 8
$\langle E_{LCs^+Pic^-/chloroform} \rangle$	-37 ± 4	-45 ± 3	-44 ± 5	-60 ± 7	-93 ± 15	-43 ± 6	-81 ± 11
$\langle E_{Cs^+/Pic^-} \rangle$				-40 ± 8	-40 ± 8	-40 ± 4	-65 ± 11

Table 8: Solutes at the chloroform / water interface: characteristics of the simulations and energy component analysis. a) Number of solvent molecules; b) size of the solvent boxes (Å); c) interaction energies averaged over the last 50 ps. Unless otherwise specified all simulations are constant volume runs of 350 ps, using the "rigid" chloroform ($k_{C\text{-}Cl}$ and $k_{Cl\text{-}Cl}$ = 500 kcal/mol/Å^2). d) For **LCs+ ||**, the simulation time is 200 ps; e) constant pressure, "flexible" chloroform model ($k_{C\text{-}Cl}$ = 216 kcal/mol/Å^2, $k_{Cl\text{-}Cl}$ =100 kcal/mol/Å^2).

The free ligand **L**. (Figure 7). As **L** moves initially from the interface into chloroform, $E_{L/water}$ drops from -20 to -3 kcal/mol while $E_{L/chloroform}$ increases from -20 to -50 kcal/mol, *i.e.* **L** displays much larger attractions with chloroform than with water (about -50 and -10 kcal/mol, respectively). At about 160 ps, **L** has returned to the interface and captured two water molecules in its crown cavity, provoking a 20 kcal/mol increase of $E_{L/water}$. Then, until the end of the simulation, **L** has similar interactions with water as with chloroform (-33 and -37 kcal/mol, respectively). The major contribution to $E_{L/water}$ comes from the crown ether moiety whereas the major contribution to $E_{L/chloroform}$ comes from the calixarene moiety.

The **LCs^+** *complex.* With the two starting orientations, it has nearly equal interaction energies with the two solvents along the dynamics (about -45 kcal/mol per solvent). This is similar to the interaction energy of the solvents with **L** uncomplexed after equilibration. The interaction with chloroform arises mostly from **L** (**LCs^+** $\perp$: -41 kcal/mol) while the interaction with water is due almost exclusively to Cs^+ (**LCs^+** $\perp$: -43 kcal/mol).

The **LCs^+Pic^-** *complex.* In contrast to **L** or **LCs^+**, the **LCs^+Pic^-** complex is more attracted by water (-92 kcal/mol) than by chloroform (-60 kcal/mol). The **LCs^+** part, initially set at the interface, interacts about equally with both solvents, but the anion interacts more with water (-54 kcal/mol) than with chloroform (-24 kcal/mol). It remains however sitting in the plane of the interface, instead of diffusing to water. A first explanation comes from the fact that the attraction energy between Pic^- and its surrounding at the interface (including Cs^+) is 11 kcal/mol more attractive than the Pic^-/water interaction calculated in bulk water (-107 kcal/mol). Secondly, migration to water would cost also a significant "cavitation energy"[28-32]. The stationary state of Pic^- at the interface allows to compromise between partial hydration of its oxygen atoms and the hydrophobicity of the aromatic ring (Figure 8). In the simulation started in bulk chloroform, the Pic- anion ended up on the water side of the interface, and displayed larger attraction with water (Table 8).

In "polar" chloroform, the **LCs^+Pic^-**/chloroform interactions are about 30 kcal/mol larger than with OPLS chloroform and display larger fluctuations e.g. the Pic^- contribution fluctuates between 0 and -80 kcal/mol depending on the position of Pic^- (see Figure 6).

The chloroform/water interaction energy is another index of interest, which may reflect the solvent mixing induced by the solute. It is attractive and depends on the size of the interface, on the nature and location of the solute. With the standard OPLS chloroform model, it increases from -101 kcal/mol with **L** free to -176 kcal/mol with the **LCs**$^+$Pic$^-$ complex. With the **LCs**$^+$ solute, the attraction is intermediate (-111 to -124 kcal/mol). These numbers cannot be simply compared, because the surface $V_x.V_y$ (defined in Figure 5) increases from the **L** (990 Å^2) to the **LCs**$^+$ Pic$^-$ (1365 Å^2) solutions. Scaling these energies to a surface of 100 Å^2 gives -9.8±1.5 kcal/mol for **L**, -12.5 and -11.6 for **LCs**$^+$ (two different starting orientations), -12.9±2.0 kcal/mol for **LCs**$^+$Pic$^-$. These numbers may be indicative of increased solvent - solvent interactions from **L** to **LCs**$^+$ and **LCs**$^+$Pic$^-$. They are smaller than the interaction energy of the neat interface (-13.6 ±1.9 kcal/mol/100Å^2), presumably because the solutes act as a small "wall" between water and chloroform. Calculations on the **LCs**$^+$Pic$^-$ complex using the polar chloroform model ($V_x.V_y$ = 1365 Å^2) show a marked increase in the chloroform/water interaction (-501 kcal/mol, *i.e.* -36.7 kcal/mol / 100 Å^2), as a result of enhanced solvent mixing and electrostatic interactions.

5. Discussion and Conclusion

We have reported theoretical investigations on the stereochemical dependent binding of alkali cations by dimethoxy-calix[4]crown6 **L**, which displays in the 1,3-alternate form a remarkable Cs^+ / Na^+ binding selectivity. Computations in chloroform, an apolar solvent used in liquid-liquid extraction systems, allow to gain insight into the complexation in the pure liquid, to simulate the thermodynamics of extraction and to model the chloroform / water interface.

5.1. CONFORMATION DEPENDENT BINDING SELECTIVITY OF CALIX[4]CROWN6 IN NON-AQUEOUS SOLVENTS.

In pure chloroform solution, the three conformers of **L** are calculated to display the largest affinity for Na^+ and the weakest affinity for Cs^+, in contrast to what was calculated previously in pure water [10], and has been measured later in methanol solution[8]. The sequence ($Na^+ > K^+ > Rb^+ > Cs^+$) is the same as the one calculated *in vacuo* for **L**, tb-**L** [10], 18-crown-6, the *222* cryptand [33], or measured for M^+/H_2O interactions in the gas phase[34]. Intrinsically, none of the conformers prefers Cs^+.

Taking into account the Pic^- anion to model the complexation process retains the Na^+ / Cs^+ selectivity for the cone form, but for the 1,3-alternate conformer, the difference drops close to zero, with the two electrostatic models of the solvent. To our knowledge, no experimental complexation data have been obtained in pure chloroform, due to the insolubility of the salts of alkali cations. Complexing ligands like calixarenes dissolve these salts in organic solvents, but ion selectivity and related thermodynamic data are, to our knowledge, not available.

In relation with extraction experiments, we modelled the ion extraction from water to chloroform and find a clear *conformation dependent binding selectivity*. The ligand **L** in its cone conformation is calculated to *extract* preferentially Na^+ over Cs^+. Conversely, the 1,3-alternate form is calculated, like the conformationally rigid 1,3-alternate calix[4]-*bis*-crown6 hosts [35] to *extract* Cs^+ better than Na^+. This is in full agreement with experiment.

5.2. CONFORMATION OF THE FREE CALIXCROWNS AND OF THEIR COMPLEXES IN SOLUTION: ON THE *IMPORTANCE OF CONFORMATION DEPENDENT SOLVATION ENERGIES*.

In solution, the conformational states of a given molecule depend on the interplay between intrinsic (gas phase) stability and solvation effects[36]. For instance, the population of C_i / D_{3d} forms of 18-crown6 differs in water, acetonitrile and apolar solvents[37,38]. Specific interactions like hydrogen bonding, due to the solvent granularity contribute to the solvation energy[38]. In polar solvents, it is accepted that stabilization of the cone form of calixarenes results from its large dipole moment [39, 40]. We show that, even with weakly polar organic solvents like chloroform, the conformational equilibria of calixarenes are modulated by the solvent. In particular, the cone form is best solvated in chloroform, as in water or acetonitrile. First, because it can complex specifically a solvent molecule. Second, it has the largest surface accessible to the solvent (1118, 1064 and 998 Å^2, respectively for the cone, partial cone and 1,3-alternate forms of **L** free [10]), leading to largest stabilization by van der Waals contacts. For the LM^+Pic^- complexes, the cone is also preferred because it is best solvated. Gas phase studies [41,42] may therefore not be sufficient to study conformational properties of calixarenes in solution. In the solid state, the conformation may still be different, as packing forces also contribute to the conformations[43,44].

5.3. COUNTERION EFFECTS ON COMPLEXATION IN SOLUTION AND LIQUID-LIQUID EXTRACTION.

The computations point out two aspects of counterions in the organic phase. First, they modulate the structure of the complex ("pulling effect") and prevent direct contacts between the complexed cation and the solvent. From the energy point of view, this modulates the ΔG_4 free energy difference between two complexes. The second and most important effect concerns the uncomplexed states, since counterions strongly increase the differences in solvation free energies of the cations (via the ΔG_3 energies). Related experimental data are worth to be noted. Extraction experiments of 18-crown6 derivatives with halide salts of alkali cations showed that the order of extraction does not simply follow the order of ionic radii, hydration enthalpy or softness parameters of the anion [45,46]. As pointed out here for the **LM**$^+$ complexes, it is likely that interactions between the complexed cation and the counterion contribute, together with the anion lipophilicity, to the cation extraction selectivity.

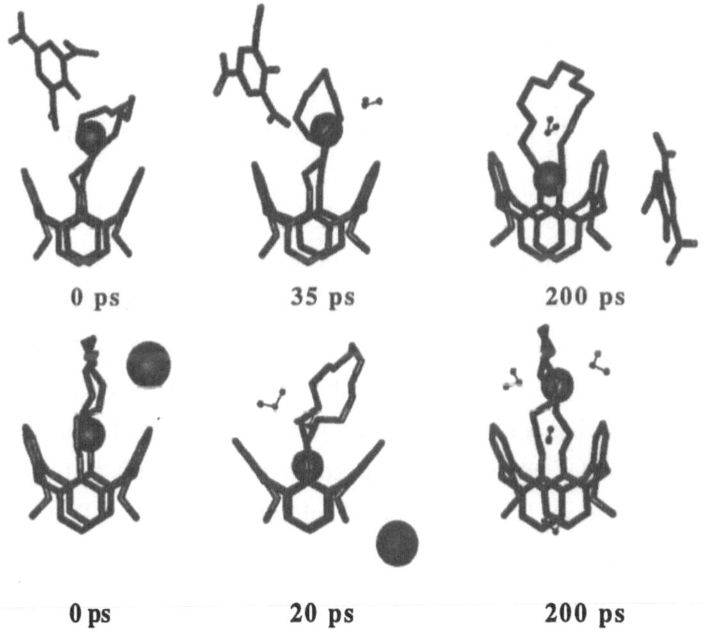

Figure 10 : Snapshots of the 1,3-alternate **LNa**$^+$**Pic**$^-$ (top) and **LNa**$^+$**Cl**$^-$ (bottom) complexes in water.

In aqueous solution, the complexed cation, likely separated from its counterion, was neglected in the previous calculations on calix-crowns [10], since cation complexation is known to separate ion pairs[47]. They may however perturb the structure of the complex, as shown in the following on the 1,3-alternate **LM**$^+$Pic$^-$ (M^+= Cs^+, K^+, Na^+) [10] and **LNa**$^+$Cl$^-$ complexes, simulated in water (Figure 10). The simulations start with intimate contacts, as in chloroform. As expected, rapid dissociation takes place, but the evolution is different with the two anions. In all **LM**$^+$Pic$^-$ complexes, the anion translates in such a way to stack over a phenolic group of **L**, until the end (200 ps) of the simulation. The Cs^+ coordination remains the same as without counterion. In the Na^+ complex, the cation attracted by Pic$^-$ down to the calix-cavity does not decomplex, as it did when no counterion was included [10] (Figure 10). The dissociation of Cl^- starts similarly and after 20 ps the **LNa**$^+$Cl$^-$ complex looks like the Pic$^-$ complexes (Figure 10). After, however, Cl^- dissociates completely and Na^+ moves back to the crown-cavity, coordinated by three water molecules, instead of decomplexing (Figure 10). These computer experiments suggest thus that *the counterion may somewhat change the nature of the complex, in water as in other solvents, especially with cations too small for the host.* The anion may also, via medium range electrostatic interactions, enhance the binding of cations which display a good fit with the host.

5.4. IONOPHORES FREE AND COMPLEXED AT THE WATER CHLOROFORM INTERFACE.

The first simulations of liquid / liquid interfaces concerned neat liquids. The water / benzene [48], water / hexanol [49] interfaces were simulated by Monte Carlo calculations. Molecular Dynamics was used to investigate the interface between water and hexane [50], nonane [51], decane [52], lipids [53], dichloroethane [54], an extended hydrophobic phase[55]. The mixing / demixing of solvents have also been modelled by Lennard-Jones particles. [56] Neutral molecular solutes such as peptides [57], fatty acids and squalene [58] were considered at the water / CCl_4 interface. With ions as solutes, MD studies involve the transfer of small ions across the water / dichloroethane interface [59-61] and the probing of interfacial width and dynamics of charge transfer at the water / octanol interface.[62] A recent review can be found in reference [63].

In the field of ion extraction by ionophores, we reported MD simulations of free ionophores (18-crown6, *222* cryptand, neutral and charged calix[4]arenes) at the chloroform / water interface.[64] Simulations of complexed ionophores at this interface were performed on Cs^+Picrate$^-$ calix[4]arene-*bis*-crown6[65], Cs^+calix[4]arene$^-$[66], CMPO.$UO_2(NO_3)_2$ [68], *222*.K^+, 18-crown6.$CaCl_2$ and 18-crown-6.Ca$(Picrate)_2$ complexes[67]. In all these studies, the solutes were found to remain "adsorbed" near the interface on the chloroform side, sometimes connected to water by "water fingers", instead of migrating to the bulk chloroform. This raises the question of energy profile for migration from the interface to the chloroform phase. Three typical situations can be considered: (i) it is a slow diffusional downhill process, which required timescales beyond the ones which have been simulated; (ii) there is some energy barrier separating the interface from the chloroform phase to be overcome; (iii) the solutes are trapped at the interface in a potential well. PMF studies on the interface crossing should contribute to investigate these situations. Based on our calculations, the hypothesis (iii) seems more likely than the two others for the systems studied.

The calculated preference of free and complexed ionophores for the interface is unlikely to be biased by the choice of the starting configuration at the interface, or even in bulk chloroform. Computer experiments starting with either an hydrophilic ion (Cl^-) or an hydrophobic solute (*n*-butane) equally shared by the two solvents at the interface led rapidly to expected behaviour, *i.e.* full migration to the bulk water and to the bulk chloroform phase, respectively[64]. Conversely, increasing the lipophilicity of a dialkoxy $(RO)_2$-calix[4]arene-crown6 solute, by replacing R = Me (**L**) by *n*-Octyl does not lead to a simulated migration to the organic phase either. Thus, based on these simulations we suggest that *the free or complexed ionophores behave as surfactants.* As a result, concentration effects should play an important role in the migration to the organic phase, via a two step process. First, the interface saturates more or less with the free or complexed ionophores. Second, diffusion to the organic phase takes place. We also suggest that the *synergistic effects* [68,69] observed in the extraction or transport of cations may not only result from the cooperative binding of the ion by two hosts, but also from a modification of the interface properties, decreasing its surface tension. From the experimental side, based on surface tension measurements of the water / benzene interface in the presence of dibenzo-18-crown6, it has been suggested that this ionophore also behaves as surfactants[70]. Other kinetic and mechanistic studies on transport by derivatized crown ethers have been interpreted by a mechanism involving rate limiting adsorption of carrier or desorption of the complex, depending of the experimental conditions[72]. This is consistent with the presently calculated results.

Several questions arise, from the theoretical side and the experimental side. First, we stressed that the force field model used to depict the solvents may play a critical role, since the use of a fixed "polar" representation of chloroform leads to full migration of the complex to the organic phase, together with significant solvent-solvent mixing. We feel however that this "polar" model may not be realistic as it exaggerates the electrostatic interactions beyond the first shell of a polar solute. With the standard model of chloroform, which accounts for the properties of this bulk liquid, the interface is narrow (about 5 Å) and no water migration takes place beyond the interface, as found in related studies. Ionophores like crown ethers, tributylphosphate [71] are known however, like small polar organic molecules, to increase the water / organic solvent mixing ("water dragging effect"[73]). These results call for experimental studies, involving the questions of solvent mixing and elucidation of the mechanisms of ion extraction.

Looking back at early gas phase simulations on ionophores [74], one can appreciate the progress achieved via combined developments of computer graphics, theory, software and computer power[75,76]. After the many studies on ion recognition in aqueous solutions, computer modelling on non-aqueous solutions of polar solutes, still in its infancy, will hopefully contribute to better understand the basis of molecular recognition in processes such as ion complexation, extraction, transport and phase transfer catalysis.

Acknowledgements

The authors are grateful to the IDRIS CNRS computer center for generous allocation of computer time, to the EEC (F12W-CT90-0062 project) for supporting part of this research and to E. Engler for software developments. M.L. thanks the French Ministery of Research and Space for a grant.

References

(1) Hill, C.; Dozol, J.-F.; Lamare, V.; Rouquette, H.; Tournois, B.; Vicens, J.; Asfari, Z.; Bressot, C.; Ungaro, R.; Casnati, A. *Nuclear Waste Treatment by Means of Supported Liquid Membranes Containing Calixcrown Compounds* in "Calixarenes 50th anniversary: commemorative issue"; J. Vicens, J. Harrowfield, Z. Asfari, Ed.; Kluwer, Dordrecht, **1995**.

(2) Ungaro, R.; Casnati, A.; Ugozzoli, F.; Pochini, A.; Dozol, J.-F.; Hill, C.; Rouquette, H. *Angew. Chem. Int. Ed.* **1994**, *33,* 1506.

(3) Dijkstra, P. J.; Brunink, A. J.; Bugge, K.-E.; Reinhoudt, D. N.; Harkema, S.; Ungaro, R.; Ugozzoli, F.; Ghidini, E. *J. Am. Chem. Soc.* **1989**, *111,* 7567.

(4) Böhmer, V. *Angew. Chem. Int. Ed. Engl.* **1995**, *34,* 713.

(5) Vicens, J.; Böhmer, V. *Calixarenes: a Versatile Class of Macrocyclic Compounds.*; Academic Publishers: **1991** .

(6) Gutsche, C. D.; Iqbal, M.; Nam, K. S.; See, K.; Alam, I. *Pure Appl. Chem.* **1988**, *60,* 483.

(7) Ungaro, R.; Arduini, A.; Casnati, A.; Ori, O.; Pochini, A.; Ugozzoli, F. *Complexation of Ions and Neutral Molecules by Functionalized Calixarenes* in "Computational Approaches in Supramolecular Chemistry"; Wipff, G., Ed.; Kluwer, Dordrecht, **1994**.

(8) Casnati, A.; Pochini, A.; Ungaro, R.; Ugozzoli, F.; Arnaud, F.; Fanni, S.; Schwing-Weill, M.-J.; Egberink, R. J. M.; Reinhoudt, D. N. *J. Am. Chem. Soc.* **1995**, *118,* 2767.

(9) Cecille, L.; Casarci, M.; Pietrelli, L. *New separation chemistry techniques for radioactive waste and other specific applications.*; Communities, Commission of the European Communities; Elsevier Applied Science: London New York, **1991** .

(10) Wipff, G.; Lauterbach, M. *Supramol. Chem.* **1995**, *6,* 187.

(11) Pearlman, D. A.; Case, D. A.; Cadwell, J. C.; Seibel, G. L.; Singh, U. C.; Weiner, P.; Kollman, P. A. *AMBER4*; University of California: San Francisco, **1991** .

(12) Jorgensen, W. L.; Chandrasekhar, J.; Madura, J. D. *J. Chem. Phys.* **1983**, *79,* 926.

(13) Jorgensen, W. L.; Briggs, J. M.; Contreras, M. L. *J. Phys. Chem.* **1990**, *94,* 1683.

(14) Jorgensen, W. L. *J. Am. Chem. Soc.* **1981**, *103,* 335.

(15) Singh, U. C.; Brown, F. K.; Bash, P. A.; Kollman, P. A. *J. Am. Chem. Soc.* **1987**, *109,* 1607.

(16) Bash, P. A.; Singh, U. C.; Brown, F. K.; Langridge, R.; Kollman, P. A. *Science* **1987**, *235,* 574.

(17) Harrowfield, J. M.; Ogden, M. I.; Richmond, W. R.; White, A. H. *J. Chem. Soc. Chem. Commun.* **1991**, 1159.

(18) Gregory, K.; Bremer, M.; von R. Schleyer, P.; Klusener, P. A.; Brandsma, L. *Angew. Chem.* **1989**, *101,* 1261.

(19) Ugozzoli, F.; Ori, O.; Casnati, A.; Pochini, A.; Ungaro, R.; Reinhoudt, D. N. *Supramol. Chem.* **1995**, *5,* 179.

(20) Ikeda, A.; Shinkai, S. *J. Am. Chem. Soc.* **1994**, *116,* 3102.

(21) Kumpf, R. A.; Dougherty, D. A. *Science* **1993**, *261,* 1708.

(22) Straatsma, T. P.; McCammon, J. A. *Annu. Rev. Phys. Chem.* **1992**, *43,* 407.

(23) Jorgensen, W. L. *Acc. Chem. Res.* **1989**, *22,* 184.

(24) Beveridge, D. L.; DiCapua, F. M. *Annu. Rev. Biophys. Biophys. Chem.* **1989**, *18,* 431.

(25) Tironi, I. G.; van Gunsteren, W. F. *Mol. Physics* **1994**, *83,* 381.

(26) Lein, G. M.; Cram, D. J. *J. Am. Chem. Soc.* **1985**, *107,* 448.

(27) Helgeson, R. C.; Weisman, G. R.; Toner, J. L.; Tarnowski, T. L.; Chao, Y.; Mayer, J. M.; Cram, D. J. *J. Am. Chem. Soc.* **1979**, *101,* 4928.

(28) Tomasi, J. *Int. J. Quant. Chem.: Quant. Biol. Symp.* **1991**, *18,* 73.

(29) Guillot, B.; Guissani, Y.; Bratos, S. *J. Chem. Phys.* **1991**, *95,* 3643. For a discussion of thermodynamic aspects of the "Scaled Partition Theory", see Morel-Desrosiers, N.; Morel, J.-P. *Can. J. Chem.* **1981**, *59,* 1.

(30) Wallqvist, A. *J. Phys. Chem.* **1991**, *95,* 8921.

(31) Blokzijl, W.; Engberts, J. B. F. N. *Angew. Chem. Int. Ed. Engl.* **1993**, *32,* 1545.

(32) Pohorille, A.; Pratt, L. R. *J. Am. Chem. Soc.* **1990**, *112,* 5066.

(33) Auffinger, P.; Wipff, G. *J. Am. Chem. Soc.* **1991**, *113,* 5976.

(34) Kebarle, P. *Ann. Rev. Phys. Chem.* **1977**, *28,* 445.

(35) Wipff, G.; Varnek, A. *J. Mol. Struct. THEOCHEM* **1996**, 000

(36) Jorgensen, W. L. *J. Phys. Chem.* **1983**, *87,* 5304.

(37) Troxler, L.; Wipff, G. *J. Am. Chem. Soc.* **1994**, *116,* 1468.

(38) Wipff, G.; Troxler, L. *MD Simulations on Synthetic Ionophores and on their Cation Complexes: Comparison of Aqueous / non-Aqueous Solvents.* in "Computational Approaches in Supramolecular Chemistry"; Wipff, G., Ed. ; Kluwer, Dordrecht, **1994**

(39) deMendoza, X.; Prados, P.; Campillo, N.; Nieto, P. M.; Sachez, C.; Fayet, J.-P.; Vertut, M. C.; Jaime, C.; Elguero, J. *Recl. Trav. Chim. Pays-Bas* **1993**, *112,* 367.

(40) Shinkai, S.; Iwamoto, K.; Araki, K.; Matsuda, S. *Chem. Lett.* **1990**, 1263.

(41) Fischer, S.; Grootenhuis, P. D. J.; Groenen, L. C.; van Horn, W. P.; van Veggel, F. C. J. M.; Reinhoudt, D. N.; Karplus, M. *J. Am. Chem. Soc.* **1995**, *117*, 1611.

(42) Royer, J.; Bayard, F.; Decoret, C. *J. Chim. Phys.* **1990**, *87*, 1695.

(43) Lipkowitz, K.; Pearl, G. *J. Org. Chem.* **1993**, *58*, 6729.

(44) Perrin, M.; Oehler, D. *Conformations of Calixarenes in the Crystalline State* in "Calixarenes: a versatile class of macrocyclic compounds"; Vicens, J., Böhmer, V., Ed. ; Kluwer, Dordrecht **1991**.

(45) Marcus, Y.; Asher, E. *J. Phys. Chem.* **1978**, *82*, 1246. Marcus, Y. (1985). Ion Solvation. Chichester: Wiley.

(46) Olsher, U.; Hankins, M. G.; Kim, Y. D.; Bartsch, R. A. *J. Am. Chem. Soc.* **1993**, *115*, 3370.

(47) Lehn, J. M. *Acc. Chem. Res.* **1978**, *11*, 49.

(48) Linse, P. *J. Chem. Phys.* **1987**, *86*, 4177.

(49) Gao, J.; Jorgensen, W. L. *J. Phys. Chem.* **1988**, *92*, 5813.

(50) Carpenter, I. L.; Hehre, W. J. *J. Phys. Chem.* **1990**, *94*, 531.

(51) Michael, D.; Benjamin, I. *J. Phys. Chem.* **1995**, *99*, 1530.

(52) van Buuren, A. R.; Marrink, S.-J.; Berendsen, J. C. *J. Phys. Chem.* **1993**, *97*, 9206.

(53) Wilson, M. A.; Pohorille, A. *J. Am. Chem. Soc.* **1994**, *116*, 1490.

(54) Benjamin, I. *J. Chem. Phys.* **1992**, *97*, 1432.

(55) Lee, C. Y.; McCammon, J. A.; Rossky, P. J. *J. Chem. Phys.* **1984**, *80*, 4448.

(56) Meyer, M.; Mareschal, M.; Hayoun, M. *J. Chem. Phys.* **1988**, *89*, 1067.

(57) Guba, W.; Haessner, R.; Breipohl, G.; Henke, S.; Knolle, J.; Sandagada, V.; Kessler, H. *J. Am. Chem. Soc.* **1994**, *116*, 7532.

(58) Guba, W.; Kessler, H. *J. Phys. Chem.* **1994**, *98*, 23.

(59) Benjamin, I. *J. Chem. Phys* **1992**, *96*, 577.

(60) Schweighofer, K. J.; Benjamin, I. *J. Phys. Chem.* **1995**, *99*, 9974.

(61) Benjamin, I. *Science* **1993**, *261*, 1558.

(62) Michael, D.; Benjamin, I. *J. Phys. Chem.* **1995**, *99*, 16810.

(63) Benjamin, I. *Acc. Chem. Res.* **1995**, *28*, 233.

(64) Wipff, G.; Engler, E.; Guilbaud, P.; Lauterbach, M.; Troxler, L.; Varnek, A. *New J. Chem.* **1996**, 000.

(65) Varnek, A.; Wipff, G. *J. Comput. chem.* **1996**, 000.

(66) Wipff, G.; Sirlin, C.; Varnek, A. *Solvent and Dynamics Effects on the Structure of Alkali Cation Complexes of the p-t-Butyl-calix[4]arene Anion: MD and FEP Computer Investigations on the* Na^+ */* Cs^+ *Binding Affinity* in "Crystallography of Supramolecular Compounds."; Tsoucaris, G., Ed.; Kluwer, Dordrecht, **1996.**

(67) Troxler, L. (1995). *Simulation par dynamique moléculaire d'ions libres et complexés par des récepteurs macrocycliques en solution.* Thèse. Université Louis Pasteur, Strasbourg,

(68) Guilbaud, P.; Wipff, G. *New J. Chem.* **1996**, 000

(69) Rozen, A. M. *J. Radioanal. Nucl. Chem.., Articles* **1990**, *143,* 337.
Natatou, I.; Burgard, M.; Asfari, Z.; Vicens, J. *J. Inclus. Phenom. Mol. Recognition* **1995**, *19,* 107-117.

(70) Danesi, P. R.; Chiarizia, R.; Pizzichini, M.; Satelli, A. *J. Inorg. Nucl. Chem.* **1978**, *40,* 1119.

(71) Danesi, P. R. *Solvent Extraction Kinetics* in "Principles and Practices of Solvent Extraction"; J. Rydberg, C. Musikas, and G.R. Choppin; M. Dekker, Inc., New York,, **1992**, 157.

(72) Fyles, T. M.; Malik-Diemer, V. A.; McGavin, C. A.; Whitfield, D. M. *Can.J.Chem.* **1982**, *60*, 2259. Dulyea, L. M.; Fyles, T. M.; Whitfield, D. M. *Can.J.Chem.* **1984**, *62*, 498. Fyles, T. M. *Journal of Membrane Science* **1985**, *24*, 229-243. Fyles T. M. *J.Chem.Soc., Faraday Trans. 1* **1986**, *82*, 617-633.

(73) Fan, W.; Tsai, R.-S.; El Tayar, N.; Carrupt, P.-A.; Testa, B. *J. Phys. Chem.* **1994**, *98,* 329. Tsai, R.-S.; Fan, W.; El Tayar, N.; Carrupt, P. A.; Testa, B.; Kier, L. B. *J. Am. Chem. Soc.* **1993**, *115,* 9632.

(74) Wipff, G.; Weiner, P.; Kollman, P. A. *J. Am. Chem. Soc.* **1982**, *104,* 3249.

(75) Kollman, P. A.; Merz, J., K. M. *Acc. Chem. Res.* **1990**, *23,* 246.

(76) Wipff, G. *J. Coord. Chem.* **1992**, *27*, 7 and references cited therein.

(77) Toner, J.-L. "*Modern Aspects of Host-Guest Chemistry: Molecular Modeling and Conformationally restricted Hosts.*" in "Crown Ethers and Analogs", Pataï Ed. **1989,** pp 77-205, and references cited therein.

MOLECULAR DYNAMICS SIMULATION AND PRELIMINARY ANALYSIS OF A LIPID BILAYER NEAR THE LIQUID–CRYSTALLINE PHASE

MATTHEW P. REPASKY, RAMKUMAR RAJAMANI and JEFFREY D. EVANSECK*

Chemistry Department, University of Miami,
1301 Memorial Drive, Coral Gables, Florida 33124–0431

ABSTRACT. The construction and theoretical treatment of a membrane bilayer model consisting of 128 lipid molecules of 1-palmitoyl-2-oleoyl-*sn*-glycerol-3-phosphatidylcholine (POPC) is described in detail. The bilayer has been simulated with 1445 water molecules on each side normal to the lipid bilayer plane providing an average water layer thickness of 14 Å. A total of 25822 explicit atoms were used in the membrane simulation with the all–atom parameter representation included in the CHARMM program. Molecular dynamics was employed using the canonical ensemble, and periodic boundary conditions were implemented in all directions to equilibrate the bilayer assembly. The Verlet algorithm was used to integrate the equations of motion with a time step of 2 fs. A total of 100 ps of molecular dynamics were carried out resulting in an average temperature of 386.1 ± 2.1 K. A preliminary analysis was conducted to evaluate the membrane phase of the computed lipid bilayer. The resulting lateral dimensions of the bilayer are 75.5 × 63 Å. The average phosphorous–to–phosphorous distance normal to the bilayer surface is 52.2 ± 1.2 Å. In addition to lipid structure, other static and dynamic properties of the membrane have been computed to be consistent with that observed near the biological liquid-crystalline phase. The intention is to use this bilayer as the starting point for the collection and analysis of equilibrium and dynamic properties of the bilayer after long time simulations, as well as study the insertion of supramolecular ionic devices into the model bilayer to better understand transport phenomena of ions and small organic molecules.

1. Introduction

Biological membranes are complex, noncovalent and cooperative two–dimensional solutions of oriented amphiphilic lipid molecules and membrane bound globular proteins.[1] The thin sheet–like lipid structures (60–100 Å) serve the dual role of (1) creating highly selective and permeable barriers to establish discrete cellular compartments, and (2) providing the necessary environment to solubilize integral membrane proteins. Lipid molecules are a significant constituent of biological membranes and are composed of both hydrophilic (polar head) and hydrophobic (alkyl tail)

L. Echegoyen and A.E. Kaifer (eds.), Physical Supramolecular Chemistry, 103–113.

moieties. These amphiphilic molecules self–assemble to form the classic lipid bilayer structure, where hydrophobic interactions are considered to be the main driving force of aggregation.[2] Other interactions, such as van der Waals attractive forces between the hydrocarbon tails, and hydrogen bonding at the aqueous interfacial region, mediate the stability and structure of the lipid membranes.

Lipid membranes have complicated phase properties since they exist in a variety of different types of organized structures.[3] The supramolecular assembly which predominates depends upon the lipid molecule, degree of hydration, ionic strength, pH, temperature and pressure.[4] X–ray crystallography coupled with differential scanning calorimetry have been used to study the structure and transitions between the lipid phases.[3] At intermediate hydration, four lamellar phases occur, as shown in Figure 1.

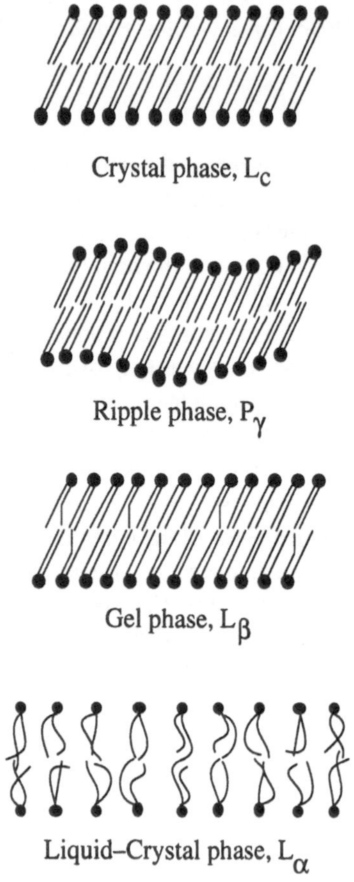

Figure 1. Four general lamellar phases of a lipid bilayer.

The first phase occurs at low temperature, which is known as the *crystal phase* (L_c). One significant characteristic of this phase is that the hydrocarbon tails are fully extended in close orthorhombic packing. The *ripple phase* (P_γ) takes on a 250–500 Å wavelength distortion. Even with the ripple, the alkyl tails retain the all *trans* conformation. The *gel phase* (L_β) involves a metastable situation with several *gauche* bonds per chain disrupting the close packing of the alkyl tails. Finally, the *liquid–crystalline state* (L_α) is defined by the large alkyl chain disorder, which has been referred to as the melted state.[5] A mix of naturally occurring lipid membranes show a main transition temperature (T_m) for $L_\beta \rightarrow L_\alpha$ just below physiological temperatures, and hence the most relevant biological state is in the liquid–crystalline form. Under some circumstances, the gel phase can be found in some biological processes. The $L_\beta \rightarrow L_\alpha$ phase change is the most intensely studied lipid phase transition.[3] Structural changes associated with the phase transitions include a lateral expansion and decrease in the bilayer thickness, yielding an overall small net volume increase. In addition, there is experimental evidence that the number of waters increase at the aqueous interface in the liquid–crystalline state.[12]

Determination of the relationship between three dimensional structure and biochemical function is crucial to the fundamental understanding of any membrane system or process. Due to the dynamic characteristics and fluidity of membranes in their common liquid–crystalline state, the three dimensional structure of only a few systems have been determined. Some small soluble peptides, such as alamethicin[6] and melittin,[7] can insert into lipid bilayers, crystallize and produce measurable X–ray diffraction. However, the only intrinsic membrane protein to be crystallized and studied by X–ray diffraction is the photosynthetic reaction center from a purple bacterium (rhodopseudomomus viridis). The photosynthetic reaction center has been crystallized in three dimensions in a detergent solution and its structure has been solved by X–ray analysis at 3 Å resolution.[8] Despite the inherent problems associated with experimental determination of membrane structure, one of the most useful starting points is to consider lipid membranes in their natural occurring configurations without proteins or small peptides. There are few known examples, such as 1,2-dilauryl–*sn*–glycerol–3–phosphatidylethanolamine (DLPE),[9] 1,2–dimyristoyl–*sn*–glycerol–3–phosphorylcholine (DMPC),[10] and 1,2–dioleoyl–*sn*–glycerol–3–phosphocholine (DOPC)[11] which have been reported by X–ray crystallography. These crystal structures provide insight into the architecture of lipid bilayers in the crystal phase.[12]

Obtaining a more detailed and complete understanding of the structure, dynamics and function of lipid membranes in the liquid–crystalline phase remains as a major challenge confronting computational studies due to two primary factors: membrane fluidity and the timescale of membrane phenomena.[3] To start, lipid membranes are inherently flexible and dynamic, thus a proper statisitical analysis should entail long simulation times and/or large supramolecular assemblies for sufficient configurational sampling and comparison with experimental observables. The timescale of cellular processes range from tens of picoseconds (water diffusion), hundreds of picoseconds (ion transport), nanoseconds (*trans/gauche* isomerization), to milliseconds (fusion and repair).[3] In addition, large numbers of explicitly defined atoms are necessary in the computational model to compute bulk properties. Recent technological advances allow for the proper simulation of large membrane assemblies, since more experimental information is being reported, theoretical methodology and force fields are being improved, and computational resources are becoming more available.

A number of research groups have built and tested explicit molecular lipid bilayer systems beyond the earlier hard sphere,[13] Lennard–Jones,[14] and mean field models[15] to understand the structural aspects of membrane lipids. The statistical properties of lipid bilayers in the gel and liquid–crystalline phases have been studied by a number of computational techniques, such as Monte Carlo and molecular dynamics simulations. Scott and other workers have reported a number of Monte Carlo simultions designed to understand protein and cholesterol interactions with lipid membranes.[16,17] Molecular dynamics have been carried out on a decane bilayer in the absence of solvent,[18] and in the presence of explicit water molecules.[19] Pastor and coworkers have used Brownian dynamics and mean field stochastic boundary molecular dynamics to investigate the long time (nanosecond) phenomena of the NMR order parameter profile of DPPC.[20] Computational studies on other membrane bilayers, such as DLPE,[21] DPPC,[22,23,36] and DMPC[24] have been completed. Recently, several large scale computations on fully solvated lipid bilayers have been reported. For instance, 32808 explicit atoms where used in defining a hydrated POPC bilayer system,[25,26] and 7100 explicit atoms in a fully solvated DMPC bilayer has been reported[27] for up to 4 ns.[28] These long time and large scale computational studies allow for the direct comparison between theory and experiment and provide details and insights missing from experimental observation.

Due to the significance of characterizing and understanding the role of biological membranes, coupled with recent advances in computational methodolgy, and improved force–field parameters, we report here on the procedure to construct a stable bilayer membrane in the liquid–crystalline phase using the all–atom representation provided by CHARMM force–field.[29] Our goal is to produce a suitable lipid bilayer model as a starting point to further study equilibrium and dynamic properties after long time simulations, as well as transport phenomena across biological membranes.

2. Computational Procedure

Phospholipids are a major class of membrane lipids that are abundant in higher organisms.[1] The type of lipid used for membrane construction is the phosphoglyceride known as 1-palmitoyl-2-oleoyl-*sn*-glycerol-3-phosphatidylcholine (POPC).[30] The schematic structure of POPC is shown in Figure 2.

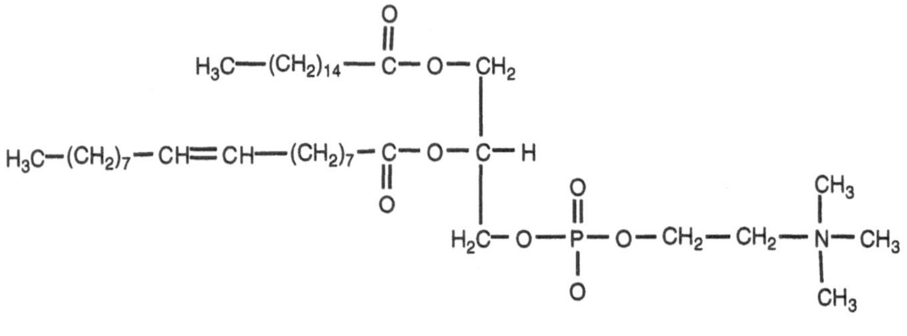

Figure 2. 1-palmitoyl-2-oleoyl-*sn*-glycerol-3-phosphatidylcholine (POPC).

POPC has one saturated palmitoyl and one unsaturated oleoyl fatty acid tail. POPC may be easily altered to DPPC by removing the double bond from the oleoyl tail and adjusting the carbon chain length. DPPC is well studied by experimental methods. The all–atom CHARMM lipid force field[29] was used to describe POPC, and the TIP3P potential was used for the waters.[31] Although the all–atom representation greatly increases the computational effort required for these types of simulations, it has been found that explicit hydrogens provide an improved degree of interdigitation which better describes the phase and properties of membranes and macromolecules.[29] From early computational studies, it has been found that the initial configuration of the lipid matrix affects the resulting stability, phase, and computed properties of membrane models. To reduce the dependence on initial starting conditions, we have employed a new strategy in the construction of the POPC lipid bilayer.

The graphical interface CHEMNOTE within the QUANTA program[32] was used to visually create the initial POPC lipid structure. The CHARMM program[33] was then employed to carry out molecular dynamics simulations and energy minimization. The starting orientation of the lipid was initally set to an all *trans* configuration with the exception of the *cis* double bond within the oleoyl lipid tail. Brief energy minimization of 100 steps of steepest descent were carried out on a single POPC molecule. The resulting change in energy was < 0.01 kcal/mol at the termination of the minimization. The partially minimized POPC structure was then replicated so that four identical copies were combined into a single building block. Another series of 100 steps of steepest descent energy minimization was carried out on the collection of the four POPC lipids. A 15 ps trajectory employing periodic conditions (planes parallel to the lipid axis) and the NVT ensemble was then carried out using velocity rescaling every 200 fs for the first 2 ps to reach the target temperature of 300 K. The default 13 Å electrostatic cut off was used for nonbond interactions in the lipid CHARMM force–field.[29] The update of the nonbond interaction list was done every 10 steps using a time step of 2 fs. From the 15 ps trajectory, four of the lowest energy configurations involving the four lipids were selected as the individual building blocks to be randomly propagated in a 4×4 matrix yielding a total of 64 lipids in the construction of the final monolayer. This method reduces computational artifacts arising from highly ordered initial conditions by introducing a measure of short range order while creating long range disorder. A stereoview of the lowest energy four POPC lipid building block is shown in Figure 3.

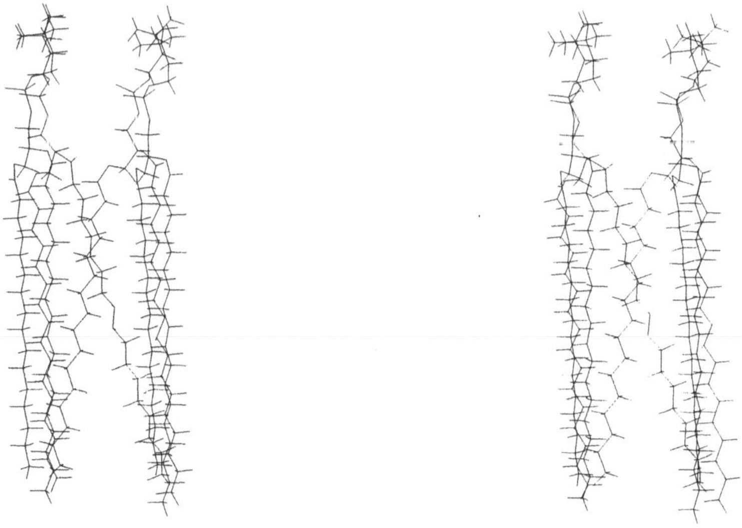

Figure 3. Stereoview of the POPC building block.

The high degree of randomness is evident and all the alkyl tails retain a pseudo *trans* conformation. The 64 POPC lipid monolayer was then copied, rotated and translated to form a crude bilayer with an approximate 1 Å separation between the hydrocarbon tails of each monolayer. In an effort to reduce the high energy van der Waals overlap between the two monolayers, an additional 100 steps of steepest descent energy minimization was carried out to establish the initial configuration of the lipid bilayer. The bilayer was then solvated by stacking several pre–equilibrated boxes of TIP3P waters on each polar head group surface to completely cover the lateral dimensions of the bilayer. The waters beyond the lateral dimensions and those 14 Å away from the surface normal were removed. This thickness corresponds to roughly six layers of water molecules. The result of this procedure was to place 1445 TIP3P water molecules on each side normal to the lipid bilayer plane. In order to equilibrate the bilayer system from the rough construction, the lipids were constrained and the water molecules allowed to relax and sample the membrane surface through molecular dynamics simulation. A total of 25 ps were carried out with the constrained lipids to allow for water relaxation with respect to the lipid bilayer. Water penetration into the polar head group region was observed. In the early stages of the simulation, the velocities of the atoms were scaled every 400 fs in order to adjust the temperature of the system. Finally, the lipid constraints were released and the simulation was carried out for 100 ps without velocity rescaling. A sample lipid configuration is shown in Figure 4 after 100 ps of molecular dynamics simulation.

Several MD studies reproducing phospholipid bilayers in the fluid phase have been recently reported.[20–28] The two primary ensembles which have been used in modeling membrane systems are the canonical (NVT) and isothermal–isobaric (NPT) ensembles. Only recently has the NPT ensemble been implemented and used in computer simulations.[34–36] It has been found that use of the NPT ensemble reduces the system dependence on initial conditions while more accurately modeling short range nonbond forces. However, these conditions are more difficult to implement and there is little control over the phase of the membrane. Because of the difficulties using the NPT ensemble, NVT conditions utilizing periodic boundary conditions have been employed within this study.

3. Discussion

In general, a short 100 ps simulation is not sufficient to study the mesoscopic properties of this POPC lipid bilayer. In long time simulations (200 ps or greater), the computed results are typically divided into static and dynamic properties.[20–28] The static properties usually considered are the behavior of order parameters, dihedral angle isomerization, presence of overall tilt and extent of its cooperativity, radial distribution of head groups and its correlation with molecular tilt, and distribution across the bilayer of both density and lateral pressure. Dynamic properties typically include the rate of dihedral angle transitions, the temporal behavior of molecular tilt and the translational motion of both head groups and centers of mass, as related to lateral diffusion. Within the scope of this study, the three main bilayer features included for the preliminary analysis are: (1) overall membrane structure, (2) area per lipid head, and (3) group distribution functions across the membrane, especially that of the alkyl tail region.

The resulting lateral dimensions of the bilayer are 75.5 × 63 Å after 100 ps of molecular dynamics simulation. The average phosphorous–to–phosphorous distance normal to the bilayer surface is 52.2 ± 1.2 Å, which is in the high experimental range reported for the liquid–crystalline phase,[30] and significantly greater than that computed by other long time simulations.[26] The larger phosphorous–to–phosphorous distance indicates that the short time 100 ps simulations results in a bilayer that is not well interdigitated. Further time evolution coupled with manual membrane compression will produce a lipid bilayer with a decreased membrane width. It is of interest to note that after 100 ps of simulation in this study, the bilayer width is stable using the all–atom lipid parameter set from CHARMM.[29] This is opposed to other instabilities seen in mixed parameter systems.[26]

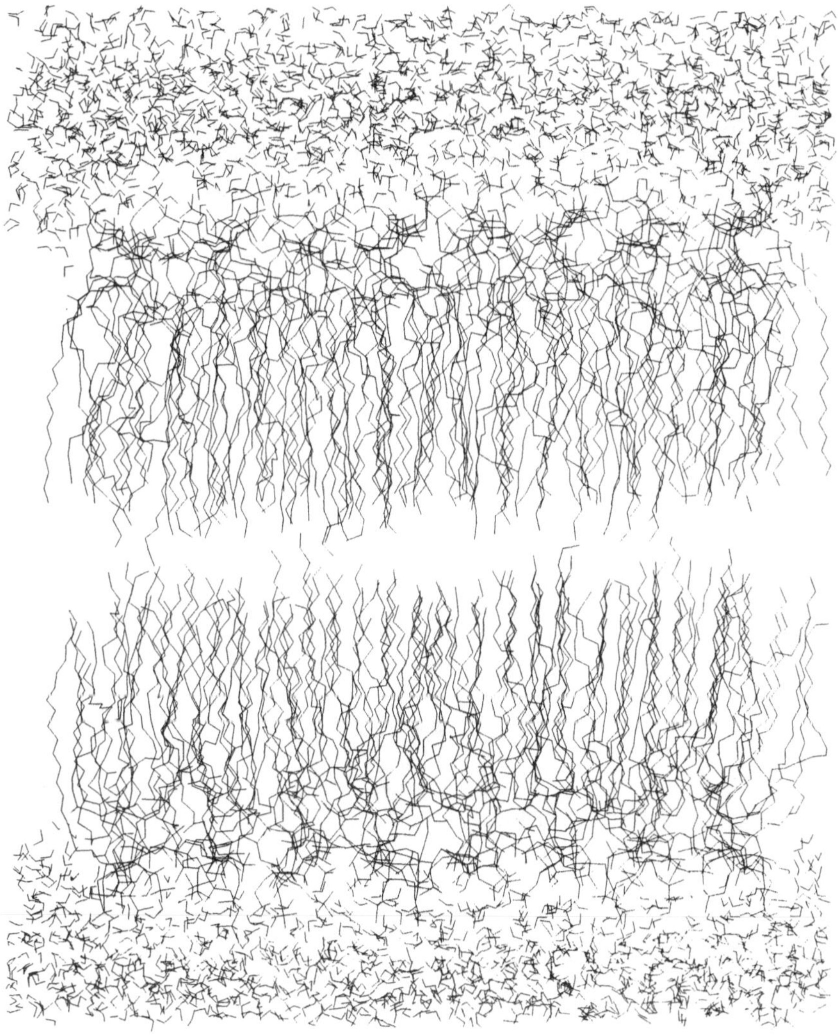

Figure 4. Sample configuration of the POPC membrane bilayer.

The area per lipid head group was determined by the method presented by Heller et al.[26] A FORTRAN program was written to compute the average area per lipid head group by determining the area of the smallest rectangle representing the lipid atoms projected onto the xz–plane. The xz–plane is defined to be the linear least square plane of the polar head groups of each monolayer. The area per lipid head group is then computed by summing the individual contributions from each lipid and dividing by the total number of lipids in each monolayer (64). Further sampling was performed over 20 random configurations, since the area was computed to have a sizable fluctuation. Our results indicate that the area per head group at the end of the simulation is 59 ± 4 Å^2. This is in good agreement with long time simulations (65.5 ± 0.04 Å),[26] and the experimental liquid crystal phase of 57–70.9 Å^2 at 50° C. The computed area per lipid head is on the lower range of the liquid–crystalline state, but definitely within the experimental range. We believe that longer molecular dynamics simulations will slightly increase the area per lipid head as the membrane model is given the chance to equilibrate.

The group distribution function is a method to describe the molecular composition across the lipid membrane. The procedure is to count the occurrences of the different lipid parts and waters, from one end of hydration to the other. A plot of the frequency of occurrences is given which shows a one dimensional view across the lipid bilayer, as shown in Figure 5.

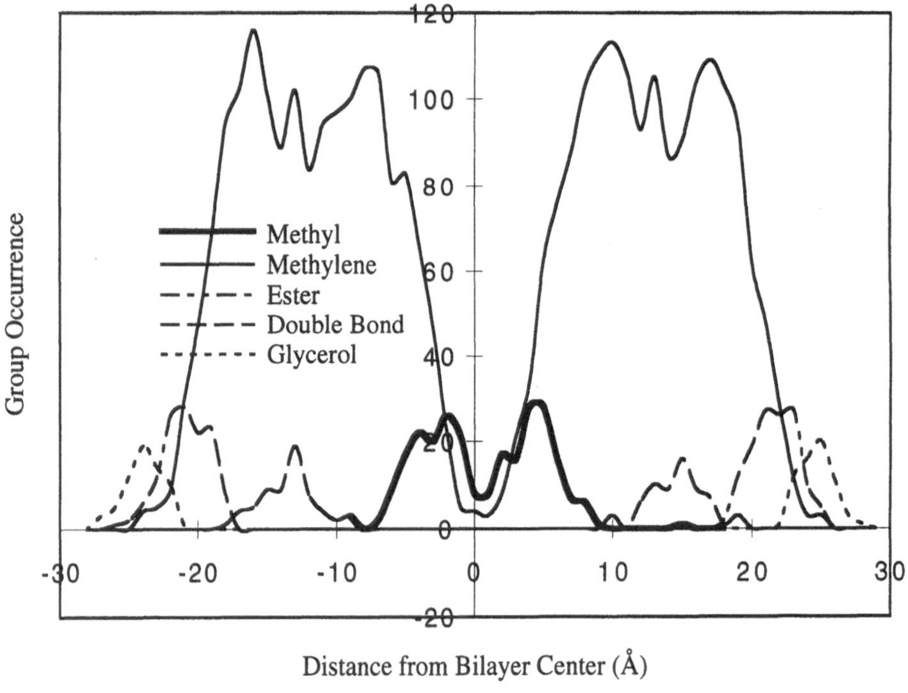

Figure 5. Group distribution function for the POPC membrane.

After 100 ps of simulation, we find good agreement with experimental results on DOPC,[11] and other long time simulations.[26] In our short time simulation, the bilayer still requires additional time to achieve a relaxed liquid–crystalline phase. This is shown by the pronounced dip in the CH_2 and CH_3 distributions at the center of the bilayer (Figure 4). The low occurrence of these functional groups at the center indicates that the interdigitation is rather poor and that more simulation time is required.

4. Conclusions

A fully solvated bilayer system made from 128 1-palmitoyl-2-oleoyl-*sn*-glycero-3-phosphatidyl choline (POPC) lipids has been made. Molecular dynamics simulations were carried out for a total of 100 ps using the canonical ensemble and periodic boundary conditions at 386.1 ± 2.1 K. The all–atom CHARMM parameter set for lipids with the computational treatment described within this report provides a stable bilayer system. A preliminary analysis on the overall bilayer geometry, area per lipid head, and group distribution function indicates that the computed bilayer is near the liquid–crystalline phase, and that further simulation time is required to collect and properly analyze the equilibrium and dynamic properties of this POPC membrane.

5. Acknowledgments

We are grateful to the Summer Award in Natural Science and Engineering and the University Research Council from the University of Miami, and the Petroleum Research Fund (PRF#30918–G4) for financial support of this research. In addition, we thank Professor Alex MacKerrell for helpful comments on the manuscript.

6. References

1. Stryer, L. (1995) *Biochemistry* 4th Ed. Freeman, New York.
2. Lehn, J–M. (1995) *Supramolecular Chemistry* VCH, Weinheim.
3. Yeagle, P. (1992) *The Structure of Biological Membranes* CRC Press, Boca Raton.
4. Jain, M. (1988) *Introduction to Biological Membranes* 2nd Ed. John Wiley & Sons, New York. Gennis, R. G. (1989) *Biomembranes. Molecular Structure and Function* Springer–Verlag, New York.
5. Venable, R. M.; Zhang, Y.; Hardy, B. J.; Pastor, R. (1993) Molecular Dynamics Simulations of a Lipid Bilayer and of Hexadecane: An Investigation of Membrane Fluidity *Science* 1993 **262**, 223–226.
6. Fox, R. O.; Richards, F. M. (1982) A Voltage–Gated Ion Channel Model Inferred from the Crystal Structure of Alamethicin at 1.5–Å Resolution *Nature* **300**, 325–330.
7. Anderson, D.; Terwilliger, T. C.; Wickner, W.; Eisenberg, D. (1980) Melittin Forms Crystals which are Suitable for High Resolution X–Ray Structural Analysis and which Reveal a Molecular 2–Fold Axis of Symmetry *J. Biol. Chem.* **255**, 2578–2582.
8. Deisenhofer, J.; Michel, H. (1989) The Photosynthetic Reaction Center from Purple Bacterium Rhodopseudomonas Viridis *Science* **245**, 1463–1473.
9. Hitchcock,P. B.; Mason, R.; Thomas, K. M.; Shipley; G. G. (1974) Structural Chemistry of 1,2–Dilauroyl–DL–phospatidylethanolamine: Molecular Conformation

and Intermolecular Packing of Phospholipids *Proc. Natl. Acad. Sci. USA* **8**, 3036–3040.
10. Pearson, R. H.; Pascher, I. (1979) The Molecular Structure of Lecithin Dihydrate *Nature* **281**, 499–501.
11. Wiener, M. C.; White, S. W. (1992) Structure of a Fluid Dioleoylphospatidylcholine Bilayer Determined by Joint Refinement of X–Ray and Neutron Diffraction Data: III. Complete Structure *Biophys. J.* **61**, 434–447.
12. McIntosh, T. J.; Simon, S. A. (1986) Area per Molecule and Distribution of Water in Fully Hydrated Dilauroylphosphatidylethanolamine *Biochemistry* **25**, 4948–4952.
13. Alder, B. J.; Wainwright, T. E. (1957) Phase Transition for a Hard Sphere System *J. Chem. Phys.* **27**, 1208–1209.
14. Rahman, A. (1964) Correlations in the Motion of Atoms in Liquid Argon *Phys. Rev. A* **136**, 405–411.
15. Marcelja, S. (1973) Molecular Model for Phase Transition in Biological Membranes *Nature* **241**, 451–452.
16. Scott, H. L. (1977) *Biochim. Biophys. Acta* **469**, 264. Scott, H. L. (1986) Monte Carlo Calculations of Order Parameter Profiles in Model of Lipid–Protein Interactions in Bilayers *Biochemistry* **25**, 6122–6126. Scott, H. L. (1991) *Biophys. J.* **59**, 445.
17. Edholm, O.; Nyberg, A. M. (1992) Cholestrol in Model Membranes: A Molecular Dynamics Simulation *Biophys. J.* **63**, 1081–1089.
18. Ploeg, P. v. d.; Berendsen, H. J. C. (1982) *J. Chem. Phys.* **76**, 3271. Ploeg, P. v. d.; Berendsen, H. J. C. (1983) Molecular Dynamics of a Bilayer Membrane *Mol. Phys.* **49**, 233–248.
19. Egberts, B.; van Gunsteren, W. F.; Berendsen, H. J. C. (1988) Molecular Dynamics Simulation of a Smectic Liquid Crystal with Atomic Detail *J. Chem. Phys.* **89**, 3718–3732.
20. Pastor, R. W.; Venable, R. M. Karplus, M. (1988) Brownian Dynamics Simulation of a Lipid Chain in a Membrane Bilayer *J. Chem. Phys.* **89**, 1112–1127. Pastor, R. W.; Venable, R. M. Karplus, M. (1988) A Simulation Based Model of NMR T1 Relaxation in Lipid Bilayer Vesicles *J. Chem. Phys.* **89**, 1128–1140.
21. Raghavan K.; Reddy M. R.; Berkowitz M. L. (1992) A Molecular Dynamics Study of the Structure and Dynamics of Water between Dilauroylphosphatidyletholamine Bilayers *Langmuir* **8**, 233–240. Damodaran, K. V.; Merz, K. M.; Gaber, B. P. (1992) Structure and Dynamics of the Dilauroylphosphatidylethnolamine *Biochemistry* **31**, 7656–7664.
22. Pastor, R. W.; Venable, R. M. (1993)*Computer Simulation of Biomolecular Systems* Theoretical and Experimental Applications Vol. 2 Eds. van Gunsteren, W.; Weiner, P. K.; Wilkinson, A. J. ESCOM, Leiden.
23. Taga, T.; Masdua, K. (1995) Monte Carlo Study of Lipid Membranes: Simulation of Diparmitoylphosphatidylcholine Bilayers in Gel and Liquid–Crystalline Phases, *J. Comp. Chem.* **16**, 235–242.
24. Stouch, T. R. (1993) *Mol. Simul.* **10**, 335.
25. Zhou, F.; Schulten, K. (1995) Molecular Dynamics Study of a Membrane–Water Interface *J. Phys. Chem.* **99**, 2194-2207.

26. Heller, H.; Schaefer, M. Schulten (1993) Molecular Dynamics Simulation of a Bilayer of 200 Lipids in the Gel and in the Liquid–Crystal Phases *J. Phys. Chem.* **97**, 8343–8360.
27. Bassolino–Klimas, D.; Alper, H. E.; Stouch, T. R.(1993) Solute Diffusion in Lipid Bilayer Membranes: An Atomic Level Study by Molecular Dynamics Simulations *Biochemistry* 32, 12624–12637. Alper, H. E.; Bassolino–Klimas, D.; Stouch, T. R. (1993) The Limiting Behavior of Water Hydrating a Phospholipid Monolayer: A Computer Simulation Study *J. Chem. Phys.* **99**, 5547–5558.
28. Bassolno-Klimas, D.; Alper, H. E.; Stouch, T. R. (1995) Mechanism of Solute Diffusion through Lipid Bilayer Membranes by Molecular Dynamics Simulations *J. Am. Chem. Soc.* **117**, 4118-4129.
29. Schlenkrich, M., Brickmann, J., MacKerell, Jr., A.D. and Karplus, M. (1996) An Empirical Potential Energy Function for Phospholipids: Criteria for Parameter Optimization and Applications, in Biological Membranes: A Perspective from Computation and Experiment. Merz, K.M.; Roux, B., Eds., Birkhauser, Boston.
30. Lafleur, M. C.; Cullis, P. R.; Bloom, M. (1990) Dipalmitoylphosphatidylcholine (DPPC) Lipid Bilayer in the Fluid Phase *Eur. Biophys. J.* **19**, 55.
31. Jorgensen, W. L.; Chandrasekhar, J.; Madura, J. D.; Impey, R.; Klein, M. L. (1983) Comparison of Simple Potential Functions for Simulating Liquid Water *J. Chem. Phys.* **79**, 926–935.
32. Molecular Simulations, Inc. Quanta version 4.1.
33. Brooks, B. R.; Bruccoleri, R. E.; Olafson, B. D.; States, D. J.; Swaminathan, S.; Karplus, M. (1983) CHARMM: A Program for Macromolecular Energy, Minimization, and Dynamics Calculations *J. Comp. Chem.* **4**, 187–217.
34. Shinoda, W.; Fukada, T.; Okazaki, S.; Okada, I. (1995) Molecular Dynamics Simulation of the Dipalmitoylphosphatidylcholine (PPPC) Lipid Bilayer in the Fluid Phase using the Nosé–Parrinello–Rahman NPT Ensemble *Chem. Phys. Lett.* **232**, 308–322.
35. Feller, S. E; Zhang, Y.; Pastor, R. W.; Brooks, B. R. (1995) Constant pressure molecular dynamics simulation: The Langevin Piston method *J. Chem Phys.* **103** , 4613–4621.
36. Shinoda,W.; Fukada, T.; Okazake, S.; Okada, I. (1995) Molecular Dynamics Simulation of the Dipalmitoylphosphatidylcholine (DPPC) Lipid Bilayer in the Fluid Phase using the Nose-Parrinello-Rahman Ensemble *Chemical Phys. Lett.* **232**, 308-322.

MOLECULAR RECOGNITION OF CARBOHYDRATES: INTERACTION OF DIOLS WITH ACETATE ION

WILLIAM L. JORGENSEN, WOLFGANG DAMM,
ANTONIO FRONTERA, and MICHELLE L. LAMB
Department of Chemistry, Yale University
New Haven, Connecticut 06520-8118, USA

Abstract. Hydrogen bonding between sidechain carboxylates and sugar hydroxyl groups is a common motif in the recognition of carbohydrates by proteins. The nature of such interactions is explored here with ab initio molecular orbital and molecular mechanics calculations for complexes of acetate ion with 1,2-ethanediol and *trans*-1,2-cyclohexanediol (TCD). Monodentate, bidentate, and bifurcated structures are found as energy minima. Striking binding enhancements are obtained for monodentate complexes upon orientation of the remote hydroxyl group towards the anion. This makes the gas-phase interaction energies for the monodentate complexes competitive with those for the doubly hydrogen-bonded alternatives.

1. Introduction

Binding of carbohydrates by proteins plays a central role in important biological events including cell-cell recognition, immune response, tissue repair, and blood coagulation.[1] Structural details have been obtained through diffraction experiments, which reveal the common occurrence of hydrogen bonds between charged protein sidechains and sugar hydroxyl groups and ether oxygens.[2] Representative binding sites are illustrated in Figures 1 and 2.[3,4] The variety of interactions between carboxylate groups of aspartate and glutamate residues and sugar hydroxyl groups is particularly striking; three common motifs emerge, which can be designated as monodentate, bidentate, and bifurcated (Figure 3). In order to help clarify the energetics of the these interactions, the present computational study was undertaken on model systems consisting of acetate ion with 1,2-ethanediol and *trans*-1,2-cyclohexanediol. Both ab initio and molecular mechanics calculations were carried out for gas-phase complexes. The results are particularly relevant to the design of artificial receptors for carbohydrates. Recent efforts indicate that greater success for polyol binding in polar solvents can be achieved through the use of anionic groups including phosphonates[5] and phosphates[6] rather than neutral functionality.[7]

L. Echegoyen and A.E. Kaifer (eds.), Physical Supramolecular Chemistry, 115–126.

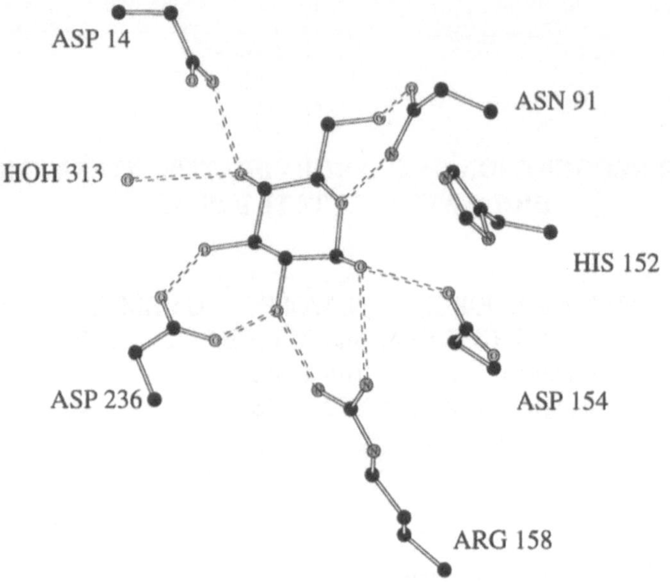

Figure 1. β–D–Glucose bound to D–galactose, D–glucose binding protein.[3]

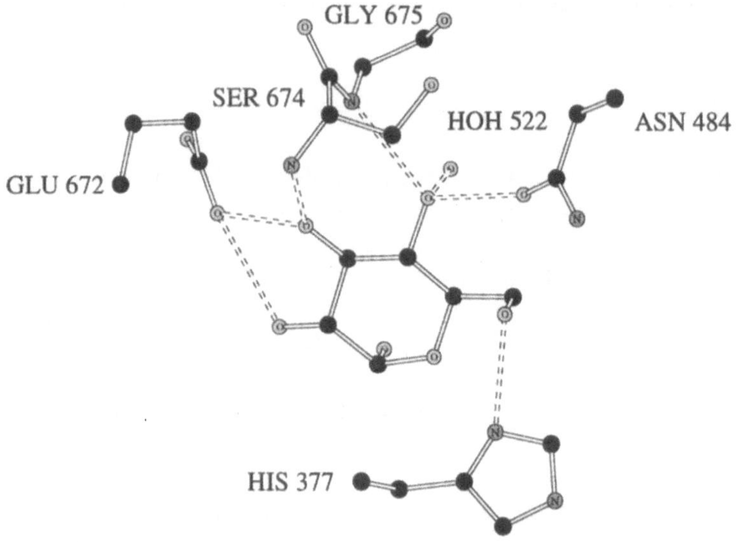

Figure 2. α–D–Glucose bound to glycogen phosphorylase.[4]

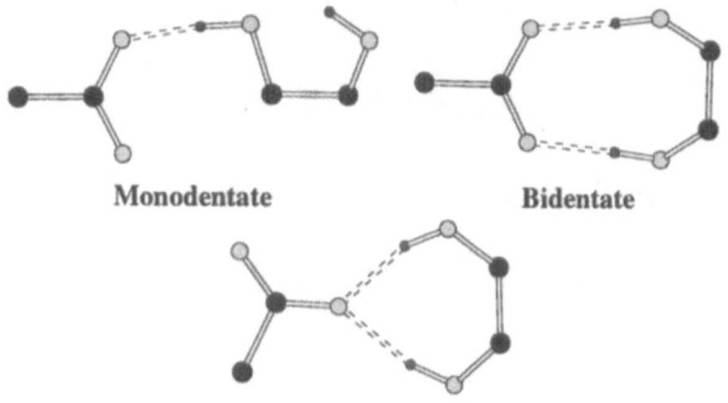

Bifurcated

Figure 3. Three motifs for carboxylate-diol interactions.

2. Computational Details

Full geometry optimizations were carried out for acetate ion (AcO^-), 1,2-ethanediol (ED), and their complexes using the Gaussian 92 program with RHF theory and the 6-31+G* basis set, which includes d-type polarization functions and a set of diffuse s and p orbitals on non-hydrogen atoms.[8] The diffuse functions are important for describing the electronic structure of anions, and results at this level have been shown to give complexation energies in good accord with experimental values.[9] Inclusion of electron correlation was considered via single-point calculations with second-order Møller-Plesset theory (MP2/6-31+G*//6-31+G*). Full optimizations were also performed for conformers of *trans*-1,2-cyclohexanediol (TCD) with the 6-31G* basis set. Vibrational frequency calculations varified the reported structures as energy minima.

Molecular mechanics calculations for these systems and AcO^-/TCD complexes were performed with the BOSS program[10] and the OPLS all-atom force field.[11] The force field parameters have been developed to give accurate results for relative energies of conformers in the gas-phase and for properties including densities and heats of vaporization of pure liquid alkanols and ED. The energy expression consists of harmonic terms for bond stretching and angle bending, a Fourier series for each torsional angle, and Coulomb and Lennard-Jones interactions between atoms separated by three or more bonds (eqs 1-4). The latter "non-bonded" interactions are also evaluated between intermolecular atom pairs, and they are reduced by a factor of 2 for intramolecular 1,4-interactions. The parameters for the present calculations are listed in Table 1. Geometrical combining rules are used for the Lennard-Jones σ's and ε's. Conformational searching for energy minima of isolated molecules and the complexes was carried out with BOSS by random variation of designated internal coordinates, followed by energy minimization and sorting for unique structures. Gradient-based procedures, particularly Fletcher-Powell and BFGS, were used for the optimizations.[12]

$$E_{bond} = \Sigma_i k_{b,i} (r_i - r_{0,i})^2 \quad (1)$$

$$E_{bend} = \Sigma_i k_{\vartheta,i} (\vartheta_i - \vartheta_{0,i})^2 \quad (2)$$

$$E_{torsion} = \Sigma_i \{V_{0,i} + V_{1,i}(1 + \cos\varphi_i)/2 + V_{2,i}(1 - \cos 2\varphi_i)/2 + V_{3,i}(1 + \cos 3\varphi_i)/2\} \quad (3)$$

$$E_{nb} = \Sigma_i \Sigma_j \{q_i q_j e^2/r_{ij} + 4\,\varepsilon_{ij} [(\sigma_{ij}/r_{ij})^{12} - (\sigma_{ij}/r_{ij})^6]\} \quad (4)$$

TABLE 1. OPLS all-atom parameters for alcohols, 1,2-diols, and acetate ion.

Bond Stretching Parameters		
Bond	r_0(Å)	k_b (kcal/mol-Å^2)
C-C	1.529	268.0
C-H	1.090	340.0
C-O	1.410	320.0
O-H	0.945	553.0
C-O$^-$	1.250	656.0
C-CO$^-$	1.522	317.0

Angle Bending Parameters		
Angle	ϑ_0 (deg)	k_ϑ (kcal/mol-rad^2)
C-C-H	110.7	37.5
H-C-H	107.8	33.0
C-C-O	109.5	50.0
H-C-O	109.5	35.0
C-O-H	108.5	55.0
C-C-O$^-$	117.0	70.0
O-C-O$^-$	126.0	80.0
H-C-CO$^-$	109.5	35.0

Non-Bonded Parameters			
Atom	q	σ (Å)	ε
O ROH	-0.683	3.120	0.170
HO ROH	0.418	0.0	0.0
C CH_3OH	0.145	3.500	0.066
C RCH_2OH	0.145	3.500	0.066
HC CH_3OH	0.040	2.500	0.030
HC RCH_2OH	0.060	2.500	0.030
O Diols	-0.700	3.070	0.170
HO Diols	0.435	0.0	0.0
O AcO$^-$	-0.800	2.960	0.210
CO AcO$^-$	0.700	3.750	0.105
CH AcO$^-$	-0.280	3.500	0.066
HC AcO$^-$	0.060	2.500	0.030

Torsional Parameters (kcal/mol)			
Angle	V_1	V_2	V_3
C-C-C-C	1.740	-0.157	0.279
C-C-C-H	0.0	0.0	0.366
H-C-C-H	0.0	0.0	0.318
C-C-C-O	1.711	-0.500	0.663
C-C-O-H	-0.356	-0.174	0.492
H-C-C-O	0.0	0.0	0.468
H-C-O-H	0.0	0.0	0.450
O-C-C-O	9.508	0.0	0.0
H-C-C-O$^-$	0.0	0.0	0.0

3. Results and Discussion

Conformers of the Diols. The three lowest-energy conformations for ED and TCD are illustrated in Figure 4. The structures can be designated by symbols, g (gauche, ca. 60°), t (trans, ca. 180°), and G (gauche*, ca. 300°), for the HOCC, OCCO, and CCOH dihedral angles. There is accord between ab initio calculations and the OPLS force field that the three lowest-energy conformers for ED are the internally hydrogen-bonded ones, Ggt, Ggg, and GgG, with relative energies of ca. 0.0, 0.3, and 1.2 at the highest levels.[13] In all, a conformational search with the OPLS force field yields 7 minima, as follows, with the relative energies in kcal/mol in parentheses: Ggt (0.0), Ggg (0.30), GgG (1.48), Gtg (1.78), ttt (1.79), Gtt (2.31), and GtG (2.88). With the two hydroxyl groups in TCD equatorial, only three minima are found with the force field corresponding again to the Ggt, Ggg, and GgG structures (Figure 4.) The energetic results from the OPLS and ab initio calculations are similar, though the Ggg structure is relatively higher in energy at the 6-31G* level. Based on the extensive ab initio results for ED,[13] it may be expected that the true relative energy for Ggg TCD is between the OPLS and 6-31G* values in Figure 4.

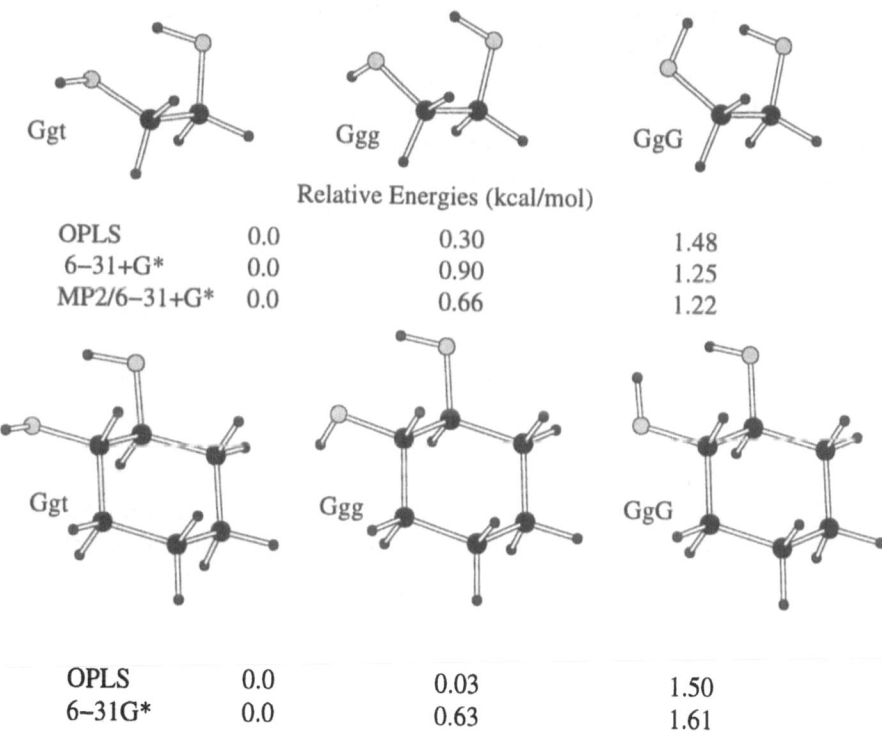

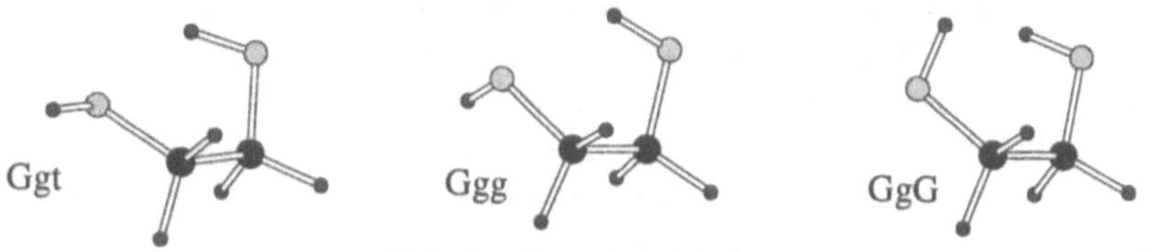

Relative Energies (kcal/mol)

	Ggt	Ggg	GgG
OPLS	0.0	0.30	1.48
6–31+G*	0.0	0.90	1.25
MP2/6–31+G*	0.0	0.66	1.22

	Ggt	Ggg	GgG
OPLS	0.0	0.03	1.50
6–31G*	0.0	0.63	1.61

Figure 4. The lowest-energy conformers of ED and TCD.

Complexes of ED with Acetate Ion. The lowest-energy structures obtained from an extensive conformational search with the OPLS force field for complexes of ED and acetate ion are illustrated in Figure 5. The optimized structure for methanol and acetate ion is also given. Structures corresponding to the three motifs in Figure 3 were found. The interaction energies, ΔE, are the energy difference for the lowest energy conformer of ED plus acetate yielding the complex, and therefore include any reorganization energies for the separated reactants. These structures were then used as starting points for the optimizations with the ab initio calculations using the 6-31+G* basis set. The results of those computations and the energetics from the MP2 single-point calculations are presented in Figure 6. The monodentate Ggt/acetate structure rearranged to the monodentate Ggg alternative, while the other OPLS minima yielded similar minima in the ab initio calculations.

There are several key observations from Figures 5 and 6. (1) The energetics from the OPLS and ab initio calculations are strikingly similar. The force field was developed without consideration of such complexation energies, so the transferability of the parameters to this application is gratifying. The OPLS interaction energies are generally between the 6-31+G* and MP2/6-31+G* values. (2) The interaction energies from the OPLS and MP2 calculations range from -26 to -31 kcal/mol. The monodentate complexes are only 0-4 kcal/mol less favorable than the bidentate and bifurcated structures. Thus, information on differential solvation and entropy effects are needed before conclusions can be reached on the populations of these competitive forms in solution. It is reiterated that the interaction energies are based on separated acetate ion and the lowest-energy conformer of ED, Ggt. The interaction energies for the complexes in Figure 5 based on separated acetate ion and the corresponding conformer of ED are -34.2, -27.7, -28.1, -25.9, and -28.4 kcal/mol for the GtG, Ggg, first GgG, Ggt, and second GgG structures. (3) The computed interaction energies for the AcO^-/methanol complex are close to the experimental enthalpy of complexation at 298 K (-17.6 kcal/mol).[14] The terms needed to convert the electronic energy change reported here to an enthalpy change at 298 K for such complexes typically amount to +1 kcal/mol.[9] (4) There is impressive, ca. 8 kcal/mol, enhancement of the interaction energies on going from the AcO^-/methanol complex to the monodentate complexes with ED. The internal hydrogen bond in the Ggt and Ggg conformers seems to increase (vide infra) the hydrogen-bond donating ability of the adjacent hydroxyl group. The charges in the OPLS model are fixed, i.e., they do not depend on conformation. Nevertheless, they show the enhancement apparent from the ab initio calculations, which do not have fixed charges.

The dependence of the interaction energies for the monodentate complexes on the conformation of ED was further investigated. The OPLS results in Figure 7 show great sensitivity to the orientation of the "backseat" hydroxyl group. This is purely an electrostatic effect as the distances between the region of ED that is varying and the acetate ion yield negligible Lennard-Jones interactions. Again, if one considers the

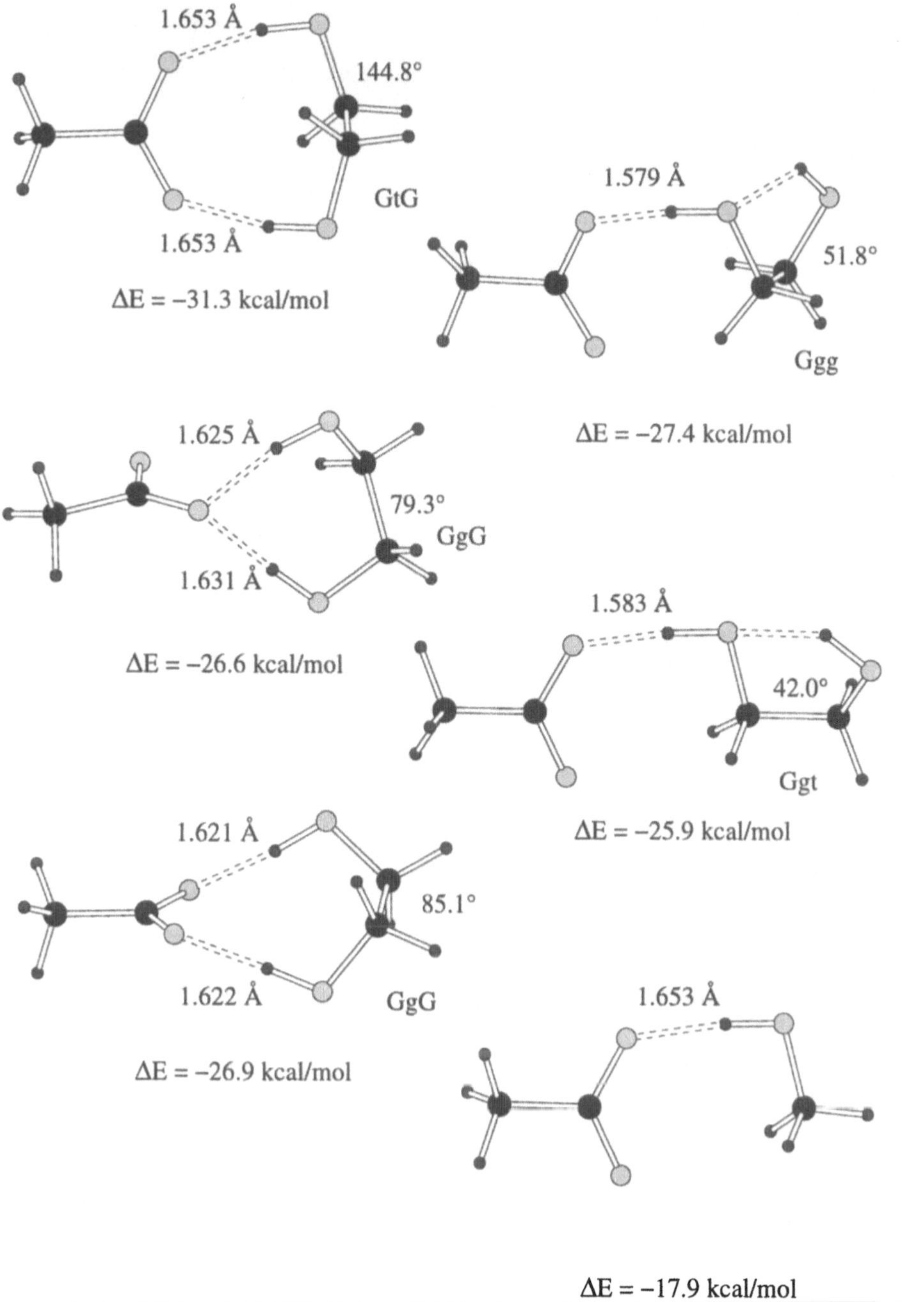

Figure 5. **Structures and interaction energies from optimizations with the OPLS force field for complexes of acetate ion and 1,2-ethanediol and methanol.**

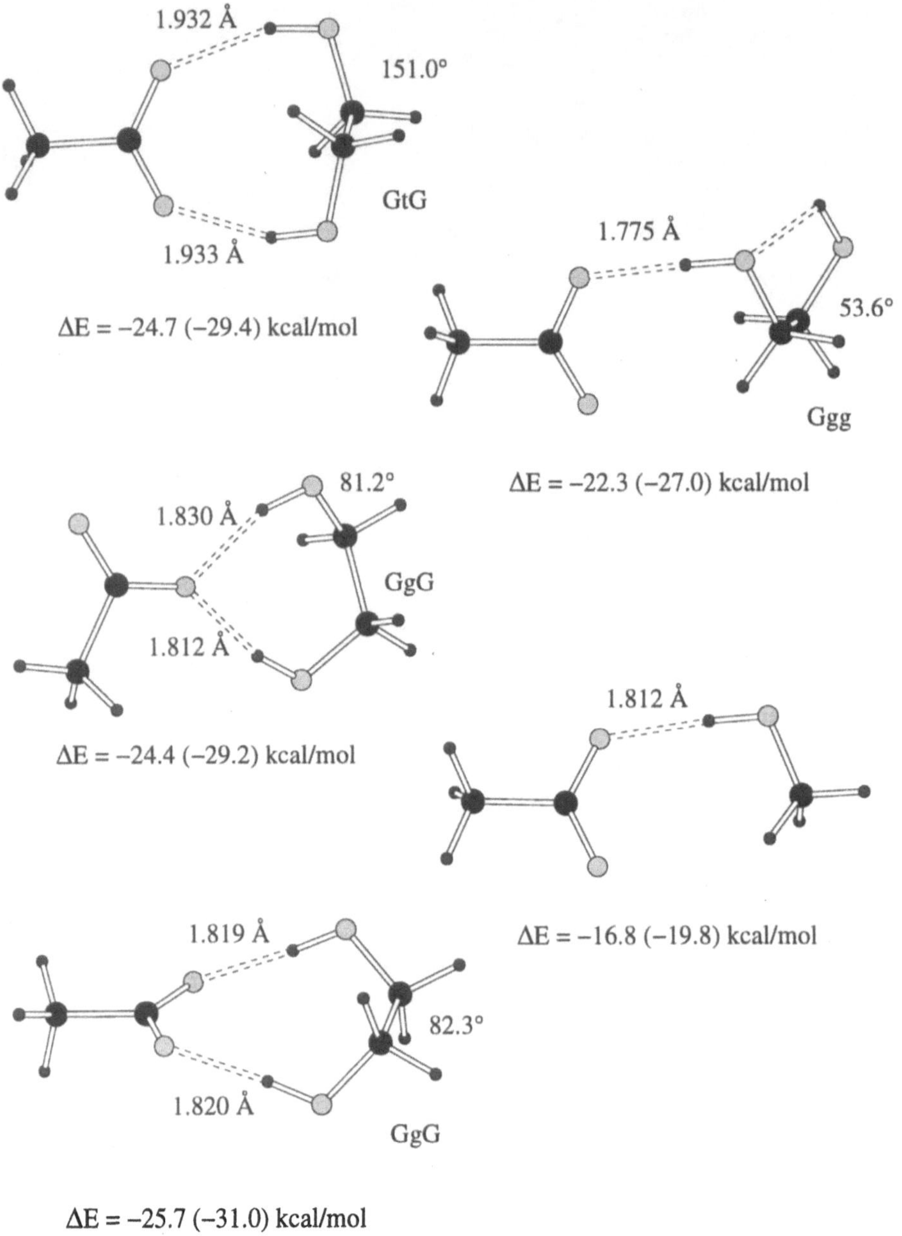

Figure 6. Structures and interaction energies from 6-31+G*//6-31+G* calculations for AcO^-/ED and AcO^-/methanol complexes. MP2/6-31+G*//6-31+G* interaction energies in parentheses.

net interaction energies relative to the corresponding conformer of ED, the ΔE values for the three structures in Figure 7 become -25.9 (Ggt), -23.9 (Gtt), and -18.3 kcal/mol (ttt). Thus, the difference for the interaction energy between the Ggt and Gtt forms is small because the distances between the remote OH and the acetate ion are similar. The internal hydrogen bond is not needed for the enhancement. However, orienting the backseat hydroxyl hydrogen away from the acetate ion in the ttt conformer is clearly electrostatically unfavorable. The dipole moments for the Ggt, Gtt, and ttt conformers of ED are 4.15, 2.91, and 0.0 D from the OPLS partial charges. The order parallels the net interaction energies, though details on the charge distributions are needed to explain the small difference in net interaction energy for the Ggt and Gtt conformers. Incidentally, MP2/6-31+G* results for the structures in Figure 7 are similar; ΔE = -25.7, -20.7, and -17.7 kcal/mol for the Ggt, Gtt, and ttt forms.

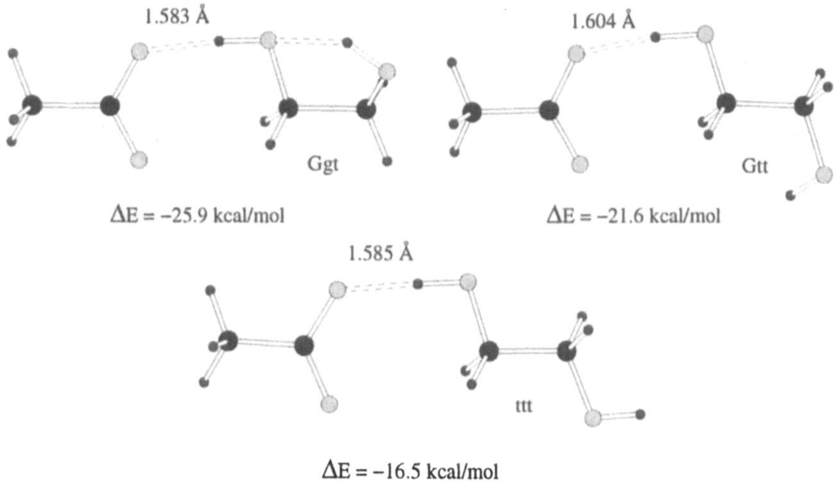

Figure 7. Dependence of OPLS interaction energies on conformation for monodentate complexes of acetate ion and 1,2-ethanediol.

<u>Complexes of TCD with Acetate Ion.</u> The conformational search for complexes of acetate ion and TCD with the OPLS force field yielded the four low-energy minima illustrated in Figure 8. The hydroxyl groups were constrained to remain diequatorial, as expected in sugars. One bidentate, one bifurcated, and two monodentate structures were found. A bidentate structure analogous to the GtG one for ED (Figures 5 and 6) would require an unrealistic OCCO dihedral angle for diequatorial TCD. The interaction energies in Figure 8 are for formation of the complex from separated acetate ion and the lowest-energy conformer of TCD, Ggt. The energetic results are very similar to those for the corresponding complexes of ED in Figure 5. The bifurcated structure for AcO^-/TCD is the lowest in energy, though the gaps between it and the bidentate and Ggg monodentate structures are only 0.4 and 1.7 kcal/mol.

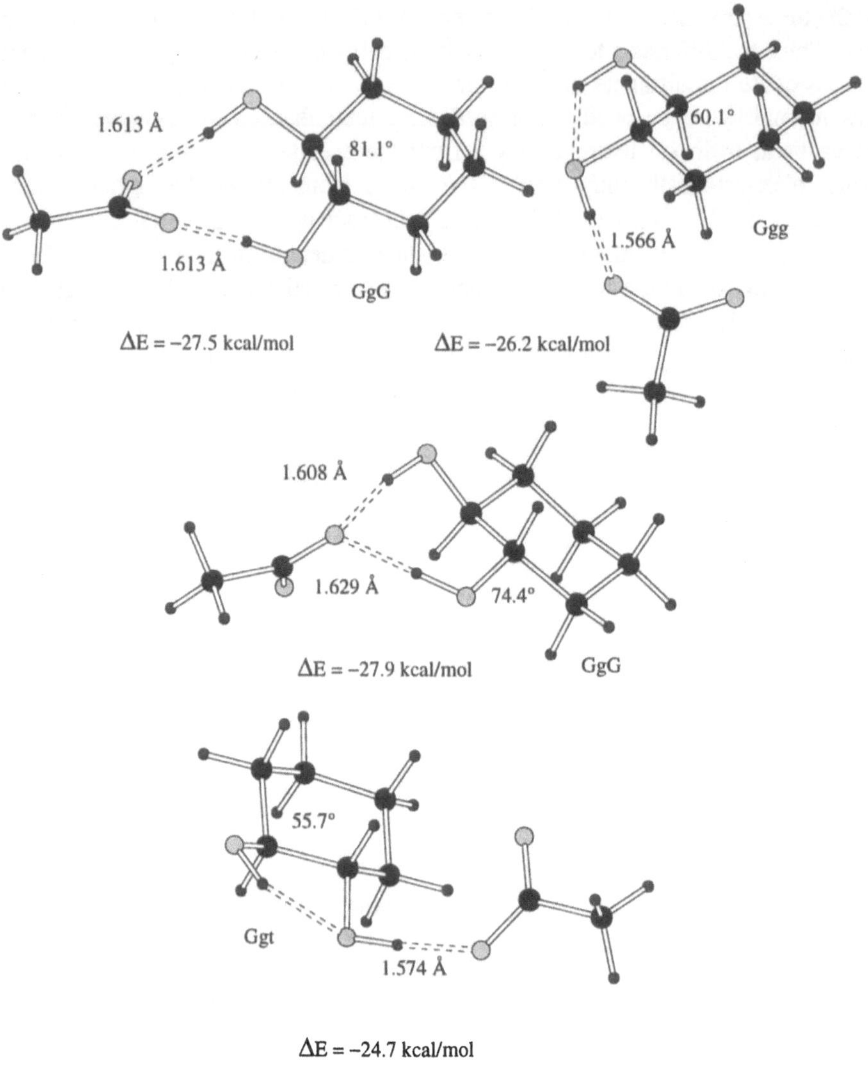

Figure 8. Structures and interaction energies from optimizations with the OPLS force field for complexes of acetate ion and *trans*-1,2-cyclohexanediol.

The differences are small enough that the conformational preference in condensed phases is uncertain. The fact that all three types of structures are observed in crystal structures of proteins with bound carbohydrates is consistent with the similarity of the gas-phase interaction energies. It also indicates that the preferences are sensitive to the details of the condensed-phase environment including, for proteins, the nature and packing of neighboring sidechains.

4. Conclusion

The present ab initio and molecular mechanics calculations on complexes of acetate ion with ED and TCD have verified the existence of three basic geometries for carboxylate/diol complexes inferred from protein crystal structures as energy minima in isolation. The OPLS all-atom force field was found to mimic well the intermolecular energetics and structures of the complexes obtained from ab initio calculations. The results revealed striking enhancement of monodentate carboxylate/alcohol binding upon orientation of the remote hydroxyl hydrogen towards the anion. This makes monodentate carboxylate/diol complexes energetically competitive with bidentate and bifurcated alternatives. Resolution of the competition is undoubtedly sensitive to environmental details such as the polarity and hydrogen-bonding ability of the solvent or, in proteins, the nature and disposition of neighboring sidechains. Monte Carlo and/or molecular dynamics simulations with the OPLS force field could be expected to further understanding of the molecular recognition of these model systems as well as carbohydrates in condensed phases.

5. Acknowledgments

Gratitude is expressed to the Office of Naval Research for support of this work. Receipt of fellowships from the Swiss National Science Foundation, Roche Research Foundation, and the government of Spain is also gratefully acknowledged.

6. References

1. For reviews, see: (a) Lee, Y.C. and Lee, R.T. (1995) *Acc. Chem. Res.* **28**, 321. (b) Bertozzi, C.R. (1995) *Chem. Biol.* **2**, 703.
2. For reviews, see: (a) Bourne, Y., van Tilbeurgh, H., and Cambillau, C. (1993) *Curr. Opin. Struct. Biol.* **3**, 681. (b) Quiocho, F.A. (1989) *Pure Appl. Chem.* **61**, 1293
3. Vyas, N.K., Vyas, M.N., and Quiocho, F.A. (1988) *Science* **242**, 1290.
4. Martin, J.L. , Johnson, L.N., and Withers, S.G. (1990) *Biochemistry* **29**, 10745.
5. Das, G. and Hamilton, A.D. (1994) *J. Am. Chem. Soc.* **116**, 11139.
6. Anderson, S., Neidlein, U., Gramlich, V., and Diederich, F. (1995) *Angew. Chem., Int. Ed. Engl.* **34**, 1596.
7. Cuntze, J., Owens, L., Alcázar, V., Seiler, P., and Diederich, F. (1995) *Helv. Chim. Acta* **78**, 367, and references therein.
8. Frisch, M. J., Trucks, G. W., Head-Gordon, M., Gill, P. M. W., Wong. M. W., Foresman, J. B., Johnson, B. G., Schlegel, H. B., Robb, M. A., Replogle, E. S., Gomperts, R., Andres, J. L., Raghavachari, K., Binkley, J. S., Gonzalez, C., Martin, R. L., Fox, D. J., Defrees, D. J., Baker, J., Stewart, J. J. P., and Pople, J. A. *GAUSSIAN 92*, Revision F; Gaussian Inc.; Pittsburgh, PA, 1993.
9. Gao, J., Garner, D.S., and Jorgensen, W.L. (1986) *J. Am. Chem. Soc.* **108**, 4784.

10. Jorgensen, W.L. *BOSS, Version 3.6*; Yale University; New Haven, CT, 1995.
11. (a) Kaminski, G., Duffy, E.M., Matsui, T., and Jorgensen, W.L. (1994) *J. Phys. Chem.* **98**, 13077. (b) Jorgensen, W.L., Maxwell, D.S., and Tirado-Rives, J., to be published.
12. Press, W.H., Teukolsky, S.A., Vettering, W.T., and Flannery, B.P. *Numerical Recipes in FORTRAN, 2nd Edition*; Cambridge Univ. Press; Cambridge, 1992.
13. (a) Cramer, C.J. and Truhlar, D.G. (1994) *J. Am. Chem. Soc.* **116**, 3892. (b) Teppen, B.J., Cao, M., Frey, R.F., van Alsenoy, C., Miller, D. M., and Schäfer, L. (1994) *Theochem* **314**, 169.
14. Meot-Ner, M. (1988) *J. Am. Chem. Soc.* **110**, 3854.

MOLECULAR MODELS BY LASER STEREOLITHOGRAPHY

C.A. VENANZI, W.J. SKAWINSKI, A.D. OFSIEVICH
Department of Chemical Engineering, Chemistry, and Environmental Science, New Jersey Institute of Technology, 323 King Blvd., Newark, NJ 07102 USA

1. Background

Visualization of complex, three-dimensional scientific data on a two-dimensional graphics screen is difficult. The image must often be continually rotated in order to get the full three-dimensional effect. In contrast, an accurate hand-held model can be useful in providing an immediate, static three-dimensional image. Presently available molecular models, such as Dreiding and CPK models, use average bond angles and bond lengths, but are unable to accurately represent torsional barriers and the resulting conformational readjustments due to torsional angle changes. Instead, it would be useful to have accurate models of optimized structures at selected incremental steps of a conformational analysis. The following describes a procedure for obtaining such models by the application of a rapid prototyping procedure, laser stereolithography [1-7]. Our previous work [8,9] has shown that the concept of constructing accurate molecular models by laser stereolithography from experimental and computational data is not only feasible but quite practical. Models were built not only of molecular structures, but also of molecular interaction pharmacophores and the transition states of chemical reactions. No other molecular model has the capability to accurately represent these latter quantities. The advantage of the stereolithography approach is two-fold:

(1) **The stereolithography models provide a revolutionary new tool for molecular modeling.** The models produced by the stereolithography technique can provide significantly more information than standard hand-held molecular models now available to chemists. Since they are derived from crystallographic data or high level quantum mechanical calculations, the models not only accurately represent structures, but can also represent steps along a reaction pathway, molecular properties, putative binding sites, and pharmacophores--i.e., whatever can be described on a molecular graphics screen.

(2) **The stereolithography models enable a blind or visually impaired scientist to participate more fully in the pursuit of scientific research.** This was clearly demonstrated during the amiloride studies (see below) in which one of us, W.J.S., who is blind, was able to use the models of amiloride, amiloride analogues, and the amiloride pharmacophore to interpret experimental structure-activity data. The models greatly facilitated W.J.S.'s communication with the sighted members of the

L. Echegoyen and A.E. Kaifer (eds.), Physical Supramolecular Chemistry, 127–142.

research team, especially the graduate and undergraduate students whose work he directed.

2. Methodology

The methodology has been described in detail elsewhere [9] and is summarized below.

2.1. DESIGNING THE MODEL

The first step in the model-building process is to create a molecular model image in the computer-aided design (CAD) program, I-DEAS [10], version 6. At present, there is no capability for transferring images directly from molecular modeling software into a CAD program -- a step which is necessary in order to produce files capable of being used by the stereolithography apparatus. Therefore, a short program was written within I-DEAS to allow automatic input of the structural data from a text file. This input data consists of five parameters for each atom: the atomic symbol, the van der Waals radius, and the x-, y- and z-coordinates in Ångstroms. The molecular structure data used as input was obtained from the Cambridge Crystallographic Database [11] or from quantum mechanical calculations, thus insuring an accurate representation of the molecular structure by the model. The SLICE [12] program was used to prepare thin "slices" of molecular structure data which was sent to the computer which controls the stereolithography apparatus. The BRIDGEWORKS [13] program was used to provide temporary supports as needed for parts of the structure.

2.2. STEREOLITHOGRAPHY

The stereolithography apparatus (model SLA-250; 3D Systems, Valencia, California, Fig. 1) consists of a cubic container approximately 25.4 cm on a side which holds the liquid resin, an acrylate ester blend consisting of aliphatic urethane acrylates, dimethylacrylate ester and diacrylate ester. Within the cubic container is a computer-controlled table which moves vertically. At the beginning of the building process, the table is elevated to a level which allows a specified thickness of resin to lie above the surface of the table. The computer-controlled laser then traces out, in the top layer of liquid resin, the shape of the first slice of the object being built. This is a curing process which converts the liquid into a solid plastic material in the exact shape of the first slice of the object. The table with the first slice now atop its surface moves downward by a specified increment until another thin layer of liquid resin lies above the surface of the first slice. The laser then traces out the shape of the next slice, again curing the liquid into a solid plastic and simultaneously bonding this second slice to the first. These operations are repeated until the entire object is built slice-by-slice. Upon completion of the building process, the table is again elevated to raise the model above the liquid. The model is removed, washed with isopropyl alcohol, and placed in an ultraviolet oven to complete the curing.

Figure 1. Stereolithography apparatus. (Photo credit: Bill Wittkop)

2.3 TIME REQUIREMENTS

Once the procedures were established, the time required to fabricate a large model of 164 atoms was approximately 31 hours. Smaller models required as little as one to two hours. Several models may be made simultaneously to further reduce the overall time required. Approximately 40 models have been made to date.

3. Evaluation of Models

A range of models representing several types of different molecular structures was produced using a variety of scales, atomic radii, and surface textures. The resulting models were evaluated to identify the most effective combination of these variables that could provide the most informative tactile representation of the structural information. The molecular models were made using spheres to represent atoms while atomic bonds were formed by the overlap of these spheres.

3.1 SCALING AND CHOICE OF ATOMIC RADII

Models have been produced using six different scales (in cm/Å): 0.19, 0.64, 0.76, 0.89, 1.25 (same scale as CPK models), and 1.78. It is most appropriate to model atoms using radii representing some physical property rather than some arbitrary value. The first models built used van der Waals radii for atomic radii. This resulted in considerable overlap of the spheres representing atoms but they were clearly distinguishable. Later models utilized atomic radii used in standard CPK models. Many of these latter radii are slightly smaller than the atomic van der Waals radii, resulting in less overlap of the atomic spheres and significantly more distinct atoms. Several models have been built using the CPK radii and proved to most clearly convey the spatial information. The use of the CPK radii ratios is particularly suited for use in education where the blind or sighted students would be initially unfamiliar with the details of specific molecular structures. In these models individual atoms are more readily recognized and located and overall structural features are clearly discernable.

3.2 ADDITION OF SURFACE FEATURES TO AID TACTILE RECOGNITION

Version 6.0 of I-DEAS represents spheres with facetted rather than smooth surfaces. The degree of facetting can be specified vertically and horizontally and this feature was utilized to provide different atom types with distinctive surface features. Maintaining a constant number of facets while changing the scale of the models results in smaller models tactually resembling spheres more closely, while the models built on the larger scales resemble polyhedra. As the scale of the model increases, it is appropriate to increase the number of facets in order to present a more spherical and distinct shape for each sphere. For models built using the scales 0.64 and 0.76 cm/Å, the facetting assists in the identification of atom types. The facets are not useful in models built on a smaller scale where all the atoms appear roughly spherical. On the larger scales some atoms are too angular which seems to make identification and general tactile examination slightly more cumbersome than necessary, essentially reducing the "signal to noise" ratio. A number of models were built on the larger scales without facets and proved to be quite distinguishable. The best combination of variables was shown to be the use of CPK radii with smooth spheres.

3.3 INTERNAL CAVITIES

In order to conserve the expensive resin, the models were eventually designed to be hollow. The internal cavities were designed and then cut from the image of the molecule in I-DEAS. This results in light, strong, hollow models which require only 40% as much material as do the solid models. Both cubic and spherical internal cavities were used. A hollow model was dropped onto a hard floor and demonstrated considerable resilience and sustained no damage. A program was written to create an optimum internal spherical cavity centered within each sphere representing an atom, resulting in a uniform wall thickness for each atom type.

4. Applications

4.1 INTERPRETATION OF STRUCTURE-ACTIVITY DATA FOR AMILORIDE ANALOGUES, POTENT SODIUM CHANNEL BLOCKERS

The first application of these models to the study of structure-activity relationships was in the interpretation of kinetic data for the blocking of epithelial sodium channels by the diuretic drug amiloride, **1**, and two of its analogues, **18** and **19**. The two analogues differ from amiloride by the insertion of an -O- or -NH- group, respectively, between the carbonyl carbon and the guanidinium group of the sidechain. Extensive semiempirical (AM1) and ab initio (6-31G*) molecular orbital calculations were carried out on these three species to determine the minimum energy conformations and rotational barriers about two critical sidechain bonds. Results showed that, unlike the planar amiloride [14,15], **18** and **19** assume nonplanar conformations [16]. The ring of **18** is approximately 30° out of the plane of the sidechain, whereas that of **19** is closer to 90°. These results were used to fabricate molecular models of these species using the 0.64 cm/Å scale. All the atoms of the molecular structures were clearly and completely recognizable. These models proved to be an excellent tool for comparing the structural differences between these three species.

The ab initio molecular electrostatic potential was used, along with the structural information, to define the steric and electrostatic features of the amiloride pharmacophore [17]. The distance between the minimum in the electrostatic potential off the chlorine at position 6 and the guanidinium hydrogens was identified as an important feature in all the planar amiloride analogues with pyrazine ring modifications that form a stable blocking complex with the ion channel [18]. The electrophysiological results of Li, et al. [19,20] showed that, in comparison to amiloride, **18** forms a slightly less stable blocking complex with the ion channel, whereas **19** is a better blocker than amiloride. These results could not be understood if **18** and **19** were assumed to be planar, since they would simply be too long to fit the steric requirements of the pharmacophore. However, structural superposition of the optimized conformers of **18** and **19** onto amiloride (by matching the location of the chlorine and guanidinium hydrogens of the analogues to those of amiloride) and subsequent calculation of the molecular electrostatic potential showed that the analogues could fit the putative amiloride pharmacophore [16].

This analysis was aided by the fabrication of a model of the steric requirements of the pharmacophore. An impression of the planar amiloride molecule in a solid block was fabricated (Fig. 2a,b). This clearly showed the relative spatial relationship of the chlorine and guanidinium hydrogens of amiloride and, therefore, served as a first approximation to the minimum steric requirements which may be present at its binding site on the surface of the sodium channel protein. The models of the two analogues were placed within this impression to determine how well their chlorine and guanidinium hydrogens could fit the locations occupied by those of amiloride. Analogue **19** (Fig. 2c) fit the spatial requirements better than **18** (Fig. 2d). The stereolithography models, in conjunction with the results from high level molecular orbital calculations, proved to be very useful in demonstrating how structural analogues with quite different conformations may be able to fit the steric and electostatic requirements of the putative amiloride pharmacophore. The unique features of the stereolithography models, compared to Dreiding or CPK models, are their ability to precisely represent the molecular structure information obtained from accurate quantum mechanical calculations and to represent complementary spatial relationships, as in the pharmacophore model.

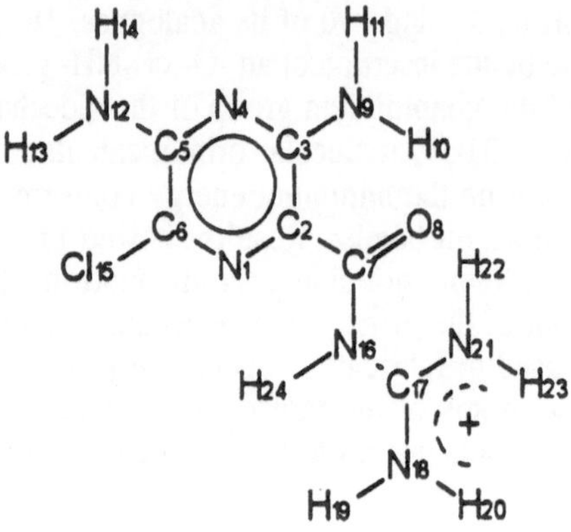

1

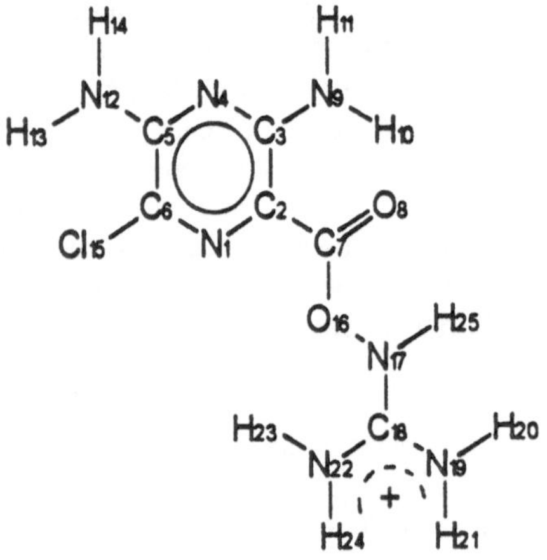

18

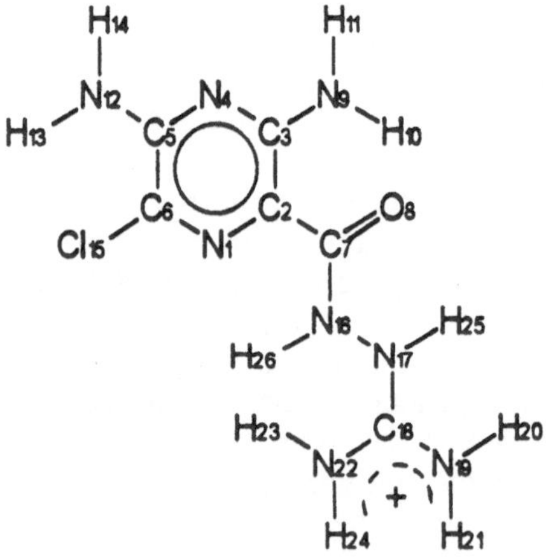

19

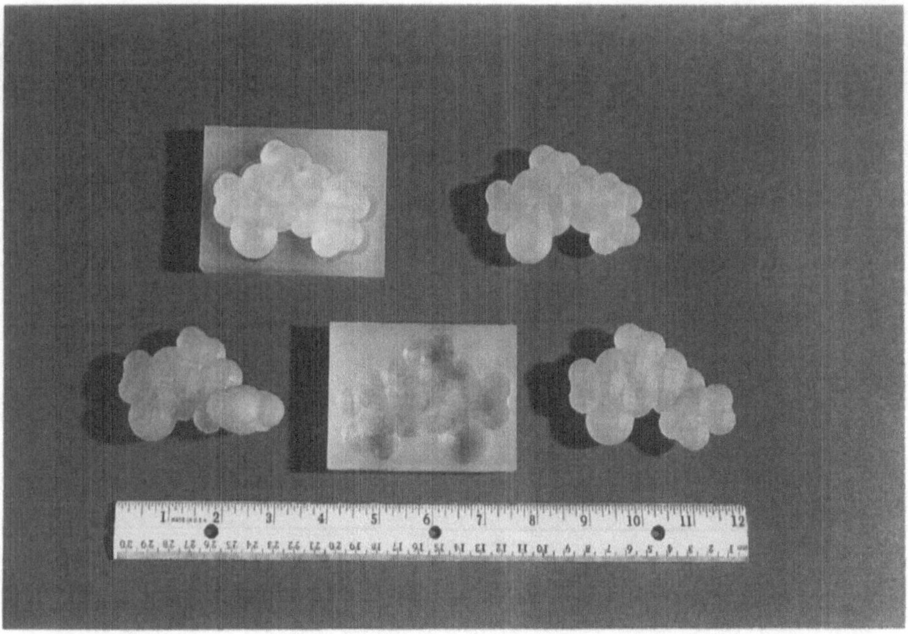

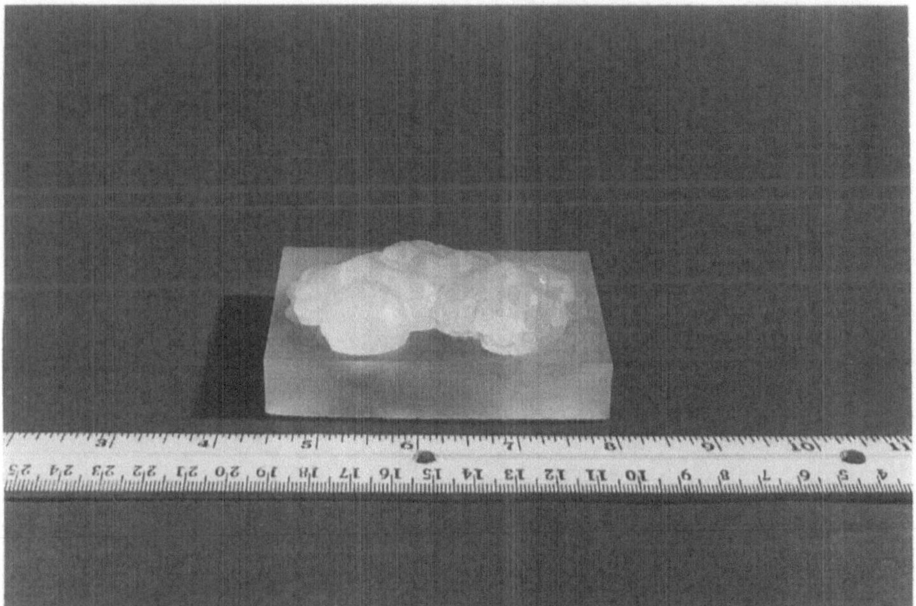

Figure 2. Solid models of amiloride analogues (approximately 7.5 cm x 6.25 cm in size) and amiloride complementary site. a. Top row, from left: amiloride in complementary site, amiloride. Bottom row, from left: **19**, mirror image of complementary site, **18**. b. Solid model of amiloride in complementary site. c. Solid model of **18** in complementary site. d. Solid model of **19** in complementary site. (Photo credit: Bill Wittkop)

4.2 INTERPRETATION OF STRUCTURE-ACTIVITY DATA FOR NOVEL AMINO ACID ANALOGUES ACTIVE AT THE CATFISH TASTE RECEPTOR

Amino acid recognition is an important element of biological systems [21-23]. Amino acid receptors have been identified in the Colorado beetle [24], the channel catfish [25,26], and bacteria [21,22]. Amino acid receptors in the catfish have been particularly well-characterized [23,25-27,28]. The important molecular features for recognition and binding at the L-alanine receptor of the channel catfish include: (a) the presence of the zwitterionic component in the L-configuration and (b) a small residue at the ß-carbon (i.e. L-alanine binds to the receptor and activates it, whereas L-phenylalanine does neither). Although the interaction of the zwitterionic moiety with the receptor is obviously ionic in nature, analysis of the data [25] for those analogues with the highest neural response seems to indicate that the interaction between the sidechain residue and the receptor is most likely nonpolar in nature. Molecular modeling techniques were used to investigate further the nature of the analogue-receptor binding and activation at the L-alanine receptor in the channel catfish [29]. Quantum mechanics (AM1, AM1-SMx, and ab initio SCRF) were used to determine the low energy conformations of the amino acid analogues in the specific L-alanine receptor environment. By superposition of the low energy conformations, a model of the steric requirements for binding was formulated. Facetted models (Fig. 3) were made of the g;lobal minimum energy conformer of several of the analogues at the scale of 0.76 cm/Å. In each case the ammonium and carboxylate ions of the zwitterion were used as reference points relative to which the remainder of the model was examined. This allows the models to be oriented to compare the space occupied by their sidechains. The three-dimensional models were useful in supplementing the information obtained from the superposition of the analogues which indicate that the steric properties of the various residues may be correlated to the relative neural activity produced by these species.

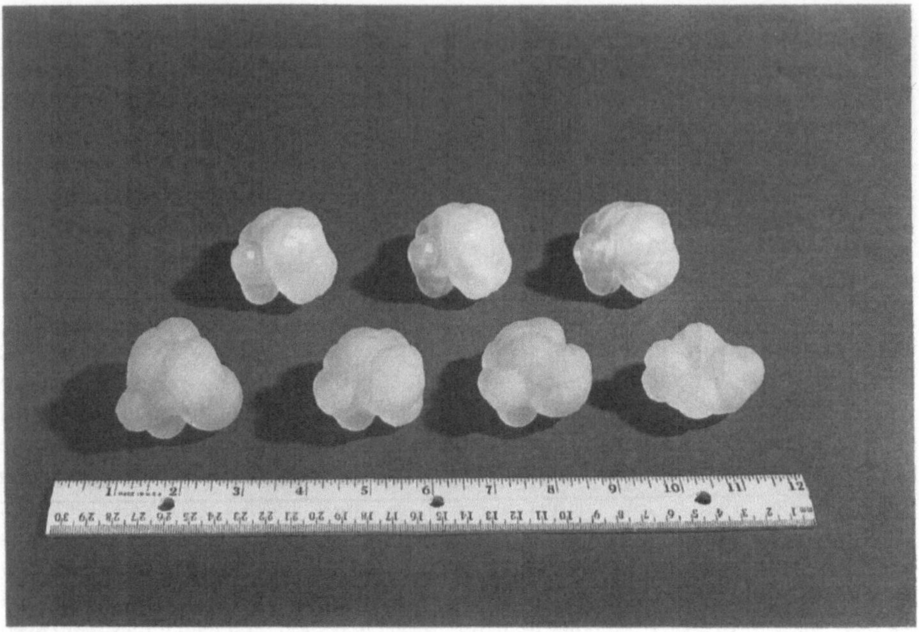

Figure 3. Models of amino acids and analogues (approximately 5 cm across), all oriented with the amino group on the top left side and the carboxylate group on the top right side. Top row from left: 1-amino-1-cyclopropane carboxylic acid, solid model, hollow model, hollow model. Bottom row, from left (solid models): L-alanine, glycine, ß-chloro-L-alanine, L-serine. (Photo credit: Bill Wittkop)

4.3 ANALYSIS OF THE REACTIVITY OF ß-CYCLODEXTRIN, AN ENZYME MIMIC

The cyclodextrins, carbohydrate macrocycles, are known to accelerate the rate of acylation of a number of different substrates. Experimental observations show that ß-cyclodextrin reacts with phenolic esters via an alkoxide ion derived from the secondary hydroxyl groups with the formation of a covalent intermediate and a subsequent release of corresponding phenols [30-32]. However, it is not clear whether the reaction occurs at the 2'- or 3'-hydroxyl group [33,34]. AM1/Langevin Dipole Solvent Model calculations [35] were carried out to study the reaction of ß-cyclodextrin with phenyl acetate in order to investigate this problem and to probe the source of the catalytic activity. The largest models yet constructed are ß-cyclodextrin (147 atoms; Fig. 4, left), consisting of seven glucose rings, derived from neutron diffraction studies [36] and the transition states of the reaction of ß-cyclodextrin with phenyl acetate at the 2'-hydroxyl (164 atoms; Fig. 4, right) and 3'-hydroxyl oxygen derived from the semiempirical molecular orbital calculations [35]. Two of these models are facetted while the third consists of smooth spheres. The scale is 0.89 cm/Å. In the transition state models, the phenyl acetate lies within the ß-cyclodextrin cavity and its orientation is clearly discernable. A model of phenyl acetate was also constructed on this scale. The larger scale was selected to allow access to the cavity for tactile examination of the orientation of the phenyl acetate. The nonstandard bond lengths and angles present in the transition states were depicted accurately in these models. It should be noted that this type of transition state model could not be made with any accuracy using standard molecular model kits since knowledge of the orientation and bonding of the substrate is required.

Facetted models of ß-cyclodextrin (Fig. 4, center) and one transition state were also built at the 0.19 cm/Å scale. All atoms appear nearly spherical and quite distinct. Tactile identification of non-surface atoms was difficult. However, models of such large molecules made at these small scales provides excellent tactile and visual representations of the overall topology and salient surface features of these molecules.

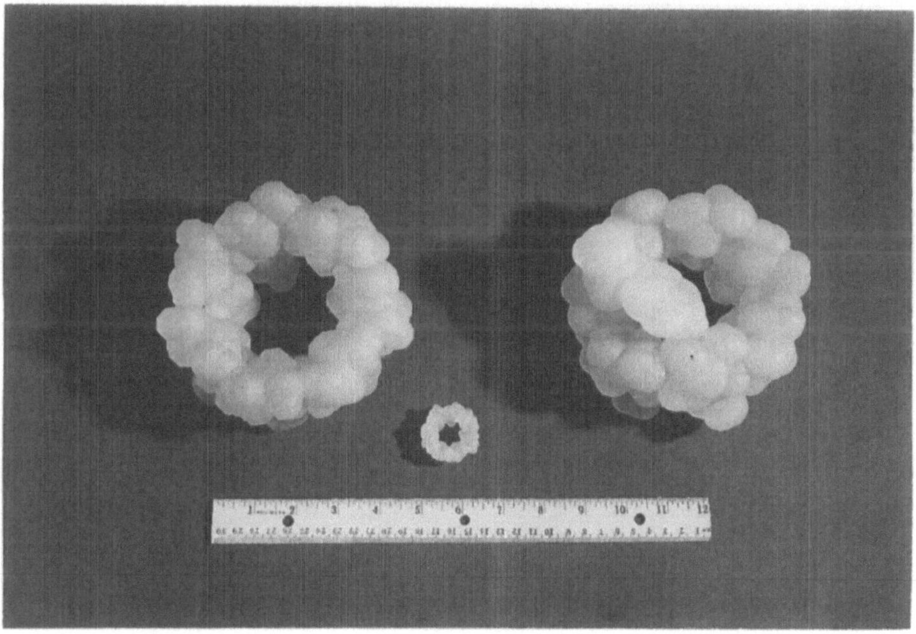

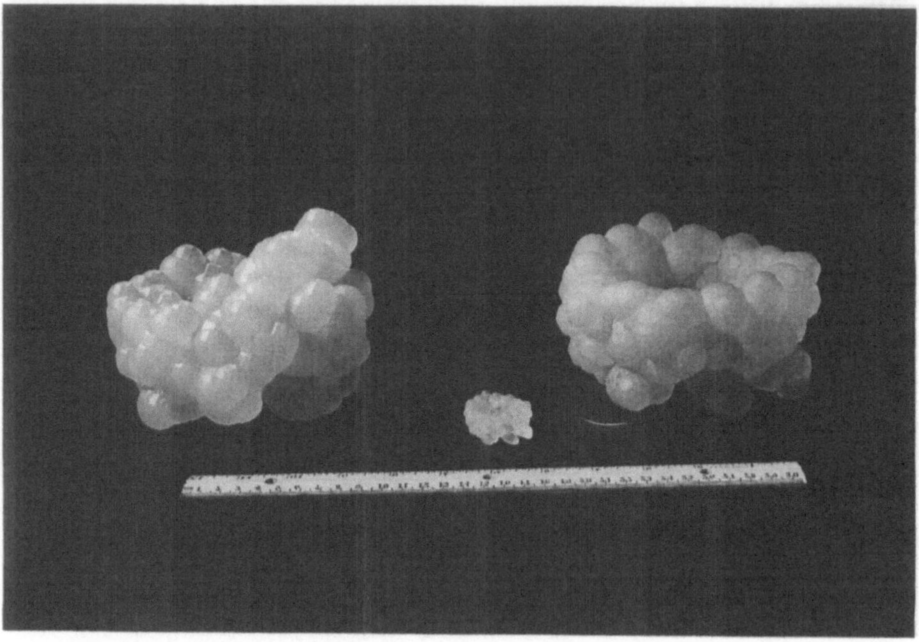

Figure 4. Solid models of cyclodextrins. From left: neutron diffraction structure (15 cm in diameter, 7.5 cm high), scaled down neutron diffraction structure (3.7 cm in diameter, 1.8 cm high), transition state structure for phenyl acetate cleavage by 2'-hydroxyl oxygen. a. Top view. b. Side view. (Photo credit: Bill Wittkop)

4.4 SIMPLE MODELS FOR EDUCATIONAL PURPOSES

A series of models of small molecules (water, methane, methyl chloride, cyclohexyl chloride and phenyl acetate) was built to be used for educational purposes by blind students. The models very clearly depict the structures of the organic molecules including the axial and equatorial substituents on the cyclohexyl ring. The models of small molecules have been used in science classes at the Washington State School for the Blind. C.A.V. has used the amiloride and cyclodextrin models in her graduate course, Computational Aspects of Physical Organic Chemistry, taught to sighted students. The feedback from both blind and sighted students has been very positive.

5. New Directions

Initial work has demonstrated that the concept of constructing accurate molecular models by stereolithography, from experimental and computational data, is not only feasible but quite practical. Future efforts will extend this work into four areas:

(1) Molecular Property Models: The encoding of molecular properties, such as the electrostatic potential, on the surface of the model. Complementarity in molecular electrostatics as well as molecular shape is important to the host-guest recognition process. Models which incorporate tactile coding of the molecular electrostatic potential or other properties on the molecular surface could be useful in the understanding of the molecular recognition process.

(2) Macromolecular Models: The fabrication of large models, such as proteins and DNA. Because of the importance of interactions between small molecules and biological macromolecules, it is useful to have models to examine structural features which control such interactions. Accurate models of macromolecules are difficult and time-consuming to construct using standard molecular model kits. More important is the difficulty of accurately specifying torsional angles in such an extensive system. Due to the limitations of the size of the stereolithography tank, adapting the procedure to deal with macromolecules is a nontrivial problem.

(3) Dynamic Pharmacophore Models: The modelling of dynamical features, such as the volume occupied by a molecule when consideration of rotation around bonds is taken into account. Since molecular structures are not static, it is useful to know the region of space occupied by constituent atoms during bond vibrations or internal rotations. This information can be used to understand the region of space specific atoms can occupy during these motions and to provide insight into the possible range of conformers which can exist under specific conditions.

(4) Applications: The construction of models of ligands in binding sites using examples where the coordinates of each are known, such as amiloride/DNA binding [37], amiloride/anti-amiloride antibody binding [38,39], and host-guest interactions in biomimetic models of chymotrypsin [32].

In each case, the models will be evaluated as to their effectiveness in providing information for blind and sighted researchers and students.

6. Acknowledgements

This work was funded by a grant to C.A.V. and W.J.S. from the National Science Foundation. W.J.S. thanks Paul Strauss, Texas Instruments Corporation, for bringing the sterolithography technique to his attention as a potential aid for the visually-impaired. The authors thank Alan Bondhus, Ram Reddy, David Lubliner, and Seymour Dreizen of the Center for Manufacturing Science at New Jersey Institute of Technology for assistance with the stereolithography technique.

7. References

1. Emery, J. (1994) Stereolithography models and prototypes, US DOE Technical Report KCP-613-5429, National Technical Information Service, Springfield, VA.
2. Burns, M. (1993) *Automated Fabrication*, PTR Prentice Hall, Englewood Cliffs.
3. Hull, C. (1988) Stereolithography: plastic prototypes from CAD data without tooling, *Mod. Cast* . **78**, 38.
4. Heinzmann, H. (1993) Stereolithography-the fast way from 3D CAD model to prototypes, *Kautsch. Gummi, Kunstst.* **46**, 19-21.
5. Schmitt, H., Geiger, M., and Steger, W. (1992) Stereolithography in original model production, *Konstr. Giessen* **17**, 13-19.
6. Bisschop, L. and Jagt, J.C. (1992) Stereolithography: a combination of CAD, laser technology, and uv-polymerization, *Kustst. Rubber* 11-18.
7. Neckers, D.C. (1990) Stereolithography: an introduction, *CHEMTECH* **20**, 615-19.
8. Skawinski, W.J., Busanic, T.J., Ofsievich, A.D., Luzhkov, V.B., Venanzi, T.J., and Venanzi, C.A. (1994) The use of laser stereolithography to produce three-dimensional tactile models for blind and visually impaired scientists and students, *Information Technologies and Disabilities* **1**, issue 4, article 6.
9. Skawinski, W.J., Busanic, T.J., Ofsievich, A.D., Venanzi, T.J., Luzhkov, V.B., and Venanzi, C.A. (1995) The application of stereolithography to the fabrication of accurate molecular models, *J. Mol. Graphics* **13**, 126-135.
10. *I-DEAS.*, SDRC, Inc., Milford, Ohio.
11. Allen, F.H., Kennard, O., and Taylor, R. (1983) Systematic analysis of structural data as a research technique in organic chemistry, *Acc. Chem. Res.* **16**, 146-153.
12. *SLICE*, 3D Systems, Valencia, CA.
13. *BRIDGEWORKS.*, Solid Concepts, Valencia, CA.
14. Venanzi, C.A., Plant, C., and Venanzi, T.J. (1991) A molecular orbital study of amiloride, *J. Comput. Chem.* **12**, 850-861.
15. Buono, R.A., Venanzi, T.J., Zauhar, R.J., Luzhkov, V.B., and C.A.Venanzi (1994) Molecular dynamics and static solvation studies of amiloride, *J. Am. Chem. Soc.* **116**, 1502-1513.
16. Skawinski, W.J. and Venanzi, C.A. (in preparation) A molecular orbital study of amiloride analogues with sidechain elongations.
17. Venanzi, C.A., Buono, R.A., Skawinski, W.J., Busanic, T.J., Venanzi, T.J., Zauhar, R.J., and Luzhkov, V.B. (1995) From Maps to Models: A concerted computational approach to analysis of the structure-activity relationships of amiloride analogues, in C.H. Reynolds, M.K. Holloway, and H.K. Cox, H. (eds.), *Computer-Aided Molecular Design: Applications in Agrochemicals,*

Materials, and Pharmaceuticals, ACS Symposium Series **589**, American Chemical Society, Washington, DC, pp. 51-63.

18. Venanzi, C.A., Plant, C., and Venanzi, T.J. (1992) Molecular recognition of amiloride analogs: a molecular electrostatic potential analysis. 1. Pyrazine ring modifications, *J. Med. Chem.* **35**, 1643-1649.

19. Li, J.H.-Y., Cragoe, E.J., Jr., and Lindemann, B. (1985) Structure-activity relationship of amiloride analogs as blockers of epithelial sodium channels: I. Pyrazine-ring modifications, *J. Membrane Biol.* **83**, 45-56.

20. Li, J.H.-Y., Cragoe, E.J., Jr., and Lindemann, B. (1987) Structure-activity relationship of amiloride analogs as blockers of epithelial Na channels: II. Side-Chain Modifications, *J. Membrane Biol.* **95**, 171-185.

21. Milburn, M.V., Privé, G.G., Milligan, D.L., Scott, W.G., Yeh, J., Jancarik, J., Koshland, J., D.E., and Kim, S.-H. (1991) Three-dimensional structures of the ligand-binding domain of the bacterial aspartate receptor with and without a ligand, *Science* **254**, 1342-1347.

22. Jeffery, C.J. and Koshland Jr., D.E. (1993) Three-dimensional structural model of the serine receptor ligand-binding domain, *Protein Science* **2**, 559-566.

23. Caprio, J. (1982) High sensitivity and specificity of olfactory and gustatory receptors of catfish to amino acids, in T.J. Hara (ed.), *Chemoreception in Fishes,* Elsevier, Amsterdam, pp. 109-134.

24. Mitchell, B.K. (1985) Specificity of an amino acid-sensitive cell in the adult colorado beetle *Lepinotarsa Decemlineata, Physiological Entomology* **10**, 421-429.

25. Bryant, B.P., Leftheris, K., Quinn, J.V., and Brand, J.G. (1993) Molecular structural requirements for binding and activation of L-Alanine taste receptors, *Amino Acids* **4**, 73-88.

26. Caprio, J. (1978) Olfaction and taste in the channel catfish: an electrophysiological study of the responses to amino acids and derivatives, *J. Comp. Physiol.* **123**, 357-371.

27. Krueger, J.M. and Cagan, R.H. (1976) Biochemical studies of taste sensation. IV. Binding of L-[^{3}H]alanine to a sedimentable fraction from catfish barbel epithelium, *J. Biol. Chem.* **251**, 88-97.

28. Brand, J.G., Bryant, B.P., Cagan, R.H., and Kalinoski, D.L. (1987) Biochemical studies in taste sensation. XIII. Enantiomeric specificity of the alanine taste receptors in the catfish *Ictalurus Punctatus, Brain Research* **416**, 119-128.

29. Venanzi, T.J., Bryant, B.P., and Venanzi, C.A. (1995) Computational analysis of binding affinity and neural response at the L-alanine receptor, *J. Comp.-Aided Mol. Des.* **9**, 439-447.

30. Bender, M.L. and Komiyama, M. *Cyclodextrin Chemistry*, Springer-Verlag, Berlin, 1978.

31. Bender, M.L., Bergeron, R.J., and Komiyama, M. *The Bioorganic Chemistry of Enzymatic Catalysis*, John Wiley & Sons, New York, 1984.

32. Breslow, R. (1980) Adjusting the lock and adjusting the key in cyclodextrin chemistry: an introduction, *Adv. Chem. Ser.* **191**, 1-15.

33. Fujita, K., Tahara, T., Imoto, T., and Koga, T. (1986) Regiospecific sulfonation onto C-3 hydroxyls of ß-cyclodextrin. Preparation and enzyme-based assignment of 3A, 3C and 3A, 3D disulfonates, *J. Am. Chem. Soc.* **108**, 2030-2034.

34. Ueno, A. and Breslow, R. (1982) Selective sulfonation of a secondary hydroxyl group of ß-cyclodextrin, *Tetrahedron Lett.* **23**, 3451-3454.

35. Luzhkov, V.B. and Venanzi, C.A. (1995) Computer modeling of phenyl acetate hydrolysis in water and in reaction with ß-cyclodextrin: molecular orbital calculations with the semiempirical AM1 method and the Langevin dipole solvent model, *J. Phys. Chem.* **99**, 2312-2323.

36. Saenger, W., Betzel, C., Hingerty, B., and Brown, G.M. (1982) Flip-flop hydrogen bonding in a partially disordered system, *Nature* **296**, 581-583.

37. Bailly, C., Cuthbert, A.W., Gentle, D., Knowles, M.R., and Waring, M.J. (1993) Sequence-selective binding of amiloride to DNA, *Biochemistry* **32**, 2514-2524.

38. Kleyman, T.R. (1992) Generation, characterization, and utilization of anti-amiloride antibodies, in E.J. Cragoe, Jr., T.R. Kleyman, and L. Simchowitz, L., (eds.), *Amiloride and Its Analogs: Unique Cation Transport Inhibitors,* VCH, New York, pp. 209-217.

39. Kleyman, T.R. and Zebrowitz, J.R. (1991) Distinct epitopes on amiloride. II. Variable-restricted epitopes defined by monoclonal anti-amiloride antibodies, *Am. J. Physiol.* **260**, C271-C276.

PHOTOCHEMICAL AND ELECTROCHEMICAL PROBES OF STRUCTURE IN SELF-ASSEMBLED MONOLAYERS

M.A. FOX,* M.O.WOLF, and R. S. REESE
Department of Chemistry and Biochemistry
University of Texas at Austin
Austin TX 78712

Abstract: The electrochemical and photochemical characteristics of substituted alkane thiols bearing electrophores or chromophores are altered by their attachment as self-assembled monolayers to a flat gold surface. The efficiency of electron tunneling through the layer and the photosensitivity of the appended groups correlate with surface packing.

1. Introduction

Important advances have been made in preparing dimensionally ordered arrays on metal and metal oxide surfaces [1,2], but relatively little attention has been paid to characterizing photochemistry in these organized layers [3]. Such experiments would be particularly interesting because comparisons with the solution phase and solid state photochemistry of the same molecules should provide an interesting means for learning about local organization and about the character of electronic interactions between the attached molecules and the surface in a self-assembled monolayer [4]. These investigations would be complementary to the more extensive array of electrochemical measurements that has been used for similar characterization of the organized monolayers [5]. For example, previous work in our laboratory has employed cyclic

L. Echegoyen and A.E. Kaifer (eds.), Physical Supramolecular Chemistry, 143–162.

voltammetry to define the structures of adsorbed multilayers [6] and to characterize the kinetics for exchange in alkane thiols functionalized at the chain end with electroactive probes [7].

Specifically, we focus in this paper on how the structure and properties of a monolayer might be changed by either photochemical or electrochemical activation. To address this general problem, we have explored the photochemistry and electrochemistry of several stilbene-functionalized alkyl thiol monolayers on gold as model monolayer arrays. With monolayers consisting of the *cis*- (**1**) and *trans*- (**2**) isomeric stilbenes *p*-substituted with a polar cyano group and an alkoxyl chain

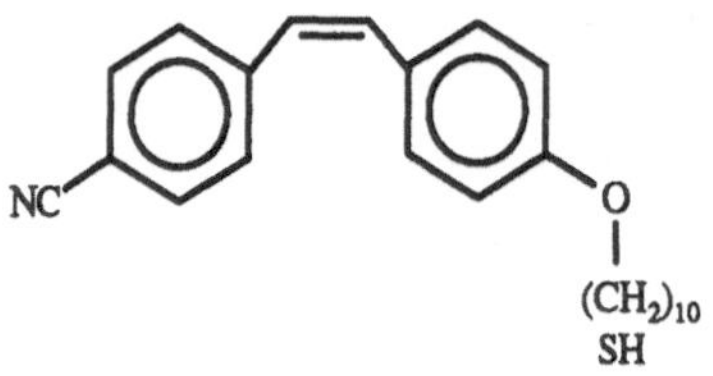

1

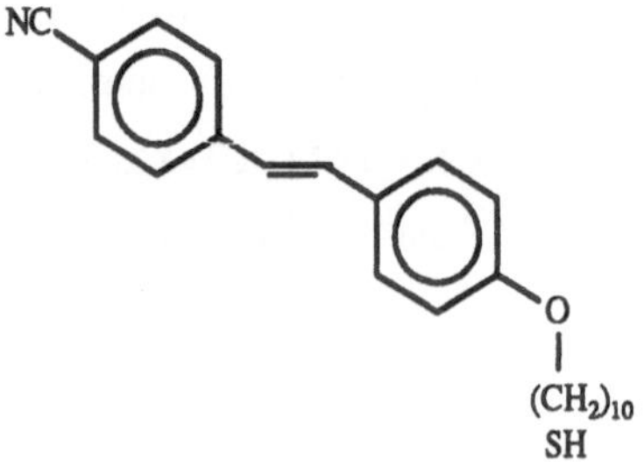

2

terminated with a thiol group for surface anchoring, we have discovered a new method for visual imaging based on photoinduced changes in surface polarity and, hence, in the resulting sessile drop contact angle made with water droplets [8]. The resulting image, which is easily discernible to the naked eye, requires selective *cis-trans* isomerization as a method for switching surface hydrophilicity and, consequently, for controlling polarity-based molecular recognition at the photochemically modified surface. Prolonged irradiation of these monolayers induces a formal [2+2]-cycloaddition of the aligned chromophores, causing a permanent alteration in the absorptivity of the component molecules chemisorbed on the assembled surface[9]. This cycloaddition, eqn 1, takes place in solution with a lower quantum efficiency than the geometric

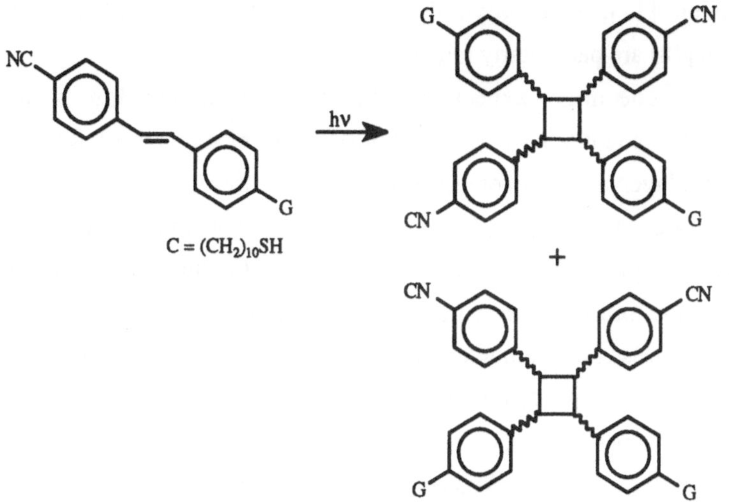

(1)

isomerization and produces both regioisomers. Nonetheless, this photoreaction disrupts conjugation between the aryl rings in the stilbene core and causes a substantive change in the absorption spectrum of the product. The observation of parallel reactivity when the corresponding molecules are confined to a surface thus would constitute a new method for permanent optical information recording.

A key question is whether these photo-altered monolayers will display altered conductivity or different physical properties in either the ground or excited state when covalently bound or adsorbed to a metal electrode surface. Conduction through a self-assembled monolayer requires thermodynamically driven electron tunneling. It been demonstrated in many low molecular weight model compounds that long distance electron transfer can be induced by irradiation of molecules bearing donor-acceptors pairs at specific sites along a covalent framework [10,11], and recent investigations in our laboratory have shown that the same principles can be used to understand the photophysical properties of chromophore-labeled macromolecules [12]. For example, we have shown that the positions of interacting chromophores in block copolymers must be precisely defined by attachment to a rigid scaffold if one is to minimize energy loss through excimer formation and/or fast back-electron transfer [13]. These systems allowed us to probe the effect of the chemical composition of the scaffold on the magnitude of the electronic coupling between the attached photoactivated donor and acceptor pair. In this connection, descriptions of the photophysical properties of rigid polymers whose frameworks include polypeptides [14,15], ladder polymers [16], polyisocyanides [17], and polyolefins obtained by ring-opening metathesis polymerization [18] are particularly relevant.

Similarly, one might expect that the chemical composition of the spacer separating the attachment point from the photoactive or electroactive termini in a self-assembled monolayer would likewise influence the observed dark or light-induced conductivity through such a layer. Significant electronic coupling between the metallic electrode surface and the probe chromophore might effect rapid excited state quenching and obviate the possibility of observing non-ablative photochemical conversions within a well-ordered self-assembled monolayer. In this connection, we therefore compare and contrast the cyclic voltammetry observed with a monolayer of thioalkylstilbene (4-(3-thiopropoxy)-stilbene) **3** (thus containing both unsaturated moieties and two heteroatoms along the skeletal backbone) with that seen with a saturated alkyl thiol

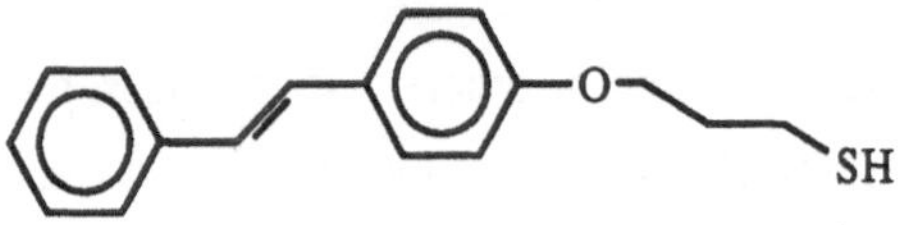

3

(dodecanethiol) that produces a well-organized monolayer estimated to be of comparable thickness.

This paper will therefore discuss: a) the synthesis and characterization of self-assembled monolayers on gold in which the organic thin layer is composed of derivatized alkane thiols bearing a photo- active (stilbene) group; b) the effect of unsaturation on observed potential-dependent electron tunneling through these self-assembled monolayers to redox-active species present in solution; and c) the effect of the metal surface on the excited state behavior of the monolayer-appended groups.

2. Synthesis of Stilbene-functionalized Alkane Thiol Monolayers on Gold

The syntheses of the thioalkylated stilbenes **1** and **2** followed the sequence outlined in Scheme 1 [8]. Thus, *p*-cyanobenzyltriphenylphosphonium chloride was coupled with *p*-hydroxybenzaldehyde under Wittig conditions to access a mixture of geometric isomers

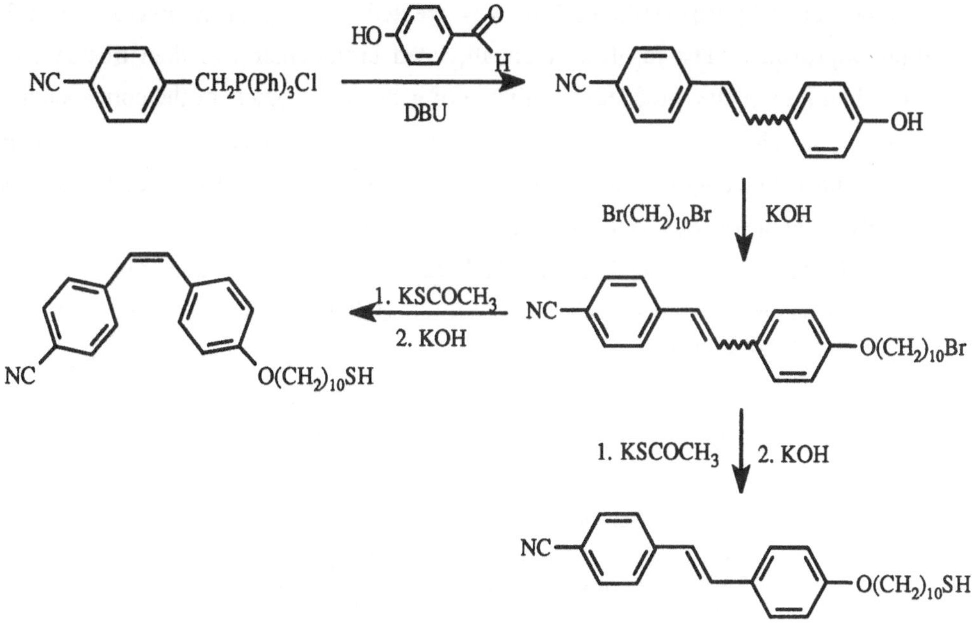

Scheme 1. Preparation of thioalkylated stilbenes **1** and **2**

of 4-cyano-4'-hydroxystilbene. The corresponding potassium salt, prepared by treatment with KOH, was alkylated with a large excess of 1,3-dibromodecane. Chromatographic separation of the *cis*- and *trans*- isomers, followed by thioacetylation and hydrolysis, yielded pure samples of **1** and **2**.

Similarly, 4-(3-thiopropoxy)-stilbene **3** was prepared by the series of reactions illustrated in Scheme 2 [19]. A mixture of the potassium salts of the *cis*- and *trans*-

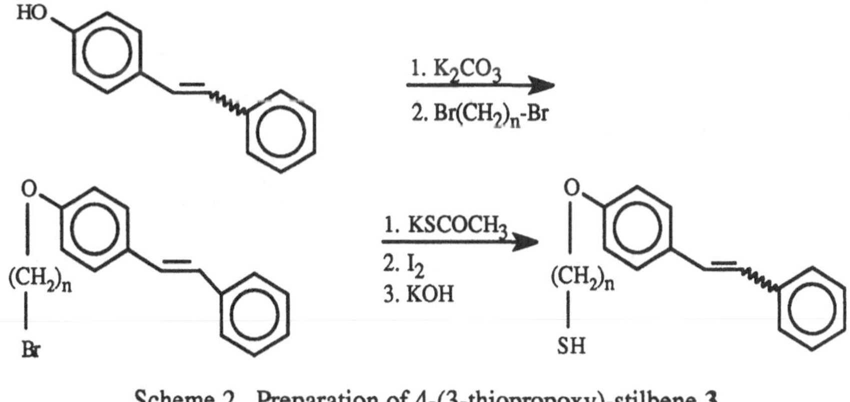

Scheme 2. Preparation of 4-(3-thiopropoxy)-stilbene **3**

isomers of 4-hydroxystilbene[20] was treated with a large excess of 1,3-dibromopropane. The resulting monoalkylated bromoether was then treated with potassium thioacetate, producing a mixture of geometric isomers of the corresponding thiol esters. The mixture of isomers was equilibrated to the predominant *trans*-isomer by heating with catalytic amounts of iodine. The resulting *trans*-thiol ester was then hydrolyzed with aqueous KOH to yield the desired thiol **3**.

Three general methods are available for spontaneous self-assembly of organic molecules on metal or metal oxide surfaces [6,7], as illustrated in Scheme 3. In the first

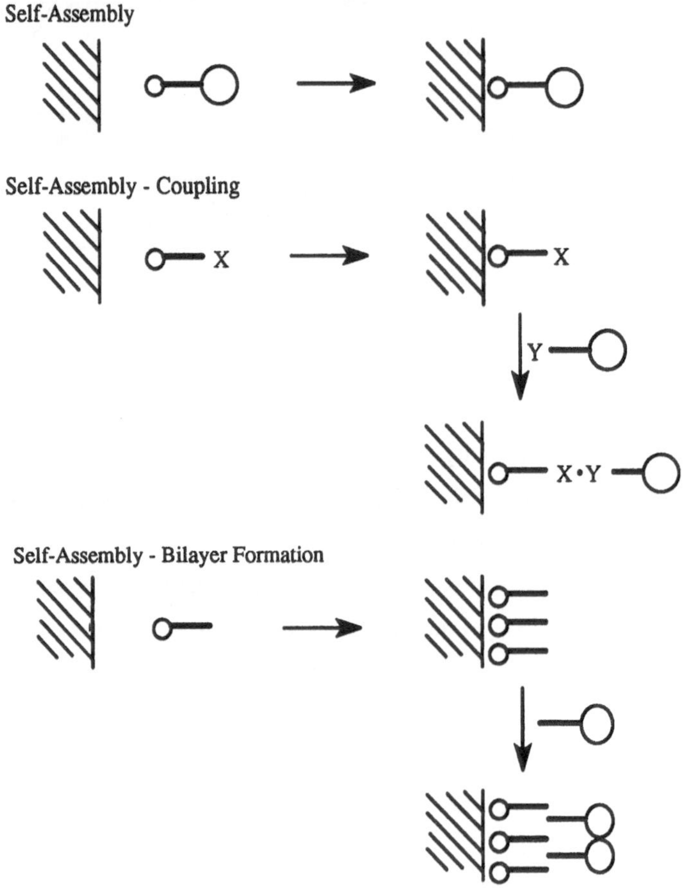

Scheme 3. Conceptual approaches to self-assembly on metal or metal oxide surfaces [6,7]

approach, direct association of the functionalized adsorbate (e.g., an alkane thiol on gold, silver, copper or mercury, an alkyl(trialkoxysilane) on a metal oxide, or an aryl thiol on a metal chalcogenide) is accomplished by dipping a freshly cleaned, or newly deposited, surface into a dilute solution of the adsorbate. A thermodynamically equilibrated assembly of the monolayer is typically accomplished in a few hours at room temperature for simple unbranched chains, although an overnight soak is typically employed to assure complete equilibration. With more highly branched or functionalized adsorbates, longer times and/or annealing at a higher temperature (either by heating the suspended surface while in contact with the adsorbate solution or by heating the dry modified surface under nitrogen to improve ordering of the adsorbed chains) may be needed for formation of a well-ordered monolayer.

Diffraction measurements show that the resulting array is densely packed and surface infra-red measurements demonstrate that the adsorbed chains are highly ordered. The adsorbed chains tilt from the surface at an angle defined by the simultaneous requirements for space-filling packing within the monolayer and lattice matching of the underlying atoms displayed on the associative surface. Well-ordered arrays produced in this way virtually completely block the resulting surface (when configured as electrodes in a standard electrochemical cell) to redox reactions of species dissolved in a contacting electrolyte solution. Terminal functional groups are easily tolerated, as are heteroatoms incorporated as components of the chain.

The second approach illustrated in Scheme 3 would promote surface functionalization by combining a surface self-assembled modification with an ω-terminated chain (as in the first approach discussed above) with a chemical coupling between the ω- functional group and a compatible reagent. For example, amides have been produced by treating an ω-aminoalkylthiol on gold with an acid chloride[7]. The resulting modified surface will then incorporate a "buried" coupling site within the resulting layer. Because the surface necessarily constrains molecular motion of the ω-functional group (and thus inhibits the structural rearrangement typically encountered in approaching the transition state of the coupling reaction), chemical transformations of this type are significantly slower than the analogous coupling reactions when conducted in homogeneous solution.

A third approach would modify the self-assembled monolayer produced in the first method by non-covalent association with one or more additional components [6]. For example, as illustrated in Scheme 3, the hydrophobic surface presented by a well-ordered monolayer of simple alkyl thiol on gold could spontaneously form a bilayer upon interaction with an added surfactant. Depending on the identity of the ω-functional group present in the surfactant (shown here as a large white ball), further hydrophobic–hydrophilic interactions could be promulgated by subsequent serial depositions, not unlike those employed for the production of Langmuir-Blodgett films.

It is also possible that mixed monolayers can be formed by exposure of a freshly prepared surface to mixtures of potential adsorbates. For example, alkane thiols terminated with either methyl groups or hydroxyl groups spontaneously self-assemble with little phase-separation into reagent-specific domains [21]. It is likely that larger domains would be produced, however, if the attached chains were functionalized with groups of appreciably different size or hydrophobicity [7]. Chemical reactions like those proposed in the second approach outlined in Scheme 3 are equally possible in these mixed monolayers, Scheme 4. Thus, esterification can be employed as a means

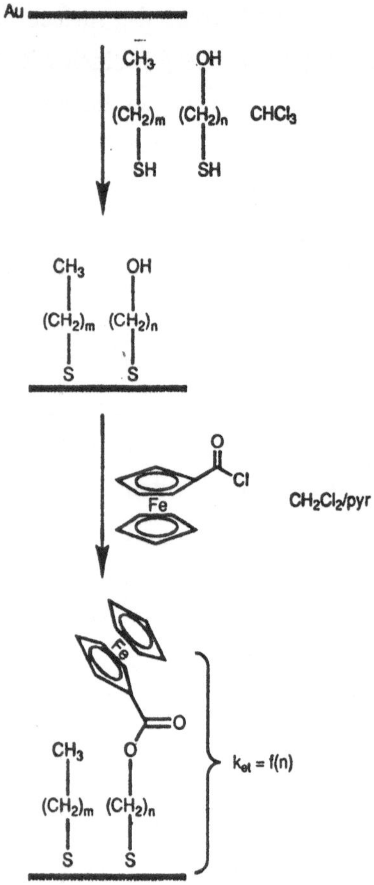

Scheme 4. *In situ* formation of redox-active modified surfaces

for attachment of functionalized acid chlorides to mixed monolayers containing ω–hydroxy-terminated alkyl thiols. With a redox-active acid chloride, the local organization of the resulting self-assembled monolayer can be probed with standard electrochemical techniques. Partial coverage of the surface is necessary if very large redox-active groups are attached to the termini of the absorbing thiol. Under such circumstances, mixed monolayers with unsubstituted alkane thiols can be used to fill binding sites on the electrode surface, while simultaneously achieving high degrees of local order and permitting exposure to bulk electrolyte of the appended redox-active groups constrained at the outer surface of the self-assembled monolayer. We envision the resulting layer to structurally resemble the simplified organization illustrated in the cartoon shown as Figure 1.

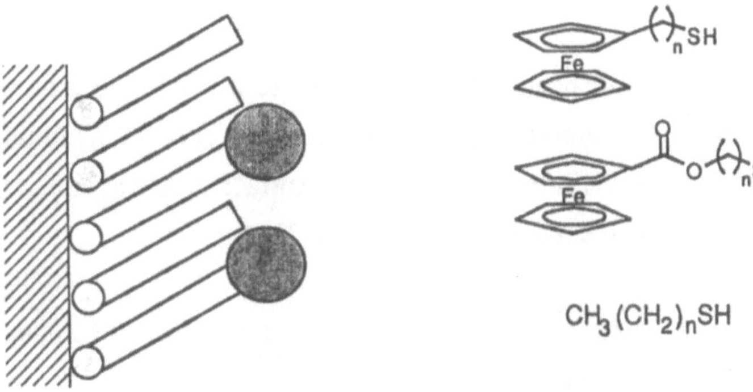

Figure 1. A simplistic view of a self-assembled mixed monolayer in which one component is an ω-terminated redox couple whose molecular diameter exceeds the surface lattice spacing [7].

For the present study, monolayers of thiols **1 – 3** were prepared by spontaneous self-assembly induced by immersing an atomically flat gold surface into a 0.01 – 0.1 M solution of the thiol in dry, deaerated solvent (either ethanol or chloroform). The nearly atomically flat gold surface was prepared by evaporative deposition of metal onto an atomically smooth Si (111) wafer. Deposition of about 50 Å of chromium was followed by the deposition of about 200 Å of gold, the chromium having served as an adhesion promoter. X-ray diffraction measurements demonstrated that a predominantly Au (111) surface was obtained by this procedure. These gold electrodes were immersed into a dilute solution of the desired thiol in either ethanol or chloroform for a period greater than 2 h (often overnight). The modified surfaces were then rinsed repeatedly with ethanol or chloroform and dried under a slow stream of nitrogen. The presence of an organic thin film was evident from the similarity of the surface infra-red spectra for these monolayer-modified electrode surfaces and that obtained for solutions or mulls of the component thiols.

Electrochemical measurements were conducted on a Bioanalytical Systems (BAS-100) electrochemical analyzer, with a saturated calomel electrode (SCE) or a Ag/AgCl/KCl electrode as reference. The modified electrodes described above were employed as working electrodes with a Pt flag (about 1 cm^2 geometric area) as the counterelectrode. Solvents employed for the electrochemical measurements were either Millipore (Milli-Q) purified water or reagent grade anhydrous acetonitrile cannulated directly from the Sure Seal container in which they were received. The electrolyte (tetrabutylammonium hexafluorophosphate) was recrystallized twice from absolute ethanol and dried *in vacuo* for 30 h at 100 °C. Upon scanning several modified electrodes prepared independently by response to a reversible redox couple in the contacting electrolyte solution, cyclic voltammetric responses reproducible to within 5% were attained. A variation in microscopic surface area was found to be less than 15% as determined by the reduction of Au_2O_3 [22].

3. Contrasting Electrochemical Behavior of Self-Assembled Monolayers of *trans*-4-thiopropoxystilbene and dodecanethiol on Gold

The structural integrity of self-assembled monolayers has frequently been probed by electrochemical measurements, often by observing redox chemistry of an electroactive species present in a contacting electrolyte solution. The electrochemical blocking behavior of these self-assembled monolayers could then be tested by comparing the cyclic voltammetry observed for the oxidation of $Fe(CN)_6^{4-}$ on gold electrodes with and without the adsorbed monolayer. The attenuated current response observed upon surface modification has previously been shown to be a useful means to characterize the homogeneity of the thin coating and to test for the presence of pinholes [23,24]. As shown in Figure 2, a monolayer of **3** on a gold electrode reduces the observable current

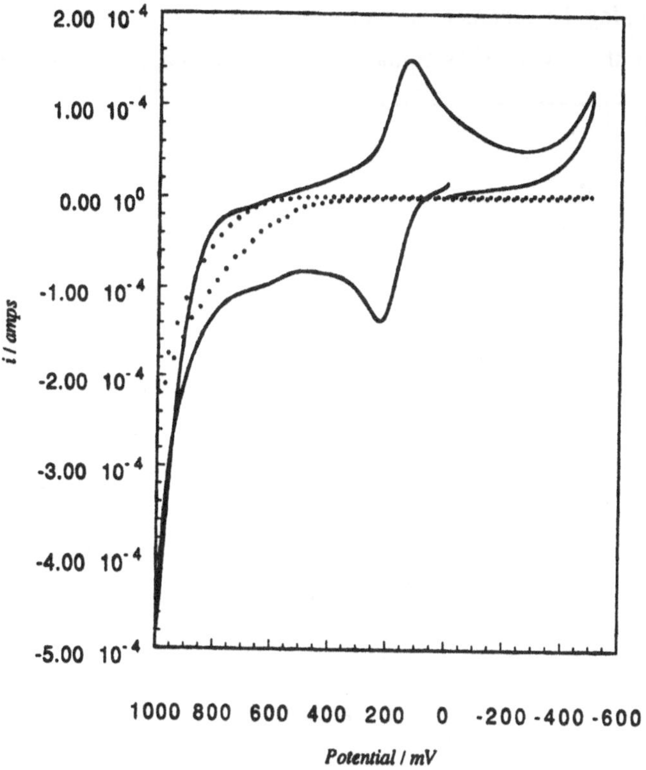

Figure 2. Cyclic voltammetric response of a 1mM solution of $Fe(CN)_6^{4-}$ in deaerated 0.1 M aqueous KCl on bare gold (solid line) and on gold modified by an overlaid self-assembled monolayer of **3** (circles). Reported potentials are referenced to SCE with a platinum flag as the counterelectrode, the curve having been measured at a scale rate of 100 mV/sec.

response to a 1mM solution of $[Fe(CN)_6^{3-}$ in 0.1 M aqueous KCl by about three orders of magnitude, compared with a unmodified gold electrode of identical surface area. This is indicative of a well-ordered, pinhole free monolayer through which electrons must be transferred by tunneling in order to effect a redox reaction of the electroactive species present in solution. Because of the known distance dependence for electron tunneling through insulating hydrocarbon layers, it is likely that electron transfer will occur more readily in these relatively thick layers at domain boundaries and defect sites where the separation between the metal electrode and the adsorbed redox species is minimized. It is therefore reasonable to expect that electrodes modified with pinhole-free self-assembled monolayers of different chemical compositions should elicit different electrochemical responses because of differences in packing morphologies. A

Tafel diagram comparing the apparent rate constants for this oxidation as a function of potential is shown for these two self-assembled monolayer, Figure 3. Both electrodes

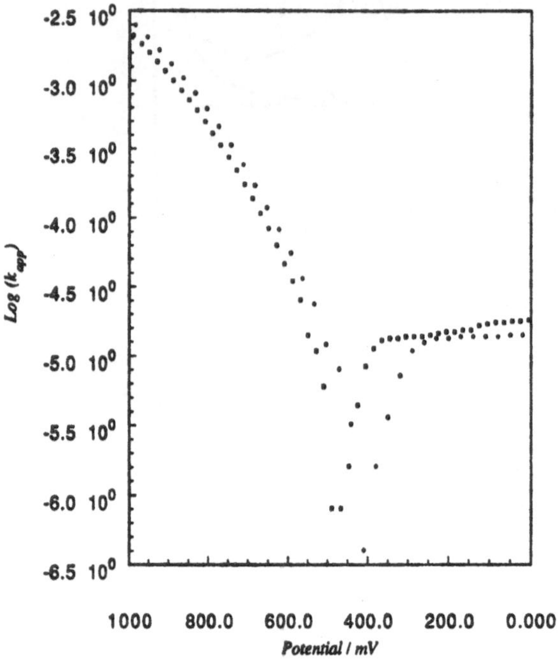

Figure 3. Tafel plot for a 1mM solution of $Fe(CN)_6^{4-}$ in deaerated 0.1 M aqueous KCl on gold modified with dodecanethiol (open squares) and on gold modified by an overlaid self-assembled monolayer of **3** (filled circles). Conditions as in Figure 2.

show appreciable curvature in the high potential region, as had previously been reported for alkanethiol monolayers [25].

A self-assembled monolayer of similar thickness was prepared by analogous treatment of a freshly prepared gold electrode surface with a dilute solution of dodecane thiol. The magnitude of the observed current response in the potential region for the $Fe(CN)_6^{3-}/Fe(CN)_6^{4-}$ couple is about twice that observed at the analogous **3**-modified surface. That is, the presence of heteroatoms and unsaturated moieties in the monolayer of **3** appears to give rise to an array which is even more well-ordered than the long straight alkyl chains protruding from the dodecanethiol-modified gold surface. Presumably, this observation implicates better monolayer packing with **3** than with the dodecanethiol.

With both modified electrodes (coated with either **3** or dodecanethiol), much less efficient blocking behavior was observed when the modified surfaces were exposed to non-aqueous electrolytes. Such behavior in similar electrodes has been previously ascribed to monolayer instability [26]. Nonetheless, the nearly ideal behavior exhibited in solvents whose hydrophobicity differs appreciably from that of the monolayer should permit confident assignment of high degrees of order to the adsorbed overlayers. The stability of such layers in the absence of contacting liquid solvent, as evidenced by invariance of the observed sessile drop contact angle over extended time periods (*vide infra*), leads us to conclude that the same order responsible for the suppressed electroactivity when in contact with an electrolyte is preserved when a thin adsorbate layer is in contact with air.

4. Photochemistry of a Stilbene-containing Self-assembled Monolayer.

Although several studies of surface patterning of monolayers have been reported [27-29], most such methods rely on the physical removal of one or more components of the monolayer in the irradiated (or otherwise modified) region, Figure 4. Typically, this

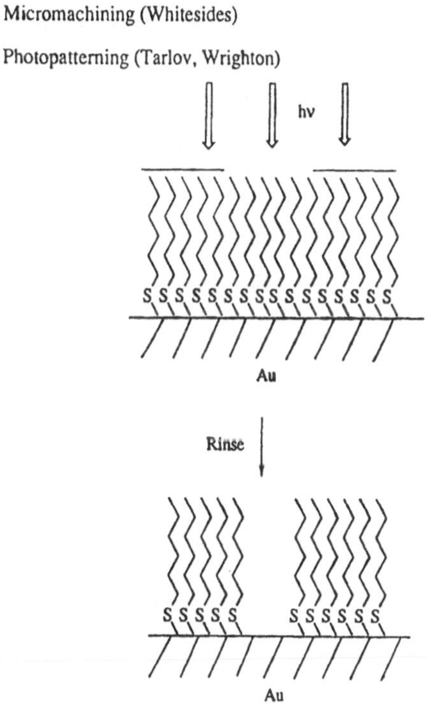

Figure 4. A schematic representation of photooxidative stripping of a self-assembled monolayer as a means for suface patterning.

is accomplished either by micromachining or by photooxidative conversion of the covalently bound Au–S–C moiety to the corresponding sulfonic acid, which can be physically displaced because of its much weaker binding to the metal surface. In contrast, we are more interested in addressing the photoconversions of light-responsive functional groups bound to the electrode surface that retain the adsorbed monolayers but with altered surface properties or spectra. We are particularly interested in those conversions in which the photochemical event can be later reversed by irradiation in a shifted wavelength region or in which thermal reversal of the photoreaction is possible in a temperature range in which the structural integrity of the adsorbed monolayer is unaffected.

The photoinduced geometric isomerization of stilbene, eqn 2, is among the most

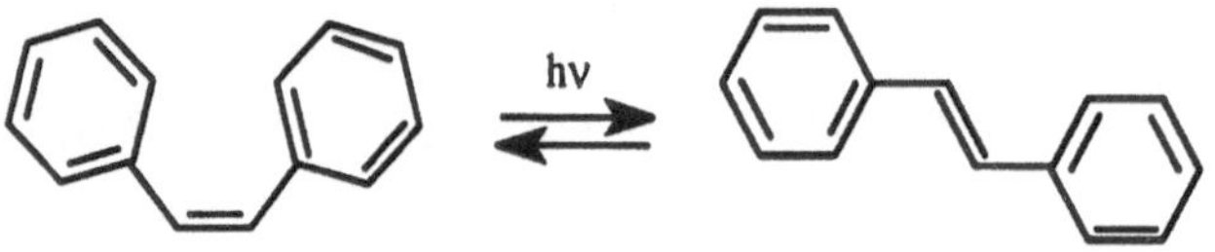

(2)

rigorously studied of all organic photochemical reactions [30]. Our initial goal was to demonstrate that parallel reactivity is attained when the stilbene core is structurally modified by the attachment of a polar probe (the *p*-cyano group) and by the surface-ordering and -anchoring ω-thioalkyl group of **1** and **2**. Indeed, steady state ultraviolet irradiation of a solution of either **1** or **2** led to production of the other geometric isomer [8]. As with the parent, the differing intensities of the two major absorption bands permitted an easy experimental method for depleting the more highly absorptive geometric isomer and, hence, shifting the wavelength-dependent equilibrium toward the less highly absorptive partner, eqn 3.

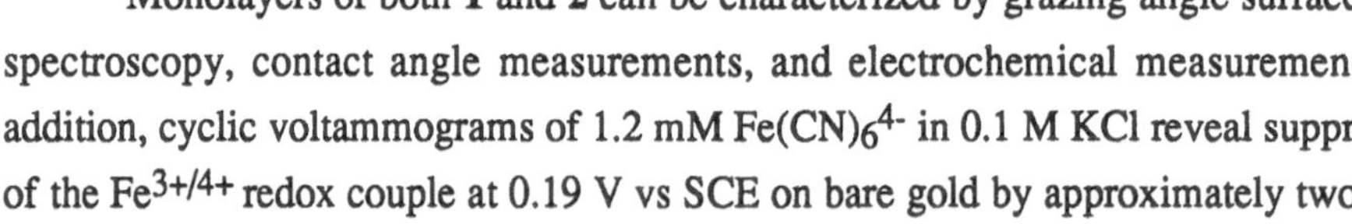

(3)

cis trans

Monolayers of both **1** and **2** can be characterized by grazing angle surface FTIR spectroscopy, contact angle measurements, and electrochemical measurements. In addition, cyclic voltammograms of 1.2 mM $Fe(CN)_6^{4-}$ in 0.1 M KCl reveal suppression of the $Fe^{3+/4+}$ redox couple at 0.19 V vs SCE on bare gold by approximately two order of magnitude at surfaces derivatized with **1** or **2**. When dispersed as self-assembled arrays on a clean, flat gold surface, monolayers of **1** and **2** display distinctly different spectral features (e.g., FT-infrared bands and ultraviolet absorptions) as well as distinguishing physical properties. For example, a monolayer of **2** is both more blocking to redox reaction of a couple dissolved in the contacting electrolyte (*vide supra)* and more hydrophilic than a monolayer of **1** (as evidenced by a smaller sessile drop contact angle of water for **2** than **1**: a surface modified by a monolayer of **2** has $\Theta = 44 \pm 2^{o}$), whereas that formed from **1** has $\Theta = 60 \pm 1^{o}$). These contact angle measurements suggest that surface hydrophobicity is significnatly influenced by the rather open exposure of the terminal polar cyano group in a monolayer of **2,** in contrast to a monolayer of **1,** where this same functional group is partially buried within the thin layer. Presumably, these effects result from the tighter packing afforded by the extended chain conformation possible with the *trans* isomer, as is suggested by molecular mechanics calculations [31].

Irradiation of polycrystalline gold surfaces derivatized with monolayers of **1** or **2** induced little change in the contact angle of water for the **2**-modified surface, whereas the gold surface derivatized with **1** shows a significant decrease in contact angle (to Θ = 45 – 50°) in the irradiated area [8]. Accordingly, photoconversion (geometric isomerization) of a self-assembled monolayer of *cis*-4-cyano-4'-(10-thiodecoxy)-stilbene **1** on a polycrystalline gold surface results in production of the spectral characteristics and physical properties of a monolayer of the corresponding *trans*-isomer **2** [9]. This photochemical isomerization, when conducted through a patterning mask, results in production of local surface features that can be made visible to the human eye by exposure of the cooled monolayer surface to atmospheric moisture. This new kind of "image development" employs the condensation of small water droplets to differentiate the hydrophilic regions where **2** is present from the hydrophobic domains where **1** is present. Because of variation of the sessile drop contact angle for these domains, an image immediately visible to the human eye is produced [8,9].

Although photochemical excitation of **2** produces **1** in solution, irradiation of a monolayer of *trans*- 4-cyano-4'-(10-thiodecoxy)stilbene **2** on gold under identical conditions does not result in any parallel change in the contact angle made with a contacting water droplet. We interpret the suppression of changes in surface properties when the geometric photoequilibration is attempted with a monolayer of **2** as indicative of the local packing difficulties encountered as a well-ordered monolayer is isomerized. That is, the more disordered monolayer of **1** on gold can be more easily converted to the more tightly packed monolayer **2**, but the reverse process is significantly more difficult.

Continued irradiation of monolayers of **1** or **2** eventually results in the disappearance of the characteristic ultraviolet absorptions of these chromophores, with appearance of a spectrum identical to that expected from the conversions shown in eqn (1) [9]. Thus, the surface hydrophobicity changes that result in a developed image can also be elaborated for permanent imaging as a photochromic local domain. The induction of local photochromism on a chromophore-laden monolayer surface is also observed when monolayers tethered to anthracene, e.g., **4**, and coumarin, e.g. **5**, are similarly employed [32].

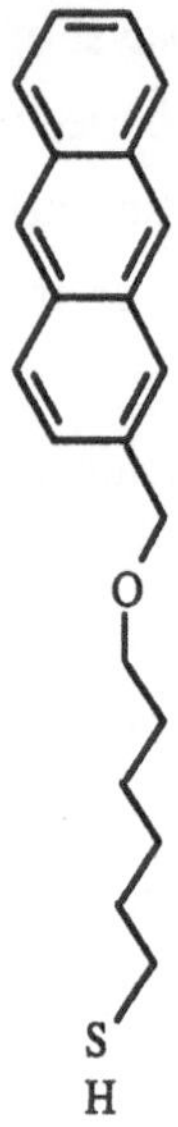

4

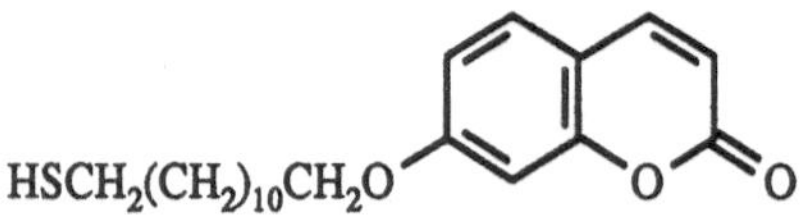

5

An important challenge for the future will be determining whether the written image produced by these photochromic changes can be erased, without net surface damage, by either photochemical or thermal methods. If so, these monolayers could be employed as easily accessible arrays for write, read, store, or erase materials.

5. Conclusions

Self-assembly on a flat gold surface of alkane thiols functionalized with electroactive or photoactive groups at the chain end provides an interesting means for preparing anisotropic thin layers. The disposition of such molecules into this highly controlled environment significantly influences their physical properties and their ability to respond to an applied bias or to the stimulation afforded by the absorption of photons.

The magnitude and shape of the overpotentials observed for oxidation or reduction of redox couples present in the contacting electrolyte provides a rough measure of the structural integrity of the thin layer. The photoproducts formed in irradiated regions of the monolayer constitute effective means for information storage in discrete domains, in which the latent image is either developed by exposure to water (producing domain-dependent optical scattering consequent to local surface hydrophobicity changes) or by measurement of surface photochromism.

6. Acknowledgments

This work was supported by the U.S. Department of Energy, the Texas Advanced Research Program, and the Robert A. Welch Foundation. We are grateful to Professor David M. Collard for preparing several of the schemes included in this paper. M.O.W. gratefully acknowledges the National Science and Engineering Research Council of Canada for a postdoctoral fellowship.

7. References

1. (a) Ulman, A. (1991) *An Introduction to Ultrathin Films: From Langmuir-Blodgett to Self Assembly*, Academic Press, San Diego. (b) Dubois, L.H. and Nuzzo, R.G. (1992) *Ann. Rev. Phys. Chem.* **47**, 437.
2. For example, see Huang, J. and Wrighton, M.S. (1993) Surface modification of n-molybdenum disulfide electrodes wtih a viologen based redox polymer: persistent attachment of a polysiloxane via a thin tin dioxide adhesion layer, *Langmuir* **9**, 3291.
3. Schlenoff, J.B., Li, M., and Ly, H. (1995) Stability and self-exchange in alkane thiol monolayers, *J. Am. Chem. Soc.* **177**, 12528 and references cited therein.
4. Ramamurthy, V. (1991) *Photochemistry in Organized and Constrained Media*, VCH Publishers, New York.
5. For example, see Bard, A.J., Abruña, H.D., Chidsey, C.E.; Faulkner, L.R., Feldberg, S.W., Itaya, K., Majda, M., Melroy, O., Murray, R.W., Porter, M.D., Soriaga, M.P, and White, H.S. (1993) The electrode/electrolyte interface - a status report, *J. Phys. Chem.* **97**, 7147; and Smalley, J.F., Feldberg, S.W., Chidsey, C.E.D., Linford, M.R., Newton, M.D., and Liu, Y.-P. (1995) The kinetics of

electron transfer through ferrocene-terminated alkanethiol monolayers on gold, *J. Phys. Chem.* **99**, 13141.

6. Creager, S.E., Collard, D.M., and Fox, M.A. (1990) Mediated electron transfer by a surfactant viologen bound to octadecylmercaptan on gold, *Langmuir* **6,** 1617.
7. Collard, D.M. and Fox, M.A. (1991) The use of electroactive thiols to study the formation and exchange of alkanethiol monolayers on gold, *Langmuir*, **7**, 1192.
8. Wolf, M.O. and Fox, M.A. (1995) Photochemistry and surface properties of self-assembled monolayers of *cis*- and *trans*-4-cyano-4'-(10-thiodecoxy)stilbene on polycrystalline gold, *J. Am. Chem. Soc.* **117**, 1845.
9. Wolf, M.O. and Fox, M.A. (1996) Photoisomerization and Photodimerization in Self-Assembled Monolayers of *cis*- and *trans*-4-cyano-4'-(10-thiodecoxy)stilbene on Gold, *Langmuir*, in press.
10. For a pioneering example describing the effect of thermodynamic driving force on the observed efficiency of long distance photoinduced electron transfer across an extended saturated sigma framework, see Miller, J.R., Calcaterra, L.T., and Closs, G.L. (1984) Intramolecular long-distance electron transfer in radical anions. The effects of free energy and solvent on reaction rates, *J. Am. Chem. Soc.* **106**, 3047.
11. Although listing of all of the key experiments in this important area of physical chemistry lies far beyond the scope of this brief article, several useful reviews describing some of these contributions are available: Fox, M.A. (1992) Electron transfer reactions, *Chem. Rev.* **92**, 365.
12. Fox, M.A. (1992) Polymeric and supramolecular arrays for directional energy and electron transport over macroscopic distances, *Accts. Chem. Res.* **25**, 569.
13. Watkins, D.M. and Fox, M.A. (1995) Synthesis and photophysical characterization of aryl-substituted polynorbornenediol acetal and ketal multiblock copolymers, *Macromolecules* **28**, 4939.
14. Galoppini, E. and Fox, M.A. (1996) Effect of the electric field generated by the helix dipole on photoinduced intramolecular electron transfer in dichromophoric α-helical peptides, *J. Am. Chem. Soc.* **118**, in press.
15. Batchelder, T., Fox, R.J., III, and Fox, M.A. (1996) Photoinduced electron transfer along an α-helical peptide, *J. Org Chem.*, submitted for publication.
16. Li, W.J. and Fox, M.A. (1996) Photoinduced electron transfer in ladderane polymers bearing pendant aryl groups, unpublished observations.

17. Hong, B. and Fox, M.A. (1994) Arene-functionalized polyisocyanides: A kinetic study of polymerization to prepare homopolymers and block copolymers, *Macromolecules* **27**, 5311.

18. Watkins, D.M. and Fox, M.A. (1994) Rigid, well-defined block copolymers for efficient light harvesting, *J. Am. Chem. Soc.* **116,** 6441.

19. Reese, S. and Fox, M.A. (1996) Electron tunneling through self-assembled monolayers on gold, unpublished observations.

20. Ruasse, M.F. and Duboise, J.E. (1972) Electrophilic bromination of aromatic conjugated olefins. I. evaluation of a competitive path mechanism in bromination of *trans*-monosubstituted stilbenes, *J. Org. Chem.* **37**, 1770.

21. Folkers, J.P., Laibinis, P.E., Whitesides, G.M., and Deutch, J. (1994) Phase behavior of two-component self-assembled monolayers of alkane thiolates on gold, *J. Phys. Chem.* **98**, 563.

22. Oesch, U. and Janata, J. (1993) Electrochemical study of gold electrodes with anodic oxide films I. Formation and reduction behavior of anodic oxides on gold, *Electrochim. Acta* **28,** 1237.

23. Chidsey, C.E.D. and Loiacono, D.N. (1990) Chemical functionality in self-assembled monolayers: structural and electrochemical properties, *Langmuir* **6,** 682.

24. Finklea, H.O., Snider, D.A., Fedyk, J., Sabatani, E., Gafni, Y., and Rubenstein, I. (1993) Characterization of octadecanethiol-coated gold electodes as microarray electrodes by cyclic voltammetry and AC impedance spectroscopy, *Langmuir* **9,** 3660.

25. Miller, C.J., Cuendet, P. and Grätzel, M. (1991) Adsorbed ω-hydroxy thiol mMonolayers on gold electrodes: evidence for electron tunneling to redox species in solution, *J. Phys. Chem.* **95,** 877.

26. Groat, K.A. and Creager, S.E. (1993) Self-assembled monolayers in organic solvents: electrochemistry at alkanethiolate-coated gold in propylene carbonate, *Langmuir* **9**, 3668.

27. Tarlov, M.J., Burgess, D.R.F., Gillen, G. (1993) UV photopatterning of alkanethiol monolayers self-assembled on gold and silver, *J. Am. Chem. Soc.* **115**, 5305.

28. Tarlov, M.J. and Newman, J.G. (1992) Static secondary ion mass spectrometry of self-assembled alkanethiol monolayers on gold, *Langmuir* **8**, 1398.

29. Lewis, M., Tarlov, M., and Carron, K. (1995) Study of the photooxidation process of self-assembled alkanethiol monolayers, *J. Am. Chem. Soc.* **117**, 9574.

30. Saltiel, J.; Charlton, J.L. (1980) in *Rearrangements in Ground and Excited States*, DeMayo, P. ed., Academic Press, p. 25.

31. Fox, M.A., Wolf, M.O., and Stewart, G.N. (1996) Photopatterning to create new structures on surfaces, *NATO Adv. Stud. Ser.* in press.

32. Thompson, H. and Fox, M.A. (1996) Photodimerization of anthracene- and coumarin-labelled monolayers on gold, unpublished observations.

SUPRAMOLECULAR PHOTOCHEMISTRY: RECENT ADVANCES

VINCENZO BALZANI, ALBERTO CREDI
Dipartimento di Chimica "G. Ciamician"
Università di Bologna
Via Selmi 2, 40126 Bologna
Italy

NICOLA ARMAROLI
Istituto FRAE-CNR
Via Gobetti 101, 40129 Bologna
Italy

1. Introduction

One of the most interesting aspects of the chemistry of supramolecular species is their interaction with light and the great variety of processes that may ensue. This is the realm of *supramolecular photochemistry* [1] that in the last few years has grown very rapidly [1-7]. The aim of this paper is to point out some latest developments, taking examples particularly, but not exclusively, from the work carried out by our research group.

We report examples of photoinduced processes in three classes of supramolecular species: (i) host-guest systems; (ii) incarcerated molecules; (iii) catenanes, rotaxanes and related species. The studies on another important class of supramolecular species, i.e., polynuclear metal complexes, have been recently reviewed [8].

2. Host-guest systems

In host-guest systems, the photochemical and photophysical properties of each partner can be profoundly modified on adduct formation. Such changes can be useful to obtain information on the structure of the adduct and, at the same time, to design photochemical molecular devices for a variety of light-induced functions.

The cylindrical macrotricyclic receptor **M**, synthesized by Lehn and coworkers [9], is an interesting host for photophysical studies because it contains two potentially luminescent naphthalene units linked to the tertiary amine (electron donor) groups of the two 18-N_2O_4 macrocycles (Fig. 1). In CH_2Cl_2 solution **M** shows a broad and relatively weak emission band with maximum at 438 nm, assigned to a charge-transfer (CT) transition from the non-bonding orbitals of the amine groups of the macrocycles to π^* orbitals of the naphthalene units (Fig. 1) [10]. Addition of CF_3COOH causes the successive protonation of the four amine units. As a consequence, the broad and weak CT band is replaced by a structured, about 800 times more intense band with maximum at 342

L. Echegoyen and A.E. Kaifer (eds.), Physical Supramolecular Chemistry, 163–177.

nm (Fig. 1), very similar to that exhibited by 2,6-dimethylnaphthalene (DMN). Therefore **M** is a fluorescent sensor highly responsive to protons. It is also an example of a "yes" and "not" fluorescent gate [11]. Upon adduct formation with $NH_3^+(CH_2)_5NH_3^+$ (cadaverine cation), for which a molecular inclusion into the cavity of **M** is demonstrated by X-ray structure [9a], the CT emission band is again quenched, which is consistent with formation of hydrogen bonds between the host and guest moieties [10].

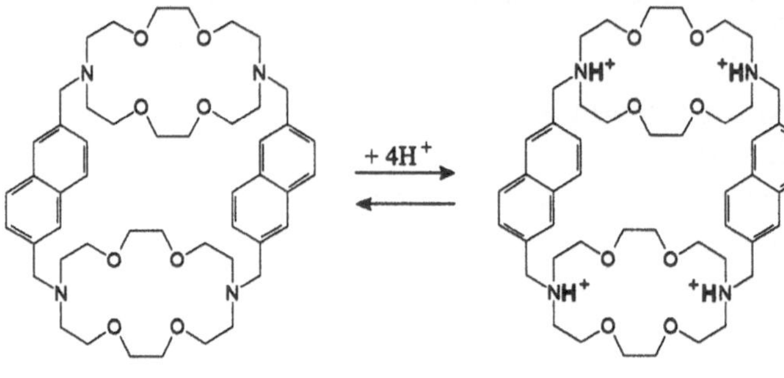

M **M·4H⁺**

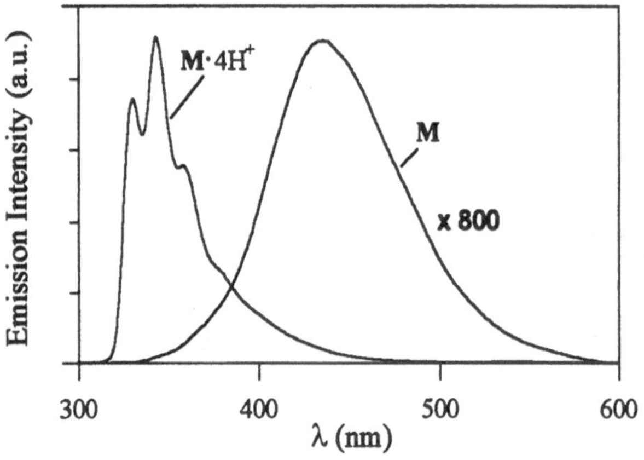

Figure 1. Structure and room temperature luminescence spectra of the cylindrical macrotricyclic receptor **M** and of its protonated form **M·4H⁺**

The $[Pt(NH_3)_2(bpy)]^{2+}$ complex, which can be used as a guest for a variety of crown ethers [12], forms a 1:1 adduct with unprotonated **M** [10]. The absorption spectrum of the adduct is noticeably different from that expected for the sum of the two separated components, particularly because of the presence of an intense tail in the 340-420 nm region assigned to a host-guest CT transition. At room temperature, the luminescence bands exhibited by the two separated components are no longer observed in the adduct. In rigid matrix at 77 K, the phosphorescence band of **M** can be observed in the adduct regardless of the excitation wavelength, but its lifetime (0.8 ms) is much shorter than that (2.6 s) of the phosphorescence of **M**. The above results indicate that $[Pt(NH_3)_2(bpy)]^{2+}$ is hosted in the cylindrical cavity of **M** with an amine ligand interacting with a 18-N_2O_4

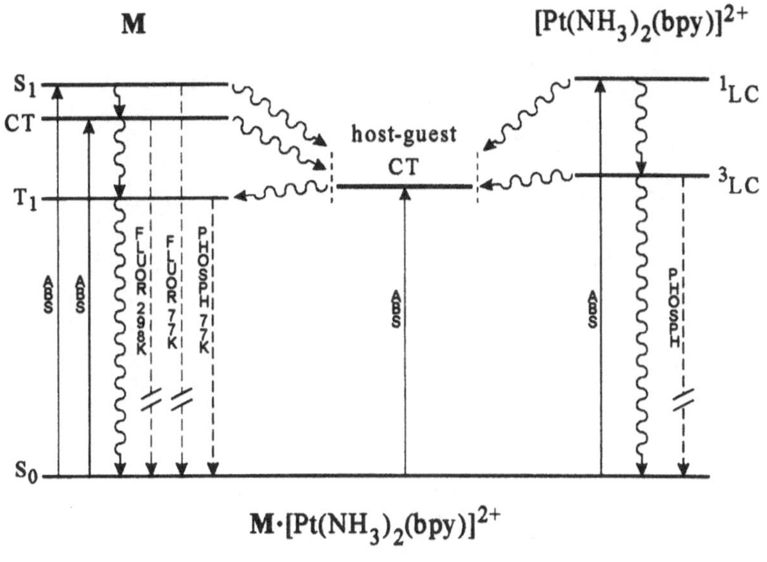

Figure 2. Schematic energy level diagram for the adduct $M{\cdot}[Pt(NH_3)_2(bpy)]^{2+}$

macrocycle unit via hydrogen bonds and the $Pt(bpy)^{2+}$ electron deficient moiety involved in a CT interaction with the DMN chromophoric units (Fig. 2).

The cage-like species **C** (Fig. 3), synthesized by Vögtle and coworkers [13], is an endoreceptor with a large cavity, made of three convergent 2,9-dianisyl-1,10-phenanthroline (DAP) units connected by two 1,3,5-trimethylbenzene-type spacers . It is well known that DAP is an excellent ligand for Cu^+ [14]. In the presence of 2,9-dimethyl-1,10-phenanthroline (DMP) and Cu^+ ions, **C** is able to coordinate three $Cu(DMP)^+$ units, forming a trinuclear metal complex of quite unusual shape [15] (Fig. 3). The three metal-containing units of this novel polynuclear complex are equivalent and exhibit interesting luminescence properties. From the viewpoint of host-guest chemistry, **C** is a receptor capable to host in its large cavity as many as six chemical species. A logic development of these studies has been the construction of mono-, di-, and tri-catenates [16].

The cage compound **C** is also particularly suitable for protonation studies because the DAP units are known to undergo profound changes in the absorption spectra and luminescence properties upon addition of acid [17]. In CH_2Cl_2 solution at room temperature **C** shows absorption and luminescence properties (absorption: λ_{max}=284 nm; fluorescence: λ_{max}=400 nm, τ=2.2 ns, Φ=0.26) very similar to those exhibited by DAP; by addition of CF_3COOH to a CH_2Cl_2 solution of **C** strong and complex changes in the absorption and fluorescence spectra and in the excited state lifetime and quantum yield are observed [18]. Analysis of the spectral changes *vs* acid concentration indicates the occurrence of two successive, well separated and reversible protonation reactions. The first protonation product is a monoprotonated species, $C{\cdot}H^+$ (absorption: λ_{max}=275 nm;

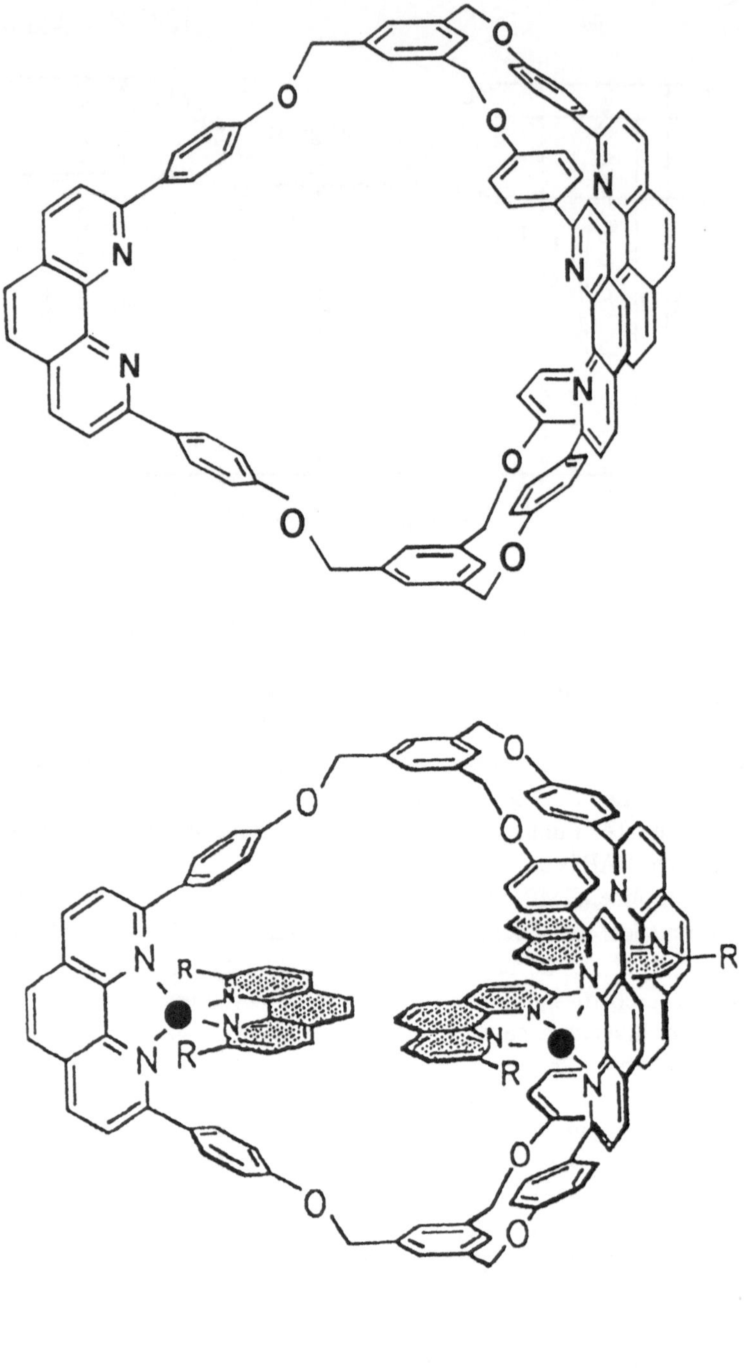

Figure 3. Tris-phenanthroline cage compound **C** and its tricopper complex (R=CH_3)

fluorescence: λ_{max}=544 nm, τ=37 ns, Φ=0.09), where the flexibility of the structure allows the interaction of a protonated DAP-type unit with the two unprotonated ones via donor-acceptor interactions. The second (and final) protonation step does not lead to a diprotonated species, but to a fully protonated **C**·$3H^+$ species (absorption: λ_{max}=303 nm; fluorescence: λ_{max}=580 nm, τ=7.6 ns, Φ=0.03), where the three independently protonated DAP-type components do not interact with one another.

3. Incarcerated molecules

Carcerands and hemicarcerands are rigidly hollow hosts synthesized by Cram and coworkers [19]. Carcerands are spherical-type molecular cavities where small molecules (particularly, solvent molecules) can be irreversibly imprisoned when the two halves of the structure are covalently linked by proper bridges in the last step of the synthesis. Hemicarcerands are cage-type molecules with larger voids and with "portals" through which a variety of molecules can enter at high temperature and then remain imprisoned at room temperature for a more or less long period of time ("constrictive" binding) [20]. Carcerands and hemicarcerands are species of great photochemical and photophysical interest because they offer the opportunity to study the behavior of guest molecules isolated in an environment constituted by a specific, discrete molecular inner phase. Furthermore, it should be possible to prepare and investigate highly reactive species like radicals or unstable isomers in the interior of such "microreactors".

The first, very exciting example of this new branch of photochemistry was the "taming" of cyclobutadiene performed by Cram and coworkers [21]. They studied the photoisomerization/photofragmentation of α–pyrone within a carcerand-type species (**H1**, Fig. 4) and found that of the two photoproducts cyclobutadiene remains imprisoned into the cage while CO_2 escapes. The cyclobutadiene is thermally stable in the inner phase at room temperature, and exists as a singlet. On further irradiation, cyclobutadiene breaks down to acetylene, as it does when is prepared in argon matrix at 8 K. Interesting questions arises as to know how the exciting light (200-250 nm) required for the formation of cyclobutadiene and then acetylene is effective. Absorption by the dimethoxybenzene-type units of the hemicarcerand and photosensitization is a possibility.

The photophysical behavior of small molecules imprisoned into hemicarcerands has recently been investigated in some detail [22-24]. In CH_2Cl_2 solution 9-cyanoanthracene, when imprisoned into the octaimine hemicarcerand **H2** (Fig. 4), undergoes noticeable changes in its spectroscopic and photophysical properties: the fluorescence spectrum shifts by about 350 cm^{-1} to the red, the quantum yield of fluorescence decreases by a factor of about 50, and the fluorescence lifetime decreases from 15 ns to 350 ps. The energy-transfer efficiency from the dimethoxybenzene chromophoric units of the cage to 9-cyanoanthracene is negligible, whereas the quenching (presumably by an electron-transfer mechanism) of the fluorescent excited state of 9-cyanoanthracene by the cage occurs with a rate constant $>10^{10}$ s^{-1}. Since the fluorescent excited state of imprisoned 9-cyanoanthracene has a short lifetime (350 ps), its quenching by external species

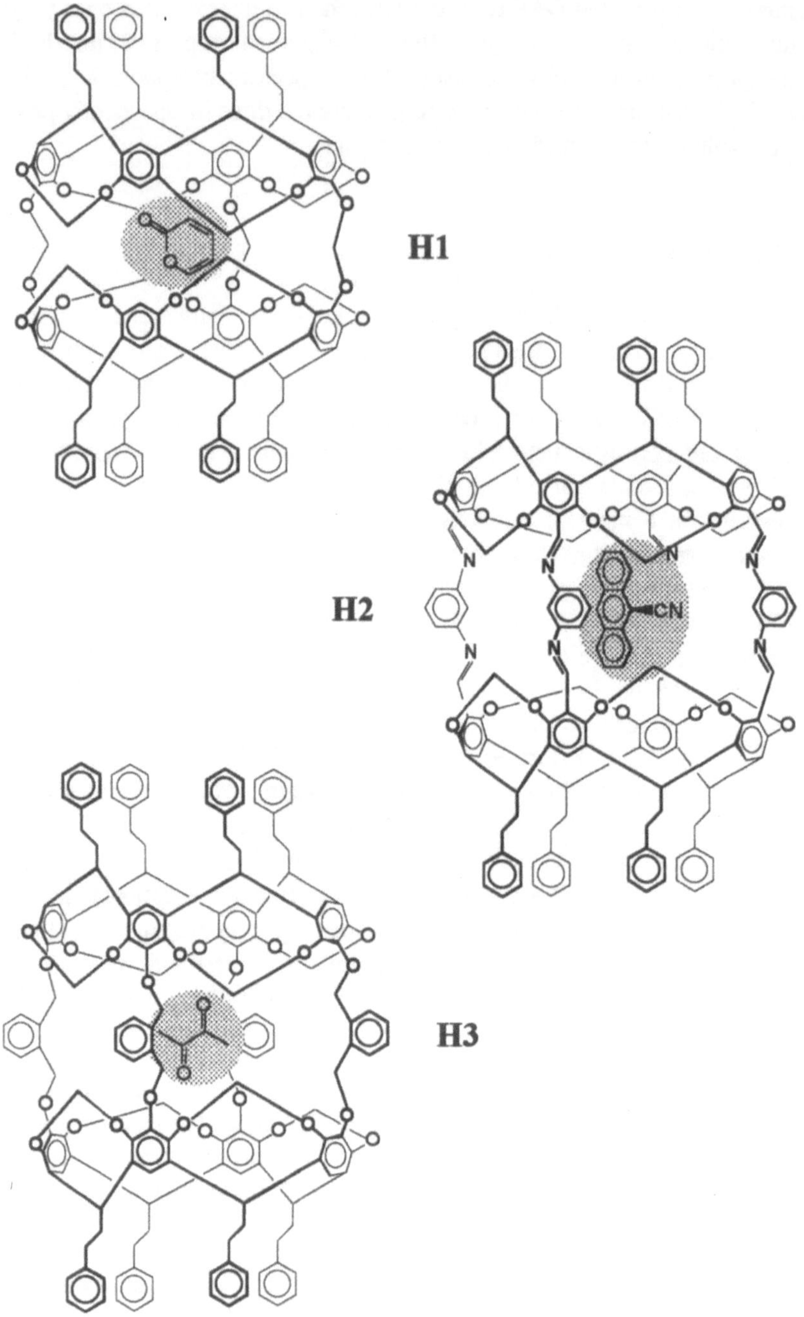

Figure 4. **Structure of the hemicarciplexes of the hemicarcerands H1, H2 and H3. For more details, see text**

through the cage walls could not be investigated.

In a successive study Pina et al. [23] have enclosed biacetyl in the hemicarcerand **H3** (Fig. 4). A first advantage of this system is that biacetyl absorbs and emits in the visible spectral region, with no interference with absorption and emission by the cage. The most important feature of biacetyl is a strong, long lived and intense phosphorescence (in deoxygenated CH_2Cl_2: λ_{max}=518 nm, τ=0.43 ms) which originates from the lowest triplet excited state, T_1, of n-π* orbital character. Both energy position and lifetime of the phosphorescence band are solvent dependent. In air equilibrated solution, however, the phosphorescence band cannot be observed because of a very efficient quenching by dioxygen (k_q= 8x10^9 $M^{-1}s^{-1}$ in benzene). When biacetyl is imprisoned into hemicarcerand **H3**, its spectroscopic and excited state properties do not depend on solvent [24]: the phosphorescence maximum of incarcerated biacetyl lies at 533 nm, and the lifetime of the T_1 excited state is 0.84 ms. Furthermore, neither the excited state lifetime nor the phosphorescence intensity are quenched by dioxygen (k_q<10^4 $M^{-1}s^{-1}$). These results show that inclusion into hemicarcerand **H3** (i) completely shields biacetyl from interaction with the solvent molecules and (ii) prevents deactivation of its long lived T_1 state by energy transfer to dioxygen. The former result is not surprising since CPK molecular models show that solvent molecules cannot penetrate into the cavity of **H3** when it is occupied by biacetyl. O_2, however, because of its smaller size should contact biacetyl through the portals of the hemicarcerand, but apparently the formation of an encounter complex of suitable distance/geometry to allow efficient energy transfer is prevented.

Since imprisoned biacetyl is completely shielded from interaction with solvent molecules, its spectroscopic and excited state properties must reflect the environment constituted by the internal cavity of the hemicarcerand **H3**. Comparison of the position of the phosphorescence maximum for the imprisoned molecules (533 nm) with the values obtained for "free" biacetyl in a variety of solvents shows that the perturbation provided by the cavity is much smaller than that provided by even an "innocent" solvent like cyclohexane. This is confirmed by the lifetime of the T_1 excited state which is longer for the imprisoned molecules than for "free" biacetyl in any solvent. The consequent picture [23] is that of a biacetyl molecule which is contained in a not too tight, rigid cavity where no specific host-guest interaction takes place. In conclusion, the peculiar spectroscopic and excited state behavior of biacetyl imprisoned in hemicarcerand **H3** supports Cram's view that the inner phase of carcerands and hemicarcerands is to be considered as a new phase of matter. It should also be pointed out that the lack of oxygen quenching on the strong and long lived phosphorescence of incarcerated biacetyl could open the way to a new family of luminescent labels for practical applications.

In another investigation [24], triplet-triplet energy transfer from acetophenone imprisoned in **H3** to *cis*-piperylene is claimed to be almost diffusion controlled (k_q= 3.7x10^9 $M^{-1}s^{-1}$), a puzzling result in view of the very low rate of energy transfer (k_q<10^4 $M^{-1}s^{-1}$, vide supra) from incarcerated biacetyl to dioxygen [23].

4. Rotaxanes, catenanes, and related species

Catenanes, rotaxanes and related species (knots, helicates, etc.) are supramolecular architectures [25] very attractive from an aesthetical viewpoint. The design of such sophisticated chemical species has since long received much attention, but only recent achievements in synthetic and analytical methods have made possible their synthesis with fairly high yields. We are engaged in a systematic investigation on the photochemical and photophysical properties of catenands, catenates, and knots prepared in the laboratory of J.-P. Sauvage (University of Strasbourg), and catenanes, rotaxanes, and pseudo-rotaxanes prepared in the laboratory of J.F. Stoddard (University of Birmingham). Before describing a few examples, we would like to recall that catenanes and rotaxanes are interesting species not only for fundamental reasons, but also because, when suitably constructed, they can behave as molecular machines that operate under the action of chemical [26] photochemical [27], and electrochemical [26a, 28] stimuli.

4.1. CATENANDS OBTAINED BY METAL TEMPLATE SYNTHESIS AROUND A METAL ION

Taking advantage of the tetrahedral-type coordination geometry imposed by the Cu^+ metal ion and its affinity for the 2,9-dianysil-1,10-phenanthroline (DAP) unit, Sauvage and coworkers [29] have prepared the metal *catenates* $Cu(2\text{-cat})^+$ and $Cu_2(3\text{-cat})^{2+}$ (Fig. 5). Demetalation of these catenates leads to the corresponding free ligands (*catenands*) where the coordinating subunits are disentangled (see, e.g., 3-cat, Fig. 5). In the cases of $Cu_2(3\text{-cat})^{2+}$, partial demetalation yields the $Cu(3\text{-cat})^+$ species, which contains a catenate and a catenand moieties. Starting from the free catenands sites, a number of catenates of other metal ions have also been prepared [29].

The luminescence properties of several catenands and catenates have been investigated [30]. The $Li(2\text{-cat})^+$ and $Zn(2\text{-cat})^{2+}$ species exhibit ligand-centered (LC) fluorescence and phosphorescence, considerably more perturbed in the case of the divalent ion. The $Co(2\text{-cat})^{2+}$ and $Ni(2\text{-cat})^{2+}$ complexes are not luminescent, as expected because of a fast radiationless decay which occurs via low energy metal-centered levels [30a]. The $Cu(2\text{-cat})^+$ and $Cu_2(3\text{-cat})^{2+}$ complexes exhibit an emission band in the red spectral region, that can be assigned to the lowest-energy triplet metal-to-ligand charge transfer (MLCT) excited state. The $Cu(3\text{-cat})^+$ species, as expected, displays two luminescence bands, which originate from the catenand and catenate moieties, respectively. The luminescence of the catenand moiety, however, is strongly quenched by the adjacent Cu-catenate unit [30c]. For $Ag(2\text{-cat})^+$ no emission can be observed at room temperature, whereas in rigid matrix at 77 K a very intense band is observed at 498 nm (τ=0.012 s) that can be assigned to the lowest ^{3}LC level. The mixed metal $CuCo(3\text{-cat})^{3+}$ species does not exhibit luminescence. This indicates that the Co-containing moiety quenches the luminescence of the Cu-containing one. In conclusion, catenands and catenanes of this family display a varied and interesting photophysical behavior. In particular, their luminescence can be tuned over the whole visible region [30].

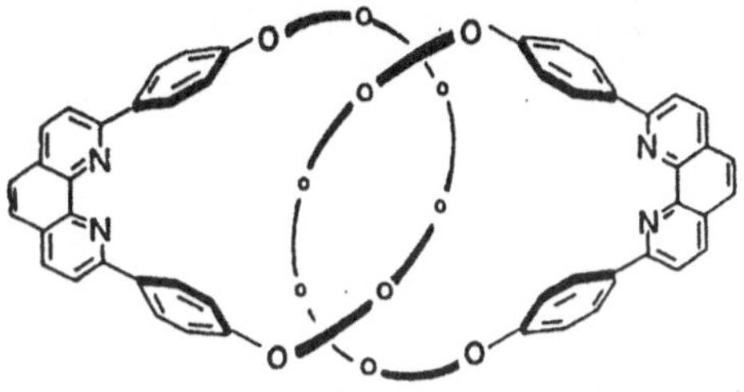

2-cat

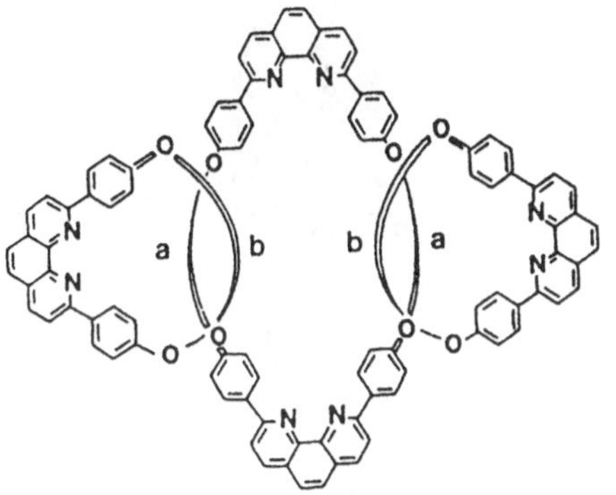

$a = -CH_2-(C\equiv C)_2-CH_2-$

$b = -CH_2CH_2-(OCH_2CH_2)_4$

3-cat

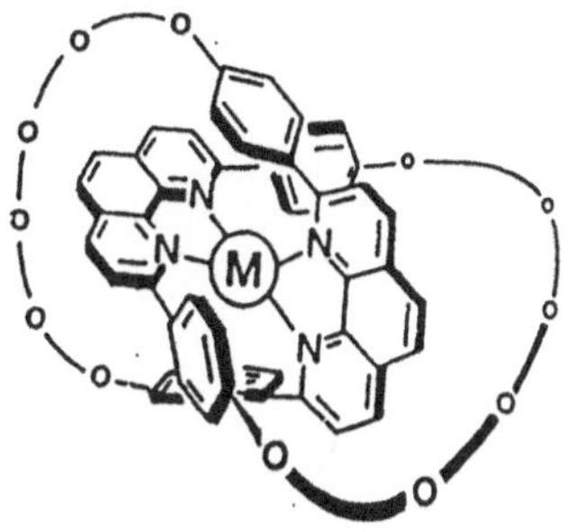

$Cu(2\text{-cat})^+$

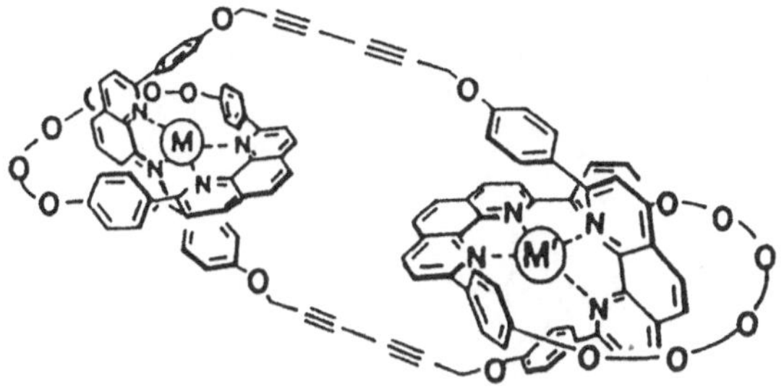

$Cu_2(3\text{-cat})^{2+}$ or $CuCo(3\text{-cat})^{3+}$

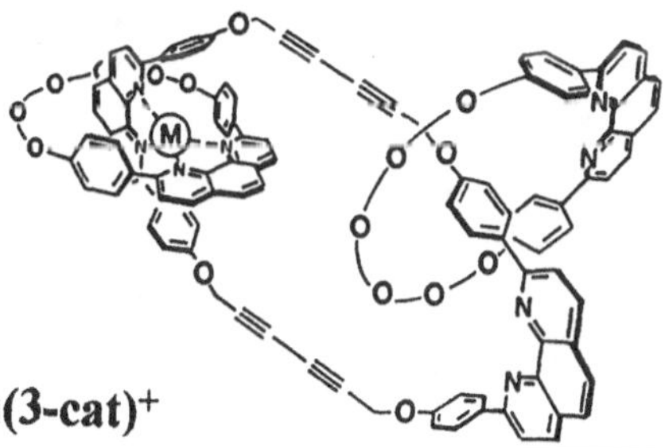

$Cu(3\text{-cat})^+$

Figure 5. Structure of catenands and catenates ($M=Cu^+$, $M'=Cu^+$ or Co^{2+})

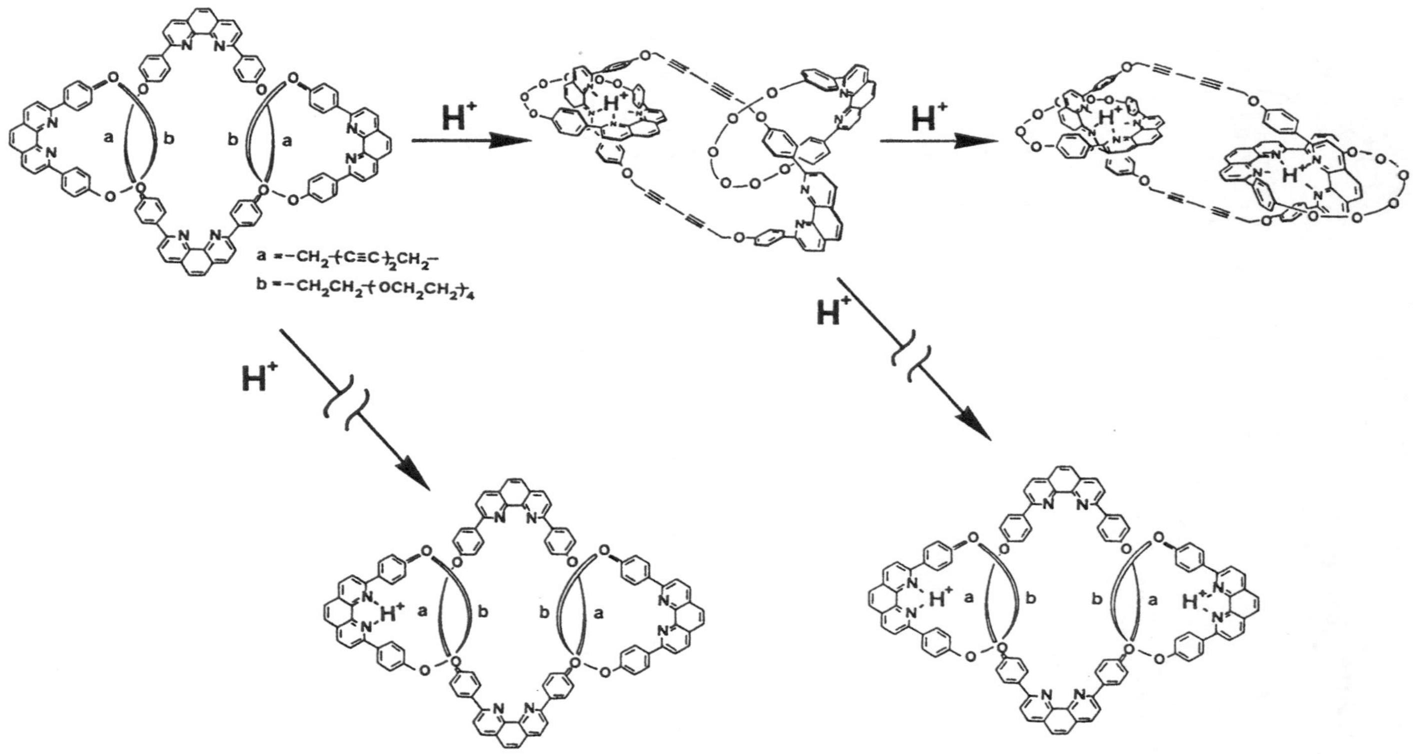

Figure 6. The first two protonation steps of 3-cat. The third and fourth protonation steps (not represented) do not lead to further structural rearrangements

As we have seen above (Section 2), DAP can be protonated. This happens also for the DAP chromophoric units of the catenands. For the [3]-catenand (3-cat), on addition of trifluoroacetic acid in CH_2Cl_2 solution four successive and distinct families of isosbestic points can be evidenced [31]. The presence of four successive protonation steps is also indicated by changes in the fluorescence spectra. Comparison of the absorption and fluorescence spectral changes observed for the protonation of isolated DAP units and of 2-cat shows that protonation of 3-cat does not involve independent DAP units, but implies a cooperative action of pairs of DAP units which organize themselves around one or two protons to form "proton catenates" (Fig. 6).

4.2. CATENANES AND ROTAXANES BASED ON DONOR-ACCEPTOR INTERACTIONS

Stoddart's catenanes and rotaxanes are based on self-assembly of electron-donor/electron-acceptor components [32]. Typical examples are shown in Fig. 7.

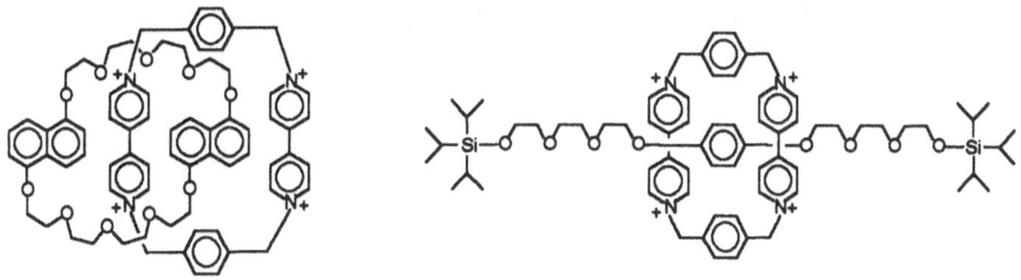

Catenane Rotaxane

Figure 7. Example of a catenane and a rotaxane based on electronic donor-acceptor interactions

Both the macrocyclic and the dumbbell-shaped components contain chromophoric, luminescent, and electroactive units. Comparative studies of the absorption spectra, luminescence properties and electrochemical behavior of catenanes, rotaxanes, macrocycles and dumbbell-shaped components, and model compounds of the chromophoric, luminescent and electroactive units are quite interesting. In fact, depending on the degree of intercomponent electronic interactions, a variety of energy- and charge-transfer processes occurs.

Consider, for example, the family of compounds shown in Fig. 8 [33]. Correlations among the reduction potentials (acetonitrile solution, room temperature) is quite instructive (Fig. 9). The different reduction behavior of 1^{4+} cyclophane and its catenanes can be accounted for by considering the charge-transfer interaction between the electron-accepting bis-(1-benzyl-4-pyridinium)ethylene (BPE) units of 1^{4+} and the electron-donating units of the aromatic crown ethers. When the cyclophane 1^{4+} is interlocked with the crown ethers, the two BPE units (i) are more difficult to reduce as a consequence of the charge-transfer interaction and (ii) are no longer equivalent since the "alongside" BPE unit interacts with only one donor moiety whereas the "inside" one interacts with two donor moieties. Therefore in the catenanes the first bielectronic wave of the free

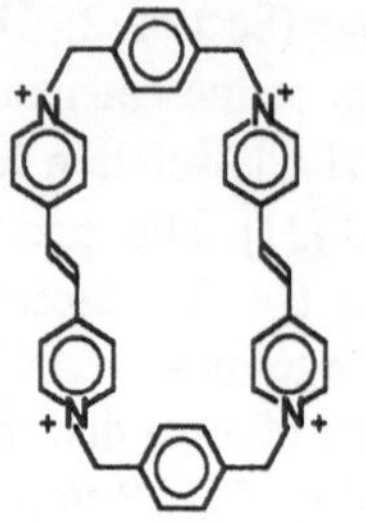

1^{4+}

BB

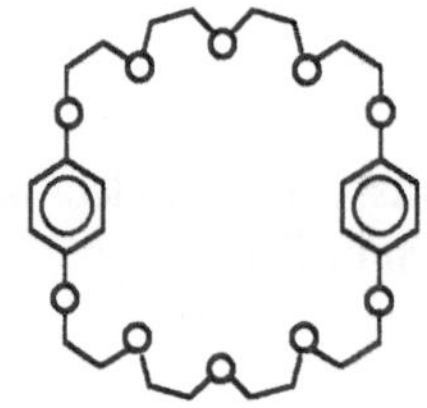

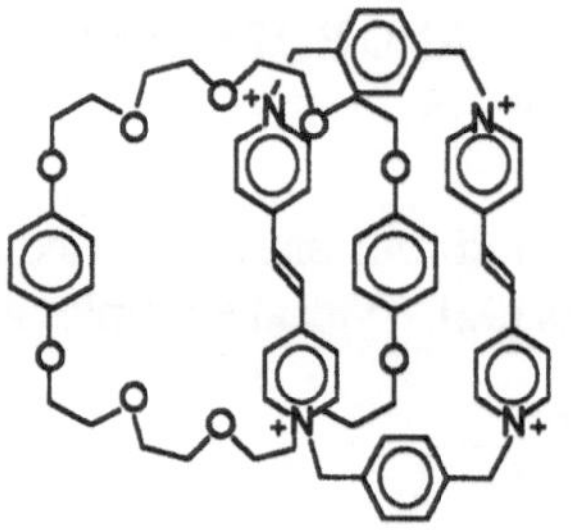

$1BB^{4+}$

BN

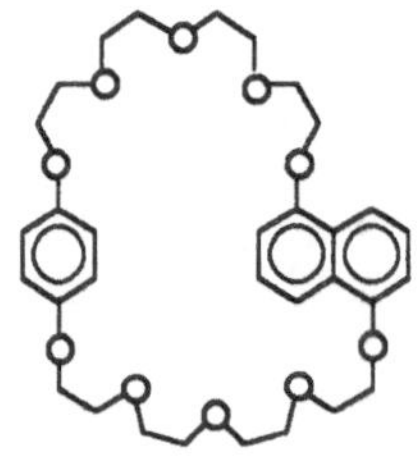

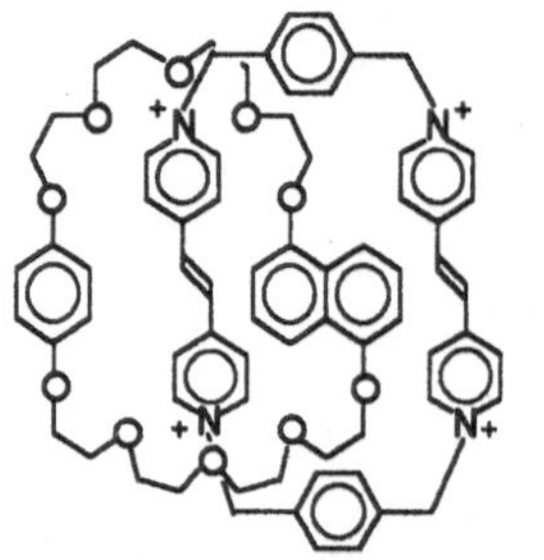

$1BN^{4+}$

NN

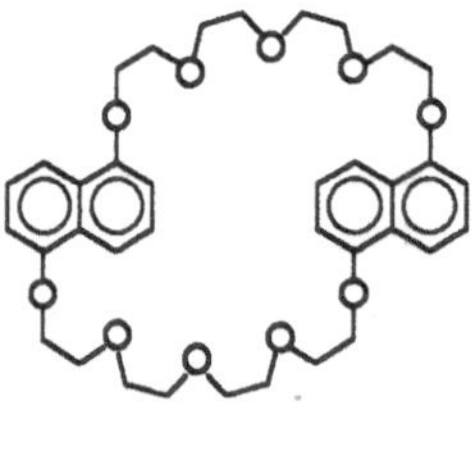

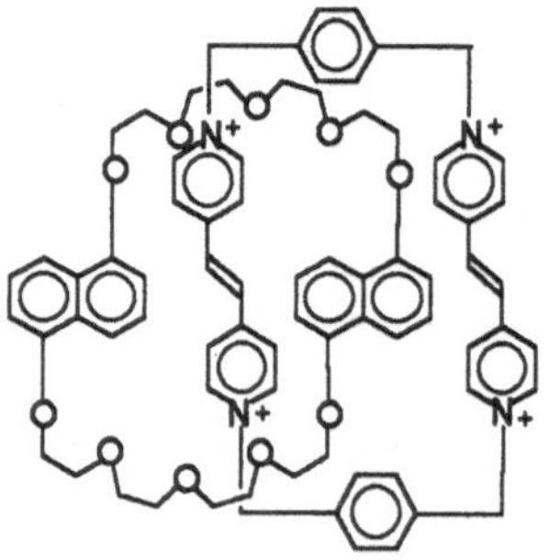

$1NN^{4+}$

Figure 8. Structure of a family of catenanes and their parent compounds cyclophane splits in two monoelectronic waves, shifted to more negative potentials. The first wave corresponds to the monoelectronic reduction of the "alongside" BPE and the second one to the monoelectronic reduction of the "inside" BPE.. The third wave in the voltammograms of the catenanes involves two electrons and must therefore correspond to the simultaneous second reduction of the two BPE units. This shows that in the catenanes the monoreduced BPE units, although they are still involved in some charge-transfer interaction (as indicated by the negative shift with respect to the second bielectronic wave of 1^{4+}), occupy equivalent positions. This is indicative of a fast rotation of the two rings. In other words, reduction of the cyclophane 1^{4+}, removing most of the charge-transfer interaction, relieves the brake which prevents free rotation.

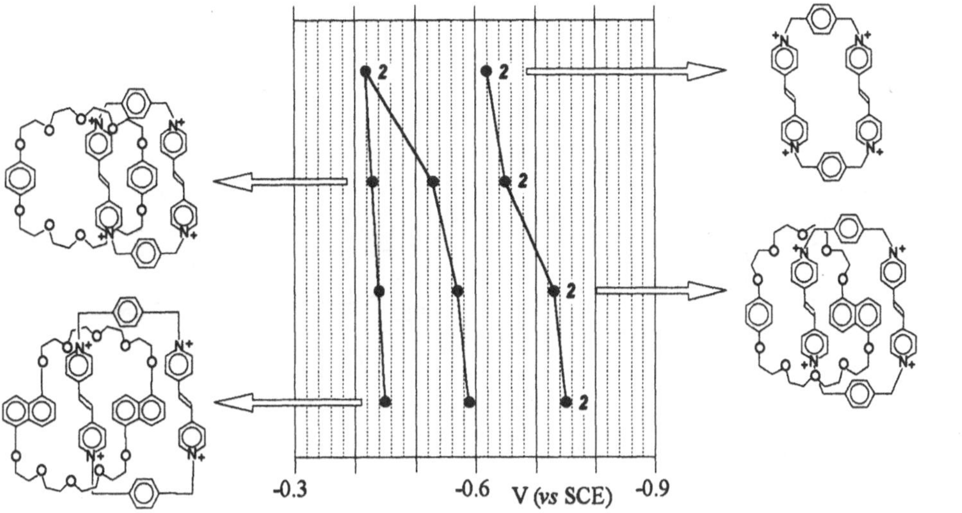

Figure 9. Correlations among the reduction potentials of 1^{4+} and its catenanes. Two-electron reduction processes are labelled with the number 2.

A detailed comparison between the behavior of the three catenanes shown in Fig. 9 evidences another interesting feature. The trend of the reduction waves of $\mathbf{1BB}^{4+}$, $\mathbf{1BN}^{4+}$ and $\mathbf{1NN}^{4+}$ shows that the BPE units become more difficult to reduce increasing the electron donor ability of the interlocked macrocycle (E_{ox}=+1.17 V for 1,5-dimethoxynaphthalene unit and +1.31 V for the 1,4-dimethoxybenzene one, *vs* SCE). Furthermore a comparison between the potentials of the first and second wave shows that the effect of the electron-donating units is more pronounced for the latter. This is due to the fact that the "inside" BPE moiety is topologically forced to experience the nature of the donor units more than the "alongside" BPE unit.

The compounds shown in Fig. 8 are quite interesting also from the photochemical and photophysical view point [33]. The results obtained (acetonitrile solution, room temperature) can be interpreted on the basis of the energy-level diagram shown in Fig. 10. The crown ethers **BB**, **BN** and **NN** exhibit a strong fluorescence. For the cyclophane 1^{4+} as well as for the BPE unit alone, direct ($^1\pi\pi^*$) excitation leads to decomposition products which originate from an intramolecular charge-transfer (CT(*intra*)) excited state. When a triplet photosensitizer (e.g., biacetyl) is used, the upper excited states are bypassed and only the $^3\pi\pi^*$ excited state is populated. Under such conditions, 1^{4+} cyclophane undergoes photoisomerization. By contrast, light excitation of the catenanes causes neither the fluorescence of the crown ether component nor the photoreaction of the cyclophane 1^{4+}. Even the use of triplet sensitizers do not cause any reaction. This behavior can be accounted for by the fact that in the catenanes low-lying intercomponent charge-transfer levels (CT(*inter*)) are present besides the levels of the two macrocyclic components. Such CT(*inter*) levels lie below the $^1\pi\pi^*$, CT(*intra*) and $^3\pi\pi^*$ levels (Fig. 10), and therefore they quench the processes which would originate from the upper states.

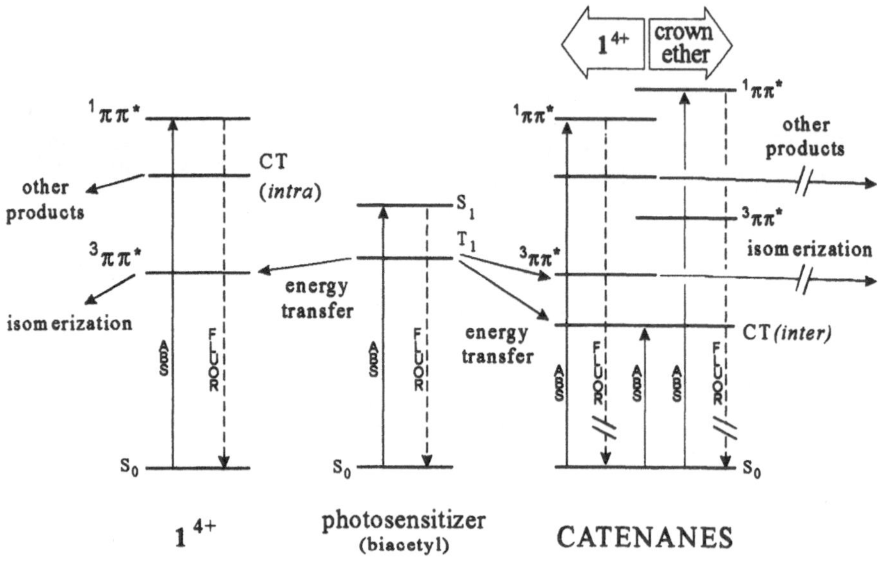

Figure 10. Schematic energy-level diagram illustrating the photophysical and photochemical processes occurring in the macrocycle 1^{4+} and in the corresponding catenanes.

5. Acknowledgments.

We would like to thank our colleagues and coworkers, whose names appear in the quoted references, for their fundamental contribution to the studies described in this paper.

6. References

1. Balzani, V., and Scandola, F. (1991) *Supramolecular Photochemistry*, Horwood, Chichester.
2. Gust, D., and Moore, T.A. (1991) *Top. Curr. Chem.* **159**, 103.
3. Balzani, V. (1992) *Tetrahedr.* **48**, 10443.
4. Wasielewski, M.R. (1992) *Chem. Rev.* **92**, 435.
5. Bissel, R.A., de Silva, A.P., Gunaratne, H.Q.N., Lynch, P.L.M., Maguire, G.E.M., McCoy, C.P., and Sandanayake, K.R.A.S. (1993) *Top. Curr. Chem.* **168**, 223.
6. Lehn, J.-M. (1995) *Supramolecular Chemistry*, VCH, Weinheim.

7. Balzani, V., and Scandola, F. (in press) in D.N. Reinhoudt (ed.), *Comprehensive Supramolecular Chemistry*, Pergamon Press, Oxford, ch. 10.

8. Balzani, V., Juris, A., Venturi, M., Campagna, S., and Serroni, S. (in press) *Chem. Rev.*

9. (a) Kotzyba-Hibert, F., Lehn, J.-M., and Vierling, P. (1980) *Tetrahedron Lett.* **21**, 941. (b) Pascard, C., Riche, C., Cesario, M., Kotzyba-Hibert, F., and Lehn, J.-M. (1982) *J. Chem. Soc., Chem. Commun.* 557.

10. Ballardini, R., Balzani, V., Credi, A., Gandolfi, M.T., Kotzyba-Hibert, F., Lehn, J.-M., and Prodi, L. (1994) *J. Am. Chem. Soc.* **116**, 5741.

11. (a) de Silva, A.P., Gunaratne, H.Q.N., and McCoy, C.P. (1993) *Nature* **364**, 42. (b) de Silva, A.P., and McCoy, A.P. (1994) *Chem. Ind.*, Dec. 19, 992. (c) Balzani, V., Credi, A., and Scandola, F. (in press) *Chim. Ind. (Milan)*.

12. Ballardini, R., Gandolfi, M.T., Prodi, L., Zappi, T., Balzani, V., Spencer, N., and Stoddart, J.F. (1991) *Gazz. Chim. Ital.* **121**, 521.

13. Lüer, I., Rissanen, K., and Vögtle, F. (1992) *Chem. Ber.* **125**, 1873.

14. Everly, R.M., and McMillin, D.R. (1991) *J. Phys. Chem.* **95**, 9071 and references therein.

15. Vögtle, F., Lüer, I., Balzani, V., and Armaroli, N. (1991) *Angew. Chem. Int. Ed. Engl.* **30**, 1333.

16. Dietrich-Buchecker, C.O., Sauvage, J.-P., Lüer, I., Frommberger, B., and Vögtle, F. (1993) *Angew. Chem. Int. Ed. Engl.* **32**, 1434.

17. Armaroli, N., De Cola, L., Balzani, V., Sauvage, J.-P., Dietrich-Buchecker, C.O., and Kern, J.-M. (1992) *J. Chem. Soc., Faraday Trans.* **88**, 553.

18. Armaroli, N., Balzani, V., Lüer, I., and Vögtle, F. (1994) *Gazz. Chim. Ital.* **124**, 17.

19. (a) Cram, D.J. (1988) *Angew. Chem. Int. Ed. Engl.* **27**, 1009. (b) Cram, D.J. (1992) *Nature* **356**, 29.

20. Cram, D.J., Blanda, M.T., Paek, K., and Knobler, C. B. (1992) *J. Am. Chem. Soc.* **114**, 7765.

21. Cram, D.J., Tanner, M.E., and Thomas, R. (1991) *Angew. Chem. Int. Ed. Engl.* **30**, 1024.

22. Parola, A.J., Pina, F., Maestri, M., Armaroli, N., and Balzani, V. (1994) *New J. Chem.* **18**, 659.

23. Pina, F., Parola, A.J., Ferreira, E., Maestri, M., Armaroli, N., Ballardini, R., and Balzani, V. (1995) *J. Phys. Chem.* **99**, 12701.

24. Farràn, A., Deshayes, K., Matthews, C., and Balanescu, I. (1995) *J. Am. Chem. Soc.* **117**, 9614.

25. Schill, G. (1971) *Catenanes, Rotaxanes, and Knots*, Academic, New York.

26. (a) Bissell, R.A., Cordova, E., Kaifer, A.E., and Stoddart, J.F. (1994) *Nature* **369**, 133. (b) Ballardini, R., Balzani, V., Credi, A., Gandolfi, M.T., Langford, S.J., Menzer, S., Prodi, L., Stoddart, J.F., Venturi, M., and Williams, D.J. (in press) *Angew. Chem. Intern. Ed. Engl.*

27. (a) Ballardini, R., Balzani, V., Gandolfi, M.T., Prodi, L., Venturi, M., Philp, D., Ricketts, H.G., and Stoddart, J.F. (1993) *Angew. Chem. Intern. Ed. Engl.* **32**, 1301. (b) Bauer, M., Müller, W.M., Müller, U., Rissanen, K., and Vögtle, F. (1995) *Liebigs Ann.* 649.

28. (a) Ashton, P.R., Ballardini, R., Balzani, V., Gandolfi, M.T., Marquis, D.J.-F., Pérez-García, L., Prodi, L., Stoddart, J.F., and Venturi, M. (1994) *J. Chem. Soc. Chem. Commun.* 177. (b) Ashton, P.R., Ballardini, R., Balzani, V., Credi, A., Gandolfi, M.T., Menzer, S., Pérez-García, L., Prodi, L., Stoddart, J.F., Venturi, M., White, A.J.P., and Williams, D.J. (1995) *J. Am. Chem. Soc.* **117**, 11171. (c) Livoreil, A., Dietrich-Buchecker, C.O., and Sauvage, J.-P. (1994) *J. Am. Chem. Soc.* **116**, 9399.

29. (a) Dietrich-Buchecker, C.O., and Sauvage, J.-P. (1987) *Chem. Rev.* **87**, 795. (b) Sauvage, J.-P. (1990) *Acc. Chem. Res.* **23**, 319.

30. (a) Armaroli, N., Balzani, V., Barigelletti, F., De Cola, L., Sauvage, J.-P., and Hemmert, C. (1991) *J. Am. Chem. Soc.* **113**, 4033. (b) Armaroli, N., De Cola, L., Balzani, V., Sauvage, J.-P., Dietrich-Buchecker, C.O., Kern, J.-M., and Bailal, A. (1993) *J. Chem. Soc. Dalton Trans.* 3241. (c) Armaroli, N., Balzani, V., Barigelletti, F., De Cola, L., Flamigni, L., Hemmert, C., and Sauvage, J.-P. (1994) *J. Am. Chem. Soc.* **116**, 5211.

31. Armaroli, N., Balzani, V., De Cola, L., Hemmert, C., and Sauvage, J.-P. (1994) *New J. Chem.* **18**, 775.

32. Anelli, P.L., Ashton, P.R., Ballardini, R., Balzani, V., Delgado, M., Gandolfi, M.T., Goodnow, T.T., Kaifer, A.E., Philp, D., Pietraszkiewicz, M., Prodi, L., Reddington, M.V., Spencer, N., Stoddart, J.F., Vicent, C., and Williams, D.J. (1992) *J. Am. Chem. Soc.* **114**, 193.

33. Ballardini, R., Balzani, V., Credi, A., Gandolfi, M.T., Prodi, L., Venturi, M., Pérez-García, L., and Stoddart, J.F. (1995) *Gazz. Chim. Ital.* **125**, 353.

PHOTOACTIVE CYCLOPHANES:

Competing with Rapid Charge Recombination

ANDREW C. BENNISTON
Chemistry Department, University of Glasgow
Glasgow G12 8QQ, United Kingdom

and

ANTHONY HARRIMAN
Laboratoire de Photochimie, Ecole Européenne des Hautes
Etudes des Industries Chimique de Strasbourg
1 rue Blaise Pascal, 67008 Strasbourg, France

1. Introduction

Illumination of a system under thermal equilibrium in the dark provides an ideal means by which to displace the equilibrium position in favour of energetic products. Of course, if the system is fully reversible the thermal equilibrium position will be re-established upon extinction of the light source and, under microscopic examination, the system will be highly dynamic. An appropriate example concerns photogalvanic cells, known since the pioneering studies of Rabinovitch in the 1930's, in which light is used to drive a reversible electron-transfer reaction in anticipation of drawing current. A second example, more successful in terms of energy storage, concerns photosensitive proteins that direct photonic energy collected by their chromophores into a photocycle that results in major conformational changes. Signal transduction by way of protein-bound photoreceptors involves a delicate balance of electrostatic, hydrophobic, and structural complementarity, often with hydrogen bonding and/or proton transfer playing an important role. We have long been concerned with combining these two aspects (namely, light-induced electron transfer and conformational exchange) into a single photosystem in which illumination drives a redox reaction that, in turn, results in a large-scale

L. Echegoyen and A.E. Kaifer (eds.), Physical Supramolecular Chemistry, 179–197.

structural change [1-3]. In this way, the time scale for restoration of the original ground-state can be prolonged, as in natural light transduction.

One approach to the design of unnatural phototactic reagents involves using cyclophanes to encapture redox-active guest molecules within their central cavity. Subsequent illumination of any redox-active chromophores appended to the cyclophane can be used to initiate an electron-transfer process resulting in transient generation of an oxidised (or reduced) guest molecule that is rapidly ejected from the cyclophane. Several such systems have been devised, based on crown ethers linked covalently to metalloporphyrin photosensitizers, with metal cations acting as the redox-active guests [1-3]. An ingenious extension of this work replaces the porphyrin with a functionalized aminobenzonitrile derivative that undergoes rapid intramolecular charge transfer upon illumination, resulting in ejection of calcium ions bound inside the azacrown ether [4]. In order for mass transfer to compete with charge recombination between the primary redox products, which is inherent in all such systems, it is necessary that this process be relatively fast. So as to construct more effective phototactic systems it is also important that we can identify and manipulate the various factors that combine to control the rate of charge recombination. We have now explored some of these factors by way of studying the photochemistry of cyclophanes retaining organic guests where the primary binding mode is of charge-transfer character.

2. Background

The relationship between the rate of charge recombination following excitation of compact charge-transfer (CT) complexes and the amount of energy to be dissipated during that process has been studied intensively in recent years. Early work by Farid, Gould and coworkers [5], using ns time resolution, was interpreted in terms of a bell-shaped energy-gap profile, as predicted by Marcus theory for non-adiabatic electron transfer after appropriate correction for quantum mechanical tunneling. Related studies by Mataga and coworkers [6], using ps and fs temporal resolution that allowed direct monitoring of the radical ion pair, showed that the rate of charge recombination decreased exponentially with increasing energy gap. Furthermore, Mataga observed only a shallow effect of energy gap on the rate of charge recombination. This behaviour was interpreted in terms of charge recombination being under vibrational control and has now been observed for many different molecular systems.

Alternative interpretations have been offered by Tachiya [7] and by Farid, Gould and coworkers [8] which are deemed to validate the application of Marcus-Jortner theory to these systems. Thus, Tachiya proposed that the rate of charge recombination in compact CT radical ion pairs was controlled by diffusional motion of solvent molecules but later studies by Miyasaka *et al.* [9] showed comparable dynamic behaviour for radical ion pairs in solution and solid states. Farid and Gould invoked the concept that the nuclear reorganization energy, an important term in Marcus-Jortner theory, increased with increasing energy gap such that the apparent exponential energy-gap profile actually arose from a series of nested bell-shaped energy-gap profiles. The latter argument seems quite reasonable when a wide energy range is studied.

With increased temporal resolution and the application of lasers operating at high repetition rates it has been demonstrated that the charge-recombination process occurring within intimate radical ion pairs might be far from simple [10]. Indeed, studies have shown the interconversion of an initially-formed Franck-Condon excited state to a thermally-relaxed radical ion pair [11]. Related studies have found non-exponential kinetics for charge recombination, perhaps because electron transfer involves solvent diffusion or takes place from high vibrational states of the radical ion pair. Time-resolved infrared spectroscopy [12] has been used to monitor the important acceptor vibrations and has indicated that certain modes can accept many quanta of vibrational energy.

Part of the problem lies with the poorly-defined stereochemistry of the parent CT complexes, at least in solution phase, and with their extremely low association constants. Indeed, the rates of charge recombination have been observed to depend on concentration of donor or acceptor and to decrease with increasing size of the reactants. To overcome these restrictions, we have incorporated the CT complexes into cyclophanes and [2]rotaxanes where the structure was clearly established by solid-state X-ray crystallography and solution-phase 2D NMR. Both donor and acceptor can be incorporated into the same cyclophane, or the macrocycle can be formed from either donor or acceptor with the complementary partner being incarcerated into the central cavity. Similarly, [2]rotaxanes can be formed with several donors incorporated into the string and with the terminal stoppers being redox-active. With such intricate molecular architectures, it seems less likely that excitation will cause substantial structural changes. Indeed, building the reactants as part of a [2]catanane provides for ae extremely compact CT complex.

3. The Reference System

N,N-Dimethyl-4,4'-bipyridinium dication (PQ^{2+}) is readily reduced, the redox potential for one-electron reduction in acetonitrile is -0.79 V vs SCE. Similarly, 1,4-dimethoxybenzene (DMB) has a low ionization potential, the redox potential for one-electron oxidation being 1.41 V vs SCE in acetonitrile. Upon mixing, PQ^{2+} and DMB form a weak CT complex, as evidenced by the development of a Gaussian-shaped absorption band having a maximum (λ_{MAX}) at 406 nm in acetonitrile. According to the results of a Benesi-Hildebrand titration, the binding constant (K) for this CT complex is *ca.* 0.5 M^{-1} while the molar extinction coefficient at the peak (ε_{MAX}) is 670 M^{-1} cm^{-1}. The half-width of this CT absorption band ($\Delta\nu_{1/2}$ = 5000 cm^{-1}), as determined by fitting the peak to a Gaussian bandshape, allows estimation of the electronic coupling matrix element (H_{DA}) as being 1750 cm^{-1} by application of Hush theory:

$$H_{DA} = \frac{0.0206\,\{\varepsilon_{MAX}\ \Delta\nu_{1/2}\ \nu_{MAX}\}^{1/2}}{R_{DA}} \quad (1)$$

Here, R_{DA} refers to the distance separating the partners within the CT complex, assumed to be 3.4 Å, while ν_{MAX} (= $1/\lambda_{MAX}$) represents the optical energy (in cm^{-1}) of the absorption transition. The value derived for H_{DA} indicates strong electronic coupling between the reactants, as might be expected for a system in which there is effective overlap of π-orbitals localized on donor and acceptor.

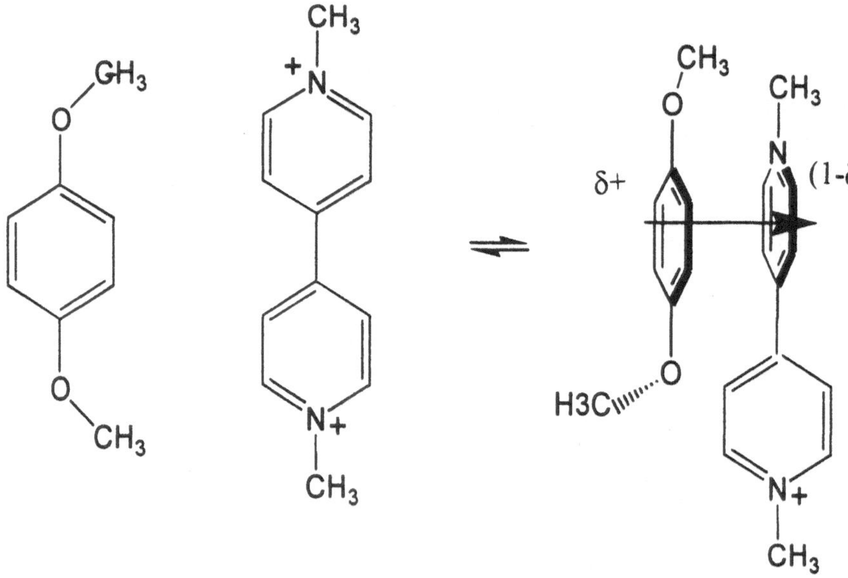

Direct illumination into the CT absorption band at 395 nm did not result in the appearance of detectable levels of fluorescence that could be supported with an excitation spectrum. However, immediately after irradiation of the CT complex with a 0.3 ps laser pulse at 370 nm the characteristic differential absorption spectrum of the reduced form of the acceptor ($PQ^{+\cdot}$) could be observed. This signal, which developed within the laser pulse, is attributed to the radical ion pair (RIP) formed by way of intermolecular electron transfer within the complex.

$$DMB + PQ^{2+} \leftrightarrow [DMB^{\delta+}...PQ^{(2-\delta)+}] \xrightarrow{h\nu} [DMB^{+\cdot}...PQ^{+\cdot}]$$

$$\swarrow \qquad \searrow$$

$$DMB^{+\cdot} + PQ^{+\cdot} \leftrightarrow [DMB^{+\cdot}.........PQ^{+\cdot}] + [DMB^{\delta+}...PQ^{(2-\delta)+}]$$

$$\downarrow$$

$$DMB + PQ^{2+} \leftrightarrow [DMB^{\delta+}...PQ^{(2-\delta)+}]$$

The RIP decayed rapidly *via* non-exponential kinetics, as monitored at 620 nm, to reform the ground-state CT complex. Despite the expected electrostatic repulsion between the radical ions within the RIP, there was minimal (i.e., *ca.* 5%) dissociation into solvated radical ions. According to the extensive studies made by Mataga and coworkers, the latter process occurs by way of a "loose" or solvent-separated RIP in which there is much less electronic coupling between the reactants. Most (i.e., *ca.* 80%) of the signal attributed to the compact RIP decayed with a lifetime of (18 ± 4) ps, although there was a very fast component having a lifetime of (1.0 ± 0.5) ps and a residual component, that decayed *via* second-order kinetics, assignable to the solvated radical cations. It is not possible, at least on the basis of these studies, to partition the derived kinetic parameters into individual rates for charge recombination and solvent penetration into the compact RIP.

While the lifetime and fractional contribution of the shortest-lived component were found to depend on the monitoring wavelength, the lifetime of the dominant component was wavelength independent. The very fast component ($\tau \sim 1$ ps) may be due to relaxation of the initial Franck-Condon excited state or to charge recombination taking place within a particularly favourable conformation of the CT complex. The main signal, however, is assigned to decay of the compact RIP and, since the yield of solvated species is rather low, it is likely that this process is dominated by charge recombination. Uncertainties about the geometry of the CT complex, together with the small binding constant, hamper further exploration of this system and attention has been turned necessarily towards better defined systems. In particular, in order to properly isolate charge recombination from solvent penetration it seems appropriate to study CT complexes where steric constraints prevent entry of a solvent molecule into the RIP.

4. A Macrocyclic Cyclophane

The next stage in development of these systems involved constraining the two redox-active reagents into a single macrocycle possessing sufficient flexibility for the reactants to form a *face-to-face* CT complex [13]. The resultant cyclophane **1** was shown by X-ray crystallography and 2D ROE to favour a closely-stacked conformation in both solid and solution phases. An intense CT absorption band, with a maximum at 397 nm in acetonitrile solution, was analysed according to eqn 1 to indicate H_{DA} = 900 cm^{-1}. Thus, electronic coupling is weaker in the cyclophane than in the CT complex formed between DMB and PQ^{2+}, presumably because of steric problems associated with adoption of the most favourable conformation. Resonance Raman spectroscopy was used to measure the nuclear re-organization energy that accompanies excitation into the CT absorption band. The derived value of 710 cm^{-1} seems rather small, consistent with minimal structural changes taking place upon irradiation, and much less than the solvent re-organization energy of 2300 cm^{-1} as determined by spectral curve fitting. The cyclophane was soluble in a small range of polar solvents, which permitted some degree of variation of the thermodynamic driving forces for forward and reverse charge transfer without seriously changing the spectral properties.

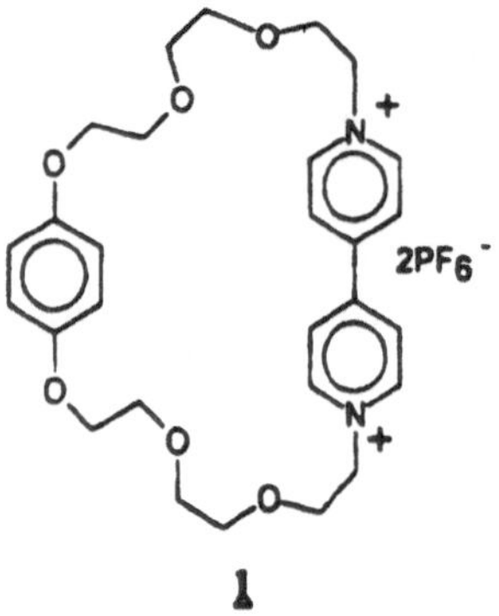

1

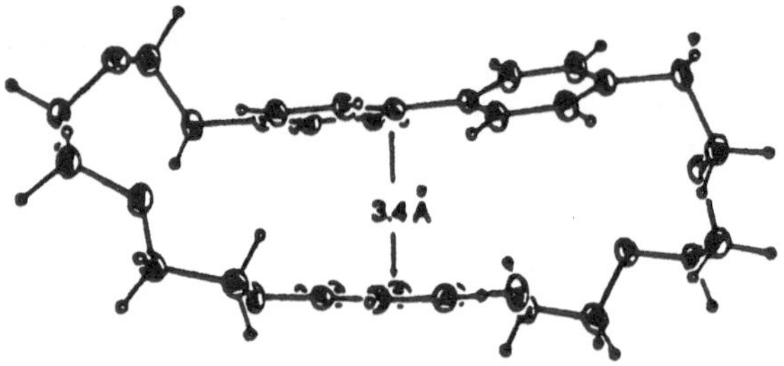

Following laser excitation into the high-energy tail of the absorption band, a transient absorption spectrum was observed that resembled the known features of the 4,4'-bipyridinium cation radical. This initial spectrum is assigned to the excited state CT complex formed upon direct excitation. Small absorption spectral shifts occur during the first few ps after excitation, followed by decay of the residual signal to the prepulse baseline. These changes are ascribed to conversion of the excited state

CT complex into an intimate RIP and subsequent charge recombination within the RIP to reform the ground-state complex [13]. The lifetime of the shorter-lived component, believed to be the Franck-Condon excited state, was (3.5 ± 1.5) ps whilst that of the RIP was (66 ± 6) ps. Whereas the lifetime of the excited state was only weakly dependent on the nature of the solvent, the rate of deactivation of the RIP was strongly solvent-dependent and increased with decreasing polarity of the solvent.

Solubility restrictions prevented detailed exploration of how the rates of formation and decay of the RIP were affected by the thermodynamic driving force. In fact, the observed rate *vs* energy gap profile could be fit equally well to several different theoretical models, each giving reasonable solutions. The overall results seem to be well-explained in terms of the potential-energy diagram shown as Figure 1. Here, excitation of the ground-state CT complex results in immediate formation of a Franck-Condon excited state. The polarity of this excited state differs from that of the ground state by *ca.* 4 D and, because of increased Coulombic repulsion between the aromatic subunits, the equilibrated conformation is more extended. Deactivation of this excited state involves, at least in part, charge transfer to form the RIP which, because of increased Coulombic repulsion, extends to the maximum separation permitted by the restraining chains. Thus, there is a stepwise separation of redox-active components and a concomitant decrease in H_{DA} from 900 cm^{-1} for the ground state to *ca.* 60 cm^{-1} for the excited state to *ca.* 14 cm^{-1} for the RIP [13].

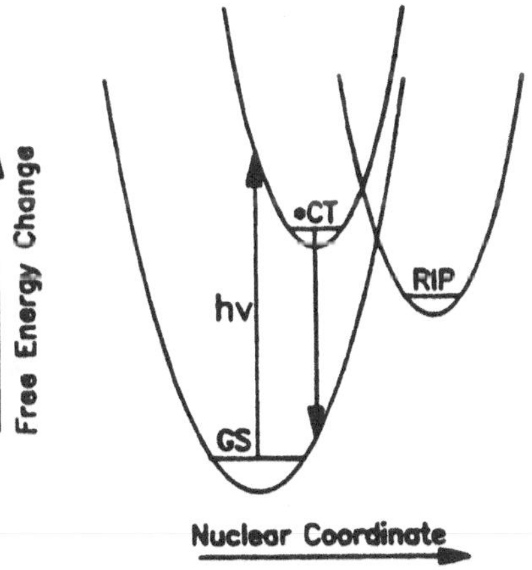

Figure 1. Potential energy diagram for direct excitation into the CT absorption band of **1**.

The time scales for these successive conformational changes must be very fast, although the actual distances are microscopic. Coherent mass transfer of this kind is promoted by induced Coulombic repulsion and it is interesting to estimate the magnitude of this effect and the size of the nuclear displacements that accompany charge transfer. In crude terms, evolution of the RIP involves each partner moving *ca.* 1 Å in 3 ps, corresponding to an apparent diffusion coefficient of *ca.* 2 x 10^{-5} cm^2/s. The approximate Coulombic repulsive energy that develops during this process is estimated as being *ca.* 1000 cm^{-1}. In order to further explore these systems, it was decided to remove the constraining chains while keeping the cyclophanic nature of the CT complex.

5. Charge-Transfer Complexes formed from Dibenzocrown ethers

It was demonstrated some years ago [14] that PQ^{2+} binds inside the central cavity of certain dibenzocrown ethers upon mixing in polar solvents and that the magnitude of the binding constant (K) was dependent on the size of the crown ether, reaching an optimum value for CE(10). We have re-examined this system in acetonitrile solution, comparing the behaviour of the crown ether to that of DMB, and thereby confirmed the relationship between cavity size and binding affinity. Thus, complexation of PQ^{2+} with the various crown ethers is manifest by the appearance of an intense CT absorption band centred around 430 nm (Table 1). The binding constant, measured by Benesi-Hildebrand analysis of the titration data, exhibits a clear correlation with cavity size while the molar extinction coefficient shows a similar dependence. The latter parameter, in part, reflects the relative orientation of donor and acceptor and it is known that, because of π-stacking between the aromatic residues, this can be optimized by matching the size of the guest with that of the cavity. For the smallest cavity size (i.e., CE(7)) the similarity between spectral properties measured with crown ether or DMB as donor is consistent with PQ^{2+} binding only to the exterior of the macrocycle.

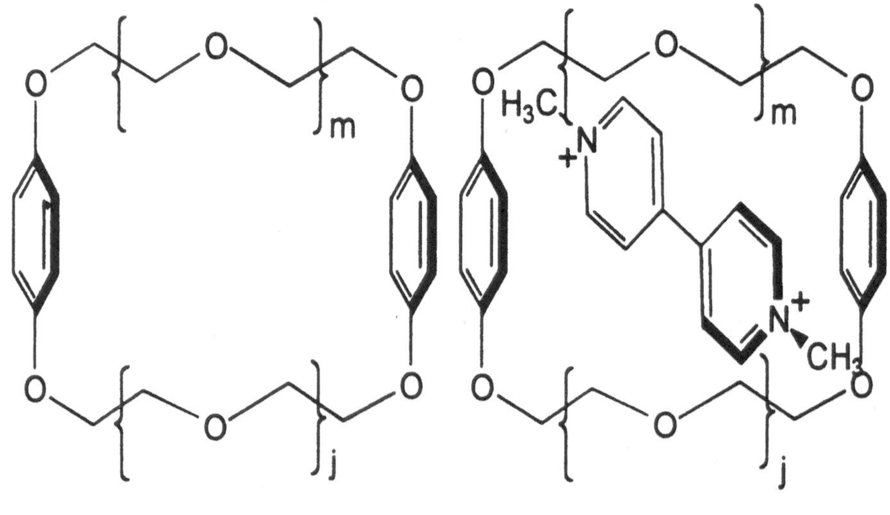

CE(7) $m = 2, j = 1$; CE(8) $m = 2, j = 2$; CE(9) $m = 3, j = 2$; CE(10) $m = 3, j = 3$; CE(11) $m = 4, j = 3$; CE(12) $m = 4, j = 4$.

TABLE 1. Spectroscopic and kinetic properties for the CT complexes formed between PQ^{2+} and dibenzocrown ethers.

Cmpd	Solvent	K (M^{-1})	λ_{MAX} (nm)	H_{DA} (cm^{-1})	λ_T (cm^{-1})	k_{CR} (10^{10} s^{-1})
CE(7)	CH_3CN	0.3	383	2065	5800	4.00
CE(8)	CH_3CN	12.0	419	1220	6600	1.03
CE(9)	CH_3CN	22.3	432	1400	5900	1.23
CE(10)	CH_3CN	48.7	440	1600	5500	1.82
CE(11)	CH_3CN	25.7	422	1445	6450	1.35
CE(12)	CH_3CN	15.5	415	1270	6850	0.95
CE(10)	CH_3NO_2	62.1	449	1690	8800	3.25
CE(10)	DMSO	35.4	418	1490	8500	2.97
CE(10)	BuCN	1235	440	1900	5100	0.90
CE(10)	$(CH_3)_2CO$	730	437	1825	4600	0.85
CE(10)	$HCONH_2$	2.9	424	1230	9300	5.50

The CT absorption bands, which correspond reasonably well to Gaussian bandshapes, were analyzed in terms of Hush theory so as to provide estimates for H_{DA} (Table 1). The derived values are seen to lie between those calculated for the unconstrained system and cyclophane **1**. As noted for the binding constants, H_{DA} depends on cavity size and reaches a maximum at CE(10). In all cases, H_{DA} is large when compared to the available thermal energy ($k_BT \sim 210$ cm^{-1}), indicating strong electronic coupling within the CT complex. For the CT complex formed between PQ^{2+} and CE(10) the binding constant was measured as a function of temperature in acetonitrile solution. The results were found to correspond to $\Delta H = -37$ kJ/mol and $\Delta S = -90$ J/mol/K. Similarly, the effect of solvent polarity was studied for complexation between PQ^{2+} and CE(10). The results, which reflect competitive solvation of the dication, are given in Table 1 and indicate that H_{DA} is quite sensitive to changes in solvent polarity. This suggests that the intimate structure of the complex might be solvent dependent.

Taking advantage of the high binding constant found in butyronitrile, cyclic voltammetry studies were made in this solvent in order to measure redox potentials for the complexed reactants. Thus, using a concentration of CE(10) = 1 mM and with a large excess of PQ^{2+} more than 99% of the crown ether is complexed. Under these conditions $E_{OX} = 1.39$ V vs SCE. Similarly, with concentrations of PQ^{2+} and CE(10), respectively, being 1 mM and 20 mM some 95% of the acceptor is complexed and E_{RED} = -

0.60 V vs SCE. The redox potential difference ($\Delta E° = 1.99$ eV) is much less than that found for the isolated species ($\Delta E° = 2.20$ eV) but still much higher than that measured for cyclophane **1** ($\Delta E° = 1.86$ eV).

The absorption maximum noted for the CT absorption band with different crown ethers and in different (polar!) solvents remains fairly constant at around 430 nm. The optical energy at the maximum (E_{OP}) corresponds to the sum of the thermodynamic driving force for reverse charge transfer (ΔG_{CT}) and the total re-organization energy (λ_T):

$$E_{OP} = [hc / \lambda_{MAX}] = \{\lambda_T - \Delta G_{CT}\} \quad (2)$$

The re-organization energy consists of contributions from a solvent-derived term (λ_S) and from a term associated with any nuclear displacement (λ_N) that accompanies CT.

$$\lambda_T = \lambda_S + \lambda_N \quad (3)$$

There seems no obvious reason why either ΔG_{CT} or λ_S should vary with the size of the crown ether; cyclic voltammetry studies showed that the respective redox potentials were independent of cavity size in butyronitrile solution. Thus, small changes in the spectral characteristics noted for the various CT complexes are attributed to changes in the nuclear re-organization energy. This seems reasonable since similar complexes are known [15] to adopt different conformations, induced by the polyoxy chains, that maximize interaction between the aromatic subunits. As such, the major effects of changing cavity size relate to variations in H_{DA} and λ_N. It is clear, however, that λ_T depends on the nature of the solvent which must indicate significant changes in λ_S and, since the absorption maximum is fairly constant, ΔG_{CT} as the solvent polarity changes.

Laser excitation into the various CT complexes caused absorption spectral changes characteristic of the RIP to be formed and then to decay over relatively fast time scales. Rate constants for charge recombination (k_{CR}) within the RIP were measured at 620 nm (Table 1), ignoring conversion of the Franck-Condon excited state into the RIP. In these systems there was no observable dissociation of the RIP into solvated radical ions and deactivation of the RIP resulted in complete restoration of the ground state CT complex. The derived rates are seen to depend weakly on cavity size, reaching an optimum at CE(10). In fact, the rates are well described in terms of the Fermi golden-rule expression

$$k_{CR} = (2\pi / \hbar) |H_{DA}|^2 \rho \quad (4)$$

which relates the rate constant to the extent of coupling between the reactants and to the density of initial and final states (ρ). Changes in the nuclear re-organization energy appear in the density term but since they mirror changes in H_{DA}, the rate decreasing with decreasing structural distortion, they tend to diminish the effect of increased electronic coupling. Also, it is important to realize that the H_{DA} terms estimated by optical absorption spectroscopy might not apply directly to the RIP.

6. Charge-Transfer Complexes with Electron-affinic Cyclophanes

So as to enlarge the range of thermodynamic driving forces for charge recombination taking place in such cyclophane-derived CT complexes and to provide a more rigid CT complex the crown ether was replaced with an electron-affinic cyclophane (*c*-DXPY) formed from two xylyl-4,4'-bipyridinium dications. Following from the extensive studies of Stoddart and coworkers [16], it is well known that this cyclophane can accommodate an unusually large number of benzenoid electron donors in its central void. With DMB as donor, the system once again resembles the basic structural unit used throughout this study. In this case, the CT absorption band is centred at 475 nm and H_{DA} = 1300 cm^{-1}. The complex is more compact than that formed by interdigitation of PQ^{2+} into CE(10) while the more positive reduction potential of *c*-DXPY lowers $\Delta E°$ to 1.91 eV. Also, the overall electronic charge on the molecule is increased such that there is likely to be strong Coulombic repulsion within the RIP. This might promote ejection of the oxidized donor [17]. The properties recorded for a series of such CT complexes in acetonitrile solution are given in Table 2.

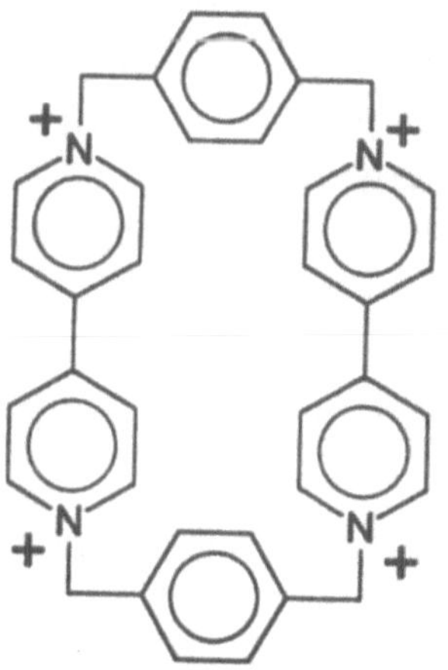

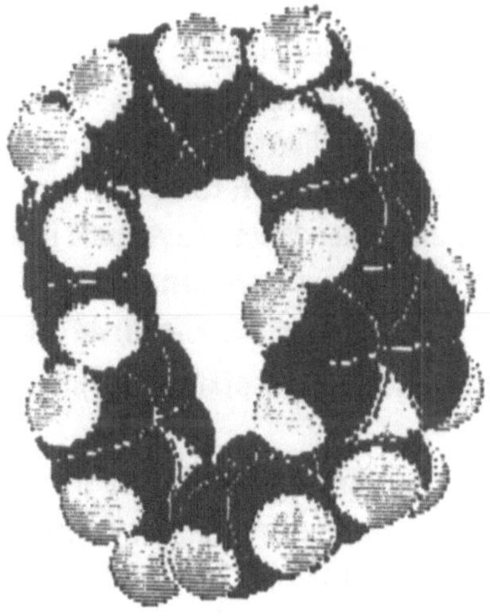

TABLE 2. Spectroscopic and kinetic properties for the CT complexes formed between *c*-DXPY and various benzenoid donors in acetonitrile.

Donor	K (M^{-1})	λ_{MAX} (nm)	H_{DA} (cm^{-1})	λ_T (cm^{-1})	k_{CR} (10^9 s^{-1})	k_{ESC} (10^8 s^{-1})
TMPD	42	820	1800	3630	248	*na*
PD	112	655	1570	5400	230	*na*
MTOL	15	601	1660	3550	145	*na*
BHQ	18	473	1670	4520	56	*na*
MEOPH	24	469	1410	4200	65	*na*
DMB	18	475	1300	3310	46	*na*
xylene	16	355	1400	4520	9.3	2.0
tBBENZ	5	333	1380	5160	7.3	10.0
FTOL	10	320	1410	5810	6.0	4.0
toluene	16	319	1280	5730	5.5	4.0
DCB	3	312	1470	5900	5.1	3.0
benzene	17	290	1400	>5600	2.6	8.0

for key to abbreviations refer to ref. 17

As above, laser excitation into the CT absorption band in acetonitrile solution results in rapid generation of the thermally-relaxed RIP, by way of a Franck-Condon excited state. There is little opportunity for small structural changes to accommodate the increased Coulombic repulsion that accompanies formation of the RIP. Rates of decay of the RIP vary over a fairly wide range and correlate surprisingly well with the optical absorption energy (Figure 2). This latter finding, which persists over a remarkably large energy gap range, suggests that the nuclear re-organization energy remains comparable for the various complexes. It also suggests that the charge-recombination process is controlled by vibrational modes. In this case, the rate of charge recombination depends on the amount of energy that has to be dissipated amongst acceptor (notably, medium frequency -C=C- skeletal) vibrations. This situation might, in fact, be the normal case for charge recombination between strongly-coupled reactants that are in close physical contact and where $-\Delta G_{CT} \gg \lambda_T$.

One disturbing feature of this system is that the total re-organization energy appears to increase systematically with increasing thermodynamic driving force (Table 2). Similar behaviour was noted previously for a small series of exciplexes in nonpolar solvent [8] and attributed to systematic changes in λ_S. This could be due to the variation in size of the electron donor used in this study but, in our case, the size and shape of the CT complex remains constant throughout the series. It is possible, therefore, that λ_N increases with increasing oxidation potential of the donor in the cyclophane-derived systems. Compared to other CT complexes λ_T seems small, probably reflecting the compactness and relative inaccessibility to solvent, and somewhat comparable to that found for CT complexes in the solid state [9].

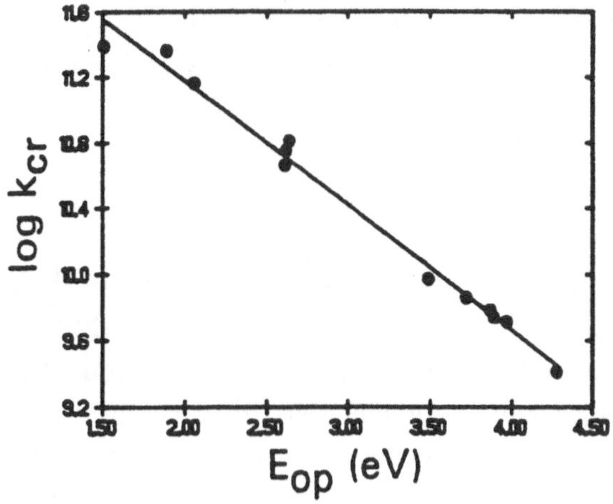

Figure 2. Energy-gap dependence for the rate of charge recombination in RIPs formed from *c*-DXPY and various donors in acetonitrile.

Of considerable interest in these systems is the observation that, at $k_{CR} < 10^{10}$ s^{-1}, the oxidized donor is ejected from the central void. Doubtless this process is favoured by strong Coulombic repulsion within the RIP. The average rate constant (k_{ESC}) for dissociation of the RIP is *ca.* 4 x 10^8 s^{-1} in acetonitrile solution at 20 °C. This rate is quite similar to those reported for dissociation of RIPs formed from simple systems, suggesting that the reduced dimensionality associated with the cyclophane is offset by the strong electrostatic forces for separation of the positively-charged reactants.

7. Restricted Motion with [2]Rotaxanes

The realization that migration of the oxidized donor could compete with rapid charge recombination within the RIP led us to consider the possibility that phototactic molecules might be constructed by incorporating the CT complex into a stoppered [2]rotaxane [18]. Appropriate systems were synthesized but, in order to be able to excite cleanly into the CT absorption band, it is necessary to ensure that the optical absorption energy is less than *ca.* 2.5 eV. This feature, in turn, means that charge recombination will be fast. In fact, for the [2]rotaxanes built from a dialkoxybenzene donor it was found that $k_{CR} \sim 5 \times 10^{10}\ s^{-1}$ which is much too fast to allow dissociation of the RIP [19].

An alternative strategy for promoting dissociation of the RIP involves incorporating redox-active terminal groups into the [2]rotaxane. If these groups associate with the exterior of the cyclophane, by way of van der Waals forces [20], secondary electron transfer can be made competitive with charge recombination, even on these fast time scales. In the absence of pre-organized structures of this type, secondary electron transfer is too slow and the RIP simply undergoes charge recombination. Other means of attaching redox-active groups to the exterior of the RIP are available [21].

8. Increased Rigidity with [2]Catananes

In order to further our understanding of the charge-recombination process taking place in compact RIPs, the synthesis of corresponding [2]catananes was undertaken. It was anticipated that such structures would be extremely compact, that absorption of light would cause only slight structural distortion, and that the total re-organization energy accompanying charge transfer would be small. Aliphatic side-chains were added in an effort to increase solubility in less polar solvents but the benefit was minimal. The X-ray crystal structure confirms the proposed interdigitation between redox-active groups. It is also clear from this structure that there is, in fact, an elongated π-stack extending over all the aromatic units. In particular, reduction potentials for the "inside" and "outside" acceptors differ by *ca.* 100 mV; variable temperature NMR studies have shown that migrational equilibration of the *c*-DXPY unit is much slower than the cyclic voltammetry time scale.

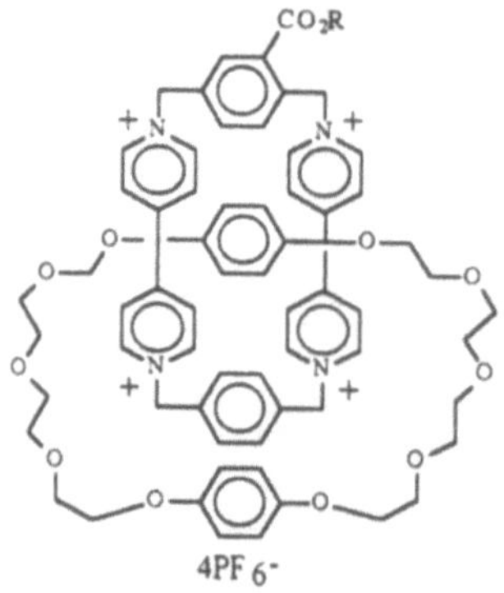

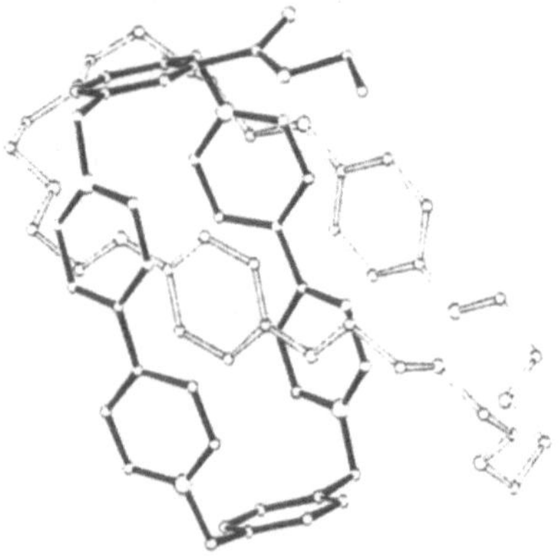

The CT absorption band is broader (half-width ~ 8700 cm^{-1}) than for the simpler analogues, perhaps because of the π-stacking that occurs. The band maximum is located at 475 nm in acetonitrile solution. Direct excitation into this absorption band generates the RIP which decays back to the ground-state CT complex with a first-order rate constant of 1.4 x 10^{10} s^{-1}. Obviously, there can be no solvent penetration into the RIP and we expect that electronic coupling remains very high despite the anticipated electrostatic repulsion. A currently unanswered question with regard to these [2]catananes concerns the extent of electron delocalization in the RIP and what role, if any, this plays in controlling the rate of charge recombination. In principle, such structures could operate as miniturized π-conductors.

9. External photosensitization

A long-term goal of this work concerns the development of well-defined, molecular phototactic systems in which absorption of light causes the coherent ejection of an organic guest from its complementary cyclophanic host. Aside from direct excitation of a host-guest CT complex, it should be possible to sensitize eviction of the guest using a photosensitizer linked covalently to either host or guest. Sensitization might be by way of energy or electron transfer from the sensitizer, although with highly electron-affinic cyclophanes it is likely that electron-transfer reactions will predominate. In order to learn about the initial photodriven steps we have studied the effects of external photosensitizers on CT complexes formed between DMB and *c*-DXPY under conditions where rapid reduction of the cyclophane takes place.

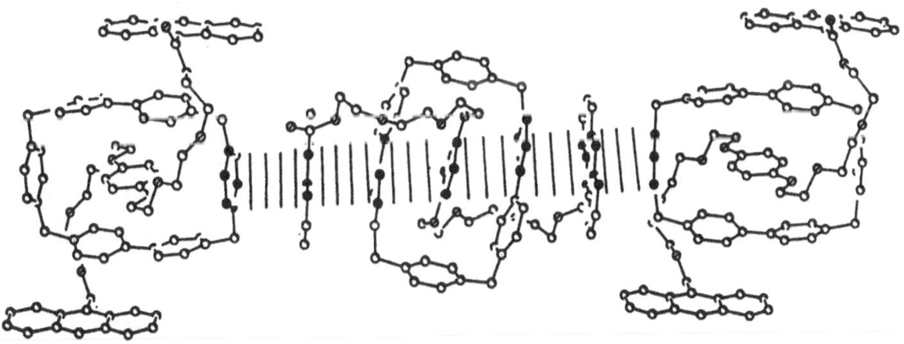

It was shown by X-ray crystallography and detailed NMR studies [18] that anthracene associates with the exterior of *c*-DPBY when the latter is part of a [2]rotaxane for which the anthracene molecules act as terminal stoppers. Excitation into the anthracene absorption bands at 355 nm results in generation of the localized excited singlet state that quickly transfers an electron to the cyclophane. The close proximity of the reactants, together with the favourable driving force for electron transfer ($\Delta G° \sim -0.71$ eV), ensures that this reaction is very fast ($k \sim 4 \times 10^{10}$ s^{-1}). Charge recombination ($\Delta G° \sim -1.97$ eV), which might take place between somewhat separated reactants, occurs on a slightly slower time scale ($k \sim 7 \times 10^9$ s^{-1}). The dialkoxybenzene donor plays no direct role in these electron-transfer reactions.

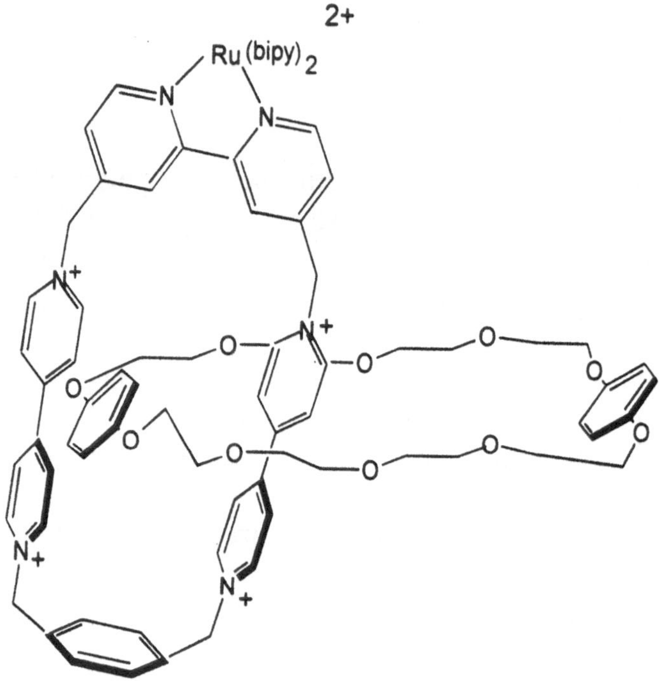

Sensitizers can also be attached to the cyclophane by way of covalent bonds. We have studied two examples; namely, the porphyrin-containing systems synthesized by Gunter and coworkers [22] and a catanane functionalized with a ruthenium(II) tris(2,2'-bipyridyl) complex. Both systems are characterized by very fast forward and somewhat slower reverse electron transfer between sensitizer and cyclophane. The catanane having the ruthenium(II) complex as external sensitizer is particularly interesting because of the inherent dielectric asymmetry introduced by the catanane. Thus, of the two appended acceptor moieties the one in direct contact with surrounding solvent is easier to reduce by *ca.* 150 mV. Since the rate of forward electron transfer in such systems increases with increasing driving force [23] we can anticipate preferential reduction of the outer acceptor. This represents the first model system to mimic this aspect of the bacterial reaction centre complex.

10. Concluding Remarks

These various structures, bearing comparable redox-active subunits built into somewhat different molecular architectures, show similar behaviour under illumination in that they are converted rapidly into a radical ion pair. The complexes differ in their internal flexibility and it is clear that minor structural changes occur within the RIP in order to provide for maximal stability. Comparison of all the available kinetic data leads us to conclude that charge recombination in these RIPs is a form of internal conversion and is controlled by vibrational processes. The most appropriate means by which to analyse the kinetic data is in the form of an "exponential energy-gap law" that allows for differences in the size of the nuclear displacement. In this respect, our findings are in exact accord with the prior work of Mataga and coworkers [6, 11].

11. Acknowledgement

We thank Alan Cowley, Vincent Lynch, Philip Magnus, and Ben Shoulders, University of Texas at Austin, for their advice and kind co-operation. Struan Gardner and Louis Farrugia, University of Glasgow, were involved in the synthesis of the catananes. We are very grateful to Fraser Stoddart, University of Birmingham, and Max Gunter, University of New England, for providing samples of some of the cyclophanes studied here. It is our pleasure to acknowledge many helpful discussions regarding the dynamics of charge recombination in CT complexes with Noboru Mataga, Hiroshi Miyasaka, and Tadashi Okada. Financial support was received from N.I.H., N.S.F., S.E.R.C., and U.L.P.

12. References

1. Blondeel, G., Harriman, A., Porter, G., and Wilowska, A. (1984) Photoredox processes in metalloporphyrin crownether systems, *J. Chem. Soc., Faraday Trans. II* **80**, 867-876.
2. Gubelmann, M., Harriman, A., Lehn, J.M., and Sessler, J.L. (1988) Photoinduced charge separation within a polymetallic supramolecular system, *J. Chem. Soc., Chem. Commun.* 77-79.
3. Gubelmann, M., Harriman, A., Lehn, J.M., and Sessler, J.L. (1990) Quenching of porphyrin excited state by silver(I) ions and charge separation in bimolecular systems and in macropolycyclic coreceptors, *J. Phys. Chem.* **94**, 308-315.
4. Mathevet, B., Jonusauskas, G., Rullière, C., Létard, J.F., and Lapouyade, R. (1995) Picosecond transient absorption as monitor of

the stepwise cation-macrocycle decoordination in the excited singlet state of 4-(N-monoaza-15-crown-5)-4'-cyanostilbene, *J. Phys. Chem.* **99**, 15709-15713.

5. Gould, I.R., Ege, D., Moser, J.E., and Farid, S. (1990) Efficiencies of photoinduced electron-transfer reactions: Role of the Marcus inverted region in return electron transfer within geminate radical-ion pairs, *J. Am. Chem. Soc.* **112**, 4290-4301.
6. Mataga, N. and Miyasaka, H. (1994) Photoinduced charge transfer phenomena: Femtosecond-picosecond laser photolysis studies, *Prog. Reaction Kinetics* **19**, 317-430.
7. Tachiya, M. and Murata, S. (1994) Non-Marcus energy gap dependence of back electron transfer in contact ion pairs, *J. Am. Chem. Soc.* **116**, 2434-2435.
8. Gould, I.R., Noukakis, D., Gomez-Jahn, Goodman, J.L., and Farid, S. (1993) Explanation of the driving force dependence of return electron transfer in contact radical-ion pairs, *J. Am. Chem. Soc.* **115**, 4405-4406.
9. Miyasaka, H., Kotani, S., and Itaya, A. (1995) Energy gap dependence of charge recombination rates of ion pairs produced by excitation of charge-transfer complexes adsorbed on the porous glass, *J. Phys. Chem.* **99,** 5757-5760.
10. Wynne, K., Galli, C., and Hochstrasser, R.M. (1994) Ultrafast charge transfer in an electron donor-acceptor complex, *J. Chem. Phys.* **100**, 4797-4810.
11. Asahi, T., Ohkohchi, M., and Mataga, N. (1993) Energy gap dependences of charge recombination processes of ion pairs produced by excitation of charge-transfer complexes: Solvent polarity effects, *J. Phys. Chem.* **97**, 13132-13137.
12. Doorn, S.K., Dyer, R.B., Stoutland, P.O., and Woodruff, W.H. (1993) Ultrafast electron transfer and coupled vibrational dynamics in cyanide bridged mixed-valence transition-metal dimers, *J. Am. Chem. Soc.* **115**, 6398-6405.
13. Benniston, A.C. and Harriman, A. (1994) Dynamics of charge transfer and recombination in a covalently-linked, face-to-face electron donor-acceptor complex, *J. Am. Chem. Soc.* **116**, 11531-11537.
14. Ashton, P.R., Slawin, A.M.Z., Spencer, N., Stoddart, J.F., and Williams, D.J. (1987) Complex formation between bisparaphenylene-(3n+4)-crown-n-ethers and the paraquat and diquat dications, *J. Chem. Soc., Chem. Commun.*, 1066-1069.
15. Allwood, B.L., Spencer, N., Shahriari-Zavareh, H., Stoddart, J.F., and Williams, D.J. (1987) Complexation of paraquat by a

bisparaphenylene-34-crown-10 derivative, *J. Chem. Soc., Chem. Commun.*, 1064-1066.

16. Bernardo, A.R., Stoddart, J.F., and Kaifer, A.E. (1992) Cyclobis(paraquat-p-phenylene) as a synthetic receptor for electron-rich aromatic compounds: Electrochemical and spectroscopic studies of neurotransmitter binding, *J. Am. Chem. Soc.* **114**, 10624-10631.
17. Benniston, A.C., Harriman, A., Philp, D., and Stoddart, J.F. (1993) Charge recombination in cyclophane-derived, intimate radical ion pairs, *J. Am. Chem. Soc.* **115**, 5298-5299.
18. Benniston, A.C., Harriman, A., and Lynch, V.M. (1995) Photoactive [2]rotaxanes: Structure and photophysical properties of anthracene- and ferrocene-stoppered [2]rotaxanes, *J. Am. Chem. Soc.* **117**, 5275-5291.
19. Benniston, A.C. and Harriman, A. (1993) A light-induced molecular shuttle based on a [2]rotaxane-derived triad, *Angew. Chem., Int. Ed. Engl.* **32**, 1459-1461.
20. Benniston, A.C., Harriman, A., and Lynch, V.M. (1994) Photoactive [2]rotaxanes formed by multiple π-stacking, *Tetrahedron Lett.* **35**, 1473-1476.
21. Benniston, A.C. and Harriman, A. (1993) Synthesis of functionalized cyclophanes *via* a self-templating effect, *Syn. Lett.* **3**, 223-225.
22. Gunter, M.J., Hockless, D.C.R., Johnson, M.R., Skelton, B.W., and White, A.H. (1994) Self-assembling porphyrin [2]-catananes, *J. Am. Chem. Soc.* **116**, 4810-4823.
23. Yonemoto, E.H., Riley, R.L., Kim, Y.I., Atherton, S.J., Schmehl, R.H., and Mallouk, T.E. (1992) Photoinduced electron transfer in covalently linked ruthenium *tris*(bipyridyl)-viologen molecules: Observation of back electron transfer in the Marcus inverted region, *J. Am. Chem. Soc.* **114**, 8081-8087.

PHOTOPHYSICAL PROPERTIES OF SUPRAMOLECULAR ASSEMBLIES IN ORGANIZED FILMS

R. M. LEBLANC
Department of Chemistry
University of Miami
Coral Gables, Florida - 33124, USA

1. Abstract

There is a growing interest in field of supramolecular biophysical chemistry in organized films. In this paper, I present some of our recent observations on molecular organization of different systems such as proteins (cytochrome *f*), and pigment-protein (chlorophyll *a*-cytochrome *c*) interactions at air-water interface and in monolayers using Langmuir-Blodgett (L-B) film, fluorescence and Fourier transform infrared, spectroscopic techniques.

2. Introduction

Recent developments in the field of supramolecular biophysical chemistry are explosive. Several studies in this area are attempting to bridge wide-range areas of science and technology. The supramolecular organization is best studied in organized monolayers. Langmuir-Blodgett technique is one of the most powerful tools to obtain supramolecular architecture. We have been applying this technique to understand the pattern of pigment organization in photosynthetic native membranes. The photosynthetic pigments such as chlorophylls and carotenoids are bound to the proteins and understanding of the pigment-protein interaction and the organization in native membranes is an important objective of our studies. Among the photosynthetic pigments, specifically chlorophyll *a* localized in the raction center and in the antenna complex of photosystems is highly efficient in energy transfer and conversion in photosystems. We have been working on photophysical and photochemical properties of chlorophyll *a*, photosystem II polypepetides and peptide-pigment assemblies in Langmuir and L-B films. Besides this, we are also working on chemical electron donor and acceptor interactions in Langmuir films. In this paper, I describe some of our recent observations on conformational stability of cytochrome *f*, and chlorophyll *a* and cytochrome *c* interactions at air-water interface.

L. Echegoyen and A.E. Kaifer (eds.), Physical Supramolecular Chemistry, 199–212.

3. Materials and Methods

Cytochrome *f* and cytochrome *c* were purchased from Sigma chemical Co. (St. Louis, MO) and were used after verifying their purity by electrophoresis in the presence of sodium dodecylsulfate. Chlorophyll *a* was extracted from fresh barley leaves and purified chromatographically according to [1]. Benzene (Accusolv, Anachemie Ltd. , Montreal) was used as a spreading solvent system for chlorophyll *a*; and the spreading solvent system for cytochrome *c* was ethanol:water (2:1, v/v). Organic solvents were distilled prior to use on a glass bead-filled column (60 cm high). For Fourier Transform Infrared (FTIR) and fluorescence studies, the monolayers were deposited on silicon and quartz slides, respectively. The FTIR spectra were recorded on a Bomem DA3-.02 Fourier Transform Infrared instrument., equipped with a nitrogen cooled HgCdTe detector and KBr beam splitter. The spectra were cooled with resolution of 1 cm^{-1} and 1000-500 scans averaged over a period of 1 hour. Triangular apodization and baseline correction were used for the infrared spectra presented here. Absorption spectra were measured with a Perkin-Elmer 553 UV-Vis Spectrometer. The fluorescence emission and excitation spectra were recorded on a Spex-Fluorolog 2 spectrofluorometer (Metuchen, NJ, Spex Industries Inc.) with a Datamate DM1 data acquisition system. Fluorescence lifetime measurements were carried out with Photochemical Research Associates Inc. International Model 3000 instrumentation (PRA, London, Ontario).

4. Results and Discussion

4.1. CYTOCHROME *F* AT AIR-WATER INTERFACE

The effect of interfacial concentration of cytochrome *f* on its Π-A and ΔV-A isotherms is presented in figure 1. Protein was deposited from an aqueous solution of 20% ethanol. the results show that, as the quantity of cytochrome *f* increase from 40 μg to 160 μg, the limiting area, obtained by extrapolating the linear part of the Π-A isotherm to zero pressure , decreases from 1300 Å^2 (curve 4 in fig 1b) to 580 Å^2 (curve 1) respectively. At the top of the linear portion of the Π-A isotherm (curve 4), the change in slope corresponding to the surface pressure of 18 mN m^{-1}, defines the collapse of monolayer of cytochrome *f*.

From the figure 1, it can be seen that, the lower the number of molecules of cytochrome f deposited the greater is the shift observed in the Π-A and ΔV-A isotherms towards larger molecular areas. This suggests that, at larger initial concentrations, the protein molecules exert a presssure on each other and reducing their spreading capability, rewsulting their desorption from the monolayer. Considering the dimensions of the trough we used, a complete coverage of the surface would require 30 μg of the protein. Thus cytchrome *f* in quantities appreciably larger than 30 - 40 μg, would result in molecules exerting pressure on each other and causing their desorption, leading to a decrease in molecular area. Our studies on surface radioactive measurements of labeled cytochrome *c* have also shown that approximately 25% of the

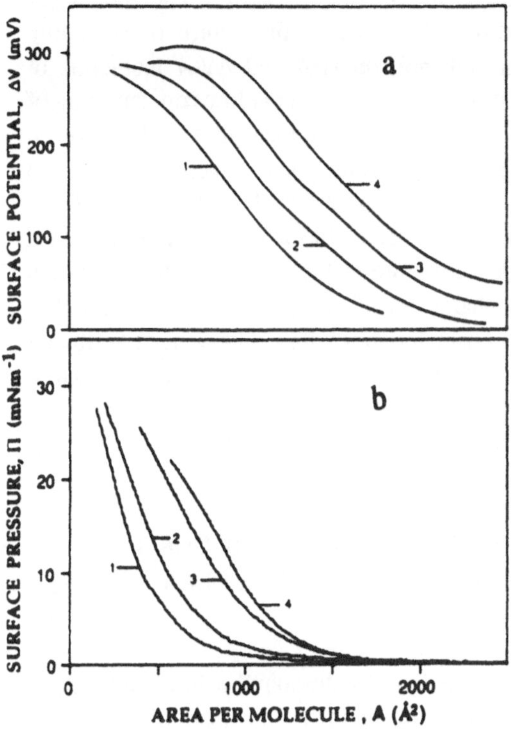

Figure 1. Surface potential ΔV-molecular area A (a) and surface pressure Π-A isptherms (b) as a function of cytochrome *f* concentration. The quantities of cytochrome *f* deposited are 160 μg (curve 1), 120 μg (curve 2), 80 μg (curve 3), and 40 μg (curve 4). The subphase was 1mM Tris-HCl buffer, pH 8.0 and temperature T = 20 ± 1°C.

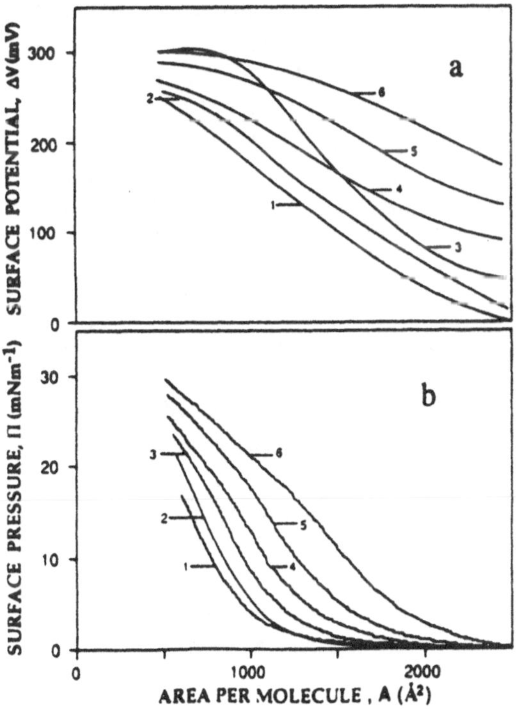

Figure 2. Surface potential ΔV-molecular area A (a) and surface pressure Π-A isptherms (b). The cytochrome *f* (40 μg) is deposited from an aqueous solution containing 0% (curves 1), 10% (curves 2), 20% (curves 3), 30% (curves 4), 40% (curves 5) and 50% (curves 6) ethanol. The subphase was 1mM Tris-HCl buffer, pH 8.0 and temperature T = 20 ± 1°C.

molecules initially deposited are solubilized in the subphase [2]. The monolayer properties of cytochrome f are also greatly influenced by the nature of spreading solvent. The effect of spreading solvent, ethanol, on Π-*A* and Δ*V*-*A* isotherms of cytochrome f are presented in figure 2. These results show that at low concentration of ethanol, less than 20%, a net solubilization of cytochrome *f* takes place. Similarly we also studied the Π-*A* and Δ*V*-*A* isotherms of cytochrome *f* as a function of pH. we noticed that the molecular areas vary depending on the pH of the subphase medium. The molecular area determined at pH 4 or 5 (close to the isoelectric point) is at a maximum, which suggests that all the molecules are at the surface. The isoelectric point corresponds to an average of zero charge for all the protein surface of water. The variations observed in Π-*A* and Δ*V*-*A* characteristics with pH have been attributed to the conformational changes in the structure of cytochrome *f* at the air-water interface. Thus, the monolayer study of cytochrome *f* at air-water interface shows that Π-*A* and Δ*V*-*A* isotherms are very sensitive to experimental parameters. Further studies on cytochrome *f* are in progress.

4.2. COMPLEXATION OF CHLOROPHYLL *A* AND CYTOCHROME *C* : ABSORPTION SPECTRA AND SURFACE PRESSURE-MOLECULAR AREA (Π-A) ISOTHERMS.

The absorption spectrum in the red region of chlorophyll *a* in ethanol [3] is characterized by a maximum at 665 nm. This corresponds to the electronic transition polarized along Y axis of the Π-electron conjugated system of the porphyrin ring, noted as QY [4]. When a small amount of ethanolic solution of chlorophyll *a* is injected into water in order to obtain an ethanol/water ratio of 3/97, v/v ($\approx 10^{-5}$ Mof chlorophyll *a*), the QY transition of chlorophyll *a* decreases in energy giving rise to a broad absorption band with maximum at 675 nm and a shoulder at 700 nm (figure 3, curve 1). These spectral changes can be assigned to the formation of a new spectral form of chlorophyll *a* molecules upon aggregation [5,6,7]. On the other hand, when the same experiment is repeated using water cytochrome *c* solution ($\approx 10^{-7}$ M), a single absorption band appeared in the spectrum at 672 nm (figure 3, curve 2). This feature can be attributed to chlorophyll *a*-cytochrome *c* complexation, since the removal of aggregated excess of chlorophyll *a* by centrifugation does not affect the maximum absorption (figure 3, curve 3). The stability of chlorophyll *a*-cytochrome *c* complex in solution is highly affected by the increasing of ethanol/water ratio. This clearly shows a complete destruction of the chlorophyll *a*-cytochrome *c* complex, since at 60% of ethanol, the absorption spectrum (figure 3, curve 4) corresponds to that of chlorophyll *a* in pure ethanol [3]. However, this solution which represents the deposition solvent for monolayer, preserves the chlorophyll *a* / cytochrome *c* molar ratio.

Surface-pressure molecular area isotherms of the chlorophyll *a*-cytochrome *c* complexes are shown in figure 4. Since the remaining number of chlorophyll *a* molecules in solution was unknown (see above), the molecular area of the complexes was calculated according to the number of cytochrome *c* molecules only. The shift of the isotherm of the mixed film compared to that of the pure cytochrome *c* (see below),

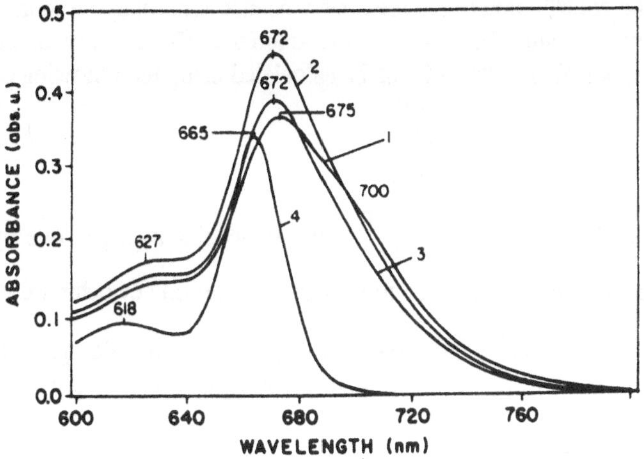

Figure 3. Absorption spectra in the red region of chlorophyll *a* in different solutions: (1) ethanolic solution of chlorophyll *a* (10-5 M) was injected to water; the water/ethanol ratio was 3/100, v/v; (2), ethanolic solution of chlorophyll *a* (10-5 M) was injected to water solution of cytochrome *c* (10-7 M); water/ethanol ratio was 3/100; (3), sample NO 2 after 10 min of centrifugation at 6000 g in order to eliminate the excess of chlorophyll *a* in aggregated form; and (4), sample NO 3 after ethanol addition upto a final concentration of 60%.

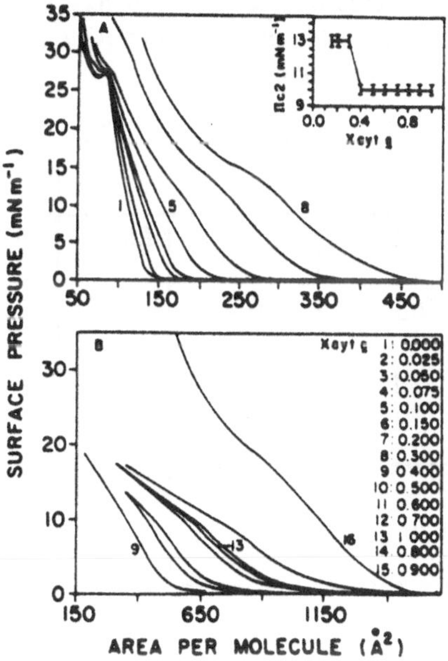

Figure 4. Surface pressure-molecular area isotherms of chlorophyll *a* and cytochrome *c* mixtures (A: 1 to 8 and B: 9 to 16) at the air-water interface. The molar fraction of cytochrome *c* is indicated in part B. The curve 16 represents the isotherm of the chlorophyll *a*-cytochrome *c* mixture prepared before deposition.
Inset: The variation of the collapse pressure, Π_{c2}as function of cytochrome *c* molar fraction X.

is related to the contribution of the previously bonded chlorophyll *a* molecules (before addition of ethanol and deposition at the interface).The number of chlorophyll *a* molecules, responsible for the shift can be calculated using the following equation:

$$n = \frac{A_c^{\Pi} - A_p^{\Pi}}{A_a^{\Pi}} \qquad (1)$$

where at the surface pressure Π, A_c^{Π} is the molecular area per one cytochrome *c* molecule in the mixture; A_p^{Π} is the molecular area per one cytochrome *c* molecule in the pure form; and the A_a^{Π} is the molecular area in the pure chlorophyll *a* monolayer. The results presented in the Table 1 show a stoichiometry of 4:1 for chlorophyll *a*-cytochrome *c* complexes.

The Π-A isotherms of pure cytochrome *c* at the air/water interface are presented in figure 4. The limiting area extrapolated to zero pressure in the linear part of the isotherm (just before the change of the slope at ≈ 9 mN m^{-1}) corresponds to 940 Å^2 [8]. The difference in the molecular area at a surface pressure of 6 mN m^{-1} reaches a value of 213 Å^2. Such a large difference has been interpreted previously as a result of sample purity [9]. However, comparing our results with those of Lamrche et al. [9,10], the molecular areas reported here are approximately 17% higher for all surface pressures less than the collapse pressure. According to our experiments, this observed difference is related to the initial surface density of molecules, known to strongly affect

TABLE 1. The mean molecular areas of one chlorophyll *a* molecule S_A, one cytochrome molecule in the mixed monolayer S_P, and the mean area per one protein molecule in chlorophyll *a*-cytochrome *c* monolayer S_C

	Surface pressure (mN m^{-1})				
	1	3	5	7	Mean
S_C[a]	1447.8	1338.4	1266.2	1203.4	
S_P[a]	1016.3	887.2	751.1	683.9	
S_A[a]	134.6	126.7	120.8	115.6	
n	3.21	3.96	4.26	4.49	3.98

Note: n is the number of the chlorophyll *a* molecules immobilized on one protein molecule, calculated by equation (1).
[a] Units Å^2 molecule^{-1}
[b] Corresponds to a protein molar fraction of 20.08%

the protein adsorption at the air-water interface at a high initial density, in which the solubilization of the protein in subphase can take place.

The slope change of Π-A isotherm observed at 9 mN m^{-1}, with a corresponding saturation potential (ΔV) and plateau at 375 mV (not shown), can be interpreted as a collapse of monolayer. For proteins, the collapse point is interpreted as the begining of a partial desorption of molecules or amino acid residues from the monolayer to the subphase [11]. A different type of collapse is observed in the case of pure chlorophyll *a* monolayer. The insoluble chlorophyll *a* monolayer is known to form a liquid-expanded film at the interface, with the characteristic, pointed collapse interpreted as a bulk phase formation. The Π-A isotherms obtained for chlorophyll *a*-cytochrome *c* mixed monolayers at different molar fractions are shown in figure 4. As it can be observed, at protein concentrations lower than 0.1, the mixed monolayers show only one collapse pressure (Π_{c1}) corresponding to that of the pure chlorophyll *a*. When the protein molar fraction reaches 0.15, an additional collapse (Π_{c2}) appears at a surface pressure of approximately 15 mN m^{-1}, which decreases to a constant value (corresponding to the collapse of the pure cytochrome *c*) when the protein concentration is higher than 40%.

As noted by the other authors, two collapse pressures, each corresponding to the ejection of one pure component, are obtained for immiscible systems [12,13]. Since this is the case for cytochrome *c* molar fractions higher than 0.1, it is clear that immiscibility and lateral phase separation take place in the film. As soon as it appears, at the critical protein concentration of 10% in the monolayer, Π_{c2} has a maximum value (15 mN m^{-1}), which is approximately 50% higher than the collapse pressure of the pure cytochrome *c* (figure 4). This increase in the collapse pressure of the mixed lipid-protein monolayers has been interpreted in the case of ovalbumin-phospholipids systems as a complex formation of lipid-protein which is more stable than the pure protein. However, the ideal stable lipid-protein complex is unlikely, since the system seems to be immiscible. Unfolding leading to protein denaturation is known to affect protein monolayers at low surface pressures [for review see 14]. Such phenomenon is investigated in mixed chlorophyll *a*-cytochrome *c* monolayers maintained at 0 mN m^{-1} before their compression. In such case, the results presented in figure 5 show an increase in molecular area and collapse pressure (12.5 mN m^{-1} to 16.5 mN m^{-1}), as the protein concentration is increased in the system. Furthermore, the collapse pressure Π_{c1} associated with the ejection for chlorophyll *a* molecules from the monolayer disappears. These observations suggest modifications of the surface properties of the cytochrome cmolecules in the mixed monolayers of 0.1 to 0.4 mole of cytochrome *c* caused by their unfolding leading to an extended denatured confirmation. This idea is also supported by the observation of an insoluble coagulum at high surface pressures (higher than Π_{c1}) in the case of the denatured chlorophyll *a*-cytochrome *c* monolayers. Surface coagulation is also reported for pure proteins [11] or mixed pigment-protein [9] monolayers compressed to high surface pressures. However, when the cytochrome *c* molar fraction is higher than 0.4 with an area contribution of more than 80%, the mixed monolayer is mainly protein in character and therefore show only one collapse pressure: that of pure cytochrome *c* (figure 5).

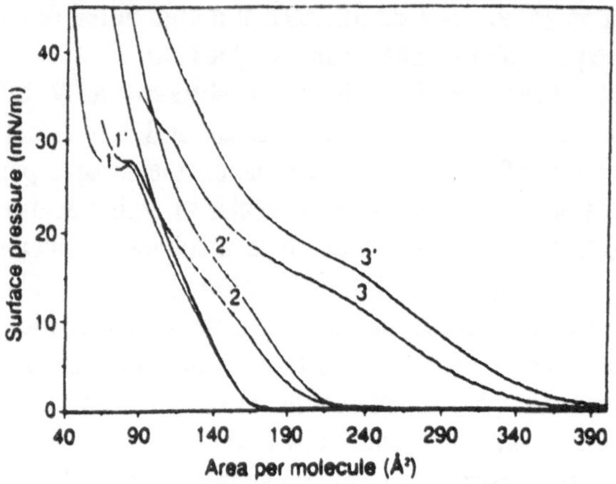

Figure 5. Π-A isotherms of mixed chlorophyll *a*-cytochrome *c* monolayers at three different cytochrome *c* molar fractions: (curve 1) 0.05, (curve 2) 0.15, and (curve 3) 0.20. 1' - 3' represent the isotherms of the same mixtures compressed 2 hours after deposition.

Further analysis of the surface pressure isotherms of the mixtures were carried out in terms of the additivity rule [12]. For the ideal mixtures the average molecular area, A_{12} at a given surface pressure, Π, in the mixed film for molar fractions X_1 and X_2, can be predicted according to the equation:

$$A_{12}^{\Pi} = X_1 A_1^{\Pi} + X_2 A_2^{\Pi} \qquad (2)$$

where: A_1^{Π} and A_2^{Π} are the molecular areas of the pure components at a surface pressure Π.

Negative deviations leading to concentrated monolayers are very common among mixed monolayers, and are interpreted as indicators of strong attractive interactions, and more thermodynamic stability in miscible systems [12,15,16], while positive deviations attributed to the weak interactions (weaker than in the pure component) in the mixed monolayer [16].

To better characterize the miscibility and have a better understanding of the thermodynamics of the interaction between the pigment and protein components, we calculated the excess free energy of mixing $\Delta G_m^{E\Pi}$ by applying the following equation [12]:

$$\Delta G_m^{E\Pi} = N \int_0^{\Pi} \left(A_{12}^{\Pi} - X_1 A_1^{\Pi} - X_2 A_2^{\Pi} \right) d\Pi \qquad (3)$$

where N is the Avogadro's constant, Π is the surface pressure below which the monolayer can be assumed to mix ideally and other parameters have the same significance as mentioned above. The excess free energy of mixing is interpreted as the energy of interaction between two components at a given surface pressure on one hand, and a complementary criterion for identification of miscibilty on the other.

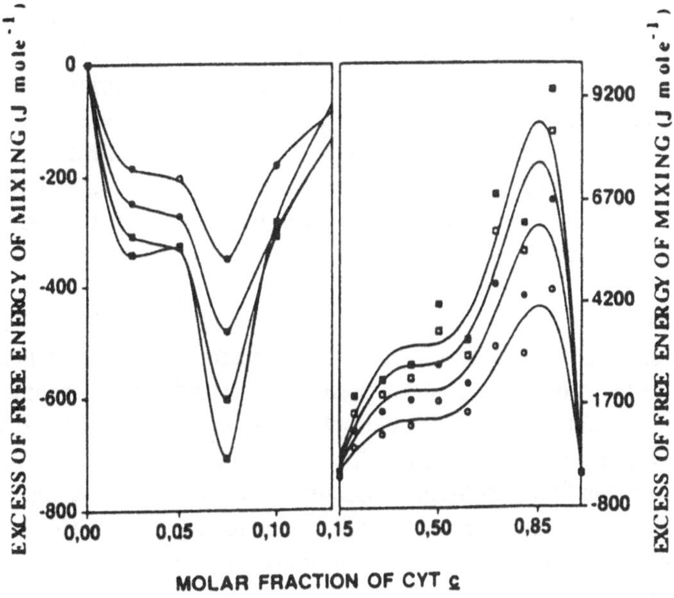

Figure 6. Excess of free energy of mixing as function of the cytochrome *c* molar fraction for chlorophyll *a*-cytochrome *c* mixed monolayers at different surface pressure: (open circles) 3 mN m^{-1}, (black circles) 5 mN m^{-1}, (open squares) 7 mN m^{-1}, and (black squares) 9 mN m^{-1}.

The results of the excess free energy of mixing deduced graphically from the Π-A isotherms of pure and mixed components are plotted against molar fractions of cytochrome *c* in figure 6. As we can see, the negative values of $\Delta G_m^{E\Pi}$ are observed for all cytochrome *c* molar fractions lower than 10% and confirm the miscibility of both components. The maximum effect observed at 0.075 of cytochrome *c* suggests that at this concentration, the pigment and protein form a complex which have the highest suitability and strongest interaction. However, this strong interaction is decreased and even inverted (positive values) when the protein molar fraction is higher than 0.15. Since at this molar fraction the system shows two well defined collapse pressures, it seems more likely that besides the pigment-protein complexes the protein rich mixed monolayer includes an excess of cytochrome *c* which may be organized as phase separated microdomains. Moreover, the positive values of

$\Delta G_m^{E\Pi}$ which increase as the protein concentration is increased beyond the threshold of miscibility, may result from repulsive forces between the two components due to the phase separation rather than to specific interactions. This idea is supported by the fact that the excess of protein is ejected from the monolayer after more than one compression-expansion cycle to high surface pressures as discussed above, and give a further evidence for the phase separation rather than stable complex formation [14,15].

From the above discussion, we can conclude that the cytochrome *c* is miscible with the chlorophyll *a* at a maximal molar fraction of 10% which corresponds to a stoichiometry of chlorophyll *a*-cytochrome *c* of 9:1. Furthermore, when the cytochrome *c* is incorporated into a chlorophyll *a* monolayer from a bulk subphase solution [10], a similar stoichiometry is deduced from the increased molecular area due to protein contribution.

4.2.1. FTIR spectra

The infrared spectra obtained in this work for cytochrome *c* in the monolayer films were similar to those of the literature reports, in solid and aqueous solution [17,18]. Similarly, the infrared spectra of the chlorophyll *a* in monolayers and L-B films were similar to those of the earlier reports [19,20]. A strong and broad band at 1652 cm^{-1} in amide I region of the free cytochrome *c* spectrum of the monolayer is attributed to the protein C=O stretching mode, in an α-helix structure (figure 7). This is consistent with the conformation of cytochrome *c*, in which the α-helical structure is more predominant, and is in good agreement with theoretical calculations [21]. However, the presence of several weak absorption bands at 1612, 1672, 1683 and 1698 cm^{-1}, in the spectrum of cytochrome *c*, can be attributed to the presence of a small amount of the β-sheet and turn structures [18]. The chlorophyll *a* monolayer also contains several infrared bands in the region of 1740-1600 cm^{-1}, which are assigned to the ester C=O, free C=O and coordinated metal C=O stretching vibrations [19,20]. Upon cytochrome *c* complexation (10%), drastic spectral changes were observed in the spectra of both cytochrome *c* and chlorophyll *a*. The amide I band of protein at 1652 cm^{-1} loses its intensity, and a new strong infrared band at 1683 cm^{-1} appears in the spectra of the chlorophyll *a*-cytochrome *c* complexes, increasing intensity, upon addition of 10%and 30% cytochrome *c*. Particularly in the 30% cytochrome *c* spectrum, the band of 1652 cm^{-1} loses its intensity, while the band at 1684 cm^{-1}, appeared as the strongest feature of the spectrum (figure 7). These spectral changes observed for the amide I region of the cytochrome *c*, are indicative of structural modifications of the cytochrome *c*. The observed changes can be interpreted in terms of an increase of the protein β-sheet and turn structures at the expense of α-helices which disappear. Such a phenomenon has been shown to characterize protein denaturation affected by pH gradient or heating. Upon binding to lipids, conformational changes from random coiled to ordered, mainly α-helical structures have been shown for several biologically active peptides and proteins including apocytochrome *c*. Such behavior confirms the transmembrane protein conformation which is α-helix. Other studies concerned with protein denaturation using a pH gradient or heating have shown a drastic conformational change mainly to β-structures

[22,23]. Such a situation is observed in the present study with molar fractions of cytochrome *c* higher than 0.1, and the spectrum of the mixed system is clearly characterized by β-structures (figure 7).

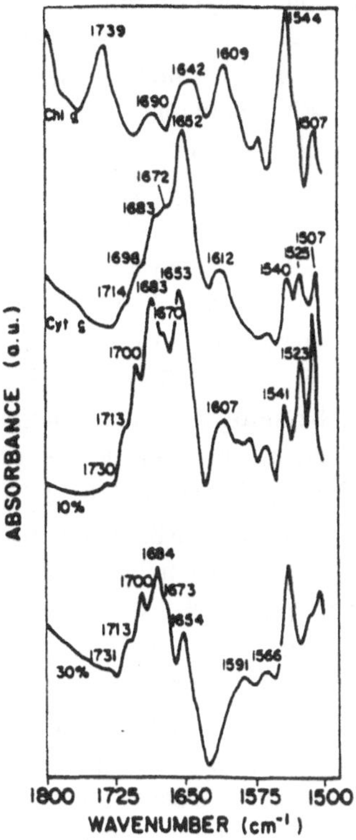

Figure 7. The FTIR spectra of the free chlorophyll *a* and cytochrome *c* monolayers, and their mixtures deposited on a silicon slide at 7.5 mN m^{-1}, at the indicated molar fraction (%) of cytochrome *c*.

Since β-structure-forms allow extensive hydrophobic interactions, protein aggregation could take place and possibly cause the positive deviations from ideality as observed above. Another infrared band at 1739 cm^{-1} in the chlorophyll *a* spectrum, related to the chlorophyll *a* ester carbonyl group, lost its intensity and shifted towards a lower frequency at about 1730 cm^{-1}. This is indicative of the participation of the carbonyl group in a stronger H-bonding network upon chlorophyll *a*-cytochrome *c* complexation. These changes can also be attributed to the participation of the carbonyl group in metal coordination and the formation of aggregated species. However, the chlorophyll *a* band at 1690 cm^{-1}, related to the free carbonyl group, shifted to about 1700 cm^{-1} in the spectra of chlorophyll *a*-cytochrome *c* mixed monolayers. A band at 1642 cm^{-1} in the spectrum of the chlorophyll *a* monolayer, which is the marker band chlorophyll aggregation, has been overlapped by the strong and broad cytochrome *c* amide I band at 1645 cm^{-1} in the infrared spectra of chlorophyll *a*-cytochrome *c* mixed monolayers. A band with medium intensity at 1609 cm^{-1} for chlorophyll *a*, and another band at 1612 cm^{-1} in the cytochrome *c* spectra are overlapped by the chlorophyll *a*-cytochrome *c* complex formation (figure 7). Since the changes observed in the IR spectra are mainly associated with the carbonyl groups of the chlorophyll and protein, it is obvious that the participation of the keto groups is very important in chlorophyll *a*-cytochrome *c* complexation. Such complex formation can occur through the participation of the protein keto groups and the magnesium ion coordination from chlorophyll *a*, or by direct H-bonding of the polar groups of the protein and the chlorophyll *a*. It is important to note that water molecules also can be involved in chlorophyll *a*-cytochrome *c* interaction.

4.2.2. Fluorescence measurements

Studying the molecular mechanism of chlorophyll *a*-cytochrome *c* interaction, we have also measured the fluorescence lifetime of the chlorophyll *a* mixed monolayer containing 20% cytochrome *c*. Measurements were performed on the monolayer at the nitrogen-aqueous buffer interface at a surface pressure of 7.5 mN m^{-1}, and for L-B films prepared at the same surface pressure. It is worth to note that our measurements of the fluorescence spectra of the L-B films deposited at 7.5 mN m^{-1} correspond well with our previous results [24]. However, in this last case the surface pressure of the deposition was 20 mN m^{-1}, and the technique of mixing was different since the cytochrome *c* was injected into the subphase underneath a monolayer of chlorophyll *a*. Furthermore, the fluorescence lifetime measured for L-B films of the chlorophyll *a*-cytochrome *c* mixture, was lower than 0.2 ns and comparable to the lifetime of the emission of pure chlorophyll *a* in L-B film. From our fluorescence lifetime measurements, in both cases, for the monolayer at the nitrogen-aqueous interface and for L-B films, we have obtained two component decay kinetics. In the first case, the lifetimes were found to be equal to $\tau_1 = 3.89 \pm 0.39$ ns (18%) and $\tau_2 = 0.33 \pm 0.13$ ns (82%), and for L-B films: $\tau_1 = 4.68 \pm 0.25$ ns (14%) and $\tau_2 = 0.33 \pm 0.07$ ns (86%). The presence of the two component fluorescence decay kinetics indicates that at least two spectral forms chlorophyll *a* are present in the system under study. We did not observe the very short lifetime (0.15 ns) reported for the pure chlorophyll *a* in monolayer [24], which is interpreted as a characteristic feature for the highly aggregated system. The major component of the chlorophyll *a*-cytochrome *c* system, which possesses the fluorescence lifetime of $\tau_2 = 0.33$ ns also can not be attributed to the contribution of aggregated form of chlorophyll *a* alone. We believe that the longer lifetime, τ_1 (14-18%), is related to some chlorophyll *a* molecules being covered by protein. This component is quite different in the measurements carried out directly on the monolayer and after deposition on the slide. The shorter component, τ_2 (82-86%) seems to be typical for the cytochrome *c* -bound chlorophyll *a* molecules, with the possibility of a common interaction decreasing the lifetime by one order of magnitude. Similar values were obtained for these components for chlorophyll *a* fluorescence decay *in vivo* [25].

In conclusion, the chlorophyll *a*-cytochrome *c* system may not directly mimic the specific functional complexes found *in vivo*. However, the findings discussed here can help to develop an understanding of the specific mechanisms involved in the chlorophyll-protein interaction.

References

1. Omata, T. and Murata, N. (1980) A rapid and efficient method to prepare chlorophyll *a* and b from leaves. *Photochem. Photobiol.* **31**, 183-185.

2. Lamarche, F. (1988) *Une approache permettant de qualifier et de quantifier les interactions lipide-protéine et chlorophylle-protéine a l'interface air-eau.* Ph. D. Thesis, Univeristy of Québec, Trois-Rivières, Canada.
3. Szalay, L., Singhal, G.H., Tombácz, E. and Kozma, L. (1973) Light absorption and fluorescence of highly diluted chlorophyll solutions. *Acta Phys. Acad. Sci. Hungaricae* **34**, 341-350.
4. Sauer, K. (1975) *Bioenergetics of Photosynthesis,* Academic Press, San Francisco.
5. Inamura, I., Ochiai, H., Toki, K., Watanabe, S., Hikino, S. and Araki, T. (1983) Preparation and properties of chlorophyll/water-soluble macromolecular complexes in water. Stabilization of chlorophyll aggregates in the water-soluble macromolecule. *Photochem. Photobiol.* **38**, 37-44.
6. Shipman, L.L., Norris, J.R. and Katz, J.J. (1976) Quantum mechanical formalism for computation of the electronic spectral properties of chlorophyll aggregates. *J. Phys. Chem.* **80**, 877-882.
7. Krawczyk, S., Leblanc, R.M. and Marcotte, L. (1988) Visible absorption, fluorescence and thermal equilibration of electronic excitation energy in Langmuir-Blodgett multilayers of chlorophyll *a* at 85 and 300 K. *J. Chim. Phys.* **85**, 1073-1078.
8. Stryer, L. (1975) *Biochemistry,* W.H. Freeman and Company, San Francisco.
9. Lamarche, F., Téchy, F., Aghion, J. and Leblanc, R.M. (1988) Surface pressure, surface potential and ellipsometric study of cytochrome *c* binding to dioleoylphosphatidylcholine monolayer at the air-water interface. *Colloids and Surfaces* **30**, 209-222.
10. Lamarche, F., Max, J.J. and Leblanc, R.M. (1988) Weak and stromg lipid protein interactions charaterized by the monolayer technique. in Ratner, B.D (ed.), *Surface Characterization of Biomaterials,* Elsevier Science Publishers B.V., Amsterdam, Netherlands, pp. 117-133.
11. Mac Ritchie, F. (1986) Spread monolayers of proteins. Adv. Colloid Interface Sci. 25, 341-385.
12. Gaines, G.L. (1966) *Insoluble Monolayers at Liquid-Gas Interfaces,* John Wiley and Sons, New York.
13. Gabrielli, G., Baglioni, P. and Maddii, A. (1981) Orientation and compatibility in monolayers I. Macromolecular compounds and fatty acids. *J. Colloid Interface Sci.* **79**, 268-271.
14. Verger, R. and Pattus, F. (1982) Lipid-protein intractions in monolayers. *Chem. Phys. Lipids* **30**, 189-227.
15. Gabrielli, G., Pugelli, M., Dei, L. and Domini, C. (1988) Mixed monolayers of polypeptides. *Coll. Polymer Sci.* **266**, 429-436.
16. Birdi, K.S. (1989) *Lipid and Biopolymer Monolayers at Liquid Interfaces,* Plenum Press, New York.
17. Yang, W.J., Griffiths, P.R., Byler, D.M. and Susi, H. (1985) Protein conformation by infrared spectroscopy: resolution enhancement by Fourier seif-deconvolution. *Appl. Spectrosc.* **39**, 282-287.

18. Byler, M. and Susi, H. (1986) Examination of the secondary structure of proteins by convolved FTIR spectra. *Biopolymers* **25**, 469-487.
19. Chapados, C., Germain, D. and Leblanc, R.M. (1980) Aggregation of chlorophylls in monolayers. part IV. The reorganization of chlorophyll *a* in multilayer array. *Biophys. Chem.* **12**, 189-198.
20. Chapados, C. and Leblanc, R.M. (1983) Aggregation of chlorophylls in monolayers. part V. The effect of water on chlorophyll *a* and chlorophyll *b* in mono- and multilayer arrays. *Biophys. Chem.* **17**, 211-244.
21. Krimm, S. and Bandekar, J. (1986) Vibrational spectroscopy and conformation of peptides, polypeptides and proteins. *Adv. Protein Chem.* **38**, 181-364.
22. Parker, F.S. (1983) *Applications of Infrared, Raman and Resonance Spectroscopy in Biochemistry*, Plenum Press, New York.
23. Surewicz, W.K. Moscarello, M.A. and Mantsch, H.H. (1987) Secondary structure of the hydrophobic myelin protein in a lipid environment as determined by Fourier-Transform infrared spectroscopy. *J. Biol. Chem.* **262**, 8598-8602.
24. Lamarche, F., Picard, G., Techy, F., Aghion, J. and Leblanc, R.M. (1991) Complex formation between chlorophyll a and cytochrome c: Surface properties at the air-water interface. Absorbance, fluorescence and fluorescence lifetime in Langmuir-Blodgett films. *Eur. J. Biochem.* **197**, 529-534.
25. Gruszecki, W.I., Veeranjaneyulu, K., Zelent, B. and Leblanc, R.M. (1991) Energy transfer process during senescence: Fluorescence and photoacoustic studies of intact pea leaves. *Biochim. Biophys. Acta* **1056**, 173-180.

CONDUCTANCE QUANTIZATION IN NANOWIRES FORMED ON MACROSCOPIC CONTACTS: FROM TABLE-TOP TO PIEZODISPLACEMENT CONTROLLED EXPERIMENTS

N. GARCÍA, J.L. COSTA-KRÄMER, P.A. SERENA,
P. GARCÍA-MOCHALES AND L. BITAR

Física de Sistemas Pequeños (CSIC-UAM)
Universidad Autónoma de Madrid, C-III
Cantoblanco, E-28049-Madrid, Spain

ABSTRACT. We have recently demonstrated (Costa-Krämer, J.L, *et al.* (1995) *Surf. Science* **342**, L1144) the existence of conductance quantization in nanowires formed on macroscopic contacts when two household wires are getting in and out of contact while they vibrate. By measuring the current through such system it is noticed that, at the last stages of breaking of the contact, the conductance varies by steps of $2e^2/h$, corresponding to the quantum of conductance. This is due to formation of nanowires with diameters of the order of the Fermi wavelength of the electrons in the wire. Experiments are presented for a wide variety of metallic contacts under different environmental conditions. These data indicate that the nanowire formation and its conductance quantization are universal and very robust phenomena. The remarkable fact is that to perform the experiment the setup reduces to two household wires, a 1kΩ resistor, a battery and a digital oscilloscope, indicating that quantum mechanical effects can be observed in a very simple way, stablishing a connection between the atomic and the macroscopic worlds. We have also performed the same type of experiments using macroscopic wires but controlling the formation and the breaking of nanocontacts with piezoelectric displacements. Under these conditions, we have studied the stability of the nanowires (reaching several hours of stability) and the I-V characteristic curves, which exhibit remarkable non-linear effects.

L. Echegoyen and A.E. Kaifer (eds.), Physical Supramolecular Chemistry, 213–227.

1. Introduction

In 1987 it was found [1] that the conductance through a small constriction in a two-dimensional electron gas (2DEG) is quantized when the constriction width varies, being the quantum of conductance equal to $G_0 = 2e^2/h$, where e denotes the electron charge and h the Planck constant. Since then, the study of the electronic transport through small metallic and non-metallic contacts has been a topic of increasing interest, due to the wide range of applications in communication and information processing technologies, where the use of nanometric and submicron integration has become essential [2].

With the advent of the Scanning Tunneling Microscope (STM) [3] and several related sophisticated techniques, it has been possible to study with extreme accuracy the transition from tunnel to contact regimes when two metals (tip and sample) approach each other. In the pioneering work of Gimzewski and Möller [4], it was shown that when the tip was brought close to the sample surface, an abrupt jump to contact at ≈ 10 kΩ was noticed. With the continuous improvement of experimental techniques, the phenomenon of the quantization of the conductance in metallic contacts has been widely studied using methods based on STM to form metallic contacts [5-9]. Conductance quantization has been also measured with the Mechanically Controllable Break Junction (MCBJ) technique [10-13]. Recently, it has been possible to induce switching behavior between tunneling and ballistic transport regimes by repeatedly bringing a sharpened nickel wire into contact with a gold surface [14]. These approaches have proven that at low and high temperatures the conductance quantization appears for several metallic species forming the contact (Au, Pb, Cu, Pt, Ni, Al, Na). However, these techniques involve the use of sophisticated equipment and control software. Very recently [15], it has been found that the same conductance quantization features appear placing two macroscopic wires in contact and making them vibrate in order to produce the formation and breaking of contacts. With this simple setup it has been possible to analyze the conductance quantization for a wide variety of metallic macroscopic contacts under very different environments (air, liquids, Ultra-High-Vacuum (UHV)) With a very similar setup, the properties of metallic liquid contacts [16] have been studied, noticing the characteristic quantization effects in these systems.

The results revealing the quantum nature of the conductance at the latest stages of the contact breaking, obtained with the use of the two macroscopic wires setup, are undistinguishable from those obtained with other experimental techniques based on the detailed control (via piezodisplacement techniques) of the separation between the two parts of the contact. This proves that, independently on the initial contact area, at the latest stages of the breaking (or first steps of the formation) of a contact, only a small contact of nanometric dimensions (nanowire) is present, with a well ordered crystalline structure. The elongation mechanism of such nanowires has been described theoretically [17] in terms of consecutive accumulation and relief (or yield) os stress stages, accompanied by structural transformations. These elongation stages are reflected as oscillations of the force during the pull-off process, and have been observed experimentally [7]. During the stress-accumulation stages, the structure of the nanowire is crystalline, determining the ballistic nature of the electronic transport through the nanoconstriction, and therefore the nanowire shape plays a predominant role when understanding the quantized nature of the conductance, as followed from different theoretical works [18-21].

Although it has been proven that with two macroscopic wires it is possible to see conductance quantization phenomena, it is very clear that the measurement procedure based on the free vibration of the wires does not permit the control of the properties of the contact at atomic level. In the present paper, we have analyzed more thoroughly the contact formed between two macroscopic wires with the help of piezodisplacement techniques characteristic of STM and MCBJ methods. With this approach, we are able to stabilize the nanometric sized contacts during time scales three orders of magnitude (several hours) higher than previously reported measurements, characterizing its electronic behaviour with unprecedented accuracy and studying the nonlinear contributions to the IV characteristic curves as a function of the number of conductance channels involved in the electronic transport process through the wire.

2. Table-top experiments

In order to record the evolution of the contact during the breaking process, we have placed the experimental setup on top of a table, and by simply hitting the table, we have induced vibrations of the metallic wires, getting them in and out of contact. The evolution of these breaking-formation processes is recorded with a digital oscilloscope. In order to measure the current flowing through the contact formed between two macroscopic wires, we have used two different experimental setups [15, 16] as shown in Figure 1. In the first case (see Figure 1a), the current is measured with an I-V converter. The conductance of the contact, G, evaluated with this setup is easily calculated with the expression

$$G = \frac{V_m}{V_{app}\, g_{I-V}} \tag{1}$$

where V_{app} is the applied voltage, V_m is the measured voltage in the oscilloscope, and g_{I-V} is the I-V converter gain. In the second setup, it is posible to

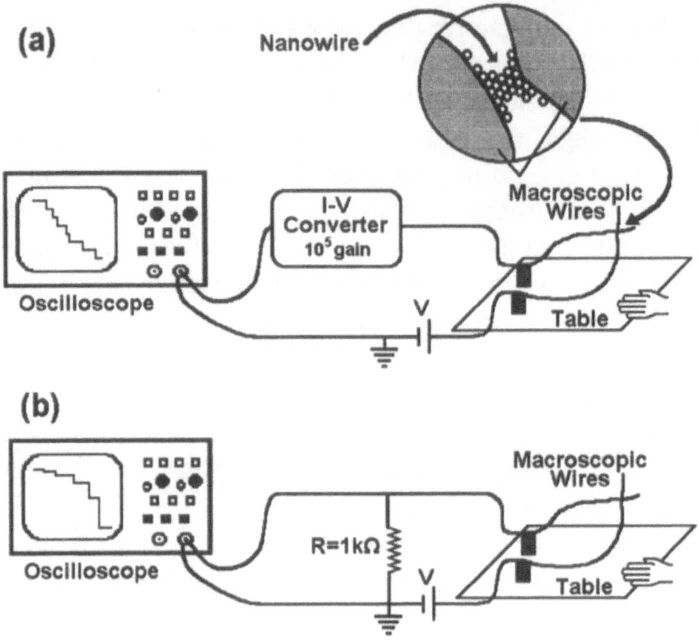

Figure 1. Schematic description of the experimental setup used to measure quantization phenomena in macroscopic contacts with: (a) an I-V converter, and (b) a resistance in series with the two wires contact.

determine the contact conductance by measuring the voltage drop across a resistor in series (Figure 1b) with the contact. With this experimental configuration, the conductance G is evaluated from,

$$G = \frac{V_m}{R_s\left(V_{app} - V_m\right)} \tag{2}$$

where R_s is the value of the resistance in series with the junction. While in the I-V converter approach, the voltage difference between the electrodes is constant, in the resistance approach, this voltage difference varies with the value of the contact resistance. In both methods, typical applied voltages range from 10 to 100 mV in order to diminish the influence of non-linear effects appearing in the I-V characterisic curves of these metallic nanocontacts for higher voltages (see Section 3).

In Figure 2 we show data obtained with the I-V setup for different macroscopic metallic wires. In all cases, the applied voltage is 90.35 mV, corresponding to a quantum current step of 7 μA. For these experiments, the rise time of the I-V converter is 3 μs at 10^5 gain. As shown, the current

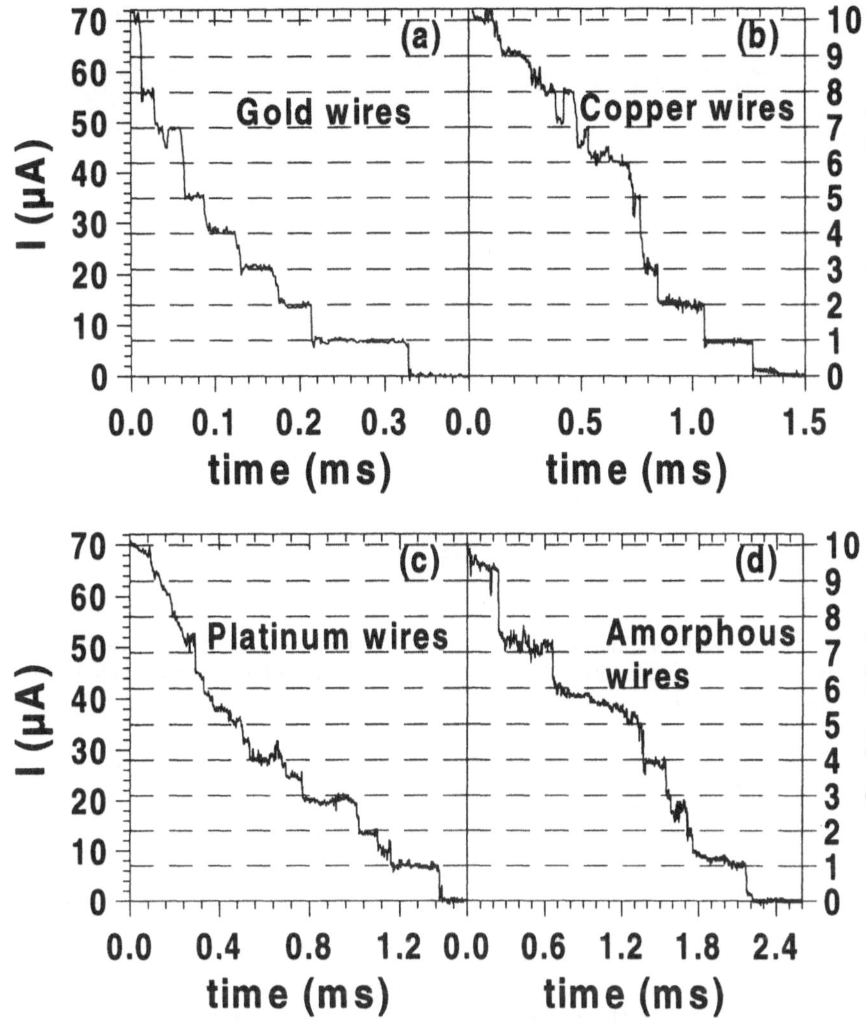

Figure 2. Conductance experiments showing the time evolution of the current and the conductance at the latest stages of the separation of two macroscopic wires. These curves were recorded using the the setup described in Figure 1a with four different metal species forming the contact: (a) gold (0.5 mm diameter wire) , (b) copper (0.7 mm diameter), (c) platinum (0.25 diameter), and (d) amorphous ($Co_{94}Fe_{6})_{72.5}Si_{12.5}B_{12.5}$ (0.125 mm diameter). The applied voltage difference between the wires was 90.35 mV.

flowing through the contact decreases as the contact breaks, until total separation of the wires is reached (zero measured current). As the macroscopic wires separate, thousands of nanocontacts stretch out, breaking eventually. At the latest stages, the remaining contacts are the strongest (from a mechanical point of view), and, like in a Darwinian selection, only the fittest survive. This final contact presents remarkable stability and mechanical

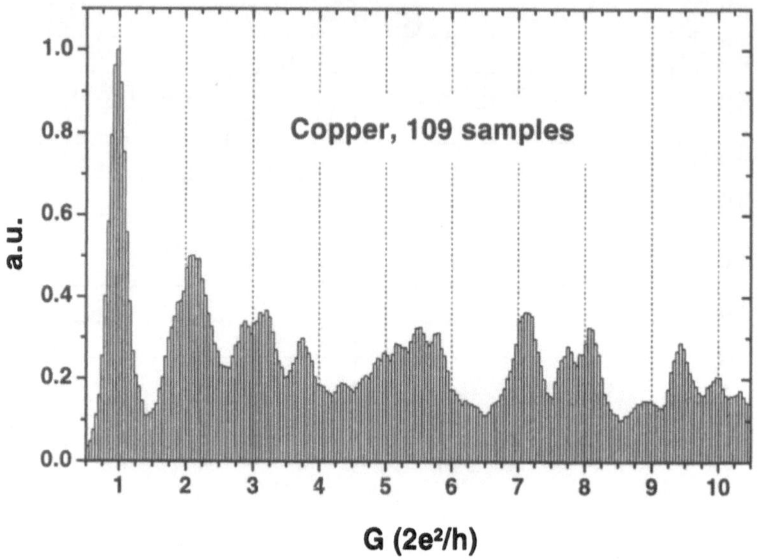

Figure 3. Histogram of conductance values for copper-copper contacts obtained with 109 different measurements. Each conductance quantum unit has been divided into 22 equally spaced intervals.

properties. As this nanowire thins out, the current decreases in well defined steps. Most of the conductance jumps appearing in Figure 2 correspond to integer values of the quantum unit of conductance ($G_0 = 2e^2/h$), reflecting the quantum nature of the electronic transport through these small constrictions. Smaller conductance jumps can be interpreted in terms of the presence of impurities, certain degree of disorder, or internal atomic rearrangements in the thicker contact areas due to stresses.

Note that for gold, copper, and platinum contacts (see Figure 2a-c) the conductance is stepped up to the 10th quantum channel, while for the amorphous metal, a more continuous dependence is observed (except the very latest stages). For gold, conductance curves show stepped behaviour up to the 40th channel, with most plateaus falling at integer values of G_0. This is in agreement with previous simulations for gold nanowires [17], where the disorder predicted was small. The importance of this behaviour is that shows that quantized conductance jumps happen in any contact between two metals, independently on the initial contact area, and therefore, the last stages of the breaking of such contacts are always characterized by the formation of a nanowire, with well defined conducatnce quantization features.

In Figure 3, we show a histogram, made with more than 100 different

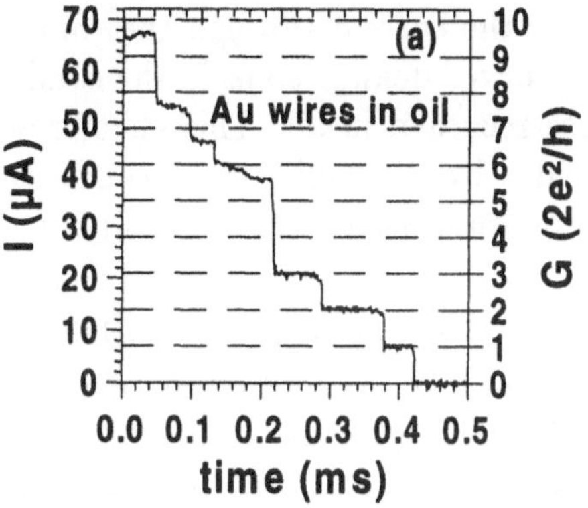

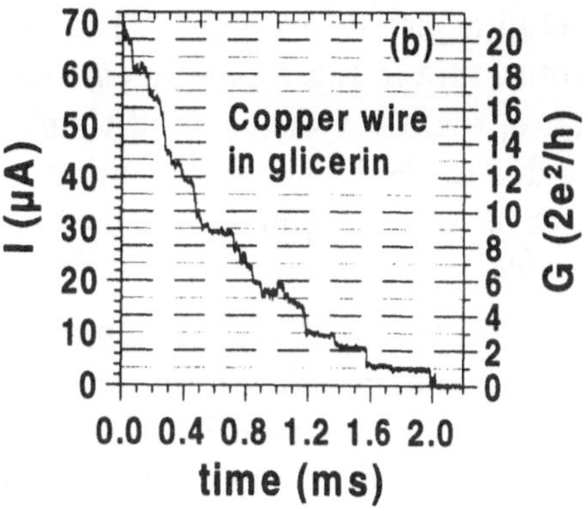

Figure 4. Conductance experiments showing the time evolution of the current and the conductance at the latest stages of the separation of two macroscopic wires immersed in a liquid: (a) gold wires in oil (Cepsa Diesel 20-40), and (b) copper wires in glycerine. The data were recorded using the setup described in Figure 1a. The applied voltage difference between the wires was 90.35 mV in (a) and 44.0 mV in (b).

experimental curves, describing the breaking of contacts formed between two copper wires. Each bar in the histogram represents the number of occurences of a given conductance interval (expressed in units of G_0) when a wide set of experimental curves (to get a reliable statistical analysis) are taken into account. It is worth to see that peaks appear close to integer values of the quantum of the conductance G_0, i.e., at well defined quantum numbers. However, this histogram cannot be compared directly with those obtained from other experimental techniques, since in these, the speed at which the junction is broken is well controlled, at variance with our kind of measurements, where the breaking process takes place many times in a single experiment but without direct control. Therefore, it is reasonable that histograms obtained with different experimental techniques present some differences, although, some common features are distinguishable. For instance, the relative high strength of the first peak (at $G = G_0$) in relation to the other peaks, indicating that the underlying structure is more stable from a mechanical point of view than those appearing for higher conductance channels. Although Figure 3 only shows the histogram corresponding to copper wires, qualitatively similar histograms, with analogous well defined peak structures, are found when studying gold, platinum, or amorphous metallic contacts [15].

To illustrate the universality of this phenomenom, we have studied the formation of these nanocontacts under different environmental conditions. In Figure 4 we present results for gold wires immersed in oil (Figure 4a) and for copper wires immersed in glycerine (Figure 4b), using an I-V converter (setup depicted in Figure 1a) to measure the current through the contact. Notice that in these experiments there are well defined steps in the conductance, slightly more marked than those recorded in air. This effect could be due to the higher stabilization provided by the liquid in the conductive neck or to the damping of the wires vibration caused by the viscosity of the liquids.

After having described some experiments performed with the I-V converter setup, in Figure 5 we show experimental curves for the time evolution of the conductance during the contact breaking process obtained with the setup depicted in Figure 1b. In this case, the current is measured with a resistor ($R_s = 1\text{k}\Omega$) in series with the metallic junction. The conductance curve for mercury (Figure 4a) was obtained at room temperature while the one for liquid tin (Figure 4b) was measured at ≈ 300 °C. In both cases, a copper wire was immersed in the liquid metal and other one was driven in and out of the liquid. It is easy to see that in this case there are fewer plateaus, in comparison with the previous experiments, although they appear more markedly. Perhaps, the initial absence of crystallinity could drive to the formed neck towards a series of metastable configurations (local minima) during the pulling process, accessing less atomic configurations in comparison with the solid-solid contacts. Another evident fact is that curves obtained with this method (using the resistor) have more noise than those obtained with the I-V converter; noticing how the noise increases as the conductance does. This can be explained in terms of Equation 2: the conductance is inversely proportional to the difference between the applied and the measured voltages, and therefore, when the voltage drop across the resistor becomes closer to the applied one, the error in the measured conductance increases considerably.

3. Piezocontrolled experiments

After describing the main results obtained in experiments without any control in the separation process of the two macroscopic wires, we proceed to design a new experiment controlling the approximation and separation between the macroscopic wires. Two Au wires (with 125 μm diameter and a purity of 99.99%) are placed within an ultra high vacuum (UHV) chamber (10^{-10} torr.), as shown in Figure 6. The separation between both wires is controlled by a comercial piezomotor inchworm which allows the movement (from nm to cm) along the z direction, whereas it is possible to control lateral displacements along the xy plane (from Å to cm) with the combined use of two quartered piezotubes and a two step motor. This UHV system has been suitably isolated from external mechanical vibrations, since it was designed as a part of a Fresnel Projection Microscope [22]. Before perform-

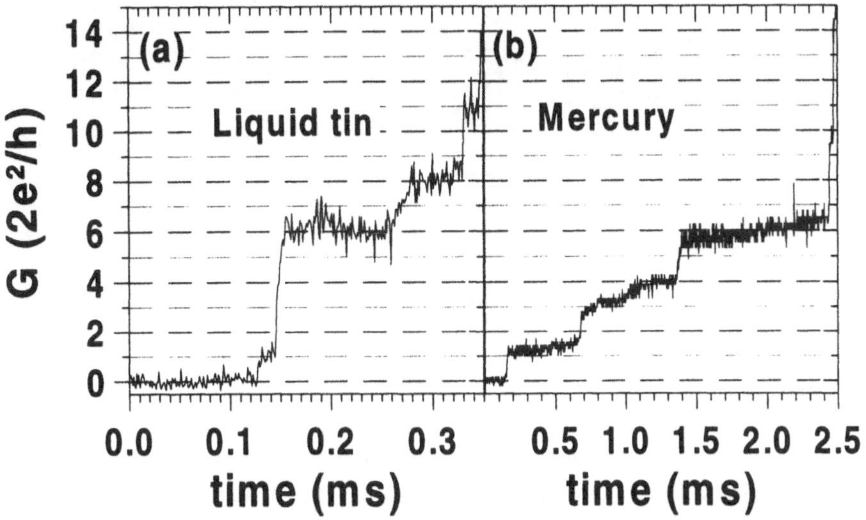

Figure 5. Conductance experiments showing the time evolution of the conductance at the first stages of the formation of a liquid contact: (a) liquid tin at 300 °C, and (b) mercury at room temperature. The data were recorded using the setup described in Figure 1b. The applied voltage difference between the wires was 50 mV in both experimnets.

ing any measurement, both wires were cleaned by heating them up to 600 °C. All the data presented in this work have been acquired at room temperature. With the help of the z-nanodisplacement control, we bring the wires into contact while the current flowing through the contact is simultaneously recorded. Once the contact is stablished, the z-displacement is reversed and the formation and later elongation of the nanoneck is monitored as change in the conductance in well defined $2e^2/h$ steps, being this behaviour similar to that observed for metallic contacts formed in air [15].

The conductance measurement is carried out by using two different techniques as it has been described in Section 1. On one hand, it is possible to measure the current through the contact by using a current-voltage (I-V) converter with 10^5 gain. However, this setup limits the accesible currents, preventing the study of the I-V behaviour at high applied bias voltages (upto 250 mV). On the other hand, there is a simpler way of determining the current through the contact, by measuring the voltage difference across a 1 kΩ resistance placed in series with the two-wires contact. With this method, we can explore the behaviour of the contact at high applied voltages, although the noise signal is higher than that measured with the first method.

When two Au wires, initially separated, are brought into contact after a

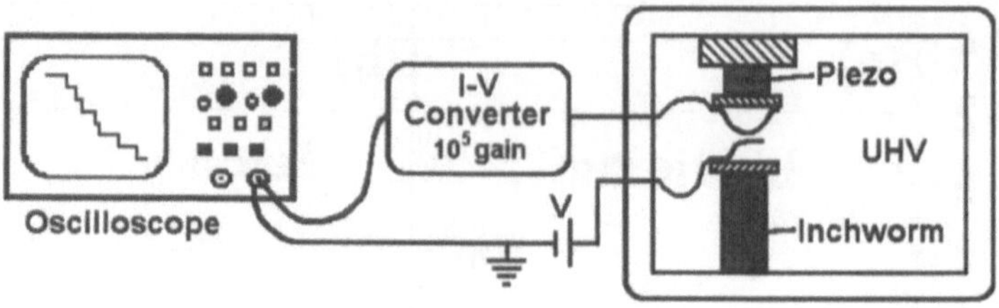

Figure 6. Schematic description of the experimental setup for studying nanocontacts under Ultra High Vacuum (UHV) conditions.

controlled approximation between them, it is possible to form a small contact corresponding to a relative low number of eigenstates through which the electronic ballistic transport takes place. Once this contact is stablished, we have observed that is stable for a long time, being able to measure a constant value of the conductance during several hours. This high stability feature is due to two factors: the excellent mechanical isolation of the UHV chamber and the intrinsic isolation provided by the macroscopic configuration of the wires. This factor is fundamental, since we have noticed that contacts formed between two rigid wires are not stable. However, when one of the wires is free to bend, the nanocontact is stable for a long time. We have been able to measure during several hours a constant current corresponding to a single metallic contact involving few conductance levels. This means that we improve by more than three orders of magnitude the stability reached by standard STM techniques [5, 6, 8, 9] at room temperature. Furthermore, this stability is remarkably higher than that obtained for contacts created at low temperatures with STM [7, 14] or with the MCBJ method [10-13]. In general, by applying a strong mechanical perturbation (for instance, a fast inchworm displacement) we have noticed that the contact can be broken or, in many cases, the conductance presents a clear jump between two quantized values [16], indicating that the contact remains with a thiner section corresponding to a lower number of conductance modes. An example of induced transition of the contact is shown in Figure 7 where a contact corresponding to 3 conductance quanta is monitored during 17 s, although due to an induced mechanical perturbation the conductance suddenly decays to one quantum of the conductance.

The high stability of the formed nanocontacts allows the measurement of its IV characteristic curves. These IV curves have been already measured for nanocontacts created with STM techniques [7, 6] and the MCBJ method [10], although the recording time used to acquire the data was usually below 0.1 s, and no systematic study of the IV curves for such

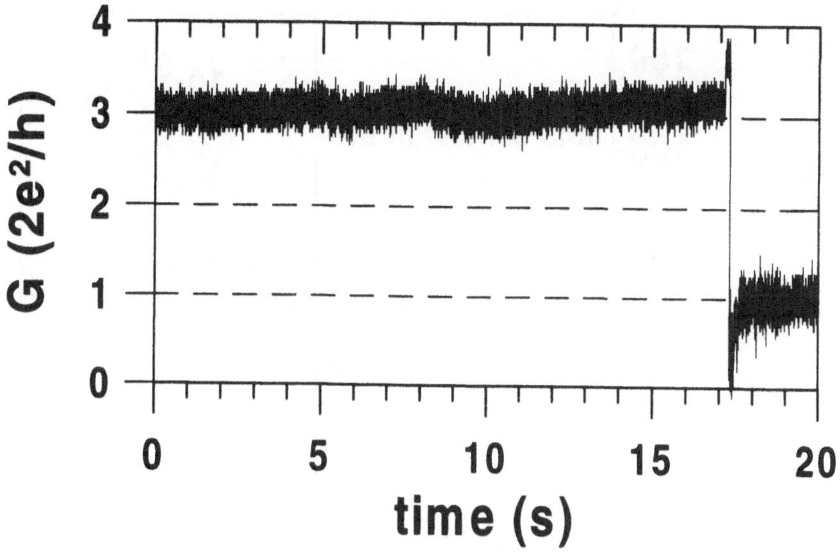

Figure 7. Mechanically induced transition between 3 and 1 conductance quanta for a contact formed between two macroscopic gold wires (0.25 mm of diameter) in UHV conditions.

nanocontacts was done. In our case, we have applied a triangular shaped potential difference to the contact with a time period of 50 s, simoultaneously recording the current through the contact. We have found that after the measurement process the contact conductance did not change to another quantized value, reflecting the contact stability above mentioned. In Figure 8, we plot a series of IV characteristic curves obtained for a determined contact at different stages (different quantized values) of the pulling process. These curves were measured with the experimental setup including the I-V converter, and they were obtained at relative low applied voltages (below 250 mV). At first glance, the IV characteristic curves present a well defined linear behaviour for contacts having a large conductances, whereas for contacts involving only few conductance channels non-linear effects become evident. In all cases, the dynamic conductance (the slope of the IV characteristic at zero applied voltage) agrees within a few percent with the expected quantized conductance of the contact having n conductance channels (n $\times$ $(2e^2/h)$).

In order to analyze the non-linear behaviour of the current-voltage curves (denoted as I(V) functions), we have substracted to such I(V) curves their linear component $I_L(V)$ (determined from fitting the linear part of the I(V) curve at low voltages) obtaining the non linear $I_{NL}(V)$ part of I(V). This non-linear component is represented in Figure 9 for some I-V curves represented in Figure 8, showing that the non-linear character does not de-

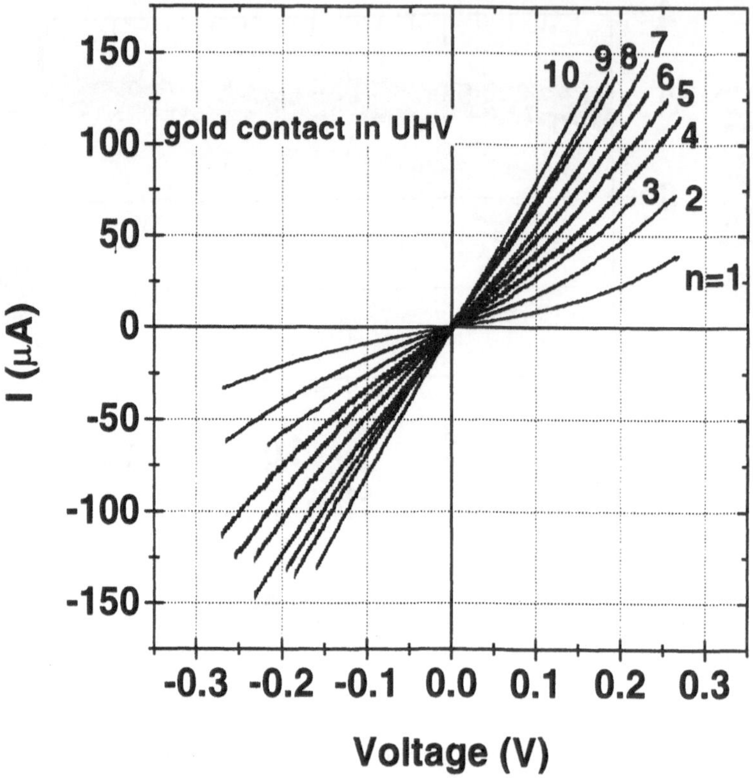

Figure 8. IV characteristic curves for stable gold-gold contacts formed under UHV conditions at RT (the number indicates the number of conductance channels characterizing the nanocontact).

pend apparently on the measured conductance within the same measurement series, i.e., without breaking the contact. When the contact is broken, and a new contact is formed by approaching the macroscopic Au wires, then it is possible to determine a new familly of IV characteristic curves for increasing conductance values. However, in this case, the non-linear part is again independent on the conductance but is different from that determined in the previously shown series. The IV non-linear behaviour can be fitted to expressions of the type $I_{NL}(V) = a_2V^2 + a_3V^3$, and we have found that within the same series $a_2 \ll a_3$ with a_3 nearly independent on the conductance. For instance, for the series depicted in Figure 8 we found $a_3 = 1500 \pm 150$ (averaging over the series). The cuadratic coefficient satisfies $a_2/a_3 < 0.1$, indicating a more cubic-like behaviour. By fitting the curve $I_{NL}(V)$ to the behaviour pV^q we notice that the value of the exponent q also determines a more cubic-like behaviour ($q \approx 2.75 \pm 0.15$). Therefore, at low applied voltages, the linear behaviour is dominant and only for contacts

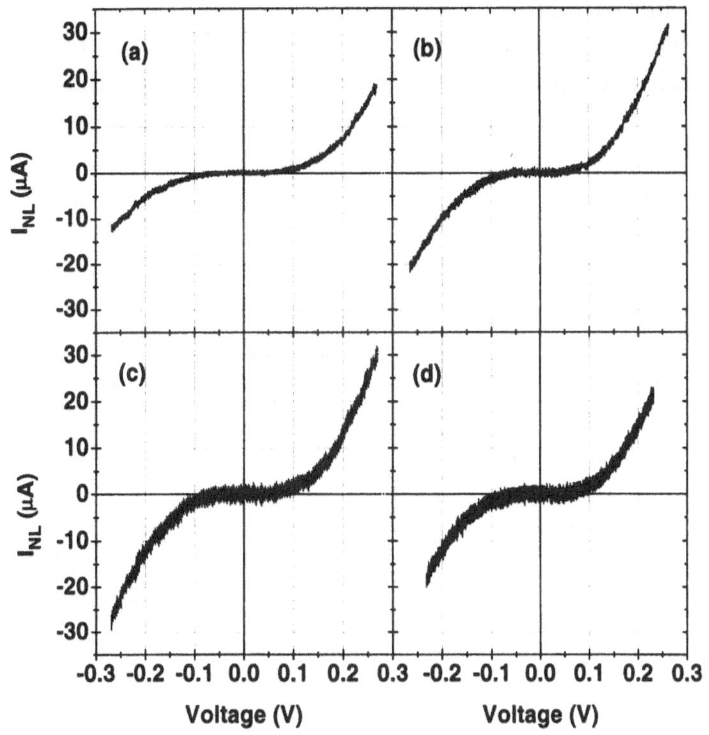

Figure 9. Non-linear component of the current in the IV characteristic curves for some contacts shown in Figure 8, corresponding to conductance values (a) n=1, (b) n=2, (c) n=4, and (d) n=6 (in units of the conductance quantum).

with the lowest conductance values, the non-linear effects (independent on the number of channels determining the conductance) become noticeable.

4. Concluding remarks

We have presented experimental evidence indicating that the study of electrical and mechanical properties of nanowires (showing well marked quantum features) might be carried out with macroscopic systems, by simply approaching two macroscopic wires. Two different setups have been used to measure the current through the nanosized metallic junction. With both schemes, the quantization of the conductance has been observed in loose macroscopic contacts, made both with solid and liquid metals. In general, conductance plateaus are more marked for liquid metals, although only few conductance steps can be obtained within a single experiment in comparison with the well defined stair-like behaviour noticed for solid-solid metallic contacts. We have analyzed from a statistical point of view the distribution of the conductance steps, noticing that most of jumps happen at integer values of the quantum of conductance G_0 for all the types of metallic junctions we have considered. Also, it is feasible to determine the distribution of the relative duration of the conductance plateaus, that presents well defined characteristics, indicating that the 2D cross section of the formed nanowires resembles the quantum chaos of the Stadium problem [23], although this point should be widely discussed elsewhere [24].

Controlled experiments with macroscopic gold wires under UHV conditions have been presented, showing that a remarkable stability is obtained for the nanocontacts, as a consequence of the absence of vibrations in the two wires system due to the ability of the macroscopic configuration to absorb mechanical noise. The nanocontact between two gold wires shows well defined I-V characteristic curves, where non-linear effects become more appreciable for thiner nanocontacts.

The authors appreciate help from M. Marqués in measuring conductance curves in liquid metals. This work has been supported partially by the Spanish DGICyT PB94-0151 and CICyT MAT94-1456-CE projects, and the EU programmes HCM and BRITE-EURAM.

References

1. van Wees, B.J., van Houten, H., Beenakker, C.W.J., Williamson, J.G., Kouwenhoven, L.P., van der Marel, D., and Foxon, C.T. (1988) *Phys. Rev. Lett.* **60**, 848; Wharham, D.A., Thornton, T.J., Newbury, R., Pepper, M., Ahmed, H., Frost, J.E.F., Hasko, D.G., Peacock, D.C., Ritchie, D.A., and Jones, G.A.C. (1988) *J. Phys. C* **21**, L209.
2. Welland, M.E. and Gimzewski, J.K. (eds.) (1995) *Ultimate Limits of fabrication and Measurement*, Kluwer Academic Publishers, Dordrecht.
3. Binnig, G., Rohrer, H., Gerber, Ch., and Weibel, E. (1982) *Phys. Rev. Lett.* **49**, 57.
4. Gimzewski, J.K. and Möller, R.M. (1987) *Phys. Rev. B* **36**, 1284.
5. Pascual, J.I., Méndez, J., Gómez-Herrero, J., Baró, A.M., García, N., and Vu Thien Binh (1993) *Phys. Rev. Lett.* **71**, 1852.
6. Pascual, J.I., Méndez, J., Gómez-Herrero, J., Baró, A.M., García, N., Landman, U., Luedtke, W.D., Bogacheck, E.N., Cheng, H.-P. (1995) *Science* **267**, 1793.
7. Agraït, N., Rodrigo, J.G., and Vieira, S. (1993) *Phys. Rev. B* **47**, 12345; Agraït, N., Rubio, G., and Vieira, S. (1995) *Phys. Rev. Lett.* **74**, 3995.
8. Olesen, L., Laegsgaard, E., Stensgaard, I., Besenbacher, F., Schiotz, J., Stoltze, P., Jacobsen, K.W., and Norskov, J.K., (1994) *Phys. Rev. Lett.* **72**, 2251; (1994) *Phys. Rev. Lett.* **74**, 2147.
9. Brandbyge, M., Schiotz, J., Sorensen, M.R., Stoltze, P., Jacobsen, K.W., Norskov, J.K., Olesen, L., Laegsgaard, E., Stensgaard, I., and Besenbacher, F. (1995) *Phys. Rev. B* **52**, 8499.
10. Muller, C.J., van Ruitenbeek, J.M., and de Jongh, L.J. (1992) *Phys. Rev. Lett.* **69**, 140.
11. Krans, J.M., Muller, C.J., Yanson, I.K., Govaert, Th.C.M., Hesper, R., and van Ruitenbeek, J.M., (1993) *Phys. Rev. B* **48**, 14721.
12. Krans, J.M., Muller, C.J., van der Post, N., Postma, F.R., Sutton, A.P., Todorov, T.N., and van Ruitenbeek, J.M. (1995) *Phys. Rev. Lett.* **74**, 2146.
13. Krans, J.M., van Ruitenbeek, J.M., Fisun, V.V., Yanson, I.K., and de Jongh, L.J. (1995) *Nature (London)* **375**, 767.

14. Smith, D.P.E. (1995) *Science* **269**, 371.
15. Costa-Krämer, J.L., García, N., García-Mochales, P., and Serena, P.A. (1995) *Surf. Science* **342**, L1144.
16. Costa-Krämer, J.L., García, N., Serena, P.A., García-Mochales, P., and Bitar, L. (1995) submitted to *Phys. Rev. B.*
17. Landman, U., Luedtke, W.D., Burnham, N.A., and Colton, R.J. (1990) *Science* **248**, 454; Landman, U. and Luedtke, D. (1991) *J. Vac. Sci. Technol* **9**, 414.
18. Landauer, R. (1989) *J. Phys. Condens. Matter* **1**, 8099.
19. Anderson, P.W. (1958) *Phys. Rev.* **109**, 1492.
20. García, N., and Escapa, L. (1989) *Appl. Phys. Lett.* **54**, 1418.
21. Torres, J., Pascual, J.I., and Sáenz, J.J. (1994) *Phys. Rev. B* **49**, 16581.
22. Vu Thien Binh, Semet, V., and García, N. (1994) *Appl. Phys. Lett.* **65**, 2493.
23. McDonald, S.W. and Kaufman, A.N. (1979) *Phys. Rev. Lett.* **52**, 1189.
24. García-Mochales, P., Serena, P.A., Costa-Krämer, J.L., García, N., and Borondo, F. (1995) submitted to *Science.*

HOST-GUEST MOLECULAR RECOGNITION WITHOUT SOLVENTS

DAVID V. DEARDEN
Brigham Young University
Provo, Utah U.S.A.

1. Introduction

Advances in mass spectrometric techniques now make it possible to study supramolecular systems in the gas phase, in the absence of solvation or counterion effects. Comparison of stability constants in various solvents(1) readily shows that solvation strongly affects the interaction between host and guest, and that counterion effects can also be large. In fact, the dependence on details of solvation, counterion, ionic strength, etc. is so strong that it is often difficult to compare data for different systems unless great pain has been taken to control all the variables.

In solution, since both the host and the guest species are solvated (particularly when a charge is also present), it is possible that many of the well-known recognition effects are actually rooted in differences in solvation energies rather than depending primarily on recognition of a host for a guest. For example, it has been argued that the observed order of crown ether - alkali cation binding constants in solution arises more from differential solvation than from specific size matching recognition.(2) In the gas phase, where neither solvent nor counterion species are present, the *intrinsic* interactions between host and guest are laid bare for experimental probing. In addition, the simplicity of the gas phase environment provides a "level playing field" on which various host-guest systems can be directly compared.

Research in our group uses laser desorption and electrospray ionization methods to introduce ions into the passive electromagnetic trap at the heart of a Fourier transform ion cyclotron resonance mass spectrometer, where the ions are allowed to react with neutrals and these reactions can be followed as a function of time. Experiments typically measure rate- or equilibrium constants. We have used these techniques to examine a number of host-guest systems in the gas phase, ranging from simple crown ether--alkali metal ion complexes to much more complex systems involving hosts such as cyclodextrins or small peptides and molecular guests such as ammonium ions. The underlying goal in all this work is to identify and probe systems where size and/or shape relationships between the host and guest species lead to measurable differences in the rates or thermochemistry of the interactions. After briefly describing our experimental techniques, gas phase results for several host-guest systems are presented, beginning with simple systems where alkali cations serve as guests and crown ethers or cryptands act as

L. Echegoyen and A.E. Kaifer (eds.), Physical Supramolecular Chemistry, 229–247.

hosts. Recognition is also observed in more complex systems involving ammonium ions as guests, and chiral crowns are observed to discriminate between enantiomeric ammonium ions in the gas phase. Finally, recent work involving much larger host species, such as cyclodextrins and small peptides, is described. This paper concentrates on work done in our laboratories, with very little description of the excellent work done elsewhere.(3-18)

2. Experimental

Our methods for generating and studying alkali cation / crown ether complexes in the gas phase using laser desorption to volatilize alkali cation salts have been described in detail in the primary literature.(19, 20) Only a brief summary will be given here.

All experiments employed Fourier transform ion cyclotron resonance mass spectrometry (FTICR/MS).(21, 22) This technique uses a large (typically 3 Tesla or greater) magnetic field to trap ions in two dimensions, with an electrostatic field providing trapping in the third dimension. The trap is suspended inside the magnetic field, within a vacuum chamber with a base pressure typically 10^{-9} Torr or lower. Ion generation typically involved painting a methanol solution of the metals of interest on the face of a stainless steel rod, which was inserted through a vacuum lock to a position inside the magnetic field, within a few cm of the ion trap. The focused output of an excimer-pumped dye laser was used to desorb ions from the salt residue on the probe, some of which were then captured inside the electromagnetic trapping cell of the instrument.

More recent experiments used electrospray ionization(23, 24) rather than laser desorption. In this technique, a solution of the species of interest is sprayed from a needle held a few kV from ground, resulting in the formation of charged droplets. Extensive differential pumping leads to evaporation of the droplets, with the charge remaining on the species dissolved within. These ions are then guided through the fringing field of the magnet and deposited in the trapping cell.

The ions are non-destructively detected through the image currents they induce on the walls of the trapping cell. The captured ions collisionally relax to the center of the trap, and react with whatever neutral background gases are present. The ion population of the cell is easily manipulated by applying RF fields which translationally excite and eject ions with selected mass-to-charge ratios. This allows reactant ions with desired mass-to-charge ratios to be isolated within the trap, ensuring that all subsequently-observed products arise from the isolated precursor(s). Neutrals are introduced either through precision variable leak valves or, when vapor pressures are low, on a direct exposure probe. Monitoring the ion population within the cell as a function of reaction time allows the determination of reaction rate constants, while observation of a constant ion population, combined with knowledge of the partial pressures of neutral reactants, allows equilibrium constants to be measured. Knowledge of the temperature allows conversion of equilibrium constants to free energy changes, and measurement of equilibrium constants as a function of temperature yields enthalpies and entropies for the reactions being examined.

Since pressure determinations typically involve the largest experimental errors in these measurements, the experiments were designed to eliminate pressure as a variable whenever possible. So, for example, rate constants for transferring alkali cations from one ligand to another were measured for all the cations simultaneously such that the pressure of the neutral was constant. Likewise, in chiral recognition measurements, stable pressures of both the chiral amine of interest and an achiral reference were established in the instrument, and measurements were made with each of the protonated enantiomeric hosts while holding the neutral pressures constant. In cases such as this, the *absolute* magnitudes of the results of course still depend on the neutral pressure measurement, but the *relative* values are independent of pressure.

3. Results and Discussion

3.1 INTRODUCTION

Three general classes of reaction are of interest for gas phase host-guest systems: i) formation of 1:1 host-guest complexes; ii) reaction of the complexes with additional host molecules to form complexes with 2:1 (or higher) stoichiometry; and iii) transfer of a (typically ionic) guest from one host to another. Likewise, guests may displace other guests from a given host in a similar exchange reaction. The formation of 1:1 complexes in the gas phase is typically quite fast, and will be touched on only briefly. The rates of the second type of reaction show very strong size effects and are faster for cyclic ligands than for their acyclic counterparts. Most of our attention will focus on the third type of reaction, involving transfer of a guest between host molecules. Conditions can often be chosen so that equilibrium is observed for these reactions, enabling the relative affinities of the two hosts for the guest to be established.

Our examination of gas phase systems begins with interactions of the simplest guests, alkali metal cations, with the prototypically straightforward crown ethers. These exhibit striking size effects in the gas phase, manifest both in reaction rates and in thermochemistry. We then move to slightly more complex hosts, cryptands, then to considerably more complex guests, ammonium ions. The recognition of enantiomers by chiral crown ethers is considered. Finally, we conclude with the application of this gas phase chemistry to much larger species, cyclodextrins and small peptides.

3.2 ALKALI METAL IONS AS GUESTS

Alkali cations are extremely convenient for gas phase host-guest studies because they are easily produced, they are closed-shell species isoelectronic with the noble gas atoms, and thus have no low-lying excited electronic states to complicate the chemistry, and their sizes span a useful range (Table 1). For our purposes, they are charged spheres whose only essential differences are their sizes.

TABLE 1. Sizes of alkali cations (from crystallographic data),(25) crown ethers, and cryptands (from models).(26)

Alkali Cations	Ionic radius, Å	Crown Ethers	Cavity radius, Å	Cryptands	Cavity radius, Å
Li^+	0.76	12C4	0.60-0.75	[2.1.1]	0.8
Na^+	1.02	15C5	0.86-0.92	[2.2.1]	1.1
K^+	1.38	18C6	1.34-1.43	[2.2.2]	1.4
Rb^+	1.52	21C7	~1.7		
Cs^+	1.67				

3.2.1 Crown ether hosts

Complexation reactions. The rates for formation of 1:1 alkali cation:crown ether complexes in the gas phase(19) are quite fast, approaching the collisional limit. Reaction efficiencies (the ratio of reaction rate to collision rate) are highest for the smallest metals, and decrease monotonically with increasing ionic radius, suggesting that cation charge density plays a role in promoting complexation. The glymes, which are open-chain analogs of the crown ethers, form complexes approximately a factor of two less efficiently than the cyclic crowns,(19) reflecting their relative lack of structural preorganization.

The efficiency of forming 2:1 crown:cation "sandwich" complexes is strongly dependent on the size relationship between the metal cation and the crown cavity.(19) In cases where the metals are much larger than the binding cavity, the reaction is generally within an order of magnitude of 100% efficiency, but when the metals are small enough to fit easily within the ligand cavity the efficiencies decrease by three or more orders of magnitude. Thus, if the metal is small enough that the first ligand can effectively occupy its coordination sphere, a second ligand will not readily attach, but if the metal is too large for the first ligand to encapsulate it, attachment of a second ligand is facile. These ideas are also consistent with experimental comparisons between crown ethers and glymes: sandwiching reactions for glymes have always been observed to be slower than those for the corresponding crowns,(19) probably reflecting the greater flexibility of the acyclic ligands which enables them to fill the metal coordination sphere more effectively than the more conformationally-restricted macrocycles.

Guest transfer reactions. Rate constants for transferring alkali metal cations from 18C6 to 12C4, 15C5, and 21C7 are given in Figure 1. 12C4 is seen to readily yield any of the alkali cations to 18C6, with little variation from one metal to another. Reaction occurs on essentially every collision. This is not surprising in that 12C4 is too small to optimally bind or encapsulate any of these metals and 18C6 easily prevails in competition for the guests.

The observed pattern for 15C5 is more complex. Li^+ and K^+ transfer to 18C6 at close to the collision rate, with K^+ being fastest, while transfer of Na^+ is about 30 times slower. Rb^+ and Cs^+ transfer quickly, but at somewhat less than the collision rate. This is consistent with an optimal fit of Na^+ inside 15C5, such that the larger ligand has difficulty in accessing and removing the metal. K^+, on the other hand, is too large for 15C5 but a good fit for 18C6, and thus readily leaves the smaller ligand. The fast rate observed for Li^+ is curious; perhaps this metal is too small to be fully coordinated by 15C5 and is therefore easily removed by the larger, more polarizable crown.

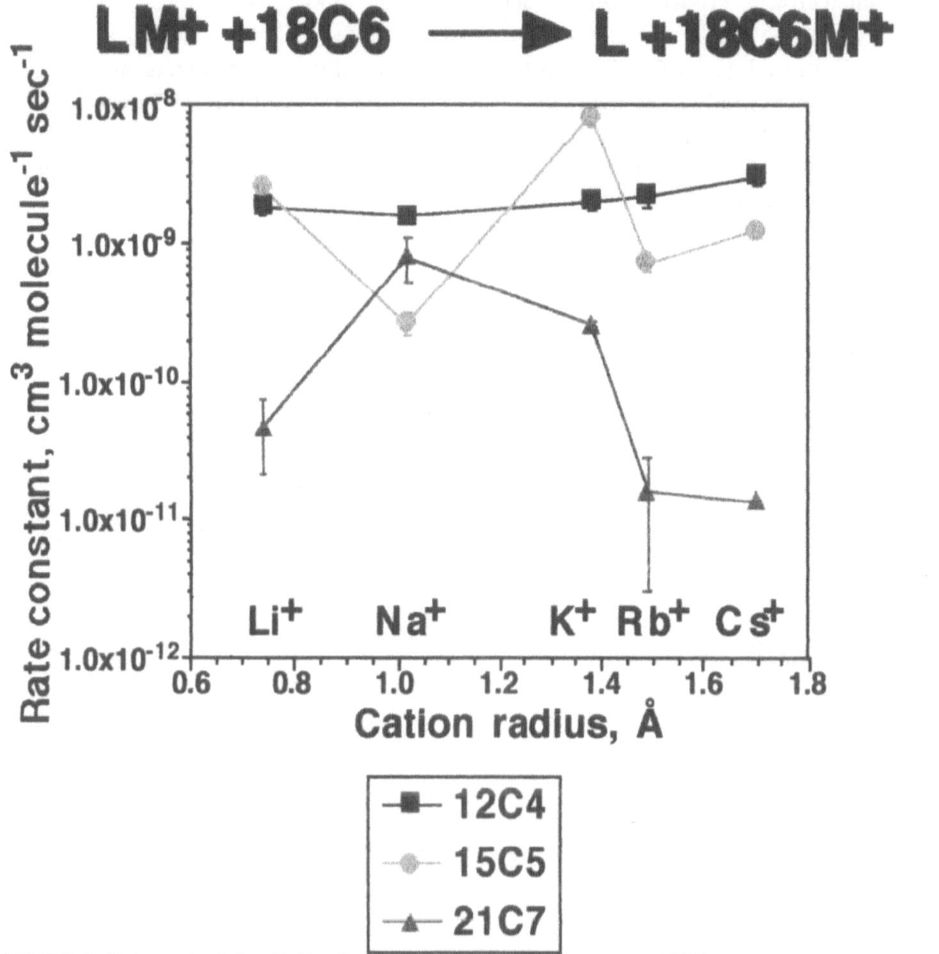

FIGURE 1. Rate constants for alkali cation transfer from the listed crowns to 18C6.

The rates for transfer from 21C7 to 18C6 are dramatically different. Only the rate for Na^+ approaches the collision rate, and rates for Li^+, Rb^+, and Cs^+ are on the order of 1% efficiency or less. The difference here is that 21C7 is larger, has an additional donor, and is more polarizable than 18C6; hence, in most collisions the metal is retained by the larger crown. This is true even for K^+, which fits optimally in 18C6. The relatively fast transfer observed for Na^+ is interesting. Perhaps, as with the 15C5-Li^+ case noted above, Na^+ is small enough that the additional donor present in 21C7 cannot effectively coordinate the metal and thus provides little advantage. The same explanation cannot apply to Li^+, which transfers very slowly. This suggests that Li^+ coordination is somehow different from that of Na^+. Molecular models of the 21C7-Li^+ complex indicate the ligand is highly folded around the small metal cation. If so, unfolding would be required for transfer to 18C6 to occur, slowing the reaction significantly.

Two cases of transfer between ligands with similar or identical cavity sizes have been examined. The first involved transfer from 18C6 to either of two isomeric alkyl-substituted crowns, *cis*-syn-*cis* and *cis*-anti-*cis* dicyclohexano-18-crown-6.(20) According to x-ray data, the substituted crowns have cavities approximately the same size as that of the unsubstituted ligand.(27) The transfer rate constants differ for the two isomers, but decrease monotonically from nearly the collision rate for Li^+ to a factor of 2 to 10 times slower for Cs^+. Similarly, in thermoneutral exchange from isotopically labeled 21C7 to unlabeled 21C7 (Figure 2), the rates are fastest for Li^+ and decrease monotonically with increasing alkali cation size, Cs^+ transfer being more than two orders of magnitude slower.

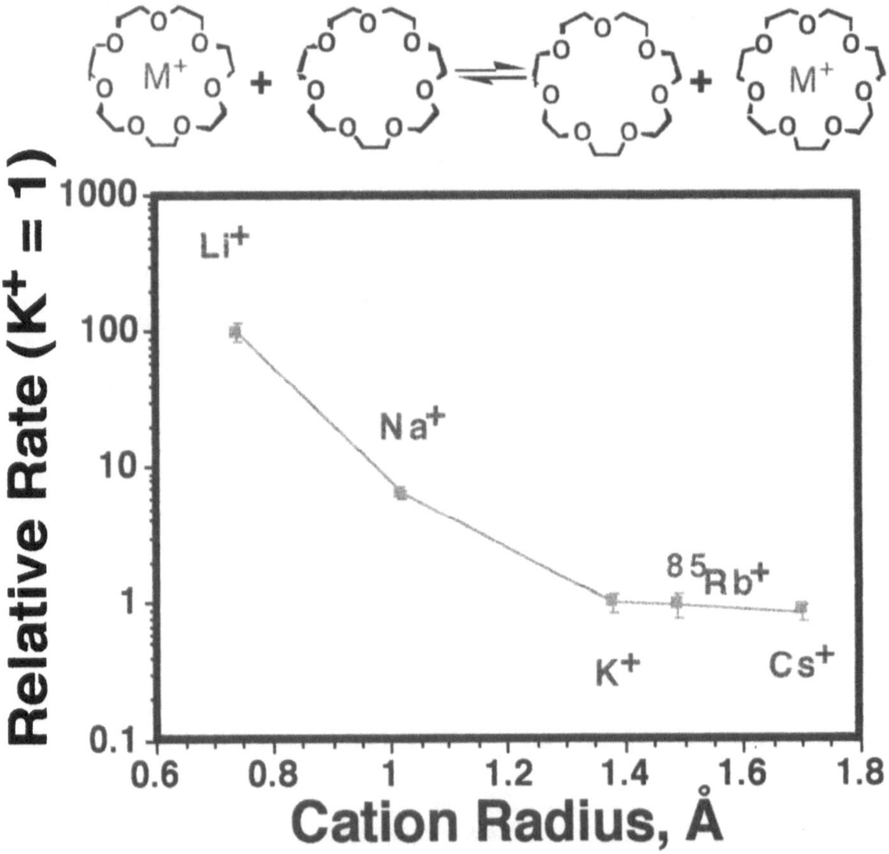

FIGURE 2. Relative rate constants for cation exchange between isotopically labeled and unlabeled 21C7.

The thermoneutral exchange reaction can be modeled in simplest terms using a double-well potential energy surface, the two wells having identical depths. Cation transfer involves hopping from one well to the other, with the rate dependent on the height of the barrier between the two wells. Although the absolute depths of the wells are not known, theory suggests(28) and it is generally believed that the metal-ligand bond strengths decrease monotonically from Li^+ to Cs^+. Thus, it is interesting that transfer is fastest when the wells are deepest, implying that the barrier is lowest for the deeper wells.

Since equilibrium is observed in alkali cation transfer from 18C6 to 21C7, the enthalpies and entropies of these transfer reactions have been investigated using van't Hoff techniques (Figure 3). In contrast to what is observed in solution,(1) the enthalpy of cation transfer from 18C6 to 21C7 is negative for all the alkali metal cations, by up to about 17 kJ mol^{-1} for Li^+ and Na^+. This gas phase behavior is reasonable, since the larger, more polarizable ligand with the greater number of donor atoms has the higher intrinsic affinity for the metals. The enthalpy is least negative for K^+, suggesting either that the 21C7-K^+ bond is especially weak, or that the 18C6-K^+ bond is especially strong. There is nothing special to suggest the former, but the size match between the cavity of 18C6 and K^+ is well known, indicating that the bond between this host and guest is atypically strong.

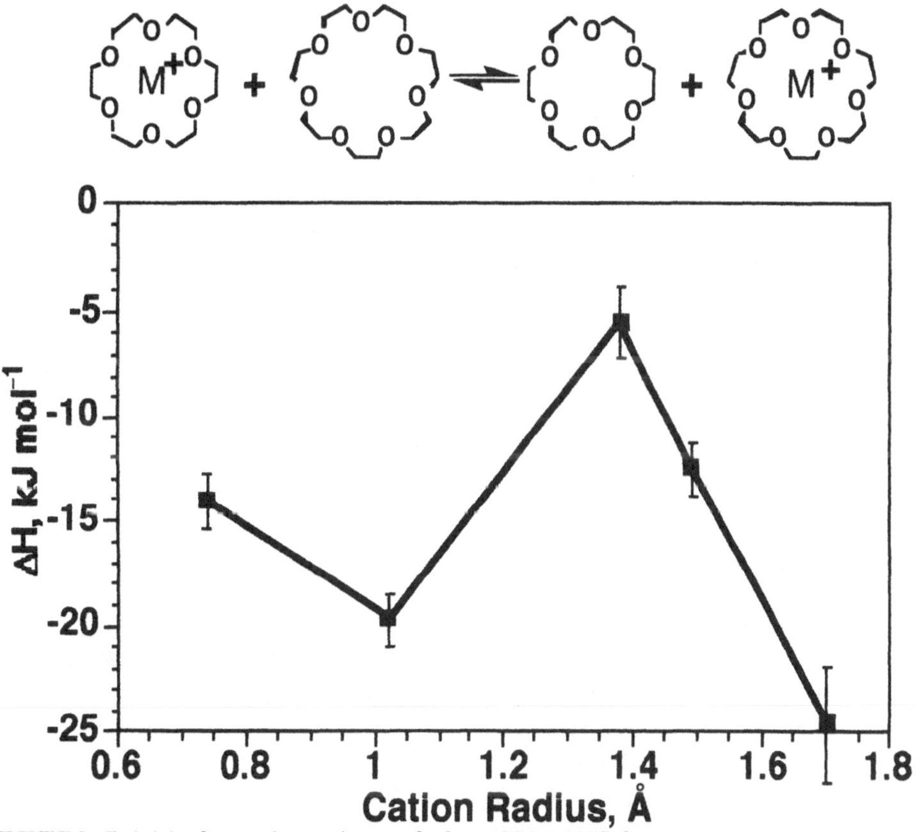

FIGURE 3. Enthalpies for gas phase cation transfer from 18C6 to 21C7, from temperature-dependent equilibrium data.

Entropically, the transfer to the larger ligand is unfavorable as expected, especially for the smallest cations, which most likely have 21C7 highly folded about them in their complexes. K^+ transfer is the least unfavorable, again probably reflecting the good fit of the metal in the cavity of the smaller ligand, which leads to a loss of vibrational degrees of freedom in the 18C6-K^+ complex.

Host modifications. Adding substituents to the ligand influences its cation binding affinity. The changes are sometimes subtle enough they are difficult to observe in solution, but are more evident in the gas phase. For example, the addition of alkyl groups to 18C6 to yield dicyclohexano-18C6 leads to increased affinity for all the alkali cations,(20) particularly for the *cis*-syn-*cis* isomer. The increase is largest for the smallest metals, and decreases monotonically with increasing metal size, consistent with the expected polarizing abilities of the metals and polarizabilities of the ligands.

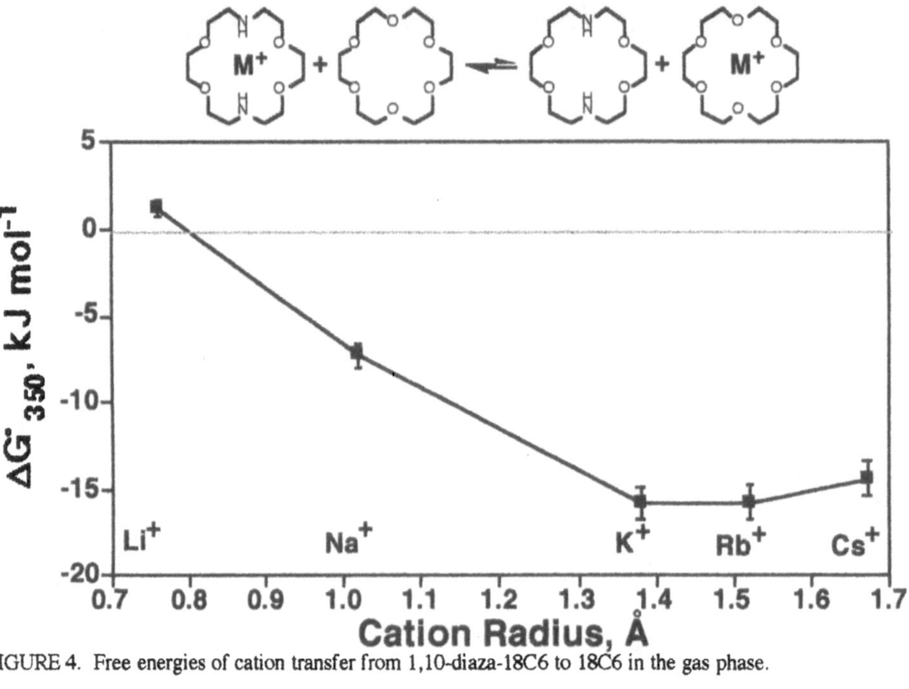

FIGURE 4. Free energies of cation transfer from 1,10-diaza-18C6 to 18C6 in the gas phase.

The binding properties of crown ethers can also be modified by changing the oxygen donor atoms to other heteroatom donors. The results of substituting two amine nitrogens for oxygens are shown in Figure 4. For most alkali cations, this substitution is destabilizing, by more than 15 kJ mol^{-1} for K^+ and larger metals. Therefore, we might expect the nitrogen atoms in cryptands to be less effective donors toward alkali metal cations than crown ether oxygens. The degree of binding preference for 18C6 over 1,10-diaza-18C6 correlates roughly with cation size, becoming larger as cation size increases. Similar effects have been noted previously in studies carried out in methanol solution, which found complexation by the nitrogen-substituted ligand to be far weaker than that involving 18C6.(29)

In contrast, for Li^+ the substitution is slightly stabilizing. These results can be rationalized by noting that each of the nitrogen donors are covalently bound to hydrogen, in addition to the two carbon atoms which join them to the macrocyclic ring. The hydrogens disrupt the binding cavity of the ligand, in effect dividing the cavity into two smaller metal binding sites rather than one large binding cavity. These "mini cavities" are too small to accommodate any of the alkali metal cations except Li^+. Models indicate Li^+ is about the right size to fit into the "mini-cavity," which may account for the increased Li^+ affinity of the nitrogen-substituted crown. Electronic effects may also be important: geometries for the Li^+-diaza-18C6 complex optimized using semiempirical AM1 calculations have the nitrogen lone pairs oriented directly toward the metal.

3.2.2 Cryptand hosts

Cryptands combine the nitrogen-for-oxygen substitution noted above with the addition of bridges, consisting of ethylene oxide groups, between the bridgehead nitrogens. By examining the relative cation affinities of cryptands and crown ethers with the same number of donor groups, cryptate effects(30) can be investigated in the gas phase. Such a comparison is shown in Figure 5, which gives free energies for alkali cation transfer

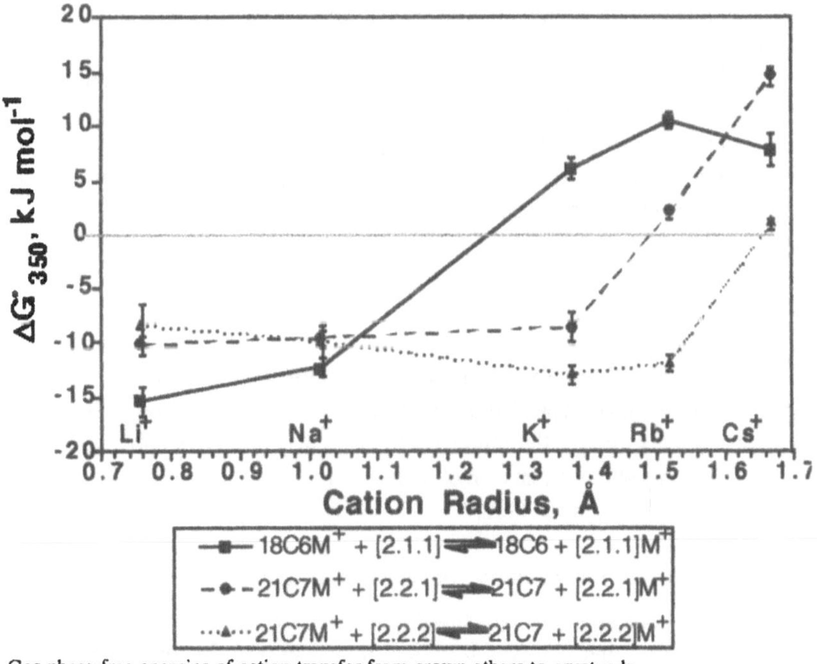

FIGURE 5. Gas phase free energies of cation transfer from crown ethers to cryptands.

from 18C6 to cryptand [2.1.1], both six-donor ligands; from 21C7 to [2.2.1], both seven-donor ligands; and from 21C7 to [2.2.2], a seven-donor crown to an eight-donor cryptand. The patterns which occur as the metal size is varied are striking. In the six-donor case, transfer from crown to cryptand is exothermic for Li^+ and Na^+, and endothermic for the larger cations. Similarly, for the seven-donor ligands the reaction is exothermic for all the alkali cations from K^+ and smaller. For the 21C7 - [2.2.2] reaction, transfer to the cryptand is exothermic for all the alkali cations studied except Cs^+, for which it is slightly endothermic.

This is probably the most striking example of size-dependent thermochemistry yet observed in the gas phase. When the cation is small enough to fit inside the binding cavity of the cryptand, the three-dimensional arrangement of the heteroatom donor groups in the cavity leads to a stronger interaction with the metal than is possible when the donor groups are constrained to a pseudo-plane, as in the crown ethers, despite the fact that two of the cryptand donors are nitrogens with intrinsically lower alkali cation binding capabilities than ether oxygens. When the cations are too large to fit within the binding cavity, the direction of equilibrium is reversed. In these cases, the cations must perch on the face of the cryptand and cannot interact effectively with the donor group(s) on the bridge. The crown, on the other hand, can still interact through all its donor atoms and therefore binds the metal more strongly than the cryptand.

This interpretation is also consistent with the kinetics of 2:1 host:guest complexation involving cryptands. For [2.1.1], 2:1 complex formation is rapid for K^+ and larger alkali cations, but too slow to observe for Li^+ and Na^+, suggesting that the latter two small cations are encapsulated within the binding cavity. Likewise, for [2.2.1] rapid 2:1 ligand:metal complexation is observed only for Rb^+ and Cs^+, implying that K^+ and smaller alkali cations fit within its binding cavity. No 2:1 complexation is observed for [2.2.2], even for Cs^+. This is consistent with NMR data in solution, which indicate that under certain solvation conditions (low temperature, with low ion pairing)(31) [2.2.2] forms "inclusive" complexes with Cs^+. Since the solvation and ion pairing forces which stabilize "exclusive" complexes are absent in the gas phase, an "inclusive" complex in this environment seems reasonable.

3.3 MOLECULAR GUESTS

Gas phase techniques can also be applied to investigate more complex, molecular species as guests. Simple prototypical molecular guests investigated in our laboratory include singly-charged alkaline earth monohalide cations and a range of substituted ammonium ions, culminating with systems involving chiral recognition. Only the work with ammonium ions is described here.

3.3.1 Interactions of ammonium ions with small, synthetic ionophores

The chemistry of ammonium interactions with crowns and cryptands is considerably more complex than that observed for alkali metal cations. In addition to complex formation, both proton transfer from the ammonium to the crown and various reactions leading to fragmentation of the ligand are sometimes observed. Herein, only the kinetics and thermochemistry of a few simple complex formation and ammonium exchange reactions will be discussed.

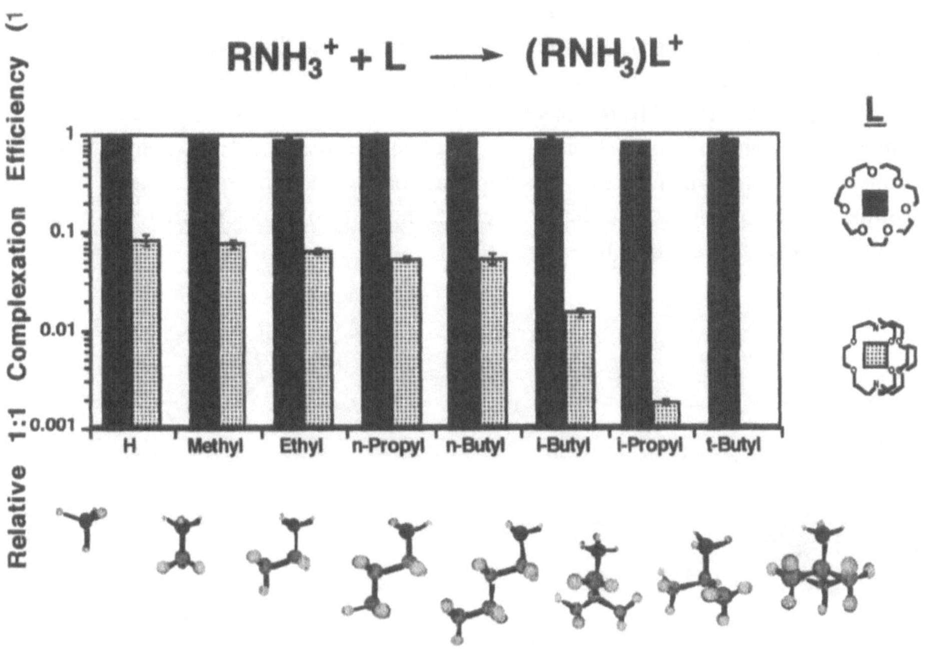

FIGURE 6. Rates for complex formation between ammonium ions and the synthetic ionophores 21C7 and [2.2.2] cryptand, relative to the rates for the same reactions with 18C6.

Figure 6 shows the rates for 1:1 complex formation between various ammonium cations and either 21C7 or [2.2.2] cryptand relative to those for forming complexes with 18C6. It is immediately apparent that the kinetics of complexation are very similar for the two crowns, although complexation to 21C7 appears to be slightly slower than to 18C6 for all the cations studied. For the cryptand, on the other hand, complexation is at least an order of magnitude slower than for 18C6.

Further, addition of the bridging group makes the rates sensitive to the substituent present on the ammonium cation. The reaction efficiencies decrease slowly as the substituent chain length increases from methyl through propyl, although additional lengthening to butyl causes little or no further decrease. The efficiencies are very sensitive to branching in the substituent: the relative efficiency for isobutylammonium is about a factor of five less than that for n-butylammonium, while moving the branch point one carbon closer to the nitrogen (going from isobutyl to isopropyl) leads to an additional decrease of an order of magnitude relative to the crown. Complexation of t-butylammonium with the cryptand is too slow to measure with our methods. Thus the steric and/or orientational requirements for binding ammonium to [2.2.2] must be much greater than for binding the cation to 18C6 or 21C7.

Although the *kinetics* of ammonium attachment to 18C6 and 21C7 are similar, equilibrium experiments show that the *free energies* of binding to the two ligands are quite different (Figure 7). Ammonium cations with small substituents, such as ammonium and methylammonium, bind more strongly to 21C7 than to 18C6, but as the size of the substituent increases the binding preference first decreases, then reverses. Thus, the equilibrium constant for t-butylammonium is more than two orders of magnitude smaller than that for unsubstituted ammonium, with the former binding 18C6 more strongly than 21C7. The reasons for these trends are not completely clear, but apparently there are additional stabilizing interactions between the ammonium cations and the larger crown which become less important or even destabilizing as the substituent on the guest becomes larger. Perhaps the conformational/steric requirements for binding 18C6 are less than those for 21C7.

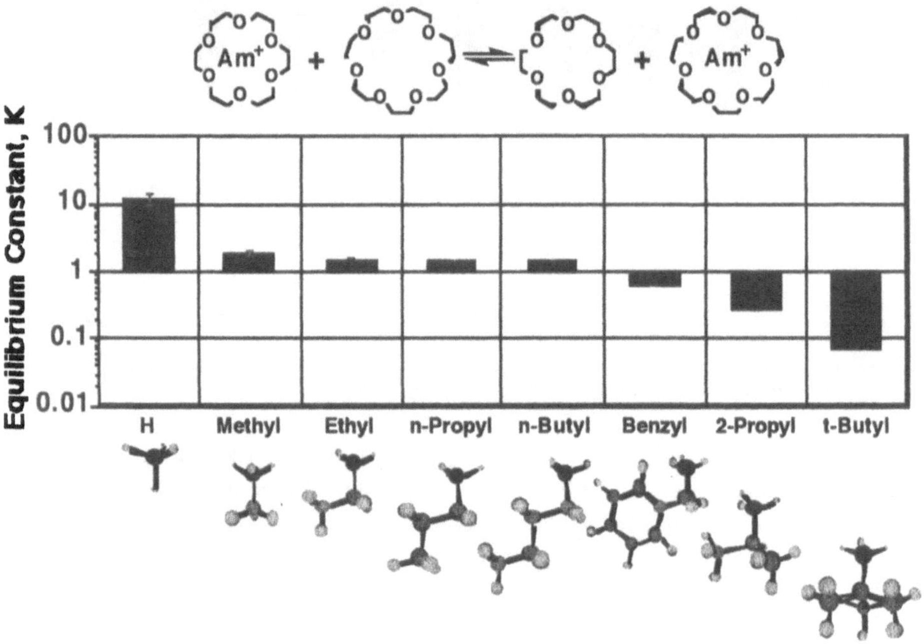

FIGURE 7. Equilibrium constants for exchanging ammonium cations between 18C6 and 21C7, all measured at 350 K.

3.3.2 *Chiral recognition*

One of the more difficult challenges in mass spectrometry is distinguishing between enantiomers, whose masses and atom connectivities are identical. Successful approaches to this problem usually involve differentiation based on different reactivities with chiral reagents. (32) Recently, fast atom bombardment (FAB) mass spectrometry has been used to examine the relative intensities of adduct peaks arising from interactions of chiral species.(33-36) Often, one of the enantiomers is isotopically labeled so that the chiral adducts can be distinguished on the basis of mass. Other experiments have examined gas phase ion-molecule reactivity directly using chiral reagents.(37-39)

Our own work in this area involves dimethyldiketopyridino-18-crown-6 (hereafter referred to as "**1**"), a crown ether with two stereocenters (Figure 8). This ligand was chosen for gas phase study because it has been extensively examined in condensed media.(40-42) Two types of interactions are likely to contribute to bonding between this crown and potential guest molecules: hydrogen bonding involving the pyridine nitrogen and/or the ether oxygens, and, for guests with π-systems, π–π stacking interactions involving the pyridino group. Our initial and most extensive studies used α-(1-naphthyl)ethylammonium ($NapEtNH_3^+$) as the chiral guest (Figure 8), while more recently we have examined a number of other chiral amines.

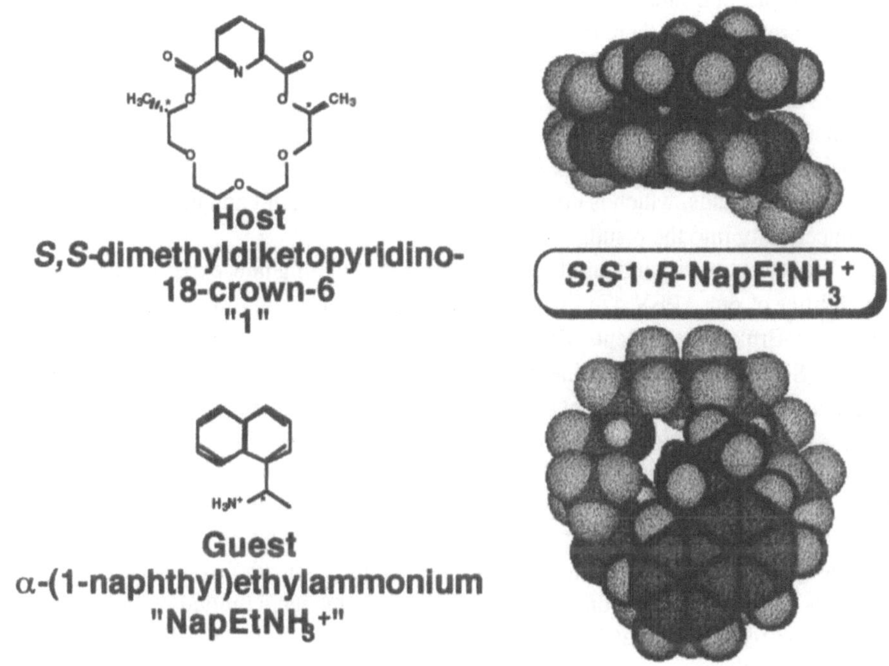

FIGURE 8. Structure of the chiral host and one of the chiral guests used in this study. The space filling models show two views of the indicated complex, with the guest outlined in bold lines. Both π-π stacking and hydrogen bonding interactions are suggested by the model.

While all of our experiments in chiral recognition use equilibrium techniques to probe free energy changes, we have approached equilibrium in several different ways. In our initial experiments,(39) the *S,S* enantiomer of the chiral host, *S,S*-**1**, and an achiral host, 18C6, were both admitted into the trapping region the instrument. The partial pressures of these two neutral ligands were carefully measured. Either *R*- or *S*-$NapEtNH_2$ was also admitted into the trapping region, and $NapEtNH_3^+$ was formed by self-chemical ionization. Reaction of $NapEtNH_3^+$ with the neutral ligands afforded the host-guest complexes. The exchange of the guest ammonium ion between the chiral and achiral hosts was allowed to proceed to equilibrium. The attainment of equilibrium was verified by monitoring ion intensities as a function of reaction time until the ratio of the 18C6•$NapEtNH_3^+$ and *S*, *S*-**1**•$NapEtNH_3^+$ ion intensities became constant. Perturbation of the system away from equilibrium by ejection of either of the complexes always resulted in re-establishment of the same equilibrium ratios after an appropriate delay, attesting to the fact that true equilibrium was reached. The difference in the equilibrium constants for the two enantiomers, expressed as a difference in free energies for the transfer reactions involving the two enantiomers, $\Delta\Delta G°$, is a measure of the degree of chiral recognition in the system.

These were difficult experiments. First, *S*, *S*-**1** is fairly involatile, so that usable vapor pressures were barely attainable when the ligand was inserted into the high vacuum region of the instrument on a direct-exposure solids probe. Second, achiral 18C6 binds $NapEtNH_3^+$ much more strongly than the chiral ligand, making equilibrium difficult to observe unless there is a large excess pressure of the less volatile, chiral *S*, *S*-**1**. Third, the lack of an external ion source on the instrument we used originally made some of the chemistry ambiguous. Could we have simply been transferring protons between neutral complexes? Fourth, most of the chiral ligand was wasted as it was slowly pumped away by the vacuum system. Finally, measurement of the partial pressures of the ligands, which is crucial to the results, is difficult and introduces a great deal of uncertainty into the results.

We set out to circumvent these problems by designing new experiments around the capabilities of our APEX 47e Fourier transform ion cyclotron resonance mass spectrometer (Bruker Instruments, Billerica, MA), which is equipped with electrospray ionization. Since this instrument has an external ion source, the ionization and complex formation aspects of the experiment can be separated such that formation of neutral complexes is not possible. Even more importantly, electrospray ionization offers a means of generating protonated host molecules which avoids the problems inherent in the low volatility of the chiral ligands, and enables us to eliminated neutral pressures as variables in these experiments. Further, with electrospray ionization only a few tenths of a milligram of the expensive chiral ligands are required for a whole range of experiments. In fact, sample consumption in our experiments is limited by the minimum mass we can accurately weigh, rather than by the amount actually needed for electrospray detection.

In our new experiments, rather than observing equilibria for exchanging an ammonium guest among neutral hosts, we observe the exchange of neutral amines by a protonated host ion. One enantiomer of the chiral amine, as well as an achiral reference amine, are introduced into the trapping cell as neutrals. One of the host enantiomers is electrosprayed in the external source, and the resulting protonated host ions are guided into the trapping cell and captured. Reaction of the protonated host with the neutral amines results in formation of complexes. Exchange of the chiral and achiral amines on the protonated chiral host is allowed to proceed to equilibrium, which is verified as before by attainment of a constant product ratio and by re-establishment of the same equilibrium ratio after the system is perturbed by ejection of either the reactant complex or the product complex. The experiments are then repeated with the other enantiomer of the host or the guest. Again, the difference in equilibrium constants for the two enantiomers yields a free energy difference which quantitatively measures the degree of recognition.

The results of these experiments are given in Table 2. For each guest, the cross chiral combination (*S*-amine with *R,R*-**1**, for example) is preferred. It is evident that the amine transfer and ligand transfer methods give answers in good agreement, indicating that the same complex is formed whether the proton is initially on the amine or on the crown. It is also apparent that π-stacking plays a role in enantiomeric recognition by this host. Cyclohexylethylamine and methylbenzylamine have substituents with similar size, but differing in that the latter is dehydrogenated, forming a π system. However, recognition for the aromatic amine is more than twice that seen for the cyclic alkylamine. The greatest recognition is observed for naphthylethylamine, which possesses a π system large enough to allow efficient π-stacking as well as hydrogen bonding between the crown and the amino group.

TABLE 2. Differences in free energies of binding *S* and *R* enantiomers by ligand *R,R*-**1**. All measurements at 300 K using amine exchange equilibria as described, except as noted.

amine	$\Delta\Delta G°$, kJ mol^{-1}
sec-butyl	0.3 ± 0.4
cyclohexylethyl	0.9 ± 0.2
methylbenzyl	2.4 ± 0.5
naphthylethyl	3.5 ± 0.6
naphthylethyl (350 K, ligand exchange)	4.2 ± 0.4

3.4 "BIG" MOLECULES

Electrospray ionization makes possible the study of much higher molecular weight host-guest systems than were accessible previously. True supramolecular systems are now becoming amenable to gas-phase study. This section describes preliminary work with two classes of relatively large molecules, cyclodextrins and small peptides.

3.4.1 Cyclodextrins

Cyclodextrins are molecules consisting of D-glucose units linked cyclically such that the overall shape of the molecule is that of a truncated cone. Electrospray of acidified methanol-water solutions of these molecules results in the formation of protonated cyclodextrin ions, which readily react with neutral amines to form complexes. We have carried out a series of experiments wherein α – (six sugar units), β- (seven sugar units), and γ -cyclodextrin (eight sugar units) were simultaneously electrosprayed and trapped. The reactions of the trapped ions with neutral amines were followed as a function of time, enabling a direct comparison of the rates of amine adduct formation for α, β, and γ -cyclodextrin.

The results of these experiments are presented in Figure 9. All three cyclodextrins adduct with all the amines studied at close to the collision rate. No clear pattern emerges in the comparison of α, β, and γ. Curiously, in most cases β-cyclodextrin forms 1:1 complexes more rapidly than α or γ. An important question yet to be answered in these experiments is whether the observed adducts are inclusion complexes or whether the amines are attached to the exterior of the polysaccharides. Improvements in the signal-to-noise which can be obtained when cyclodextrins are electrosprayed should shed further light on this question.

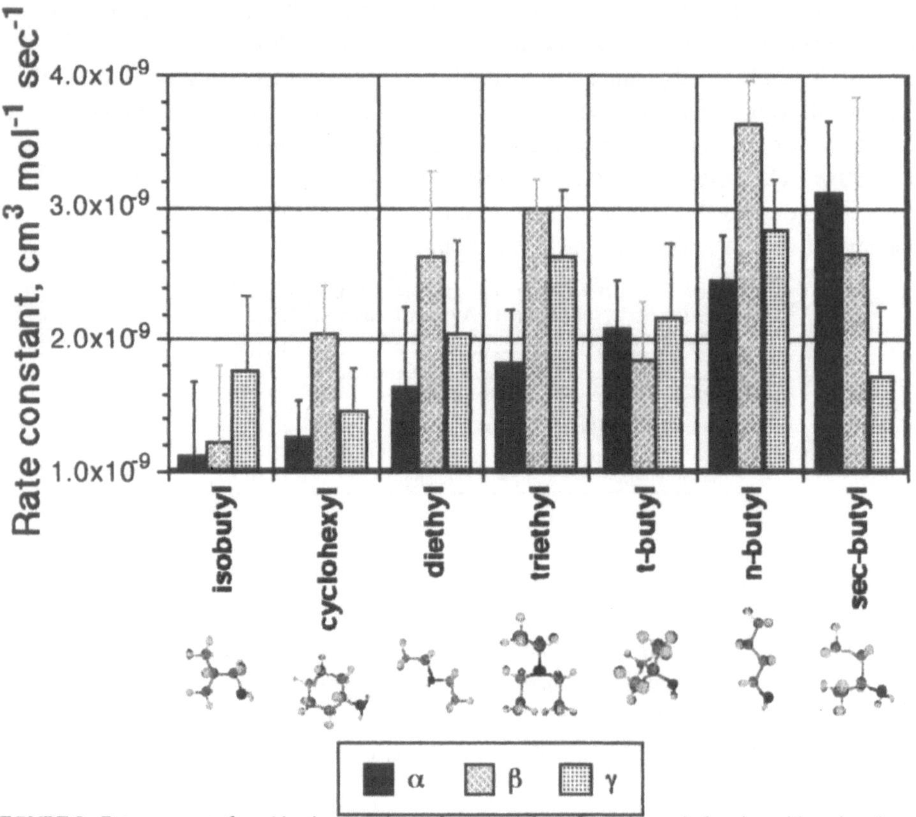

FIGURE 9. Rate constants for adduction reactions of protonated α-, β-, and γ-cyclodextrins with amines in the gas phase.

3.4.2 *Peptides*

With electrospray ionization, protonated peptide ions are readily produced, often as multiply-charged species with several protons attached. In addition, if the peptides are sprayed from solutions buffered at neutral to basic pH in the presence of alkali cations, metal cationized ions can be generated. We have used these methods to produce a number of alkali-cationized peptides, but most of the discussion which follows describes work with the small cyclic decapeptide gramicidin S. This molecule is convenient to study because it is very easily electrosprayed.

Gramicidin S possesses two basic residues, both ornithines, so it is not surprising that electrospray normally results in intense signal for the doubly protonated peptide. Similarly, one might expect that when spraying from neutral solutions it might be possible to doubly sodiate the peptide, rather than doubly protonating it. Such is indeed the case. However, it is interesting that more than two such sodium-for-proton substitutions can be accomplished, such that the number of sodium atoms attached *exceeds* the net charge on the ion. We have observed, for example, up to four sodium ions attached to doubly-charged gramicidin S. In such cases of ***hypermetallation*** of the peptide, alkali cations must replace protons on the backbone of the molecule, most likely the amide protons.

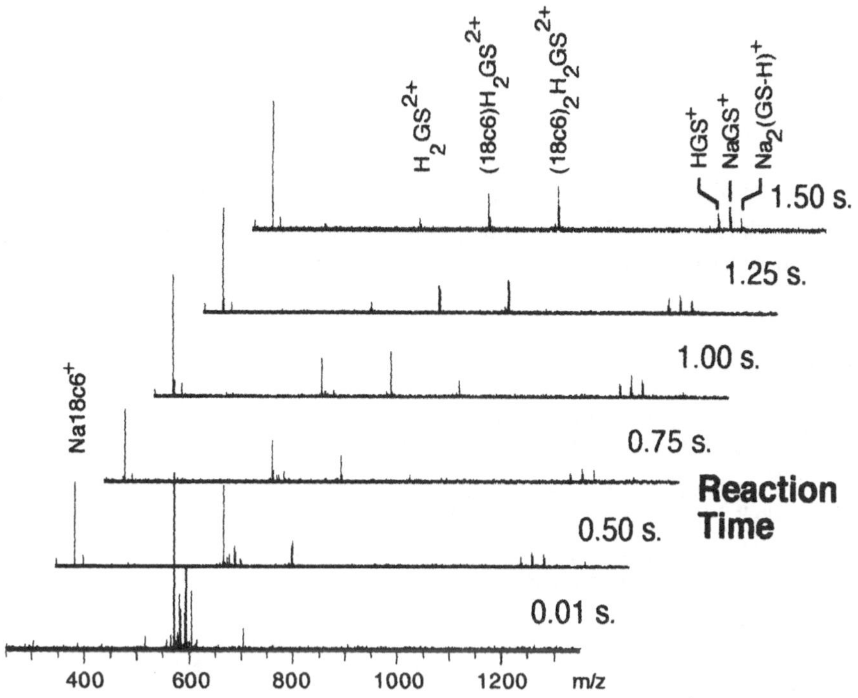

FIGURE 10. Reaction of isolated +2 gramicidin S ions cationized with protons and/or Na^+ with neutral 18C6.

Both metallated and protonated doubly-charged gramicidin S react readily with ionophores in the gas phase. Figure 10 shows the reactions which occur when doubly-charged gramicidin S ions with protons and sodium cations serving as charge carriers are trapped in the presence of neutral 18C6. Two types of reaction are evident. First, the crown rapidly removes Na^+ from HNa•gramicidin^{2+}, Na_2•gramicidin^{2+}, (Na_3-H)•gramicidin^{2+}, and (Na_4-2H)•gramicidin^{2+}, yielding 18C6•Na^+ and the corresponding singly-charged gramicidin ions. Although we have not yet obtained quantitative rate constants for these metal stripping reactions, it is clear that the reactions are quite fast. The reactions also exhibit metal selectivity: when both Na^+ and K^+ are present, K^+ is stripped more rapidly than the smaller metal. Thus, the sodium and potassium affinities of singly-charged gramicidin S are less than those of 18C6, and gas phase macrocyclic chemistry offers a means for selectively reducing the charge state of metallated peptides.

The doubly-protonated gramicidin S ion also reacts with 18C6, successively forming adducts with first one, then a second crown molecule. It is likely that the gramicidin is protonated on the two ornithines,(43) and that the crowns are forming ammonium complexes at these well-separated sites.

Singly-charged, singly-metallated gramicidin S can also be stripped by 18C6, but the reaction is roughly two orders of magnitude slower than for the +2 ion. Stripping of other singly-charged, singly-metallated ions such as singly sodiated or singly potassiated cyclosporin A is similarly slow. Again, K^+ is stripped more rapidly from this singly-charged ion than is Na^+.

4. Acknowledgments

We are grateful for support of this work by the U. S. National Science Foundation, by the donors of the Petroleum Research Fund, administered by the American Chemical Society, and by the Office of Naval Research.

5. References

1. Izatt, R. M., Pawlak, K., Bradshaw, J. S. and Bruening, R. L. (1991) *Chem Rev* **91,** 1721-2085.
2. Gokel, G. W., Goli, D. M., Minganti, C. and Echegoyen, L. (1983) *J. Am. Chem. Soc.* **105,** 6786-6788.
3. Brodbelt, J., Maleknia, S., Liou, C. C. and Lagow, R. (1991) *J. Am. Chem. Soc.* **113,** 5913-5914.
4. Brodbelt, J., Maleknia, S., Lagow, R. and Lin, T. Y. (1991) *J. Chem. Soc., Chem. Commun.*, 1705-1707.
5. Maleknia, S. and Brodbelt, J. (1992) *J Am Chem Soc* **114,** 4295-4298.
6. Maleknia, S. and Brodbelt, J. (1992) *Rapid Commun Mass Spectrom* **6,** 376-381.
7. Liou, C. C. and Brodbelt, J. S. (1992) *J Amer Soc Mass Spectrom* **3,** 543-548.
8. Liou, C. C. and Brodbelt, J. S. (1992) *J Am Chem Soc* **114,** 6761-6764.
9. Wu, H.-F. and Brodbelt, J. S. (1993) *J. Am. Soc. Mass Spectrom.* **4,** 718-722.
10. Brodbelt, J. S., Liou, C.-C., Maleknia, S., Lin, T.-Y. and Lagow, R. J. (1993) *J. Am. Chem. Soc.* **115,** 11069-11073.
11. Wu, H.-F. and Brodbelt, J. S. (1994) *J. Am. Chem. Soc.* **116,** 6418-6426.
12. Katritzky, A. R., Malhotra, N., Ramanathan, R., Kemerait, R. C., Zimmerman, J. A. and Eyler, J. R. (1992) *Rapid Commun Mass Spectrom* **6,** 25-27.
13. Vincenti, M., Dalcanale, E., Soncini, P. and Guglielmetti, G. (1990) *J. Am. Chem. Soc.* **112,** 445-447.
14. Vincenti, M., Pelizzetti, E., Dalcanale, E. and Soncini, P. (1993) *Pure Appl. Chem.* **65,** 1507-1512.
15. Timofeev, O. S., Gren, A. I., Zagorevskii, D. V., Nekrasov, Y. S., Lobach, A. V., Bogatskii, A. V. and Luk'yanenko, N. G. (1984) *Bull. Acad. Sci. USSR, Div. Chem. Sci.* **33,** 2385-2387.
16. Gren, A. I., Mazepa, A. V. and Timofeev, O. S. (1993) *Org. Mass Spectrom.* **28,** 265-266.
17. Timofeev, O. S. (1995) *Rapid Commun. Mass. Spectrom.* **9,** 837-839.
18. Kralj, B., Timofeev, O. S., Zigon, D., Kirichenko, T. I. and Marsel, J. (1995) *Rapid Commun. Mass Spectrom.* **9,** 160-166.
19. Chu, I. H., Zhang, H. and Dearden, D. V. (1993) *J. Am. Chem. Soc.* **115,** 5736-5744.
20. Chu, I.-H. and Dearden, D. V. (1995) *J. Am. Chem. Soc.* **117,** 8197-8203.
21. Marshall, A. G. (1985) *Acc. Chem. Res.* **18,** 316-322.
22. Marshall, A. G. and Verdun, F. R. (1990) *Fourier Transforms in NMR, Optical, and Mass Spectrometry: A User's Handbook,* Elsevier, Amsterdam.
23. Smith, R. D., Loo, J. A., Loo, R. R. O., Busman, M. and Udseth, H. R. (1991) *Mass Spectrom Rev* **10,** 359-451.
24. Smith, R. D., Light-Wahl, K. J., Winger, B. E. and Loo, J. A. (1992) *Org. Mass Spectrom.* **27,** 811-821.

25. Shannon, R. D. (1976) *Acta Crystallogr., Sect. A: Found. Crystallogr.* **32,** 751-767.
26. Dalley, N. K. (1978) in *Synthetic Multidentate Macrocyclic Compounds*, eds. Izatt, R. M. and Christensen, J. J., Academic, New York.
27. Dobler, M. (1981) *Ionophores and Their Structures,* Wiley, New York.
28. Glendening, E. D., Feller, D. and Thompson, M. A. (1994) *J. Am. Chem. Soc.* **116,** 10657-10669.
29. Buschmann, H. J. (1986) *Inorg. Chim. Acta* **125,** 31-35.
30. Lehn, J.-M. and Sauvage, J.-P. (1975) *J. Am. Chem. Soc.* **97,** 6700-6707.
31. Kauffman, E., Dye, J. L., Lehn, J.-M. and Popov, A. I. (1980) *J. Am. Chem. Soc.* **102,** 2274-2278.
32. Martens, J., Lübben, S. and Schwarting, W. (1991) *Z. Naturforsch* **43b,** 320-325.
33. Hofmeister, G. and Leary, J. A. (1991) *Org. Mass Spectrom.* **26,** 811-812.
34. Sawada, M., Shizuma, M., Takai, Y., Yamada, H., Kaneda, T. and Hanafusa, T. (1992) *J Am Chem Soc* **114,** 4405-4406.
35. Sawada, M., Okumura, Y., Shizuma, M., Takai, Y., Hidaka, Y., Yamada, H., Tanaka, T., Kaneda, T., Hirose, K., Misumi, S. and Takahashi, S. (1993) *J. Am. Chem. Soc.* **115,** 7381-7388.
36. Sawada, M., Okumura, Y., Yamada, H., Takai, Y., Takahashi, S., Kaneda, T., Hirose, K. and Misumi, S. (1993) *Org. Mass Spectrom.* **28,** 1525-1528.
37. Nikolaev, E. N., Goginashvili, G. T., Tal'rose, V. L. and Kostyanovsky, R. G. (1988) *Int. J. Mass Spectrom. Ion Proc.* **86,** 249-252.
38. Honovich, J. P., Karachevtsev, G. V. and Nikolaev, E. N. (1992) *Rapid Commun. Mass Spectrom.* **6,** 429-433.
39. Chu, I.-H., Dearden, D. V., Bradshaw, J. S., Huszthy, P. and Izatt, R. M. (1993) *J. Am. Chem. Soc.* **115,** 4318-4320.
40. Jones, B. A., Bradshaw, J. S. and Izatt, R. M. (1982) *J. Heterocyclic Chem.* **19,** 551-556.
41. Davidson, R. B., Bradshaw, J. S., Jones, B. A., Dalley, N. K., Christensen, J. J., Izatt, R. M., Morin, F. G. and Grant, D. M. (1984) *J. Org. Chem.* **49,** 353-357.
42. Bradshaw, J. S., Huszthy, P., McDaniel, C. W., Zhu, C. Y., Dalley, N. K., Izatt, R. M. and Lifson, S. (1990) *J. Org. Chem.* **55,** 3129-3137.
43. Gross, D. S. and Williams, E. R. (1995) *J. Am. Chem. Soc.* **117,** 883-890.

EVALUATION OF HYDROGEN-BONDING INTERACTIONS IN MOLECULAR RECOGNITION BY ION TRAP MASS SPECTROMETRY

J.S. BRODBELT
Department of Chemistry and Biochemistry
University of Texas at Austin
Austin, TX 78712

1. Abstract

Hydrogen-bonds are among the most important interactions involved in selective complexation in host-guest chemistry. In this report, a variety of hydrogen-bonded crown ether/ammonium ion-complexes are generated in the gas phase by association reactions between an amine and a polyether in a quadrupole ion trap. The nature of the hydrogen-bonding interactions of the ion-complexes are evaluated by comparison of their collision-activated dissociation spectra. Energy-resolved collisional activated dissociation is used to estimate the association energy of each complex, a parameter which reflects the net strength of the hydrogen-bonding interactions. A quantitative evaluation of proton-bound polyether/amine complexes shows that the number of available hydrogen-binding sites, the gas-phase basicities of the polyether and amine components, and the geometry of the interactions correlates with the association energies.

2. Introduction

Intermolecular and intramolecular hydrogen-bonds are arguably the most important interactions in molecular self assembly and molecular recognition in solution chemistry [1-4]. Typical hydrogen-bond enthalpies range from 1-10 kcal/mole [3]. Hydrogen-bonding interactions are also significant factors in host-guest chemistry [8-17] and site-selective complexation, such as in antibiotic functions and enzyme catalysis. The phenomena of multiple binding interactions is especially well-modelled by crown ethers which have several identical oxygen donor sites which participate as hydrogen-bond acceptors. The array of binding sites is responsible for the size-selective complexation that is characteristic of macrocycles.

L. Echegoyen and A.E. Kaifer (eds.), Physical Supramolecular Chemistry, 249–260.

Hydrogen-bonding interactions are also important in the chemistry of gas-phase ions, in particular protonated molecules and proton-bound complexes [18-23]. One recent study reported that intramolecular hydrogen-bonds in protonated polyethers could be as strong as 25 kcal/mole as determined from high pressure mass spectrometric measurements of proton-transfer reactions [22]. The total association energies of various polyether/ammonium ion-complexes were determined to be 29 - 46 kcal/mole [23], depending on the number of binding sites of the polyether and the structure of the ammonium ion. In fact, multiple hydrogen-bonds could contribute up to an additional 21 kcal/mole of binding energy to an ammonium ion/polyether complex (as compared to a complex with a single binding interaction) [23].

In the present work, a systematic evaluation of the intermolecular binding forces responsible for the association of crown ethers (hosts) with amines (guests) is presented. The nature of hydrogen-bonding interactions of crown ether/ammonium ion-complexes are examined in a quadrupole ion trap mass spectrometer with respect to macrocyclic size effects, the gas-phase basicities of the ligands, and the number of possible binding interactions [24]. Low energy collisionally activated dissociation (CAD) is used as a means to assess the relative hydrogen-bonding characteristics of the various polyether/ammonium ion complexes. In the CAD process, ions are internally energized through multiple activating collisions with helium. The activation process can be well-controlled to allow evaluation of the binding energetics of the ions. Then energy-resolved CAD [25-28] is used to estimate the association energies of the complexes.

3. Experimental

All experiments were performed in a Finnigan Ion Trap Mass Spectrometer operated in the mass selective instability mode [29]. Samples were introduced into the chamber via a leak valve or solids probe inlet at pressures between 6.0 - 8.0 x 10^{-6} torr (uncorrected). The base pressure was raised to 1.2 x 10^{-4} torr He (uncorrected) after sample pressures had stabilized. A q-value of 0.4 and an activation time of 10 ms were used for all experiments unless otherwise noted. Ion-molecule reaction times were in the range of 10 - 150 ms while a cooling time prior to activation was held at 10 ms. The trap temperature was kept at 120 °C. The range of tickle voltages applied across the endcaps were between 10-400 mV. The threshold activation voltage was defined as that voltage where the fragment ion intensity is 10% of the total ion intensity. All samples were purchased from Aldrich Chemical Co. (Milwaukee, WI) with the exception of anhydrous ammonia (MG Industries).

4. Results and Discussion

The major objective of these studies is to develop a means of evaluating the differences in hydrogen-bonding interactions of polyether/ammonium ion complexes based on interpretation of many CAD experiments done to characterize the dissociation reactions of the complexes. In the context of this study, several types of CAD experiments assist in determining the nature of the hydrogen-bonding interactions of the complexes. First, one can examine energy-, time- and pressure-resolved collisional activation data to monitor the effects of energy deposition and time scale of activation. Second, one can begin to evaluate the importance of structural effects by varying the nature of the model amine guest (floppy vs. small polyatomic vs. bulky molecule) and the polyether host (flexible acyclic ether vs. more rigid cyclic ether). In the latter case, glymes are acyclic analogs to the cyclic crown ethers, and thus they have similar types of vibrational modes involved in internal energy re-distribution. Ultimately, in this systematic study, the dissociation trends observed for the different polyether/amine complexes are interpreted as reflecting the nature of the hydrogen-bonding interactions involved in stabilizing the ions in the gas phase.

4.1 COLLISIONALLY ACTIVATED DISSOCIATION

A variety of complexes of the form (polyether + H^+ + amine) were formed in the gas phase and then subjected to collisionally activated dissociation. Examples of the CAD results obtained are illustrated in Figure 1 for the complexes formed between 18-crown-6 and N,N-diethylmethylamine or methylamine. Activation of the proton-bound 18-crown-6/methylamine ion-complex results in extensive fragmentation of the crown ether ring by losses of ethylene oxide units. In contrast, collisional activation of the proton-bound 18-crown-6/N,N-diethylmethylamine ion-complex results in formation of predominantly protonated N,N-diethylmethylamine with little fragmentation of the crown ether ring, even under a range of quite energetic conditions.

The dissociation trends described the 18-crown-6/N,N-diethylmethylamine ion-complex are characteristic of the general type of hydrogen-bonded complexes which simply disassemble after activation. These complexes are presumably weakly bound, with little evidence for substantial energy tied into the binding interactions. The dissociation spectrum of the 18-crown-6/methylamine complex is one in which multiple hydrogen-bonding interactions between the two organic substrates apparently are significant. For example, the binding interactions may include: coordination of two of the crown ether oxygens to two of the amine hydrogens and formation of a proton-bridge formation between one of the crown ether oxygens and the nitrogen atom. In this case, simple decomplexation to form protonated 18-crown-6 and protonated methylamine is not the most favored process. Instead, ions due to cleavages and rearrangements of covalent bonds in the crown ether skeleton are predominant, leading to formation of ions at m/z 133 and 177. This behavior suggests that the interaction energy of the complex may be sufficiently high such that the energy necessary to surpass the dissociation threshold results in an internally hot macrocyclic ion which may then undergo further fragmentation.

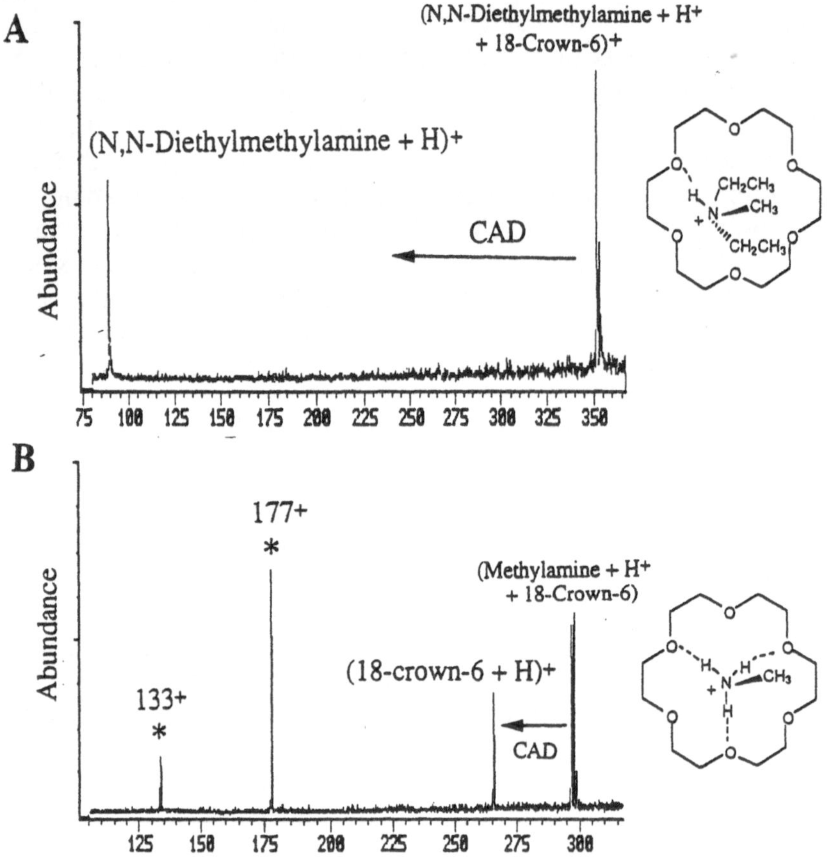

* crown ether fragment ion

4.2 POLYETHER SIZE EFFECTS ON COMPLEXATION INTERACTIONS

A comparison of the dissociation patterns of complexes involving ammonia and three different crown ethers in the quadrupole ion trap proved particularly intriguing because ammonia is the smallest model guest and it may in fact bind within the cavity of the larger crown ethers. The systematic use of ammonia as the guest means that the active vibrational modes in the amine portion of the complex remain the same throughout this set of comparative experiments, and thus cannot contribute to qualitative variations in the results. Summarized in Table 1 are the CAD spectra for the complexes generated from reactions of ammonia and pyridine with 12-crown-4, 15-crown-5, and 18-crown-6. In each case the ammonium ion complex was isolated and collisionally activated. The complex incorporating 12-crown-4 simply decomplexes after activation, and no skeletal fragmentation is observed. In contrast, skeletal decomposition is observed for the complex involving the 15-crown-5 ligand, and for the 18-crown-6 complex, the sum of the crown ether fragment ions is greater than the intensity of the intact protonated 18-crown-6. This striking variation in dissociation patterns supports an argument in which the strength of the collective hydrogen-bonding interactions of the complexes increases with the size of the crown ether and thus alters the resulting fragmentation patterns. Additionally, it should be noted that the number of internal vibrational modes of the polyether increases from 12-crown-4 to 15-crown-5 to 18-crown-6, presumably influencing the dissociation rates of the complexes. The overall slower decomposition rates of the complexes containing the larger crown ethers may allow increasing competition from pathways which involve macrocyclic bond cleavages.

Table 1. Distribution of Fragment Ion Current from CAD of Acyclic and Cyclic Polyether Ammonium Ion Complexes: (Polyether + H^+ + Amine).

	Distribution of Ion Current[a]				
Amine:	NH_3		Pyridine		
Polyether:	$(M + H)^+$	$\Sigma Frgs^+$	$(Pyr + H)^+$	$(M + H)^+$	$\Sigma Frgs^+$
diglyme	85	15	100	0	0
triglyme	60	40	100	0	0
tetraglyme	45	55	95	5	0
12-crown-4	>95	<5	100	0	0
15-crown-5	55	45	96	4	1
18-crown-6	30	70	80	16	4
21-crown-7	5	95	40	40	20

[a] $(M + H)^+$ represents the protonated polyether, and $(Pyr + H)^+$ represents protonated pyridine. $\Sigma Frgs^+$ represents the sum of all the polyether-related fragment ions.

4.3 COMPARISON OF COMPLEXES INVOLVING CYCLIC VS. ACYCLIC POLYETHERS

In order to further evaluate the significance of entropic effects in influencing the energetics and/or kinetics of the dissociation pathways, the CAD spectra of complexes incorporating acyclic polyether ligands such as glymes and glycols were compared to the spectra described for the complexes involving cyclic ethers (Table 1). Complexes involving two amine guests, pyridine and ammonia, and three acyclic ethers including diethylene glycol dimethyl ether, triethylene glycol dimethyl ether, and tetraethylene glycol dimethyl ether were examined. In terms of models hosts, the acyclic polyether ligands have similar numbers and types of vibrational modes compared to their crown ether counterparts and thus are relevant models to assist in probing the importance of the flexibility of the host structure on the dissociation behavior of the complexes. With respect to model guests, pyridine was selected because it represents a bulky guest with less propensity for multiple binding, whereas ammonia encourages multiple hydrogen-bonding interactions.

For all of the acyclic polyether/pyridinium ion complexes, dissociation results predominantly in decomplexation, forming protonated pyridine and protonated ether molecules. Fragmentation of the acyclic ether is never significant, regardless of the size or basicity of the polyether. This result re-affirmed that the formation of multiple hydrogen-bonds is not indicated for these complexes, just as noted for the crown ether/pyridinium ion complexes. For the complexes involving the larger crown ethers (21-crown-7, 18-crown-6), some macrocyclic skeletal fragmentation is observed, and the relative extent increases with the size of the ether.

In contrast, dissociation of many of the ammonium ion complexes results in extensive fragmentation of the ether ligands, especially for the larger crown ethers and glymes. The extent of ether skeletal cleavage increases with the number of oxygen binding sites of the ether. This latter result suggests that the number of possible hydrogen-bond acceptor sites influences the interaction energy in a way similar to the trend noted previously for the crown ethers. Secondly, the observation of extensive fragmentation of the acyclic ether ligands, just as was observed for the complexes involving cyclic ethers, supports the idea that the flexible acyclic ligands may participate in the types of multiple bonding interactions already described for the pre-organized crown ethers.

However and most interestingly, some of the crown ethers show less extensive fragmentation overall than their acyclic counterparts. For example, the ammonium ion complex involving tetraglyme shows an abundance ratio for $(M + H)^+$: Σ (Fragment ions) equal to 4:5, whereas the ratio is only 5:4 for 15-crown-5. The hydrogen-bonds in the tetraglyme complex may be collectively more favorable than those in the 15-crown-5 complex because tetraglyme is a more flexible ligand which can wrap around ammonia whereas ammonia is too large to fit in the cavity of 15-crown-5. Also, the kinetics for dissociation may be sufficiently slower for the floppy acyclic ligands, permitting greater competition from slower dissociation pathways. In either case, this result demonstrates that the cyclic ethers have qualitatively similar but quantitatively different binding natures than the acyclic ligands.

4.4 MEASUREMENT OF ASSOCIATION ENERGIES

Binding energies of gas-phase ions can be estimated by examination of the dissociation behavior of collisionally activated ions. The extent of fragmentation of an ion is monitored as a function of the energy deposition, then the amount of energy deposition is correlated with a calibrated scale of binding energies. Energy-resolved CAD experiments were performed for a variety of polyether/amine complexes. The energy-resolved experiments showed that the number of available hydrogen-binding sites, the gas-phase basicities of the polyether and amine components, and the geometry of the interactions correlated with the threshold values (Table 2). For example, the 12-crown-4/NH_4^+ complex dissociated at a energy value of 32 kcal/mol, while the 18-crown-6/NH_4^+ complex dissociated at 41/mol kcal. This result suggests that the geometry of the hydrogen-bonds involved in the 18-crown-6/NH_4^+ complex are enhanced by the greater flexibility of the larger polyether ring size. When a tertiary amine is involved in the complexes instead of ammonia, then the association energies are significantly reduced. For example, the association energy of the (18-crown-6 + H^+ + N,N-diethylmethylamine) complex is only 34 kcal/mole, and the (12-crown-4 + H^+ + N,N-diethylmethylamine) complex is unstable. This result is attributed to the reduction in the number of hydrogens attached to the nitrogen of the guest molecule that can serve as hydrogen-bond donors, therefore permitting fewer bonding interactions between the amine and crown ether. A similar effect is noted for the comparison of the association energies for the complexes involving butylamine vs. N-methylbutylamine. The methyl group of N-methylbutylamine reduces the number of potential hydrogen-bonding interactions to the polyether. In Table 2, the association energy for the dissociation of the (18-crown-6 + H^+ + butylamine) complex is listed as > 50 kcal/mole because its association energy exceeded the limits of the calibration curve.

Table 2. Association Energies of Hydrogen-Bonded Complexes (All values in kcal/mol)

	NH_3	N,N-DEMA	Butyl amine	N-MBA	N,N-DMEDA	EDA	1,3 DAP	Pyridine
12-crown-4	32	n/a	37	34	n/a	35	34	33
15-crown-5	35	30	44	38	34	41	39	36
18-crown-6	41	34	>50	41	39	>50	>50	35
Triglyme	32	28	37	35	n/a	36	33	31
Tetraglyme	35	29	43	39	n/a	39	35	34

n/a dimer does not form

N,N-DEMA is N,N-diethylmethylamine; N-MBA is N-methylbutylamine; N,N-DMEDA is N,N-dimethylethylenediamine; EDA is ethylenediamine; 1,3-DAP is 1,3-diaminopropane

Complexes between polyethers and diamines were also evaluated. The association energies for complexes incorporating ethylenediamine and 1,3 diaminopropane are similar for each of the crown ethers. All of the complexes are more strongly bound (34 to > 50 kcal/mole) than the simpler complexes involving NH_3. This result suggests that each amine functional group can participate in intermolecular hydrogen-bonding interactions with the crown ether oxygen atoms, and both ethylenediamine and 1,3-diaminopropane are flexible enough to achieve similar optimized interactions. However, the interaction energy is reduced for N,N-dimethylethylenediamine because one nitrogen has two methyl groups instead of hydrogens. This structural change results in a reduction of the binding energy to the crown ethers by up to 10 kcal/mole.

With respect to the association energies of the polyether/amine complexes, there is a definite interplay between the intrinsic gas-phase basicity of each component in the complex and the number of possible hydrogen-bonding interactions of each component. As the number of possible hydrogen-bonding interactions increases, the potential for a greater association energy increases. For example, NH_3 has three hydrogens that may serve as donors in hydrogen-bonding interactions, whereas pyridine only has one. From this simple view, ammonia has a more favorable structure for generating strongly-bound complexes. However, the difference in gas-phase basicities between the two components in a complex also has a great impact on the stability of the complex. For complexes in which the difference in intrinsic basicities is large, then one component will bind the proton more strongly, therefore reducing the association energy for dissociation. For instance, the gas-phase basicity of pyridine is 213.1 kcal/mole, the gas-phase basicity is only 195.6 kcal/mol for ammonia, and the gas-phase basicity of 15-crown-5 is 212.5 kcal/mol [30]. The intrinsic basicities are most similar for pyridine and 15-crown-5, indicating that the (15-crown-5 + H^+ + pyridine) complex is predicted to be thermochemically most stable, assuming formation of complexes with only one strong hydrogen-bond. The association energy for the 15-crown-5/NH_3 complex is 35 kcal/mol, whereas the association energy for the 15-crown-5/pyridine complex is 36 kcal/mole, despite the fact that the latter complex has only one active hydrogen-bond. The 15-crown-5/NH_3 complex potentially has three or four hydrogen-bonds which creates a large association energy and consequently a high association energy for dissociation, but the large difference in gas-phase basicities likely moderates the net effect. This latter factor is greatly reduced when the amine guest is changed from NH_3 to butylamine, which still permits three hydrogen-bonding interactions but has a larger intrinsic basicity overall (210.6 kcal/mol [30]). The gas-phase basicity of butylamine is substantially closer to that of 15-crown-5 relative to the difference in basicities between 15-crown-5 and ammonia. The association energy for the butylamine/15-crown-5 complex is estimated as 44 kcal/mol, showing an increase of 9 kcal/mole relative to the NH_3/15-crown-5 complex. In summary, by correlating the gas-phase basicities and number of possible hydrogen-bonding interactions, an understanding of the structural factors which influence the stabilities of loosely-bound complexes can be obtained.

5. Conclusions

The formation of multiple hydrogen-bonds can have striking effects on the dissociation behavior of polyether/ammonium ion-complexes because the multiple hydrogen-bonding interactions allow formation of strongly-bound complexes, in contrast to the loosely-bound proton-bridged complexes that are typically formed by ion-molecule association reactions. Total hydrogen-bonding energies may exceed 40 kcal/mole for strongly bound complexes. The extent of polyether skeletal cleavage can thus be correlated with the collective hydrogen-bonding interactions of the complex. Both the number of possible interactions and thermodynamic factors affect the capability of any polyether and amine for forming a strongly-bound complex. These studies have provided insight into some of the requirements for multiple binding interactions of simple model host-guest systems in the gas phase and have shown that the occurrence of such multiple binding interactions can be evaluated experimentally by application of CAD mass spectrometric techniques. Moreover, the results of this study have illustrated that model host-guest complexes formed in the gas phase may demonstrate binding properties that parallel the types of multiple interactions that govern complexation in solution.

6. Acknowledgements

The support from the Welch Foundation (F-1155), NIH, the National Science Foundation, the Dreyfus Foundation and the Sloan Foundation are acknowledged.

7. References

1. Rebek, Jr., J., Askew, R., Ballester, P., Costero, A., "Convergent Functional Groups. 5. Ternary Complexes in the Molecular Recognition of Arylethylamines", (1986) *J. Am. Chem. Soc.*, **110,** 923.
2. Etter, M.C., "Encoding and Decoding Hydrogen-Bond Patterns of Organic Compounds", (1990) *Acc. Chem. Res.*, **23,** 120.
3. Vinogradov, S.N., Linnel, R.H. (1971), *Hydrogen Bonding*, Van Nostrand Reinhold, New York.
4. Capon, B., McManus, S.P., (1976) *Neighboring Group Participation*, Plenum Press, New York, vol. 1.
5. Caffrey, M.S., Daldal, F.; Holden, H.M., Cusanovich, M.A., "Importance of a Conserved Hydrogen-Bonding Network in Cytochromes c to their Redox Potentials and Stabilities", (1991) *Biochemistry*, **30,** 4119.
6. Jeffrey, G.A.; Maluszynska, H.; (1981) *Int. J. Quantum Chem: Quantum Biology Symposium*, **8,** 231.
7. SantaLucia, J.; Kierzek, R., Turner, D.H. "Stabilites of Consecutive A-C, C-C, G-G, U-C, and U-U Mismatches in RNA Internal Loops: Evidence for Stable Hydrogen-Bonded U-U and C-C Pairs", (1991), *Biochemistry*, **30**, 8242.

8. Cram, D., "The Design of Molecular Hosts, Guests, and their Complexes", (1988)*Science*, **240,** 760.
9. Lehn, J.M., "Supramolecular Chemistry-- Scope and Perspectives: Molecules, Supermolecules, and Molecular Devices", (1988) *Angew. Chem. Int. Ed. Engl.*, **27,** 89.
10. Sutherland, I., "Molecular Recognition by Synthetic Receptors", (1986) *Chem. Soc. Rev.*, **15,** 63.
11. Vogtle, F., Weber, E., Eds.(1985), *Host Guest Complex Chemistry: Macrocycles*, Springer-Verlag: New York.
12. Takeda, Y., "The Solvent Extraction of Metal Ions by Crown Compounds", (1984) *Top. Curr. Chem.*, **121,** 1.
13. Weber, E., Vogtle, F., (1981)*Top. Curr. Chem.*, **98**, 1.
14. Cram, D.J., Trueblood, K.N., (1981)*Top. Curr. Chem.*, **98,** 43.
15. Izatt, R.M., Eatough, D.J., Christensen, J.J., "Thermodynamics of Cation-Macrocyclic Compound Interactions" (1973) *Struct. Bonding*,**16**, 161.
16. Frensdorff, H.K., "Stability Constants of Cyclic Polyether Complexes with Univalent Cations", (1971) *J. Am. Chem. Soc.*, **93,** 600.
17. Bonas, G., Bosso, C. Vignon, M.R. "Determination of Stability Constants of Macrocyclic Ligan-Alkali Cation Complexes by Fast Atom Bombardment Mass Spectrometry", (1989), *J. Incl. Phen. Mol. Recogn. in Chem.*, **7,** 637.
18. T. Morton, J. Beauchamp, "Chemical Consequences of Strong Hydrogen Bonding in the Reactions of Organic Ions in the Gas Phase. Interaction of Remote Functional Groups" (1972)*J. Am. Chem. Soc.*, **94,** 3671.
19. D. Aue, H. Webb, M. Bowers, "Quantitative Evaluation of Intramolecular Strong Hydrogen Bonding in the Gas Phase", (1973) *J. Am. Chem. Soc.*, **95**, 2699.
20. C.-C. Liou, E. Eichmann, J.S. Brodbelt, "Intermolecular Dehydration Reactions of Protonated Alkenol Adducts", (1992) *Org. Mass Spectrom.*, **27** , 1098.
21. Sharma, R.B., Blades, A.T., Kebarle, P., "Protonation of Polyethers, Glymes and Crown Ethers in the Gas Phase", (1984) *J. Am. Chem. Soc.*, **106,** 510-516.
22. Meot-Ner, M., "The Ionic Hydrogen Bond. 2. Intramolecular and Partial Bonds. Protonation of Polyethers, Crown Ethers and Diketones", (1983)*J. Am. Chem. Soc.*, **105**, 4906-4911.
23. Meot-Ner, M., "The Ionic Hydrogen Bond. 3. Multiple NH---O and CH---O bonds. Complexes of Ammonium Ions with Polyethers and Crown Ethers", (1983)*J. Am. Chem. Soc.*, **105,** 4912-4915.
24. Liou, C.-C.; Wu, H.-F.; Brodbelt, J. S. " Hydrogen-Bonding Interactions in Polyether/Ammonium Ion Complexes", (1994) *J. Am. Soc. Mass Spectrom.* **5**, 260 - 273.

25. Fetterloff, D. D.; Yost, R. A. "Energy-Resolved Collision-Induced Dissociation in Tandem Mass Spectrometry", (1982) *International Journal of Mass Spectrometry and Ion Physics.*, **44**, 37-50.
26. Dawson, P. H. "The Collision Induced Dissociation of Protonated Water Clusters Studied Using A Triple Quadrupole", (1982) *International Journal of Mass Spectrometry and Ion Physics.* **43**, 195-209.
27. Armentrout, P. B. (1992)"*Thermochemical Measurements by Guided Ion Beam Mass Spectrometry*", in N. Adams, L. Babcock, eds., "*Advances in Gas Phase Ion Chemistry*", JAI Press Inc., Greenwich, 1, 83.
28. Anderson, S. G.; Blades, A. T.; Klassen, J.; Kebarle, P. "Determination of Ion-Ligand Bond Energies and Ion Fragmentation Energies of Electrospray -Produced Ions by Collision-Induced Dissociation Threshold Measurements", (1995) *International Journal of Mass Spectrometry and Ion Processes* **141**, 217-228.
29. R. E. March, R. J. Hughes and J. F. J. Todd (1989), *Quadrupole Ion Storage Mass Spectrometry*, Wiley: NY,.
30. Lias, S. G.; Liebman, J. F.; Levin, R. D. "Evaluated Gas-Phase Basicities and Proton Affinities of Molecules: Heats of Formation" (1984) *J. Phys. Chem. Ref. Data*, **13**, 695.

DIFFRACTION STUDIES OF SUPRAMOLECULAR COMPOUNDS

Jerry L. Atwood
Department of Chemistry
University of Missouri-Columbia
Columbia, MO 65211

Small molecule crystallographers have been quite successful in attacking problems of fundamental significance to chemistry. The results of these endeavors have played a critical role in the development of almost all areas of chemistry, not just those directly related to structure and bonding. Great detail is known about the metrical parameters of more than 100,000 molecules subjected to single crystal X-ray diffraction study. In sharp contrast, structural studies of molecules of biological dimensions by means of X-ray diffraction techniques have been few indeed. The results of protein crystallography have, however, had a very great impact on the development of molecular biology and related fields. This is so, even though the metrical parameters of many proteins are not clearly revealed by the diffraction studies. Often, features of interest in protein structures are not revealed at all by single crystal studies. Of course, all scientists recognize the source of the difficulty in structurally characterizing molecules from the realm of biology: the molecules are very large and do not generally crystallize well (if they crystallize at all).

One of the major thrusts of supramolecular crystallography is to bridge the gap between small molecule and very large molecule crystallography. We seek to build up large assemblies from component parts, assemblies which will exhibit features of interest with respect to proteins. We search for the molecule or assembly of intermediate size: large enough to be relevant to biological systems but small enough to reveal details of structure and bonding.

Before discussing detailed investigations as examples of the use of diffraction in the study of supramolecular compounds, it is necessary to address that which is special about the crystallography of supramolecular compounds. It will be seen below that this area of crystallography does not present the same problems or expectations as does that of small molecules or of proteins.

The major difficulties associated with the crystallography of supramolecular compounds may be divided into five categories.

Crystal Growth. One strives for the best possible crystals, but with large assemblies crystals are often small and/or poorly diffracting. The molecular unit, or assembly, is often organized by weak interactions, interactions without strong directionality. Thermal motion may be quite high. The clear voice of experience tells one to spend the time, within reason, to get the best possible crystalline sample. A wide range of crystallization

L. Echegoyen and A.E. Kaifer (eds.), Physical Supramolecular Chemistry, 261–272.

techniques, including those utilized in the protein field, are available. However, often one is forced to proceed with data collection on a crystal of poorer quality than small molecule crystallographers normally accept.

Data Collection. If one begins with a poorly diffracting crystal, one will certainly suffer with a lack of observed data. Using conventional moving crystal, moving detector technology, it may be attractive to count each reflection for a long time in order to obtain as many 'observed' reflections as possible. This has two major disadvantages: an often overloaded diffractometer is tied up for an unacceptably long period of time and it is very difficult to gain a significant increase in the observed data. There is also a cost factor, since one pays either in funding or in access for data collection.

Structure Solution. A paucity of observed diffraction data quite often means that conventional direct methods fail to provide a solution to the problem. Many supramolecular compounds of interest do not possess heavy atoms, or, if they do, there may be too many of them (with regard to this point, making structure solution by heavy atom methods difficult). With many structures which are solved by conventional methods, there is still a time factor: the solution is not normally straightforward. Note that even for crystals which diffract well, some of the problems are beyond the reach of conventional techniques, problems with > 300 non-hydrogen atoms in the asymmetric unit.

Structure Refinement. The lack of observed data has already been recognized, as has the problem of high thermal motion. To these can often be added disorder. The non-directionality of bonding can lead to both static and dynamic disorder. (This also makes both the solution of the structure and the interpretation of the data more difficult.) The above-mentioned problems generally lead to high R values. When this author did a study of published calixarene structures some time ago, the average conventional R value was > 0.11. High R values may lead to difficulties in publication: many referees, editors, and journals now hold crystallographic structure determinations to the same standards, regardless of the compound, supramolecular or small-molecule. R values are not often discussed for protein structure determinations.

Structure Interpretation. In the view of this author, the standard for the R value should not be so rigidly enforced for supramolecular compounds. On the other hand, the standard for the interpretation of the structure should be more tightly enforced. Problems associated with a complex, complicated structure, held together by weak interactions, and displaying an array of types of disorder, lead to great difficulty in structure interpretation. To this one must add the reality that an interaction of little or no importance to one investigator may be of great interest to another.

Crystallographers realize that the problems discussed above originate in large part from initial crystal quality. If the supramolecular chemist does the best possible job in obtaining crystals, there is now new hope in the form of very sensitive area detectors, particularly those based on the CCD technology. Instruments based on this technology are capable of greatly improving the quality and quantity of diffraction data, and at a vast reduction in time for a large problem. This is illustrated in Example 2 below.

Example 1. O-H..., N-H..., and C-H...π Aromatic Hydrogen Bonding

The first example concerns the role of water which is exposed to "hydrophobic" molecular regions. Proteins, during the normal course of synthesis and activity, must expose hydrophobic surfaces to water. Indeed, proteins in the crystalline state are most certainly enshrouded by a shell of water molecules. Supramolecular complexes have revealed the action of water with a range of hydrophobic molecular entities. The O-H...π aromatic hydrogen bonding interaction of water, exemplified in Figure 1, is one such example [1]. Indeed, the subject of O-H...π hydrogen bonding has received considerable recent attention. Jorgensen and Severance used Monte Carlo methods to predict the O...centroid separation to be 3.11 Å in $C_6H_6...H_2O$ with one hydrogen atom of the water molecule directed toward the center of the aromatic ring [2]. Our group then found experimental evidence for such aromatic π hydrogen bonding to water in Na_4[p-sulfonatocalix[4]arene]·13.5 H_2O and several related complexes [1]. O-H...π hydrogen bonding has also been found in alcohols [3-5] and in a silanol [6].

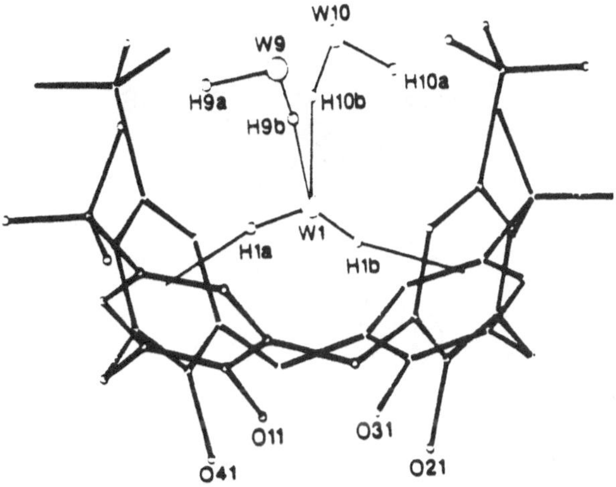

Figure 1. Environment of the water molecule W1, which displays π hydrogen bonding to two aromatic rings.

C-H...π hydrogen bonding has been found in two host-guest complexes [7] and in several other systems [8,9]. Hanton, Hunter, and Purvis have added an example of intramolecular N-H...π hydrogen bonding [10]. General crystallographic evidence for N-H...π and O-H...π hydrogen bonding has been recently discussed [11].

From these first studies of this important electrostatic interaction, it is seen that the X...centroid distances are of the order of

C...centroid = 3.5 Å,
N...centroid = 3.3 Å,
O...centroid = 3.1 Å,

as would be expected on the basis of atomic size. This is further supported by a set of *ab initio* calculations of benzene/water and benzene/ammonia. Here it is reported that the expected N...centroid distance is 0.26 Å greater than that for O...centroid [12]. However, we have now found, in a host-guest system an example of N-H...π hydrogen bonding; an abnormally short N...aromatic distance is revealed.

In the solid state, $[HNC_5H_4C_5H_4NH^{2+}]_2$[p-sulfonatocalix[4]arene]·7 H_2O displays two types of 4,4'-bipyridinium cations: one is intercalated into the bilayer structure [13] and one is partially contained in the calix[4]arene cavity. The latter one is oriented such that the N-H^+ portion is directed at the center of one of the aromatic rings of the calix[4]arene, Figure 2. The N...centroid distance is 3.06 Å. With the assumption of a normal N-H bond length, this leads to an H...centroid distance of 2.01 Å, much shorter than the H...centroid separation of 2.42 Å [14] found by Hanton, Hunter, and Purvis [10]. Indeed, the N...centroid distance is even shorter than the O..centroid distances, both observed [1] and calculated [2]. Further studies of aromatic π hydrogen bonding are in progress.

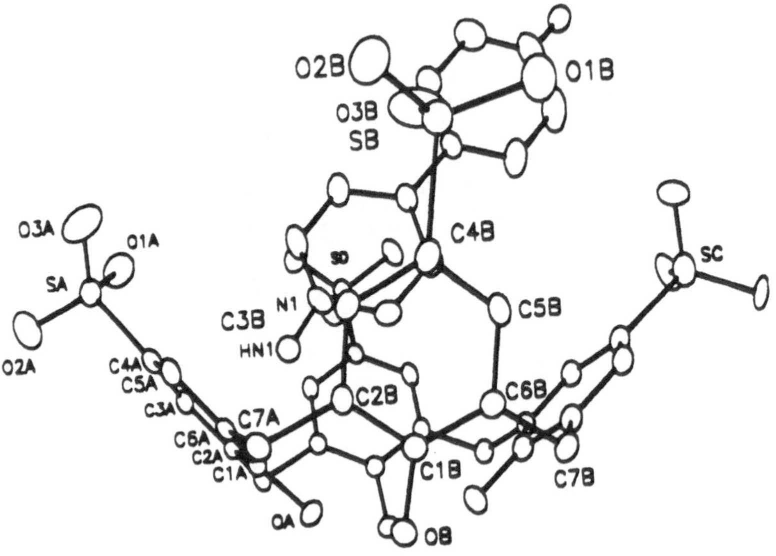

Figure 2. Structure of $[HNC_5H_4C_5H_4NH^{2+}]_2$[p-sulfonatocalix[4]arene]·7 H_2O, illustrating the N-H...aromatic π hydrogen bonding.

Supramolecular complexes in which methyl groups face layers of water molecules provide examples of C-H...O(water) interactions. Such studies are of benefit for at least two reasons: they provide proof that such interactions can exist and they afford metrical parameters which can be interpreted in terms of strengths of bonding. A model system for this interaction is found in the structure of a protonated aminomethylcalix[4]arene, [p-(dimethylaminomethyl)calix[4]arene]·$(HCl)_2$·9 H_2O [15]. Here, layers of methyl groups face onto a layer of water molecules (approximately four water molecules thick). Unfortunately, the protons have not been located with the present data, but it is clear that diligent effort will reveal the nature of the C-H...O(water) interaction.

Carbonic anhydrase provides an example of the importance of ordered water in a biological system. The crystal structure of the bio-molecule [16] shows that there is an ordered arrangement of eight water molecules in a hydrophobic pocket near the site of catalytic action. The synergism of synthesis and crystallography must be noted. The existence of the hydrophobic pocket with eight water molecules provides the synthetic supramolecular chemist with the goal of constructing such a pocket from small building blocks.

Example 2. Host-Guest Complexes Based on Fullerenes

Over the past four years our group, in conjunction with that of Prof. Colin L. Raston, has endeavored to prepare and study crystalline complexes of fullerenes with calixarenes and cyclotriveratrylenes (CTV, **1**). We have collected data on more than ten compounds. The longest low temperature data collection was six weeks on a CAD4. During this four-year period we have used approximately one year of single crystal diffractometer time. The only publication [17] to arise from this work is described briefly below.

The compound has the formula $(C_{60})_{1.5}(CTV)(C_7H_8)_{0.5}$, and presents the following crystal data: (298K, Enraf-Nonius CAD4 diffractometer): M = 1577.6, monoclinic, C2/m, a = 30.131(9), b = 17.436(5), c = 14.638(5) Å, β = 116.33(2)^o, U = 6892 Å^3, Z = 4, D_c = 1.53 g cm^{-3}, μ(Mo-K$_\alpha$) = 0.87 cm^{-1}, 5707 unique reflections with $2\theta_{max}$ = 50°, 1813 with $I > 3\sigma(I)$ used in the refinement. The structure was solved by direct methods (SHELX-86) and refined using alternating full matrix least squares and difference fourier synthesis (SHELX-76). The well-defined (C_{60})(CTV) complex resides across a mirror plane while the highly disordered second fullerene and toluene of crystallization are located on a 2/m site. Final R = 0.15.

Even though much time was devoted to the collection of the data (full data was collected on three different crystals), the best effort yielded only 32% observed data. Fortunately, this was sufficient for structure solution. However, much time and effort were required for such solution.

With regard to the interpretation of the structure, first note that the structure confirms the stoichiometry established by analysis: the asymmetric unit comprises 1/2 of a 1:1 (C_{60})(CTV) complex (the other half generated by a mirror plane), 1/4 of a highly disordered C_{60} residing over a *2/m* site symmetry, and possibly a 1/4 of a highly disordered toluene molecule of the same site symmetry. In the published account of the structure, the following observation was made: "While the structure is of low precision in consequence of this disorder, the $(C_{60}$.CTV) complex gave a meaningful refinement, clearly showing docking of the fullerene with the CTV," Figure 3. It was further noted: "The lack of disorder of this C_{60} relates to its contact with the rigid CTV molecule, and also to the matching of symmetry elements; the three-fold axis of the CTV coincides with a three-fold axis of C_{60} such that the C-9 ring of CTV lines up with a C-6 ring of the fullerene with the three immediate C-5 rings residing over the three aromatic rings of the CTV." Even at an R value of 0.15, important information was revealed by the structure. Much of the problem with this supramolecular compound doubtless results from the highly disordered C_{60} in the structure which appears to act as space filler in the packing of the ordered (C_{60}.CTV) units.

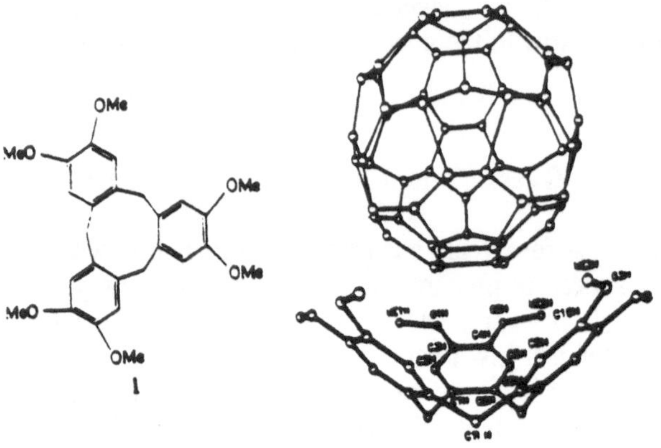

Figure 3. Ball and socket C_{60}.CTV structure (1 = CTV).

The only structure solution of a complex of a calixarene with C_{60} is for the p-phenylcalix[4]arene. Here also, a small data set eventually afforded the structure. The C_{60} is not in the calixarene cavity, but rather is packed on the exterior, among the π-surfaces. Unfortunately, the best efforts, including data collection on a CCD-based diffractometer, has led only to R > 0.20.

Example 3. Anion Complexation by Metallated Calixarenes

Our group has experienced great success in producing a new class of anion complexation agents based on calixarenes and cyclotriveratrylenes. We have shown that the binding of transition metal ions to the carbocyclic rings of the outer face of calixarenes can markedly modify their inclusion properties, resulting in the inclusion of polar and anionic guests within the macrocyclic cavity [18,19]. Similar metallation of CTV results in the formation of new inclusion species in which the guest is held within the bowl of the CTV molecule in close proximity to a redox active transition metal center. The formulations of the compounds were confirmed by FAB mass spectrometry. The structure of a key supramolecular compound based on CTV was revealed by a single crystal X-ray structure determination, Figure 4. The compound has the formula $C_{57}H_{72}O_6Ru_3.6BF_4.9H_2O$, and the crystal data is as follows: M = 1839.36, yellow plate, cubic, Pa3, a = 25.173(2) Å, U = 15953(3) Å^3, Z = 8, D_c = 1.54 g cm^{-3}, μ(Mo-K_α) = 6.61 cm^{-1}. Of 5357 data collected on an Enraf-Nonius CAD4 diffractometer (2θ = 4 - 50^0 at 20^{0}C) 3400 were independent and 1775 were judged observed [I > 3σ(I)]. The structure was solved and refined as for the fullerene complex in Example 2. The asymmetric unit was found to contain one third of one cationic molecule and a total of two tetrafluoroborate anions. The final difference electron density map also revealed the presence of three significant peaks corresponding to disordered solvent of crystallization, probably water. The remaining tetrafluoroborate anion was found to be disordered over three sites and, as a consequence, was extremely poorly defined. The final refinement converged with R = 0.079.

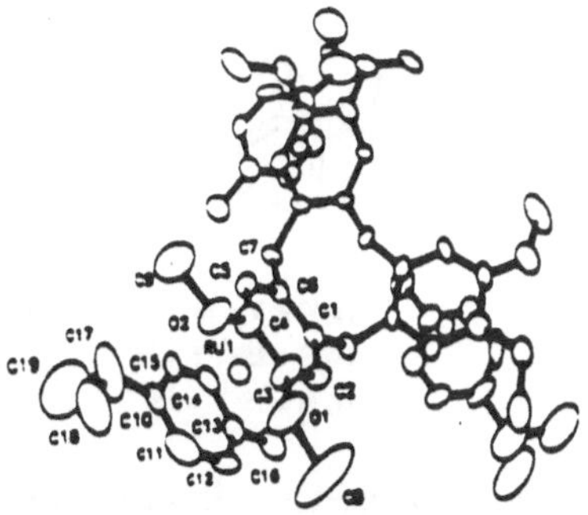

Figure 4. Cation in $[\{Ru(\eta^6\text{-}MeC_6H_4\text{-}4\text{-}CHMe_2)\}_3(\eta^6{:}\eta^6{:}\eta^6\text{-}CTV)][BF_4]_6$. The guest anion is badly disordered and is not displayed.

Within the asymmetric unit one of the two tetrafluoroborate anions is disordered over three sites; two of these sites are situated upon three-fold rotation axes while, most importantly, the final anion is distributed in a highly disordered fashion within the cavity of the CTV host. This disorder probably arises from a size mismatch between that small BF_4^- anion and the large, shallow cavity of the CTV. The structure does, however, demonstrate the principles (i) that the characteristic packing of free CTV may be disrupted such as to allow guest molecules to enter the cavity, and (ii) the electronic properties of the electron rich CTV cavity may be modified by appending of transition metal centers in order to allow the inclusion of anionic guest species, both interpretations which do not depend on precise metrical parameters.

Recent work has also been directed towards the synthesis of calixarene-based bowl-lile host molecules in which a catalytically or redox active metal center is either embedded within, or is in close proximity to the hydrophobic cavity of a calixarene. In order to achieve this goal without the interference of solvent or other potential guest molecules the possibility of direct covalent attachment of transition metals to the calixarene aromatic rings was investigated.

Shinkai and co-workers have reported the synthesis of neutral π-chromium tricarbonyl complexes of **1c** (R = H, Y = n-Pr) in which the metal binds to the outer face of the calixarene, and have demonstrated their usefulness in chemical modification of the host molecules [20]. However, these air-sensitive metal complexes are only formed under forcing conditions (130°C) in conjunction with a very limited range of especially rigid, unhindered calix[4]arenes and are unlikely to display properties such as water solubility and the ability to accommodate polar guests, which are important in mimicking conditions *in vivo* [21-23].

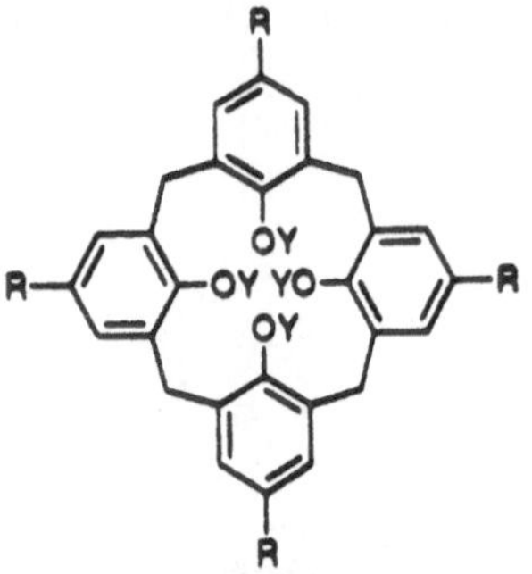

1a R = Y = H
1b R = Bu^t, Y = H
1c R = H, Y = Pr^n

Accordingly, we have synthesised a wide variety of more robust, cationic model compounds which are stable to both air and moisture and are likely to entrap a wide range of guest molecules within the hydrophobic cavity.

Treatment of the rhodium complex **2a** with Ag[BF_4] in acetone followed by refluxing with calix[4]arene **1a** in neat CF_3CO_2H resulted in the formation of the bimetallic complex [{Rh(η^5-C_5Me_5)}$_2$(η^6:η^6-$C_{28}H_{24}O_4$)][BF_4]$_4$ **3a** in 95% yield, even in the presence of excess [Rh(η^5-C_5Me_5)(acetone)$_3$]$^{2+}$. The formulation of **3a** was confirmed by a FAB mass spectrum which clearly indicated a molecular cation peak at *m/z* 898 as well as a smaller signal at *m/z* 985 corresponding to the cation in (**3a**) associated with one tetrafluoroborate anion.

Previous reports have noted that the related Cp*rhodium p-phenol complexes [Rh(η^5-C_5Me_5)(η^6-PhOH)][PF_6]$_2$ tend to deprotonate, forming [{Rh(η^5-C_5Me_5)}$_2$(η^6-PhO-H...O(η^6-)Ph)][PF_6]$_3$. Under the conditions employed in the present work, however, the calixarene oxygen atoms are fully protonated due to the presence of CF_3CO_2H (a signal due to the OH protons was located in the ^{1}H NMR spectrum of **3a** as an extremely broad resonance at *ca.* d 5.3 ppm, while the infrared spectrum demonstrated a broad band assigned to ν(OH) at 3420 cm^{-1}). The broadness and high field chemical shift of the OH NMR signal suggests that the hydroxylic protons in **3a** may be relatively acidic and only loosely associated with the cationic complex. Treatment of **3a** with excess Na_2CO_3 in acetone resulted in the loss of two protons to form the dicationic complex [{Rh(η^5-C_5Me_5)}$_2$(η^6:η^6-$C_{28}H_{22}O_4$)][BF_4]$_2$ **4a**.

Interestingly, if **1a** is refluxed in acetone with [Rh(η^5-C_5Me_5)(acetone)$_3$]$^{2+}$ in the presence of only a few drops of CF_3CO_2H a yellow precipitate of a further compound, **5a**, is obtained. More strikingly, the majority of resonances in the spectrum of **5a** were split into pairs (*e.g.* for the Cp* rings singlets at δ 2.16 and 2.15 ppm were observed) suggesting an asymmetric complex in which there is a slight difference between the two coordinated rings. This data, in conjunction with the less acidic conditions employed in the synthesis of **5a** leads us to formulate the complex as the singly deprotonated trication $[\{Rh(\eta^5\text{-}C_5Me_5)\}_2(\eta^6\text{:}\eta^6\text{-}C_{28}H_{23}O_4)][BF_4]_3$.

Reaction of the iridium complex **2b** with **1a** in acetone/CF_3CO_2H results in the formation of the iridium analogue of **3a**, $[\{Ir(\eta^5\text{-}C_5Me_5)\}_2(\eta^6\text{:}\eta^6\text{-}C_{28}H_{24}O_4)][BF_4]_4$ **3b** in 56% yield. Interestingly however, reaction of **2b** (pre-treated with $Ag[BF_4]$) with **1a** in neat CF_3CO_2H resulted in the surprising isolation of the tetrametallic species $[\{Ir(\eta^5\text{-}C_5Me_5)\}_4(\eta^6\text{:}\eta^6\text{:}\eta^6\text{:}\eta^6\text{-}C_{28}H_{24}O_4)][BF_4]_8$ **6b** in 76% yield, even in the presence of a 1 : 1 mole ratio of **2b** to **1a**. Calix[4]arene is extremely poorly soluble in CF_3CO_2H and a significant quantity of the undissolved macrocycle was recovered at the end of the reaction suggesting that the primary metalation step is rate limiting under these conditions. The FAB mass spectrum of **6b** exhibited peaks corresponding to calix[4]arene coordinated to one, two, three and four Cp^*Ir fragments whilst the symmetrical nature of the product was clearly indicated by its ^1H NMR spectrum which exhibited a single set of resonances for the calixarene ligand at similar chemical shift to those assigned to the coordinated faces of **3b**. The observation of the characteristic AB pattern for the methylinic bridges indicates a cone conformation [24].

Reaction of complexes **2** in neat CF_3CO_2H with the more sterically hindered *p*-t-butylcalix[4]arene **1b** again led, in almost quantitative yield, to the bimetallic complexes $[\{M(\eta^5\text{-}C_5Me_5)\}_2(\eta^6\text{:}\eta^6\text{-}C_{44}H_{56}O_4)][BF_4]_4$ (M = Rh **7a**, Ir **7b**), related to **3a** and **3b**. Tetrametallic species were not formed even in the presence of excess metal complex, presumably because of the highly sterically hindered nature of the *p*-t-butylcalixarenes. Complexes **7** were characterised as described above and exist solely in the cone conformation.

Consistent with the high acidity observed for **3**, slow crystallization of complexes **7** from NO_2CH_3 / Et_2O gave two new species, **8a** and **8b**, exhibiting ^1H NMR spectra similar to that of **5a** suggesting them to be the tricationic complexes $[\{M(\eta^5\text{-}C_5Me_5)\}_2(\eta^6\text{:}\eta^6\text{-}C_{44}H_{56}O_4)][BF_4]_3$. In confirmation the iridium complex **8b** as a diethyl ether/nitromethane solvate was characterised by a single crystal X-ray structure determination, Figure 5. The structure of **8b**.Et_2O.NO_2Me, obtained from crystals grown from slow vapour diffusion of diethyl ether into a nitromethane solution of the complex, is an excellent illustration of the type of host-guest interaction which make this class of compounds of interest. The calix[4]arene host adopts the cone conformation with the two iridium centers situated upon opposite sides of the macrocycle. Deeply embedded within the hydrophobic cavity is a molecule of diethyl ether. Calculated hydrogen atom positions suggest that the $-CH_3$ moiety C(1)S is orientated such as to form weak C-H...π hydrogen bonds with the unmetalated aromatic rings (B, D) while avoiding the partially positively charged coordinated decks. The hydrogen bonding network at the calixarene oxygen atoms is asymmetric with short contacts from Ob to Oa and Oc of 2.52 and 2.64 Å, respectively, whilst longer bonds are formed to Od; Od...Oa, Oc 2.71 and 2.76 Å. These distances are shorter than those observed for the longest O...O contact in the sodium salt of the *p*-t-butylcalix[4]arene monoanion (2.92 Å) where the oxygen atoms are not bridged by a proton, whilst the distances between hydrogen bonded oxygen atoms in the latter structure are significantly less than in **8b** (2.43 - 2.58 Å). This evidence suggests that crystallographic disorder results in either Oa or Oc being deprotonated without any large energetic preference between the two possibilities. Consistent with an oxocyclohexadienyl description for the coordinated decks of the calixarene ligand, the C(1)-O distance is markedly shorter for rings A and C, 1.30 Å av., than the unmetalated rings B and D, 1.37 Å av.

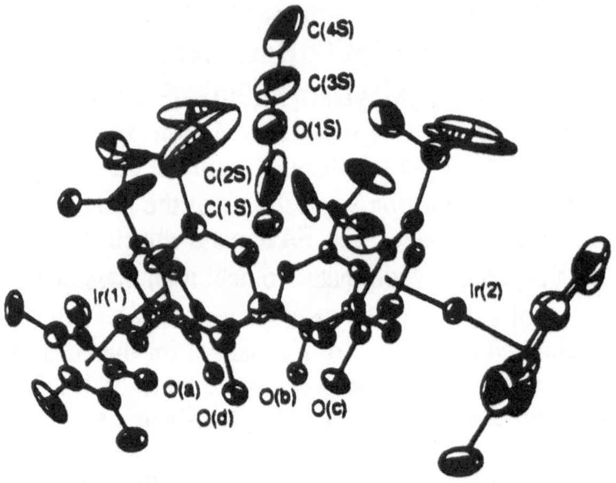

Figure 5. Structure of the cation in **8b**.$Et_2O.NO_2Me$, with the included molecule of Et_2O.

The disorder problems inherent with these systems are further exemplified with the sulfate complex of the tetrametallated, 6+ cation, as shown in Figure 6.

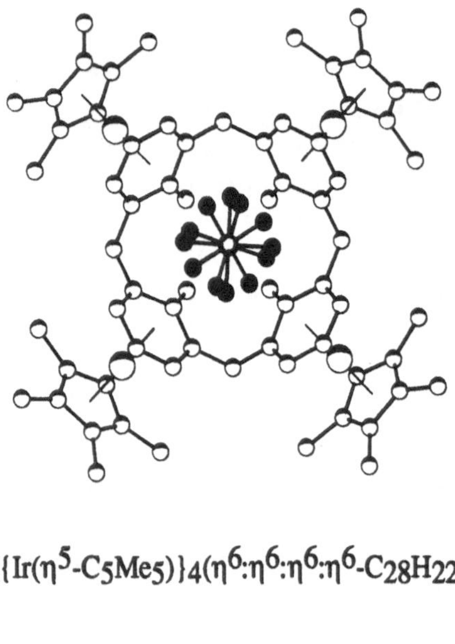

Figure 6. Structure of the $[\{Ir(\eta^5\text{-}C_5Me_5)\}_4(\eta^6\text{:}\eta^6\text{:}\eta^6\text{:}\eta^6\text{-}C_{28}H_{22}O_4)][SO_4]^{4+}$ cation.

As this chemistry develops it is clear that close attention must be paid to complementary spectroscopic techniques in order to sort out the disorder, as well as to ascertain the speciation.

This area of research has recently produced a success notable under this discussion. A compound based on a calixarene provided only very small crystals. Indeed, no reflection data could be obtained on a CAD4 diffractometer, not even one reflection. Using a CCD-based Siemens system, 60% observed data were obtained during a 12-hour data collection. The structure was apparent at the 4-hour mark.

Conclusions

Supramolecular chemistry is presenting great challenges to the crystallographer. The field of supramolecular crystallography is now seen to lie somewhere between that of small molecule and protein crystallography. The supramolecular compounds are more difficult to solve than normal small molecules, but generally not so difficult as proteins. The data obtained from supramolecular compounds does not provide such metrically precise parameters as does that for small molecules, but the data is much better than that from proteins.

References

1. Atwood, J. L., Hamada, F., Robinson, K. D., Orr, G. W. and Vincent, R. L. (1991) *Nature* **349,** 683.

2. Jorgensen, W. L. and Severance, D. L. (1990) *J. Am. Chem. Soc.* **112,** 4768.

3. Hardy, A. D. U. and MacNicol, D. D. (1976) *J. Chem. Soc., Perkin Trans. 2* , 1140.

4. Schweizer, W. B., Dunitz, J. D., Pfund, R. A.;, Tombo, G. M. R. and Ganter, C. (1981) *Helv. Chim. Acta* **64,** 2738.

5. Rzepa, H. S., Webb, M. L., Slawin, A. M. Z. and Williams, D. J. (1991) *J. Chem. Soc., Chem. Commun.*, 1136.

6. Al-Juaid, S. S., Al-Nasr, A. K. A., Eaborn, C. and Hitchcock, P. B. (1991) *J. Chem. Soc., Chem. Commun.*, 1482.

7. Atwood, J. L., Bott, S. G., Jones, C. and Raston, C. L. (1992) *J. Chem. Soc., Chem. Commun.* , 1349; Steed, J. W., Juneja, R. K., Burkhalter, R. S. and Atwood, J. L., submitted for publication.

8. Hamor, T. A., Jennings, W. B., Proctor, L. D., Tolley, M. S., Boyd, D. R. and Mullan, T. (1990) *J. Chem. Soc., Perkin Trans. 2* , 25.

9. Grossel, M. C., Cheetham, A. K., Hope, D. A. O., Lam, K. P. and Perkins, M. J. (1979) *Tet. Lett.*, 1351.

10. Hanton, L. R., Hunter, C. A. and Purvis, D. H. (1992) *J. Chem. Soc., Chem. Commun.*, 1134. The geometrical parameters given in this contribution afford N...centroid = 3.21 Å, with the assumption that the hydrogen atom lies directly over

the centroid of the aromatic ring (a difference of up to 0.15 Å is noted between the H...plane and H...centroid distances, but further details are not given).

11. Viswamitra, M. A., Radhakrishnan, R., Bandekar, J. and Desiraju, G. R. (1993) *J. Am. Chem. Soc.* **115**, 4868.

12. Bredas, J. L. and Street, G. B. (1989) *J. Chem. Phys.* **90**, 7291. However, it must be noted that these calculations over estimate the O...centroid distance by almost 0.3 Å: O...centroid = 3.40 Å; N...centroid = 3.66 Å.

13. (a) Atwood, J. L., Orr, G. W., Hamada, F., Vincent. R. L., Bott, S. G. and Robinson, K. D. (1991) *J. Am. Chem. Soc.* **113**, 2760; (b) Atwood, J. L., Orr, G. W., Hamada, F., Bott, S. G. and Robinson, K. D. (1992) *Supramol. Chem.* **1**, 15; (c) Coleman, A. W., Bott, S. G., Morley, S. D., Means, C. M., Robinson, K. D., Zhang, H. and Atwood, J. L. (1988) *Angew. Chem. Int. Ed. Engl.* **27**, 1361.

14. Estimated from the parameters given in ref 10. The results of our investigation may also be compared directly to those in ref 10: H...plane(aromatic) = 2.42 versus 1.98 Å, respectively; H...C(aromatic) = 2.72 - 2.87 versus 2.22 - 2.62 Å, respectively.

15. Atwood, J. L., Robinson, K. D., Orr, G. W., unpublished results.

16. Eriksson, E. A., Jones, T. A. and Liljas, A. (1986) in *Zinc Enzymes* (Bertini, I., Luchinat, C., Maret, W. and Zeppezauer, M., eds.), Birkhauser; Boston, p. 317; Eriksson, A. E., Jones, A. T. and Liljas, A. (1989)*Proteins* **4**, 274.;Eriksson, A. E., Kylsten, P. M., Jones, T. A. and Liljas, A. (1989) *Proteins* **4**, 283.

17. Steed, J. W., Junk, P. C., Atwood, J. L., Barnes, M. J., Raston, C. L. and Burkhalter, R. S. (1994) *J. Am. Chem. Soc.* **116**, 10346.

18. Steed, J. W., Juneja, R. K., Burkhalter, R. S. and Atwood, J. L. (1994) *J. Chem. Soc., Chem. Commun.*, 2205.

19. Steed, J. W., Juneja, R. K. and Atwood, J. L. (1994) *Angew. Chem. Int. Ed. Engl.* **33**, 2456.

20. Iki, H., Kikuchi, T. and Shinkai, S. (1992) *J. Chem. Soc., Perkin Trans. I*, 669; Kikuchi, T., Iki, H.,Tsuzuki, H. and Shinkai, S. (1993) *Supramol. Chem.* **1**, 103.

21. Tabushi, I. (1984) in Inclusion Compounds, Atwood, J.L., Davies, J. E. D. and MacNicol, D. D. (Eds), Academic Press, London, pp. 445-471.

22. Breslow, R. (1984) in Inclusion Compounds, Atwood, J.L., Davies, J. E. D. and MacNicol, D. D. (Eds), Academic Press, London, pp. 473-508.

23. Atwood, J. L., Orr, G. W., Robinson, K. D. and Hamada, F. (1993) *Supramol. Chem.* **2**, 309.

24. Gutsche, C. D. (1989) *Calixarenes*, Royal Society of Chemistry, Cambridge.

Cyclodextrins: Structures, Functions, Hydrogen Bonding

W. Saenger
Freie Universität Berlin
Institut für Kristallographie
Takustr. 6, D-14195 Berlin
Fax: +49-30-8386702;
e-mail: saenger@fu-berlin.chemie.de

1. Introduction

The cyclodextrins, Schardinger dextrins or cycloamyloses are a family of cyclic oligosaccharides. They are produced by the action of glucosyl transferases on the helical starch molecule where these enzymes cleave off one turn and reseal the two open ends so that a cyclic structure is formed. Because glucosyl transferases are not very specific with respect to the cutting sites, a family of cyclodextrins is obtained with 6 to 9 D-glucoses in α(1-4)linkages; they are called cyclohexa-, -hepta-, -octa-, -nonaamylose or α-, β-, γ-, δ-cyclodextrin (α-CD, β-CD, γ-CD, δ-CD).

The most prominent feature of the cyclodextrins (CD) is that they can form inclusion complexes in aqueous solutions with a large variety of molecular or ionic compounds that have the correct dimensions to fit into the central cavities of the CD[1-6]. For these reasons, the CD have been used as model compounds to study intermolecular interactions which are so important in biological systems. Because the central cavities are hydrophobic in character and both rims of the CD are lined by hydroxyl groups, they have modest enzymatic activity for a number of chemical reactions, and consequently CD have been widely used as model enzymes. The catalytic activity can even be enhanced if the hydroxyl groups are suitably modified so that functional groups help in the catalytic processes [7].

In contrast to the catalytic activity, CD can also prevent decomposition of sensitive molecules if these are encapsulated in the CD cavities. This is now one of the major applications of the CD especially in the pharmaceutical industry where drugs, which are sensitive to environmental conditions, can be stabilized by inclusion, and drugs which are not soluble enough can be easier-

L. Echegoyen and A.E. Kaifer (eds.), Physical Supramolecular Chemistry, 273–288.

solubilized by complex formation. Moreover, it has been shown that certain chemical reactions take a different path in presence of CD, and the CD cavity, which is chiral due to the handedness of the D-glucoses, has been used successfully as chromatographic material for the separation of enantiomers [1,3,4].

In this short review, I will focus on the inclusion process, the packing of CD in crystal structures, the structures of several CD inclusion complexes, hydrogen bonding, and the dynamics within crystal lattice of β-CD · $12H_2O$.

2. Structures of Cyclodextrins

The D-glucopyranose units in the CD adopt the $^4C_1$-chair conformation (Figure 1). They are all in the same orientation and the CD have the appearance of a torus with central cavity, where the narrower end is occupied by the O(6) hydroxyl groups and the wider end by the O(2), O(3) hydroxyl groups. The latter may form intramolecular, interglucose hydrogen bonds which stabilize the toroidal structure. The height of the torus, ~ 8 Å, is determined by the length of the D-glucose, and the outer diameter depends on the number of D-glucose units, 14.6 to 17.5 Å for α- to γ-CD. The cavity diameter is in the range α-CD 4.7 - 5.2 Å, β-CD 6.0 - 6.4 Å, γ-CD 7.5 - 8.3 Å [4,5]. The solubility in water differs significantly, being 14.5 (α-CD), 1,85 (β-CD), 23.2 (γ-CD) g/100 ml solution at room temperature.

The structure of the CD-torus is remarkably stable and well-defined due to the interglucose *O*(2)···*O*(3) hydrogen bonds which, because of the different curvatures of the molecules, are longer (and weaker) in α-CD than in β- and γ-CD. The CD-torus can adapt to the shape of the included guest molecule and, for α-CD complexes with aromatic guests, it is elliptically distorted (vide infra, Fig. 3b). This, however, does not change the $^4C_1$ chair conformation of the D-glucoses and is mainly compensated by slight variations in D-glucose puckering.

The main degree of structural freedom is in the orientation of the C6-O6 groups. They may rotate freely but prefer the (-)*gauche* form (torsion angle O5-C5-C6-O6 at -60°) so that the O(6) groups point "away" from the central axis of the CD. The (+)*gauche* form (180°) is observed if hydrogen bonding to an enclosed guest molecule occurs, and the *trans* form is sterically unfavorable and not adopted.

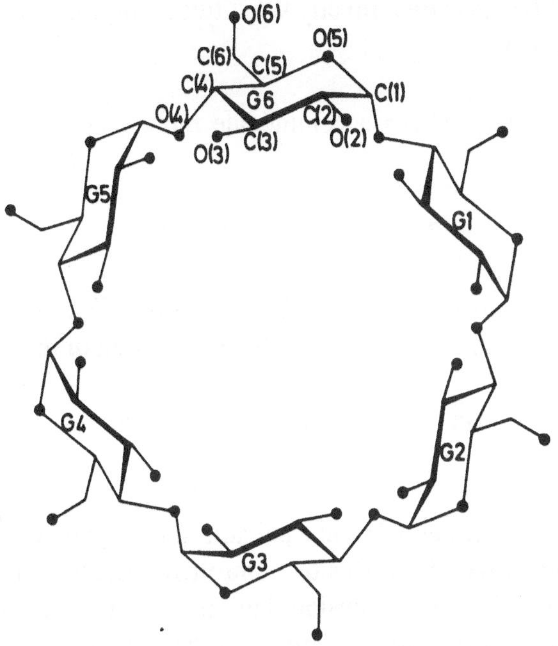

Figure 1: Chemical structure and atom numbering scheme for α-cyclodextrin. Hydroxyl groups indicated by ----•.

3. Thermodynamic and kinetic data of the cyclodextrin inclusion process

If a ligand for inclusion into CD is dissolved in water and CD is added, complex formation occurs if the ligand L as a whole or in part can be enclosed in the cavity. There is an equilibrium

$$CD + L \Leftrightarrow CD \cdot L$$

which can be followed by some suitable signal. If the ligand has an absorption spectrum in the visible region of light, there is generally a shift in the spectrum which can be monitored by spectrophotometers. If the ligand is achiral, inclusion by the chiral CD gives rise to an anisotropy which can be measured by circular dichroism. Other spectroscopic methods to follow the equilibrium are NMR-spectroscopy, calorimetry and competition experiments were one ligand that cannot be measured directly substitutes a ligand in the cavity which can be easily detected.

If a ligand is very long, it can accommodate more than one CD so that the equilibrium is

$$2\,CD + L \;\Leftrightarrow\; (CD)_2 \cdot L$$

In this case, there is a one-to-one equilibrium at low CD concentration, which is followed at higher concentration of CD by the addition of the second CD.

The equilibrium constants are in the mM range. Some of these data are collected in Table 1.

The rate of inclusion between CD and a suitable small ligand can be in the μsec range which is indicative of a diffusion controlled process. If the ligand has a certain larger geometry, the inclusion into the cavity can be slower than diffusion controlled, and occurs as slow as in the sec or min range, as shown in the series of ligands in Table 2. Since there is a dependence of the rate of inclusion on the geometry of the ligand, this suggests that there is really an inclusion into the cavity and not just an association of the ligand at the outside of the CD molecule.

In all these thermodynamic and kinetic investigations, it is important to make sure that the buffer systems used to control the pH of the solution are not themselves included in the CD cavity. It has been shown that phosphate is not included but chloride, bromide, iodide, nitrate, perchlorate and other ions or buffer substances can enter the cavity [4]. In this case, there will a competition between these substances and the respective ligand which is under investigation.

TABLE 1. Thermodynamic parameters of some cyclodextrin inclusion compounds at 25°C taken from [4]

Substrate	Cyclodextrin	K_D [mol/l]	ΔH [kcal/mol]	ΔS [cal·mol^{-1}·K^{-1}]	Method [a]
p-Nitrophenol	α	2.6 x 10^{-3}[b]	- 4.2	- 2.8	Sp
p-Nitrophenolate	α	0.27 x 10^{-3}[b]	- 7.2	- 8.7	Sp
	α		- 9.0	- 15	Ca
	α	8 x 10^{-3}	- 7.3	- 15	Ca
Perchloric acid	α	2.5 x 10^{-2}	- 7.5	- 17	Ca
Sodium perchlorate	α	5 x 10^{-2}	- 9.7	- 23	Ca
Anilinium	α	3 x 10^{-2}	- 12.3	- 35	Ca
perchlorate	α	10^{-3}	- 9.6	- 18	Ca
Benzoic acid	α	1.6 x 10^{-1}	- 11.6	- 26	Ca
4-Aminobenzoic	α	10^{-5}	- 0.3	21	Ca
acid	α	2.1 x 10^{-1}	- 7.3	- 21	Ki
2-Aminobenzoic					
acid	α	3.5 x 10^{-1}	- 1	8	Ki
Diisopropylphos-					
phorofluororidate	α	2.2 x 10^{-3}	- 4.6	- 3	Ki
m-Chlorophenyl	α	9.8 x 10^{-3} [3]	- 5.7	- 8.6	Ki
acetate					
m-Ethylphenyl					
acetate					
Benzoylacetic acid					

[a] Sp, Ca, Ki signify spectroscopic, calorimetric and kinetic methods respectively, [b] At 14°C, [c] At 50.3°C.

Substrate[a]	K, M^{b}	ΔH, kcal/mole	k_R, M^{-1} sec^{-1}	k_D, sec^{-1b}	Solvent (phosphate buffer)
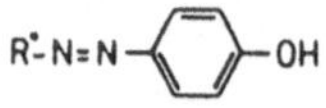	3.7×10^{-3}	7.0	1.3×10^{7}	5.5×10^{4}	$I = 0.1$, pH 3.5
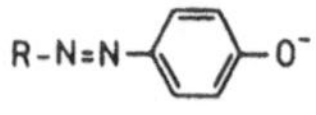	1.55×10^{-3}	6.3	1.7×10^{5}	2.6×10^{2}	$I = 0.1$, pH 11
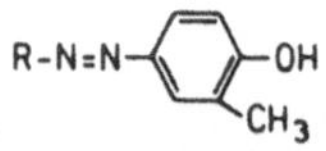	2.4×10^{-3}	6.4	1.2×10^{5}	3.5×10^{2}	$I = 0.1$, pH 3.5
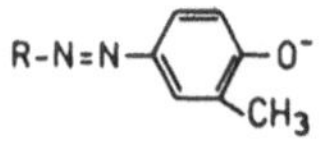	2.1×10^{-3}	5.8	1.5×10^{2}	0.28	$I = 0.1$, pH 11
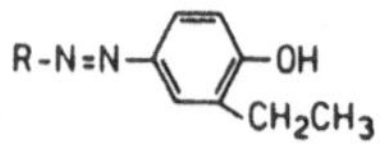	2.2×10^{-3}	6.5	6×10^{3}	19	$I = 0.1$, pH 3.5
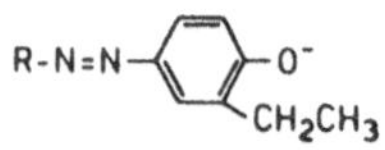	3.5×10^{-3}	7.7	2.8	1×10^{-2}	$I = 0.1$, pH 11.5
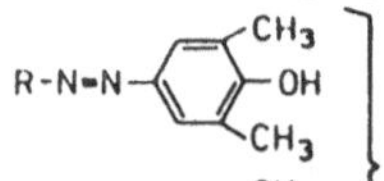	No inclusion				
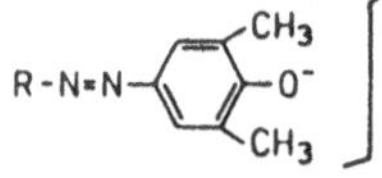	No inclusion				

[a] R= 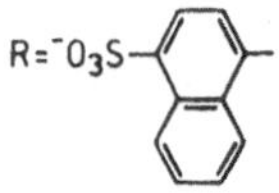
[b] All values refer to 14°.

Table 2. Thermodynamic (K_{Diss}, ΔH) and kinetic (k_R, recombination; k_D, dissociation) parameters for the inclusion of a series of azo dyes into α-cyclodextrin. Note that the dissociaiton constants K are about the same in this series but that the kinetic data vary by 5 orders of magnitude, being fast for slim and slow for bulky substituents in *ortho* position to the phenolic O-H group of the ligands. Taken from [4; see also [26]].

4. Cyclodextrin inclusion complexes crystallize in different packing modes

For the crystallization of CD inclusion complexes a hot, saturated solution of CD and an approximately tenfold molar excess of the ligand is cooled down slowly. The obtained crystals usually contain the ligand enclosed in the CD cavities. X-ray structure analysis of these crystals shows that there are certain packing motifs in which the complexes are organized, depending on the type of CD and the type of ligand [6,9] (Figure 2).

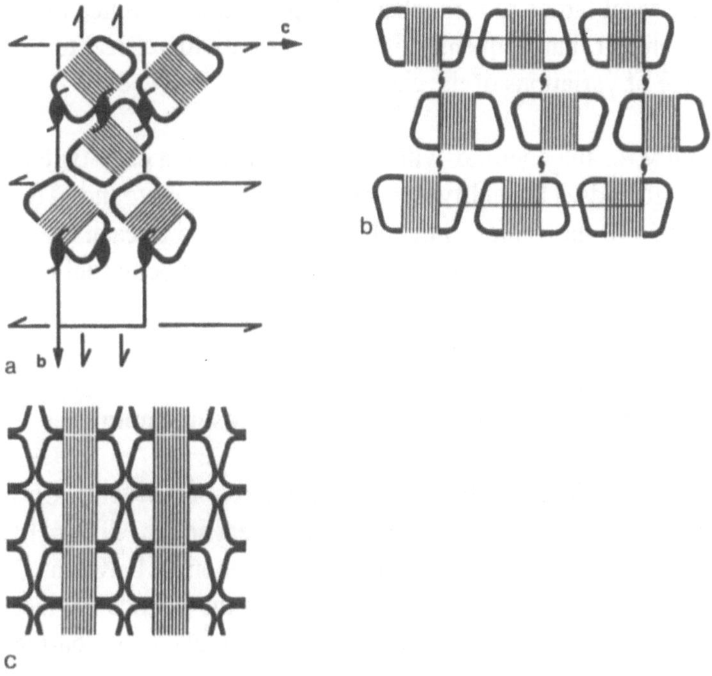

Figure 2: Packing of cyclodextrin molecules in crystal lattices with (a) cage, (b) brick, (c) channel-type arrangement. Cyclodextrin molecules are seen from the side, avities are indicated by shading.

For α-CD, small molecular guests like water, iodine, methanol, propanol, acetic acid etc. give rise to a cage-type packing; the α-CD molecules containing the guest molecules are arranged crosswise such that the cavity of each α-CD is closed on both sides by adjacent α-CD molecules so that the cavities turn into cages. If ionic guests like triiodide or long molecular guests are enclosed in the cavity of α-CD, channel-type structures are formed in which the α-CD molecules are packed on top of each other like coins in a roll; this gives rise to "infinite" cylinders which exhibit a channel-type cavity enclosing the guest molecules. The third type packing pattern, which is comparable to the organization of bricks in a wall, is obtained if α-CD forms complexes with aromatic guest molecules.

For β-CD the packing patterns are simpler than for α-CD. A cage-type pattern is formed with small ligands like water or methanol. With larger guest molecules, two β-CD molecules arrange head-to-head to form a basket-like structure which encloses the ligand, and these baskets are piled on top of each other in the crystal lattice to form channel-type structures. Since the baskets are not always colinear, but frequently tilted or rotated with respect to each other, a certain degree of variations of the channel-theme is observed.

For γ-CD, only for the complex with water we observe a cage-type pattern, but for all other ligands, channel-type packing patterns are found. The reason is that adjacent γ-CD molecules in the channel are linked to each other by strong hydrogen bonds so that the packing is more "rigid" and better defined than with α- and β-CD.

5. Examples of the crystal structures of CD inclusion complexes

In Figs. 3a-d, sections of the crystal structures of several CD inclusion complexes are shown. In all these structures except for α-CD · $6H_2O$, the CD ring is stabilized by intramolecular hydrogen bonds between O(2), O(3) hydroxyl groups of adjacent D-glucose units, and the ligands are located in the CD-cavities. Depending on the geometry of the inclusion and on the nature of the ligand, there is either no hydrogen bonding interaction, or hydrogen bonding interactions between ligand and CD hydroxyl groups are observed. It should be noted that in the complex β-CD · Methanol · $8H_2O$, there is additional $C-H\cdots O$ hydrogen bonding interaction that helps to stabilize the geometry of the complex (Figure 4). Besides hydrogen bonding, van der Waals interactions contribute to the fixation of a ligand in the CD cavity.

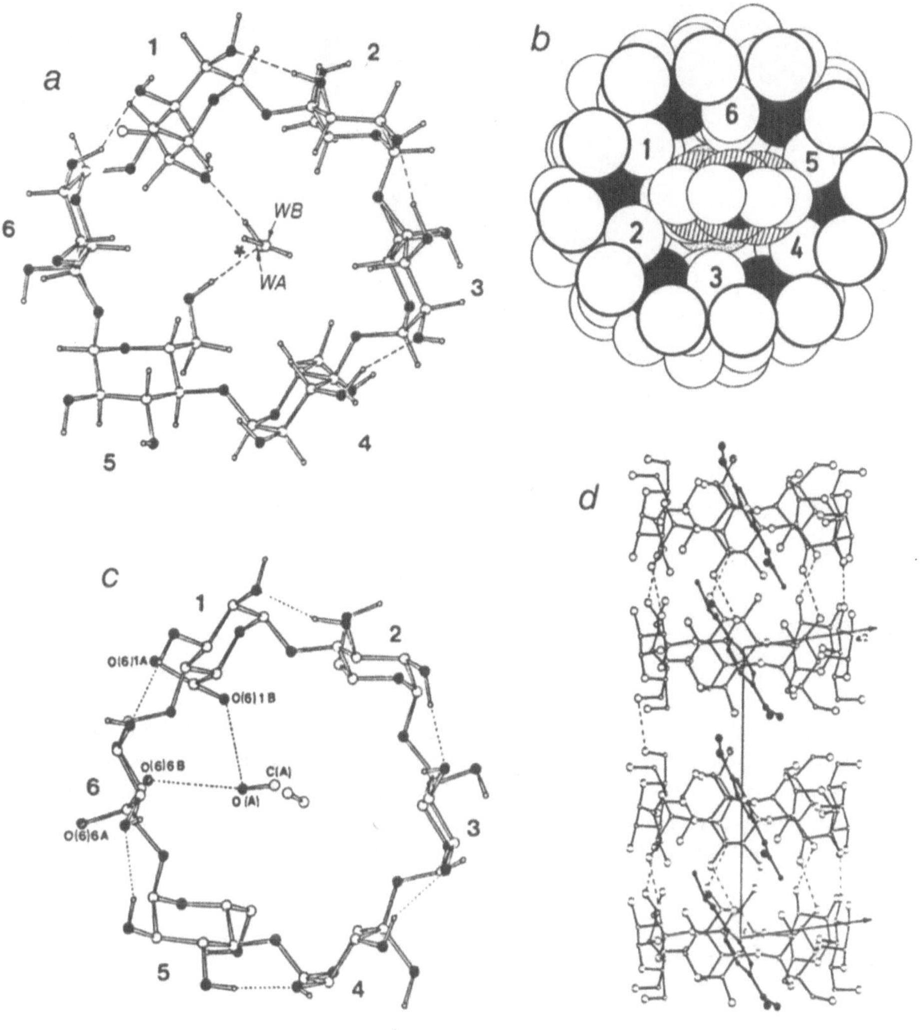

Figures 3a-d: A representative collection of cyclodextrin inclusion complexes taken from several crystal structures. Only sections of the structures are shown. (a) Neutron diffraction study of α-CD · $6H_2O$; two H_2O (WA, WB) are enclosed in the cavity, four (not shown) are located between the α-CD molecules; hydrogen bonds are shown dashed. (b) Elliptical distortion of α-CD in the complex α-CD · p - iodoaniline · $6H_2O$ [22]. (c) The complex α-CD · methanol · $5H_2O$; the two CH_3-OH molecules are disordered at 50% occupancy each, and C(6)-O(6) bonds O(6)1, O(6)6 partially occupy two positions [23]; hydrogen bonds dashed. (d) Channel-type structure of β-CD · p - nitroanilide [24].

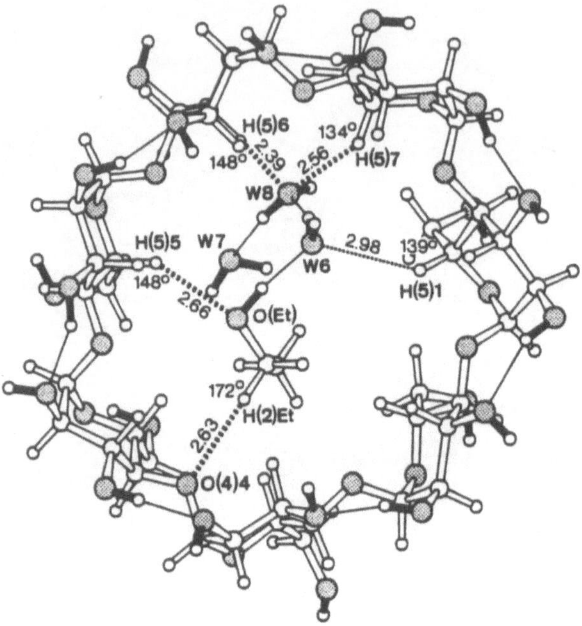

Figure 4: Stabilization of the geometry of inclusion in β-CD · ethanol · $8H_2O$ by $O-H\cdots O$ (thin lines) and $C-H\cdots O$(dotted) hydrogen bonds. One ethanol (Et) and three water molecules (W6, W7, W8) ocupy the β-CD cavity. Neutron crystal structure at 15K [25], i.e. well below the phase transistion where O-H groups, which are disordered at room temperature, become ordered (indicated by filled bonds), $H\cdots O$ distances in (Å), $O-H\cdots O$ and $C-H\cdots O$ angles in (°).

6. $O-H\cdots O$ hydrogen bonding in the CD hydrates

If CD is crystallized from water without added ligand, hydrates are formed. They are of the cage type for all three CD and contain increasing amounts of water if we go from α- to γ-CD: α-CD · $6H_2O$, β-CD · 12 H_2O, γ-CD · 15 H_2O. The first two hydrates have been studied by X-ray and neutron diffraction analyses [10,11], the latter by X-ray diffraction. The neutron diffraction data have been collected because neutrons interact with the nuclei of the atoms, and C, O, H can be located with the same accuracy. In contrast, X-rays interact with the electron clouds of the atoms, and it is notoriously difficult to locate H in the crystal structures of these large molecules, especially if some disorders occurs.

For α-CD · 6 H_2O, the neutron diffraction study shows that all the hydroxyl groups and water molecules are well ordered and engaged in a network of $O-H\cdots O$ hydrogen bonds [9,10], Figure 3a. The α-CD cavity accommodates two water molecules, and four are located between the α-CD rings. The α-CD molecule itself is distorted such that only four of the six interglucose, intramolecular hydrogen bonds are formed, and two of these hydrogen bonds are broken because the α-CD ring is collapsed to some extent to adopt to the small size of the enclosed water molecules.

If the hydrogen bonding scheme is followed from one O-H group to the next, we observe formation of four, five and six-membered or larger rings, and there are infinite chains runnning through the crystal structure [12], Figure 5. This is certainly a result of the cooperative effect, which favours structures like $\cdots O-H\cdots O-H\cdots O-H\cdots$ more than isolated $O-H\cdots O$-hydrogen bonds. Quantum chemical calculations have shown that cooperative hydrogen bonds are 25% stronger than individual ones [13], and the neutron structures indicate that they are in fact shorter [14]. Since the hydrogen bonds are all running in the same direction in cooperative systems, we call them homodromic [12].

In β-CD · 12 H_2O, the neutron diffraction study [11] has shown that *all* of the O(2), O(3) hydroxyl groups are disordered (Figure 6), and several of the water molecules are also disordered. This disorder gives rise to hydrogen bonds of the type $O\text{-}(½)\,H\cdots H\,(½)\text{-}O$ which have been interpreted as a dynamic equilibrium between two states:

$$O-H\cdots O-H \Leftrightarrow H-O\cdots H-O$$

This kind of "flip-flop" dynamic disorder [15] is found for all the seven interglucose, intramolecular hydrogen bonds in β-CD not only in the hydrate but also in the complex β-CD · ethanol · $8H_2O$ at room temperature [16]. If the crystals are cooled, differential scanning calorimetry has shown that around -46°C, a phase transition occurs [17]. Below the phase transition temperature, all the flip flop hydrogen bonds become well ordered and are of the "normal" form $O-H\cdots O$.

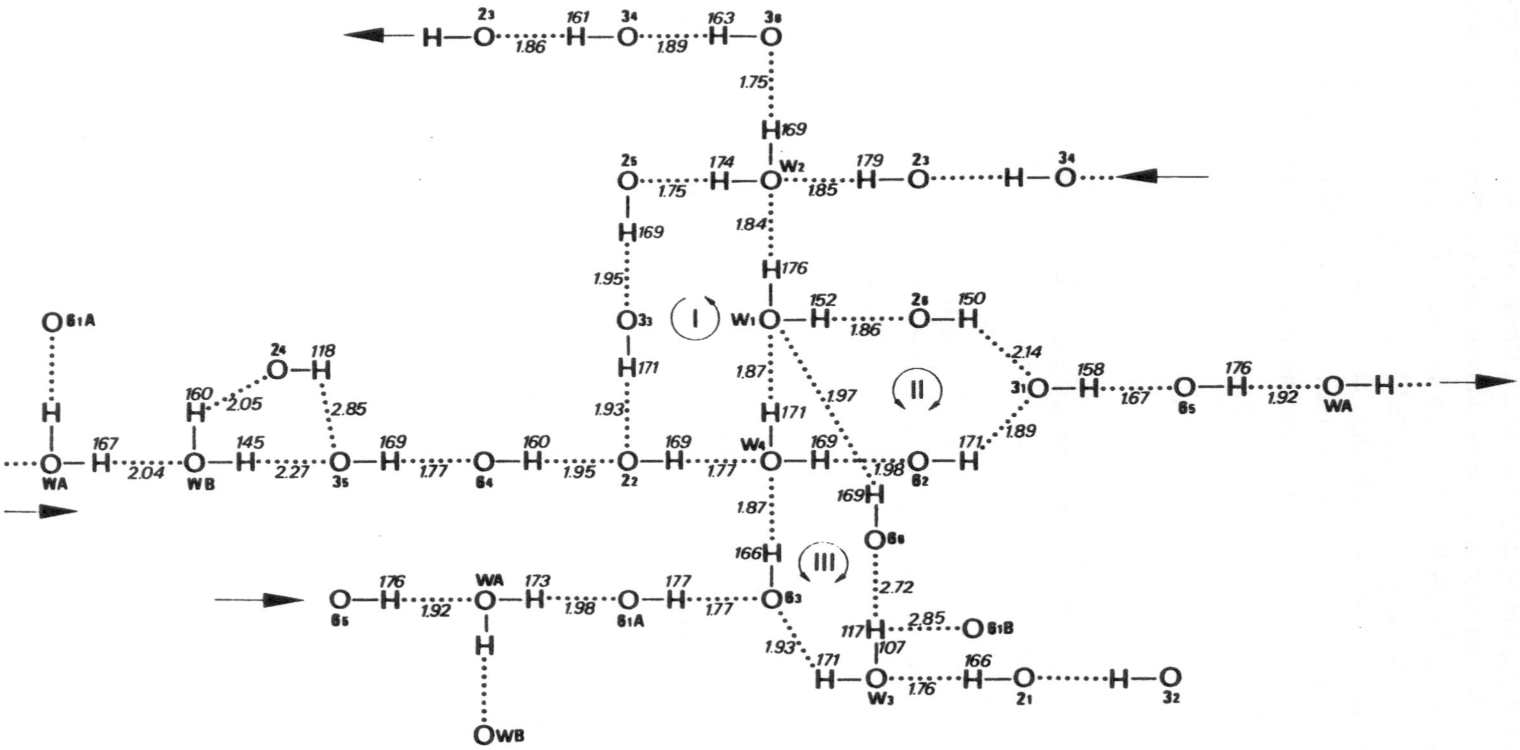

Figure 5: Hydrogen bonding scheme in α-CD · $6H_2O$ [10], taken from Ref [14]. Numbers give $H\cdots O$ distances in (Å) and $O-H\cdots O$ angles in (°). Note that homodromic arrangement $\cdots O-H\cdots O-H\cdots O-H\cdots$ predominates, circle I is also homodromic. Arrows indicate "infinite" chains.

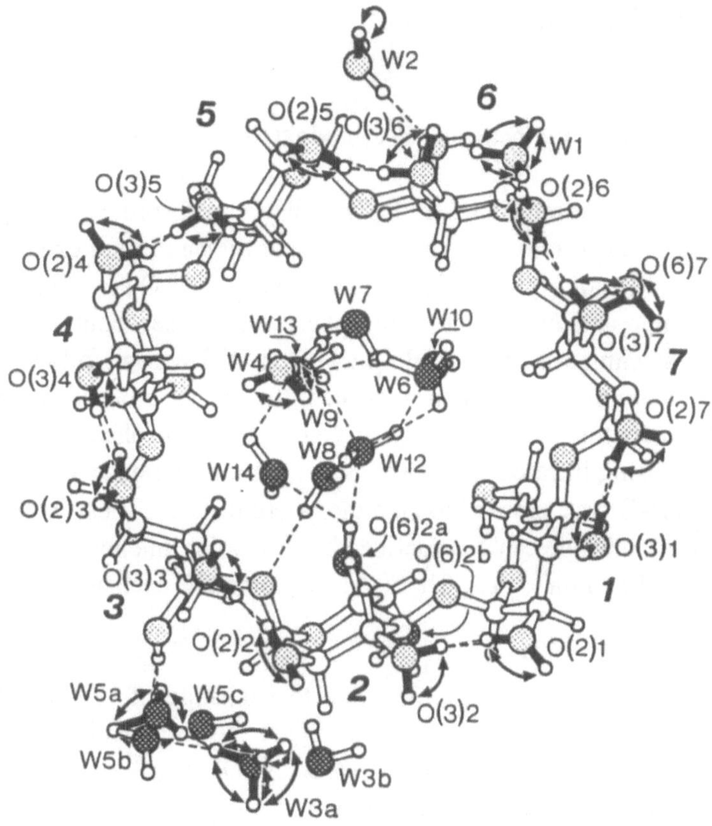

Figure 6: Section of the neutron crystal structure of β-CD · $12H_2O$ [11]. All of the O(2)H, O(3)H hydroxyl groups show twofold flip-flop disorder. The distance between H-atoms (A) and (B) in each hydrogen bond is too short (~ 1Å) to permit simultaneous, full occupation. Consequently, these positions are only ~ 50% filled; double-headed arrows indicate jump positions of O-H groups in flip-flop disorder. *light shading*: fully occupied, *dark shading*: partially occupied oxygen positions. Taken from [18].

These studies have shown that the flip-flop disorder is dynamic and not static; otherwise a phase transition and ordering of the hydrogen bonds would not have taken place upon lowering of the temperature. In a separate study, quasielastic neutron scattering experiments were conducted on β-CD · $11H_2O$ [18]. They showed clearly that the disorder occurs in the range of $2{\cdot}10^{10}$ to $2{\cdot}10^{11}$ sec^{-1}, and disordered water molecules in the β-CD cavity and between the β-CD molecules jump between the disordered positions in about the same frequency range. Below the phase transistion, these water molecules also become well ordered, and the number of sites which are only partially occupied at room temperature in the β-CD cavity are reduced from 8 to 6, with full occupation in the low temperature form.

7. Dynamics of water molecules in β-cyclodextrin · $12H_2O$

Since most of the water molecules in β-CD · $12H_2O$ are disordered over several sites, it was of interest to see whether these water molecules are exchangeable with water from the atmosphere. If crystals of β-CD · $12H_2O$ are removed from the mother liquor and exposed to atmosphere, they lose about three water molecules per asymmetric unit to yield β-CD · $9H_2O$. This was followed by X-ray structure determination which indicated the water molecules that had disappeared. Since this process is fully reversible, a H/D exchange experiment was carried out using infrared spectroscopy to monitor the replacement of H by D if β-CD · $12H_2O$ was exposed to an atmosphere of D_2O. These experiments showed that the exchange can be described by first order kinetics $c_H = c_{H,0} \cdot e^{-kt}$ where $k = 1{,}4 \cdot 10^{-2}$ min^{-1}, i.e. 50% of the H atoms are exchanged within 50 min [19].

To show that this process is not only due to exchange of protons (H^+) by deuterons (D^+), but involves exchange of intact H_2O molecules, a separate study was conducted where crystalline β-CD · H_2O^{16} was exposed to H_2O^{18} atmosphere, and the exchange was monitored by mass spectrometry. The exchange kinetics were of the same order as in the H_2O/D_2O experiment and suggested that indeed the entire water molecules are exchanged and not only H^+ and D^+. This means also that H_2O diffuses nearly freely through the crystal lattice. An assessment of the diffusion rate has shown that it is approximately 1/1000 of that in pure water. This is rapid especially in view of the cage-type crystal structure of β-CD · $12H_2O$ which does not contain channels through which the water molecules could diffuse. Since all the O-H are changed to O-D in the H/D exchange experiments, this means that there must be conformational changes of the β-CD molecules in the crystal lattice so that paths can be opened for diffusion of the water molecules. A molecular dynamics simulation has indeed shown that β-CD atoms vibrate with amplitudes around 0.4 Å so that there is enough space for water molecules to diffuse through the lattice [20].

8. Conclusions

In this review, only some of the properties of CD and their inclusion complexes could be described. There is a vast literature on this topic and there are biannual conferences where new results on CD properties, chemistry, applications are discussed [3]. I have only treated here the unmodified cyclodextrins but there are a number of CD with modifications to provide them with novel properties [5]. Besides the CD with functional groups that can have catalytic activities, the most interesting are those where the O-H groups are replaced O-CH_3. Surprisingly, these methylated CD, which should be insoluble in water due to their hydrophobic character, are very soluble in cold water, but they precipitate or crystallize if the aqueous solution is heated above 60°C. The crystals contain either no or only one water molecule [21], depending on the type of CD and type of methylation (in positions 2,6, 3,6, or 2,3,6), and they also form inclusion complexes in solution and in the solid state [5]. This appears to be a clear manifestation of the hydrophobic effect which usually is observed with biological macromolecules but can be studied again in detail using the model system CD.

9. Acknowledgements

This work was supported by Deutsche Forschungsgemeinschaft, Fonds der Chemischen Industrie, Bundesministerium für Forschung und Technologie and the EU network "Molecular Recognition".

References

1. Szejtli, J. (1988) *Cyclodextrin Technology*, Kluwer, Dordrecht.
2. Wenz, G. (1994) Cyclodextrine als Bausteine supramolekularer Strukturen und Funktionseinheiten, *Angew. Chemie* **106**, 851-870.
3. D. Duchêne (1991) E. *New Trends in Cyclodextrins and their Derivatives* Editions de Santé, Paris.
4. W. Saenger (1980) Cyclodextrin Inclusion Compounds in Research and Industry, *Angew. Chem. Int. Ed. Engl.* **19**, 344-362.
5. K. Harata (1991) Recent advances in the X-ray analysis of cyclodextrin complexes. In J.L. Atwood, J.E.D. Davies, D.D. MacNicol, Eds. *Inclusion Compounds*, Oxford Univ. Press, pp. 311-344.
6. W. Saenger (1984) Structural aspects of cyclodextrins and their inclusion complexes in J.L. Atwood, J.E.D. Davies, D.D. MacNicol, Eds. *Inclusion Compounds* Vol. 2 Academic Press London, pp. 231-259.

7. V.T. D-Souza, K. Hanabusa, T. O'Leary, R.C. Gadwood, M.L. Bender (1985) Synthesis and evaluation of a miniature organic module of chymotrypsin. *Biochem. Biophys. Res. Commun.* **129**, 727-732.
8. W. Linert, L.-F. Han, I. Lukovits (1989) The use of isokinetic relationship and molecular mechanics to investigate molecular interactions in inclusion complexes of cyclodextrin. *Chem. Phys.* **139**, 441-455.
9. W. Saenger (1985) Nature and size of included guest molecule determines architecture of crystalline cyclodextrin host matrix. *Israel J. Chem.* **25**, 43-50.
10. B. Klar, B.E. Hingerty, W. Saenger (1980) Hydrogen bonding in the crystal structure of α-cyclodextrin hexahydrate: the use of a multicounter detector in neutron diffraction. *Acta Cryst.* **B36**, 1154-1165.
11. Ch. Betzel, W. Saenger, B.E. Hingerty, G.M. Brown (1984) Circular and flip-flop hydrogen bonding in β-cyclodextrin undecahydrate: a neutron diffraction study. *J. Amer. Chem. Soc.* **106**, 7545-7557.
12. W. Saenger (1979) Circular hyrogen bonds. *Nature* **279**, 343-344.
13. B. Lesyng, W. Saenger (1981) Theoretical investigations on circular and chain-like hydrogen bonded structures found in two crystal forms of α-cyclodextrin hexahydrate. *Biochim. Biophys. Acta* **678**, 408-413.
14. G.A. Jeffrey, W. Saenger (1991) *Hydrogen Bonding in Biological Structures.* Springer Verlag New York.
15. W. Saenger, Ch. Betzel, B.E. Hingerty, G.M. Brown (1982) Flip-Flop hydrogen bonding in a partially disordered system. *Nature* **296**, 581-583.
16. Th. Steiner, S.A. Mason, W. Saenger (1991) Disordered guest and water molecules. Three-center and flip-flop hydrogen bonds in crystalline β-cyclodextrin ethanol octahydrate at T=295K: a neutron and x-ray diffraction study. *J. Amer. Chem. Soc.* **113**, 5676-5687.
17. H. Hanabata, T. Matsuo, H. Suga (1987) Calorimetric study of β-cyclodextrin undecahydrate. *J. Inclusion Phenom.* **5**, 325-333.
18. Th. Steiner, W. Saenger, R.E. Lechner (1991) Dynamics of orientationally disordered hydrogen bonds and of water molecules in a molecular cage. A quasielastic neutron scattering study of β-cyclodextrin · 11 H_2O. *Mol. Phys.* **72**, 1211-1232.
19. Th. Steiner, A.M. Moreira da Silva, J.J.C. Teixeira-Dias, J. Müller, W. Saenger (1995) Rapid water diffusion in a cage-type crystal lattice: β-cyclodextrin dodecahydrate. *Angew. Chem. Int. Ed. Engl.* **34**, 1452-1453.
20. J.E.H. Koehler, W. Saenger, W.F. van Gunsteren (1987) Molecular dynamics simulation of crystalline β-cyclodextrin dodecahydrate at 293K and 120K. *Eur. Biophys. J.* **15**, 211-234.
21. Th. Steiner, W. Saenger (1995) Crystal structure of anhydrous heptakis-(2,6-di-O-methyl)cyclomaltoheptose(dimethyl-β-cyclodextrin). *Carbohyd. Res.* **275**, 73-82.
22. W. Saenger, K. Beyer, P.C. Manor (1976) The crystal and molecular structure of α-cyclodextrin-p-iodoaniline trihydrate. *Acta Cryst.* **B32**, 120-128.
23. B. Hingerty, W. Saenger (1976). Crystal and molecular structure of the α-cyclodextrin-methanol-pentahydrate complex. Disorder in a hydrophobic cage. *J. Amer. Chem. Soc.* **98**, 3357-3365.
24. M.M. Harding, J.M. Maclennan, R.M. Paton (1978) Structure of the complex cycloheptaannylose-p-nitroacetanilide. *Nature* **274**, 621-623.
25. Th. Steiner, W. Saenger (1992). Geometry of $C-H\cdots O$ hydrogen bonds in carbohydrate crystal structures. Analysis of neutron diffraction data. *J. Amer. Chem. Soc.* **114**, 10146-10154.
26. F. Cramer, W. Saenger, H.-Ch. Spatz (1967) Inclusion compounds XIX. The formation of inclusion compounds of α-cyclodextrin in aqueous solutions. Thermodynamics and kinetics. *J. Amer. Chem. Soc.* **89**, 14-20.

DIFFRACTION, SPECTRAL AND THERMAL ANALYSIS OF MIXED CRYSTALLINE PHASES

Mechanisms of Solid-to-Solid Organic Reactions

M. A. GARCIA-GARIBAY,* A. E. CONSTABLE, J. JERNELIUS, T. CHOI, D. CIZMECIYAN AND S. H. SHIN
Department of Chemistry and Biochemistry
University of California
Los Angeles, California, 90024

1. Introduction

Crystalline organic compounds possess several unique properties that make them attractive candidates for use in non-linear optics, organic conductivity, organic ferromagnetism, sensing materials and other applications [1-4]. Their limited but fascinating chemical reactivity also makes them attractive as reaction media for solvent-free synthesis [5-8]. However, there are important challenges that must be overcome in each of these areas before their general use becomes a reality. Like other supramolecular systems, the properties of organic crystals are determined by a collection of weak but subtly balanced non-bonded interactions. The particular structures and characteristics of the crystals are difficult to predict and manipulate. Despite a long history of analytical progress in areas ranging from microscopy, calorimetry and X-ray diffraction, to nuclear magnetic resonance, atomic force microscopy and others, the role of the chemist has remained largely descriptive and passive. Many challenges regarding the *rational design* of organic crystals, the engineering of their supramolecular properties, and the understanding of their chemical reactivity remain unsolved and the study of the organic solid state will require a great deal of attention before the potential of its promises may be delivered.

In this paper we analyze several aspects of chemical reactivity in the crystalline solid state. In particular, the effects of product accumulation on the structure and properties of several crystalline compounds are examined. By focusing on the role of mixed crystalline phases and phase separation we analyze possible mechanisms for solid-to-solid reactions. It will be shown that highly selective solid state photochemical reactions may be carried out to very high conversion suggesting the possibility of developing useful and solvent-free synthetic methodologies.

2. Organic Solid State Reactivity

While everyday experience suggests that crystalline solids should be generally stable, reactions in crystals are known to occur under special circumstances [7-10]. Known examples include unimolecular and bimolecular reactions as well as polymerizations [5-

L. Echegoyen and A.E. Kaifer (eds.), Physical Supramolecular Chemistry, 289–312.

11]. For thermal reactions, activation barriers must be compatible with temperatures which do not destroy (melt) the crystal. While many examples of thermally activated reactions have been reported, most examples of solid state reactivity involve photochemical methodologies which allow for activation at temperatures that are well below the melting point. From a molecular level, it has been shown that solid state reactions can only occur when there is a strong geometric similarity between the structures of the reactant, the transition state and the final product [13]. To stress the role of the crystal structure, reactions that occur under the influence of crystalline media are known as "topochemical reactions" [11, 12]. Topochemical reactions are characterized by strict size and shape limitations that arise from the non-bonding repulsive interactions between a reacting molecule and its closest neighbors which form a "reaction cavity". The reaction cavity is a useful qualitative concept that represents the space occupied by the reactant [13]. As indicated in Scheme 1, allowed reactions must satisfy the structural constraints imposed by the reaction cavity which include limitations on the number of reaction partners (if any), and on the extent of molecular displacement and conformational motions.

Scheme 1

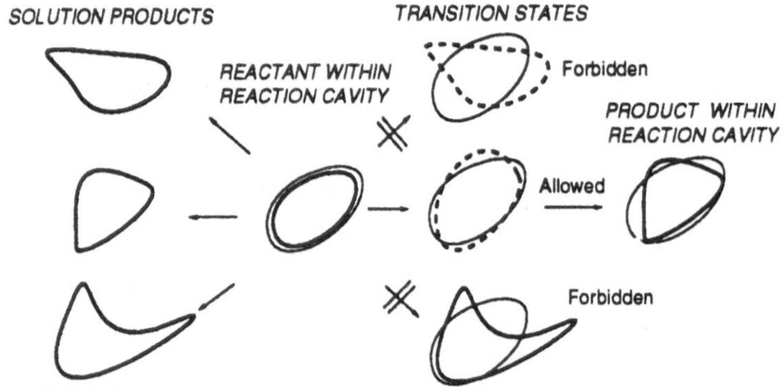

2.1. EXAMPLES OF SOLID STATE CONTROL

Unlike reactions in fluid solution media, the structure of the reactant and its environment in the solid state may be determined with precision by X-ray diffraction. The potential of mechanistic studies based on product analysis and X-ray structural data has been recognized for many years and examples involving dimerization [11] and atom transfer reactions have been reported [8, 14]. It is also known that the crystalline environment is capable of inducing highly chemo-, regio-, and stereoselective reactions [6, 7, 9, 10]. Some remarkable examples of solid state control reported in the literature include the generation of optically pure compounds from achiral reagents [15-18], the preparation of single crystalline polymers [19, 20], and the control of highly reactive intermediates [21, 22]. In general, it is not unusual that reactions in the solid state differ from those observed in solution. Recent examples from our laboratory involve reactive species such as excited states, biradicals, radical pairs and carbenes (Scheme 2). Their high solution reactivity makes them interesting candidates for the study of fundamental aspects of solid state control while their generality offers an opportunity to develop techniques and methodologies required for preparative solid state transformations. In this paper we will concentrate on changes occurring to the solid state materials hosting those reactions.

Scheme 2

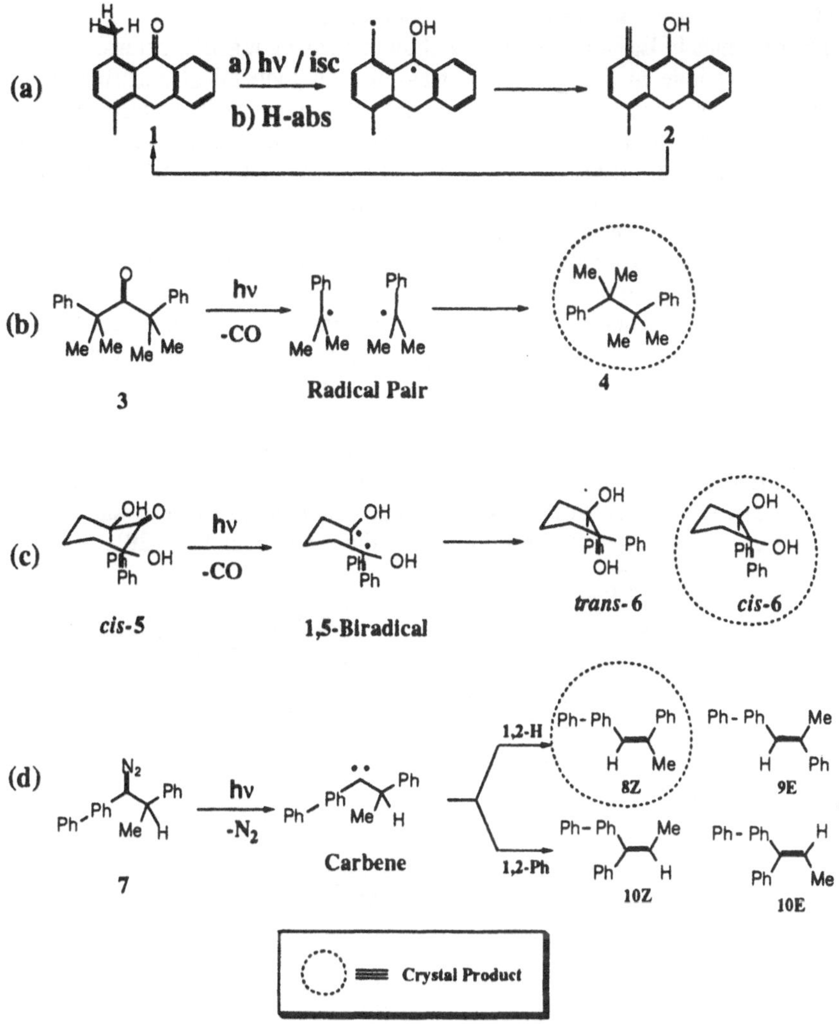

The examples in Scheme 2 represent solid state reactions with increasing complexity from the point of view of selectivity and solid state reactions demands. The first reaction is reversible and causes no net change in the structure and composition of the crystal host. The bottom three reactions are irreversible. Changes with product accumulation in reactions (b)-(d) may change the crystal structures that are responsible for the solid state reactivity that is observed in the first place. In solid **1**, an excited state H-atom transfer results in formation of *ortho*-quinodimethane **2** and the rate of reaction may be measured by direct or indirect methodologies. Because of rapid regeneration of the starting material, the structure of the crystal remains unaltered and structure-reactivity correlations based on rate measurements and X-ray structural data do not need to deal with crystal phase changes, or with product-induced defect sites [23, 24]. In contrast, irreversible chemical changes in reactions (b) through (d) may cause complications in experiments investigating structure-reactivity correlations and reactions must be carried out to low conversion values to minimize perturbations to the crystal lattice. However, if reaction selectivity is high, accumulation of the product would be very desirable from a preparative point of view.

The reactions in Eq b-d in Scheme 2 show increasing reaction control by the solid state. The example in Scheme 2b portrays the photochemical decarbonylation of tetramethyl-diphenylacetone (**3**). Compound **3** reacts both in solution and in the crystalline solid state to give tetramethyl-diphenylethane through an intermediate radical pair that is formed efficiently in both media. Scheme 2c illustrates another decarbonylation reaction involving *cis*-2,6-diphenyl-2,6-dihydroxy-cyclohexanone (**5**) [25, 26]. Compound **5** reacts photochemically to give a 1,5-biradical that combines in solution to give cyclopentanes ***cis*-6** and ***trans*-6**. The corresponding reaction in the solid state is highly stereospecific and yields the *cis* isomer in excellent yields. The final example in the scheme represents one of the most remarkable examples of solid state control encountered in our laboratory. The substrate is a biphenyl-substituted arylalkyl diazo compound (**7**) capable of forming a highly reactive carbene intermediate [27]. The corresponding carbene reacts in solution to give four main products from competing 1,2-H shifts (**8Z** and **8E**) and 1,2-Ph migrations (**9E** and **9E**). Reaction in crystals of diazo compound **7** yields exclusively compound **8Z** from an extraordinarily selective 1,2-H shift. Recent evidence with related compounds suggests that the product may also form in an excited state reaction of the diazo compound [28, 29]

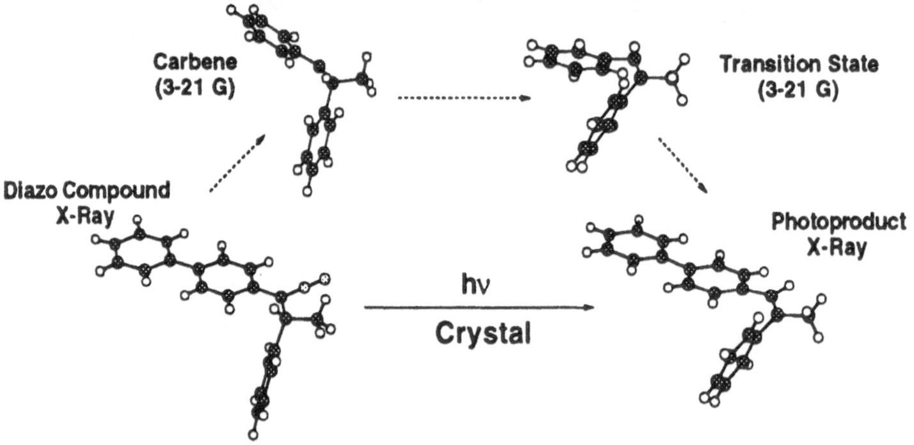

Figure 1. Structural analysis of the reaction mechanism of diazo 7 in the solid state. The structures of the carbene and the transition state were calculated with the omission of the 4-phenyl group. There are two molecules in the asymmetric unit of 8Z with different biphenyl torsion angles.

2.3. STRUCTURAL INSIGHTS INTO REACTION MECHANISMS. A REMARKABLE EXAMPLE

The potential of solid state structure-reactivity correlations based on X-ray diffraction comes from structural knowledge of the reactant and the product as well as the reaction trajectory that inter-connects them. This correlation may be illustrated for the reaction of compound **7** (Scheme 2d) by a structural comparison of the reactant, the carbene intermediate, the transition state, and the final product (Figure 1). The corresponding structures were obtained by X-ray diffraction (**7** and **8E**) and by 3-21G *ab-initio* calculations (carbene and transition state) [30]. The least motion nature of the solid state reaction and the structural similarity of all of the species along the reaction coordinate can be clearly appreciated in Figure 1. In this case, the solid state gives an unprecedented selectivity for such a highly reactive intermediate. This also provides an invaluable experimental confirmation of the theoretical reaction mechanism, which involves the overlap of the transferring hydrogen with the empty p-orbital of the singlet carbene [31]. Analysis of Figure 1 also supports the reaction cavity concept as a feasible explanation for the high selectivity observed in the crystal.

2.4. PRODUCT ACCUMULATION

The mechanistic potential of solid state reactivity relates strictly to pure crystalline phases. Deviations may occur when accumulation of the product causes phase changes or defects in the structure of the crystal [26, 31-35]. Thus, to the extent that chemical reactivity in crystals is determined by topochemical factors, phase changes may affect the mechanisms, selectivities and extent of reaction that can be observed. The ideal solid state reaction would be one that offers selectivity and specificity characteristic of the pure reactant phase while carried out to completion from one solid to another. It is clear that the consequences of structural and chemical changes are important from both mechanistic and preparative points of view. In the sections that follow, we will analyze solid state reactivity from a phase transformation point of view and will adopt a general model to describe and analyze the consequences of product accumulation on reacting crystals. In a later section, we will illustrate spectral, diffraction and thermal analyses of the examples in Scheme 2 within the context of their phase transformation properties. Finally, we will conclude by speculating on the implications of these studies on prospective synthetic applications and on material sciences.

3. Organic Solid State Reactions: Molecular and Phase Transformation Mechanism

The consequences of product accumulation on the structural and chemical properties of a reacting crystal can be analyzed in terms of a relatively simple and general model adapted from the original work of G.M.J. Schmidt on the dimerization of cinnamic acids [23]. Schmidt suggested that solid state reactions may be analyzed from two different points of view: one which addresses their mechanisms in terms of molecular reorganizations and one in terms of transformations of the solid phase.

As suggested by the arrows in Scheme 3, a separation of molecular and phase change mechanisms is somewhat artificial since accumulation of the product affects the reactant phase while changes in the latter affect the extent and type of product formation. A close look reveals a very complex, non-linear problem. Much solid state work published in recent years has dealt primarily with a molecular analysis of product formation. From a practical point of view, problems associated with phase changes can be neglected if reactions are carried out to sufficiently low conversion values. The involvement of non-topochemical reactions that occur at defect sites can be experimentally documented and sometimes avoided. Changes in reaction rates and product selectivity can sometimes be associated with internal stress [34, 35], with sample melting, or with surface effects [36]. Mechanisms and consequences of phase transformation have been studied much less. Phase changes depend on the properties of the ensemble and, as suggested in Scheme 3, they may be determined by composition, temperature, pressure and by whether or not relaxation is allowed.

Scheme 3

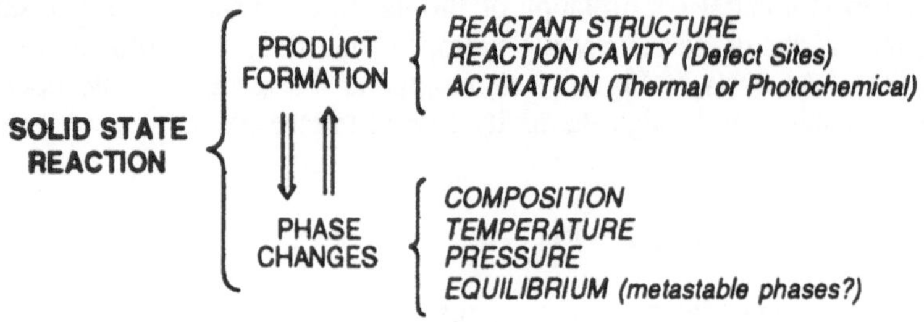

3.1. MIXED CRYSTALLIZATION: THE KEY TO EFFICIENT SOLID STATE REACTIVITY

According to the model presented here, early stages of an organic solid state reaction should take the form of a dilute mixed crystal containing small amounts of product but retaining the crystallographic properties of the reactant (Scheme 4). Mixed crystals or solid solutions are usually prepared by co-crystallization of the components from solution or from the melt under conditions of thermodynamic equilibrium. It is known that mixed crystals require the components to be chemically compatible and structurally similar [37] . Interestingly, the formation of mixed crystals by means of solid state reactions may occur under kinetic control and their composition may be subject to different constraints. However, whether or not mixed crystallization is possible should still be determined by the structural similarity of the reactant and the product [37]. Thus, it is interesting that *structural factors facilitating solid state reactivity and selectivity also facilitate mixed crystallization!*

Scheme 4

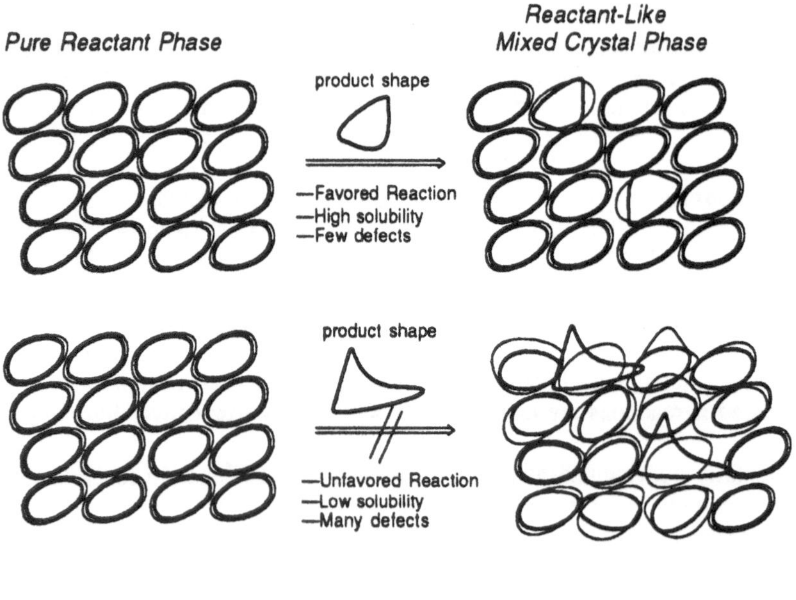

We postulate that solid state reactions having a reactant ideally predisposed to form a given product will be favored from both a molecular (reaction cavity) and a phase transformation (solid solubility) point of view. A reaction that leads to a small structural mismatch will cause a minimal number of defects, thus allowing high selectivity and high conversion values. Conversely, solid state reactions which proceed reluctantly imply poor structural similarity between the reactant and the product so that low solid solubility, severe crystal lattice perturbations and a large number of defect sites are expected. Those reactions may have low conversion limits or rapid loss of selectivity.

3.2. PHASE TRANSFORMATION MECHANISMS

Following the formation of a dilute, reactant-like mixed crystal phase, the progress of solid state reactions should depend on the total solubility of the product in the crystal phase of the reactant, on their phase separation mechanisms, and also on the influence of intermediate and final solid phases on the reactivity of the starting material. While the number of possible scenarios may be large, we distinguish three general phase transformation pathways that depend on the overall changes that occur as the composition of the crystal changes towards the product. The participation of more than one mechanism and more than one phase may be possible but Scheme 5 aims to give a simple starting point for a more detailed analysis of solid state reactions.

Scheme 5

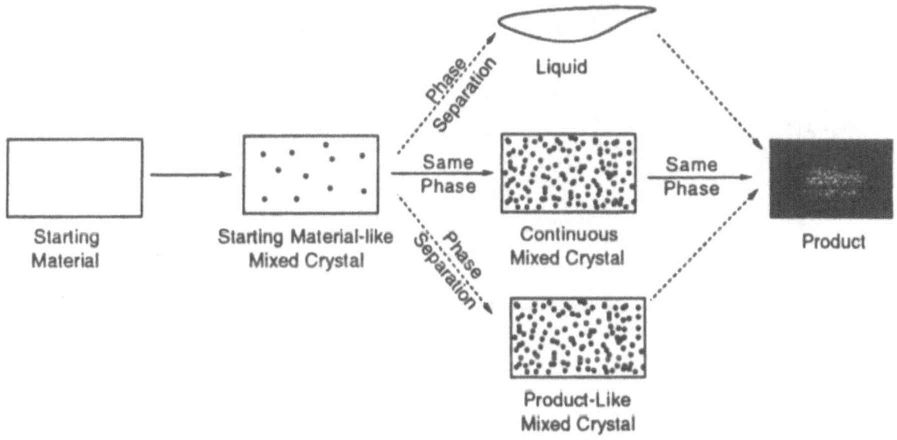

3.3. SOLID-TO-SOLID OR MELTING?

With a crude assumption that neglects possible kinetic complications associated with the relaxation of metastable mixed crystals and crystallization of prospective phases, one may postulate that solid-to-solid reactions may be guaranteed if reactions are carried out below the liquid regions of the phase diagram of the corresponding reactant and product mixtures. For instance, binary systems with limited solid state solubility may possess phase diagrams such as the one represented in Scheme 6a. As indicated by the dotted line, solid-to-solid reactions may be expected if the reaction is carried out below the eutectic point. Unfortunately, phase diagrams are generally not known in advance and reactions may be inadvertently carried out in liquid phases. Furthermore, the number of stable and metastable phases increases with the number of components (products) in the mixture. In practical terms, it may be valuable to explore solid state reactions at varying temperatures and to investigate the phase diagram of the reactant and its product(s) by following calorimetric changes as a function of reaction progress.

3.4. TOPOTACTIC REACTIONS

The ideal solid state chemical reactions are those where the reactant and the product are capable of forming continuous solid solutions with a single crystallographic phase (Scheme 6b). Since continuous solid state solubility requires a very high geometric similarity, reactants in this case have similar environmental influences as the reaction proceeds from beginning to end. The phase diagrams of such systems are characterized by a single solid phase and do not possess a temperature minimum (Scheme 6b). Several examples of such transformations have been documented in the literature[19, 38-40] and it has been recently speculated that they may be more common than expected although they may require careful reaction conditions [41]. These reactions are known as "topotactic" or "single crystal-to-single crystal" transformations and known examples include the photodimerization of benzylidene cyclopentanones [39, 40, 42], the polymerization of some diacetylenes [19], the solid state racemization of several cobaloximes [38], and the carefully carried out dimerization of cinnamic acid [29]. These reactions are all characterized by relatively small molecular and atomic motions and unlike most solid state reactions, progress may be followed by X-ray crystallographic analysis [19, 38-42].

Scheme 6

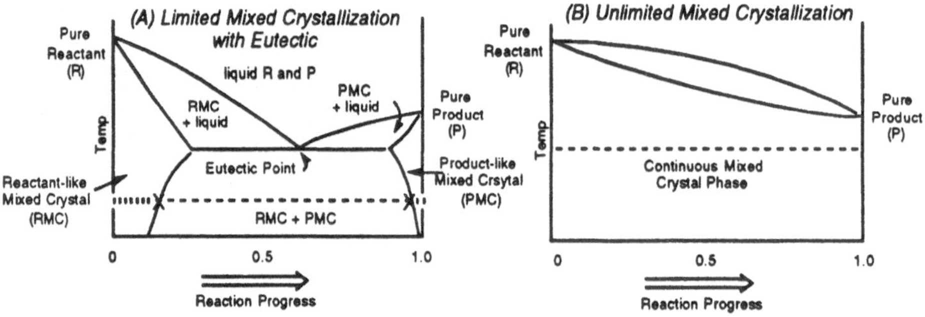

3.5. PHASE SEPARATION REACTIONS

While there seem to be few documented examples of solid state reactions that involve phase separation mechanisms, we believe that phase separation followed by product recrystallization is likely to be the most common type of solid-to-solid reaction mechanism. While it is known that phase separation may only occur after the products have exceeded their solubility limit in the phase of the reactant, the details of phase separation and recrystallization of the new phase should be quite complex. In principle, the evolution of a reacting system from a phase transformation point of view may be followed according to the phase diagram of the reactant and the product(s). A solid-to-solid reaction along the dotted line in the hypothetical phase diagram of Scheme 6a starts with formation of reactant-like mixed crystals (RMC). The reaction continues within the RMC phase until the solubility limit of the product is reached at about 20% conversion. After this composition, the RMC no longer tolerates the product, so phase separation occurs. Segregation of the reactant and the product results in re-crystallization in the RMC phase and in a "product-like" mixed crystal phase (PMC) containing about 10% of starting material. Conversion values between *ca.* 20 and 90 % involve a weighted mixture of crystals with compositions given by the solubility limits of the two allowed solid phases. At the end, the reaction reaches completion within the PMC phase.

Scheme 7

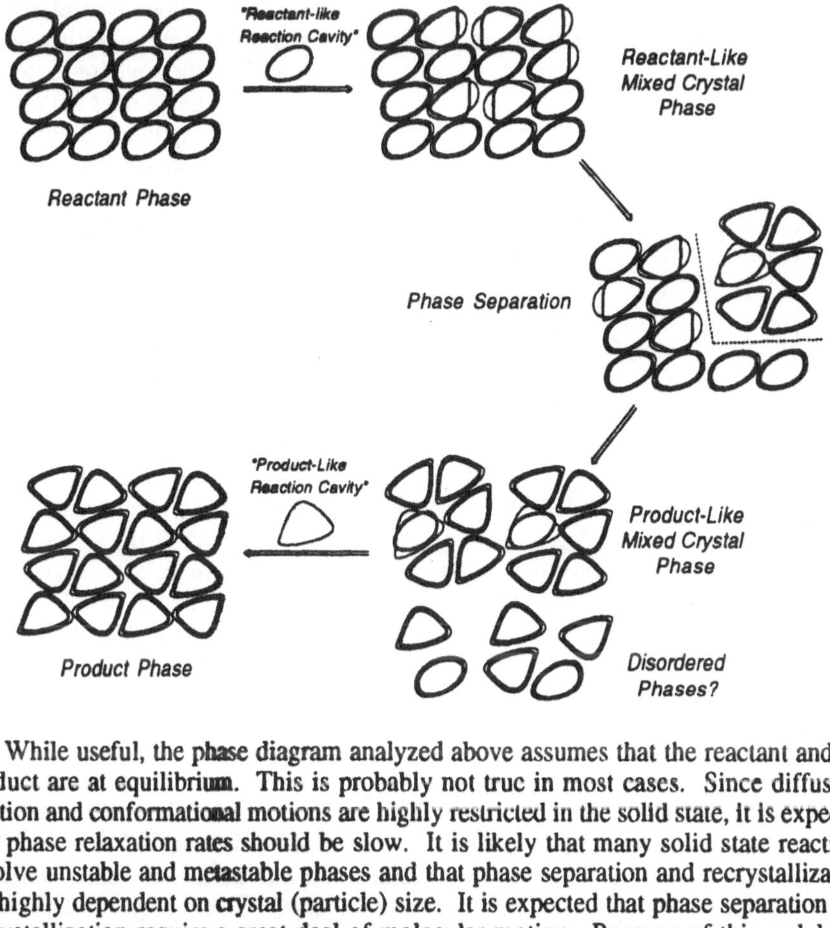

While useful, the phase diagram analyzed above assumes that the reactant and the product are at equilibrium. This is probably not true in most cases. Since diffusion, rotation and conformational motions are highly restricted in the solid state, it is expected that phase relaxation rates should be slow. It is likely that many solid state reactions involve unstable and metastable phases and that phase separation and recrystallization are highly dependent on crystal (particle) size. It is expected that phase separation and recrystallization require a great deal of molecular motion. Because of this, solubility limits in a solid state reaction may exceed those observed by co-crystallization of the components. The most important questions from a preparative point of view relate to the sensitivity of the reaction to the transformations occurring in the solid phase that is hosting it. The involvement of disordered phases should result in loss of reaction control but the influence of product-like phases may be positive or negative and the answer to these questions probably varies from sample to sample.

3.6. THE EXPERIMENTAL STUDY OF PHASE TRANSFORMATIONS IN SOLID STATE REACTIONS

To analyze the phase transformations involved in the solid state reactions shown in Scheme 2 we have explored the combined use of X-ray diffraction, solid state ^{13}C CPMAS NMR, solid state FT-IR and thermal (DSC) analysis. All of the reactions in Scheme 2 are photochemically activated. In such cases, it is important that products do not absorb at the same wavelength as the starting material to prevent filter effects and to facilitate advance of the reaction through the crystal bulk. Therefore, care was taken to irradiate at the tail of the absorption spectrum to facilitate the deepest possible penetration of the exciting UV-light [41] . For instance, the absorption spectra of ketone **5** and its photoproduct **6** are shown in Figure 2. Crystalline samples were irradiated at $\lambda \geq 350$ nm where **5** absorbs weakly and **6** does not absorb at all. Compounds **3** and **4** have absorption spectra that are comparable to those of **5** and **6**, respectively, and irradiation of compound **3** was carried out under similar conditions. Crystals of **7** were irradiated with $\lambda \geq 450$ nm which allows for selective excitation of the n,π^* transition of the reactant. Excitation was carried out by using a monochromator and a 250 W Xe lamp or cutoff filters and a 450 W Hg arc. Solid state irradiations were carried out with powdered samples between two quartz plates either at room temperature or at 0 °C. Samples were irradiated for time periods ranging from a few minutes up to 36 h.

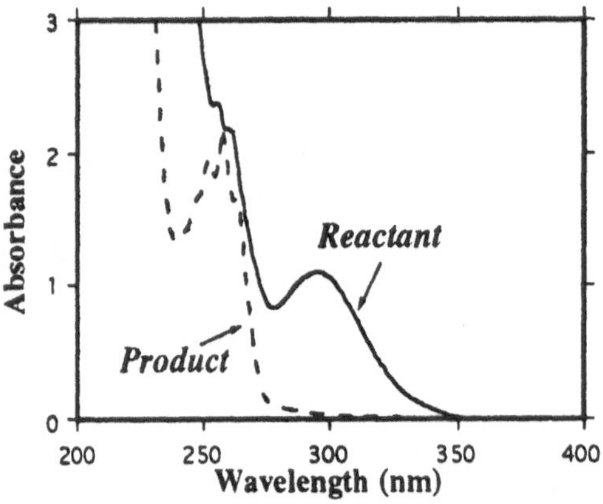

Figure 2. UV-Absorption spectra of *cis*-2,6-diphenyl-2,6-dihydroxycyclohexanone (Reactant) and 2,5-diphenyl-2,5-dihydroxycyclopentane (product)

Each of the solid state techniques in this study was used to probe different aspects of the phase transformations accompanying each reaction. For instance, X-ray diffraction reveals the molecular and crystal structures of the stable phases of the reactant and the product, and the extent of order in the medium as the solid state reaction occurs. Among several solid state spectral techniques, FT-IR and ^{13}C CPMAS NMR provide information that ranges from fingerprint-type to molecular functional group identification, to composition and dynamic aspects. Since polymorphs may be identified by either spectral technique, it may be expected that metastable or mixed crystalline phases may also have characteristic spectra. To complement the spectral data, the thermal properties of the reactant and the product and their solid state reaction mixtures were characterized by differential scanning calorimetry (DSC). In order to obtain information on reactivity changes as a function of reaction progress we explored the use of constant irradiation and FT-IR detection in dilute KBr matrices with the goal of determining *changes in quantum yields as a function of reaction progress*. The quantum yield of a reaction is defined as the number of molecules reacted per unit time per unit volume in relation to the number of photons absorbed per unit time per unit volume. In general, the quantum yield of a reaction is given by a collection of constants with absorption of light being the only variable that affects the observed macroscopic efficiency. Under conditions where excitation of the reactant does not change, photochemical reactions display apparent first order kinetics with apparent rate constants directly proportional to the quantum yield of reaction and the intensity of the exciting beam. Quantum yields can also be expressed as the rate of reaction relative to the rate of other excited state decay pathways. Thus, changes in crystal structure which affect the reaction rate will be apparent in the observable quantum yield.

Scheme 8

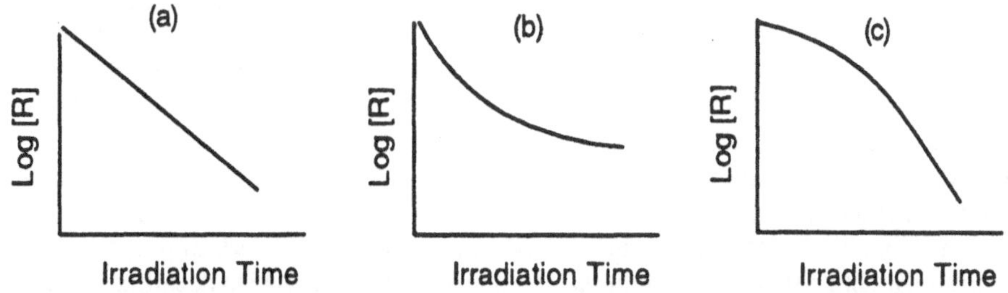

From a practical point of view, the proposed method rests on the applicability of the Beer-Lambert law, on the transparency of the photoproducts, on the optical invariability of a KBr pellet during the time of irradiation, and on the assumption that reaction in a KBr matrix is similar to reaction in a pure powdered solid. The approximate reliability of the above assumptions has been established experimentally in our laboratory. Expectations for these measurements, presented in the form of first-order plots of amount of reactant left as a function of irradiation time are indicated in Scheme 8. Scheme 8a shows an apparent single exponential behavior indicating a quantum yield of reaction that remains unaltered from the lowest to the highest conversion values. The slope of the plot is related to the quantum yield of reaction and to the intensity of the incident beam. Since reaction of the same compound in different solid phases (polymorphs) is expected to have different slopes, a continuously changing slope may reflect a continuously changing environment and/or crystal phases. Plots with upward curvature, such as that in Scheme 8b, indicate that the quantum yield of reaction decreases as a function of irradiation time. Trivial explanations may be based on factors that diminish the light absorbing properties of the reactant such as scattering and filter effects, and factors that reduce the reaction efficiency due to energy transfer. Alternatively, the behavior in Scheme 8b may be due to structural changes in the crystal host that ultimately disfavor the reaction. The third plot is the opposite of the second one and indicates that the efficiency of the reaction increases as the conversion is increased. Explanations for this behavior must involve changes that facilitate the reaction such as phase transitions into liquid or solid phases with fewer constraints on molecular motion.

4. Phase Characterization of Solid State Reactions

The experimental strategy behind this work involves the characterization of the stable phases of the reactant and the products as well as any intermediate phases that may occur along the solid state reaction. Among the goals are: to distinguish between solid-to-solid reactions and reactions that involve melting, to distinguish between topotactic single crystal-to-single crystal reactions and reactions that involve phase separation, and to find out whether phase separation reactions occur through identifiable metastable phases or through phases that are related to those of the reactant and product. As seen below, this survey roughly supports the formulation of Scheme 5 and gives an excellent starting point for addressing some interesting questions of organic solid state reactivity.

4.1. PHOTOCHEMICAL DECARBONYLATION IN THE SOLID STATE.

Reactions b and c in Scheme 2 involve solid state decarbonylation of acyclic and cyclic ketones, respectively. The formation of radical pairs and non-conjugated biradicals requires the cleavage of at least one sigma bond. In solution and in the gas phase, diffusional separation of the two radicals or extended conformations of a biradical chain allow for reactions to occur with external reagents as well as with themselves. However, in the crystalline solid state, due to the close proximity of the two radical centers, reactions from biradicals and radical pairs can only be accomplished if they are faster than bond reformation between the two radical termini. Work in progress in our group suggests that this requirement may be satisfied in the stepwise decarbonylation of certain crystalline ketones. This reaction also offers a general route to non-conjugated biradicals in the crystalline solid state. As indicated in Scheme 9, the reaction occurs by two successive cleavage reactions on the triplet state surface. A successful solid state reaction requires substitution patterns that facilitate excited state cleavage and the loss of carbon monoxide relative to recombination of the acyl-alkyl radical pair. We have recently found that solid state reactivity and stability may be tuned on and off by a simple and judicious choice of substituents [25, 43].

Scheme 9

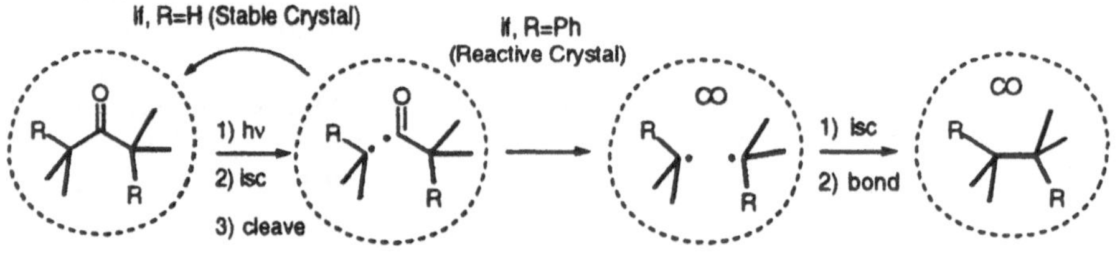

4.2. DECARBONYLATION OF TETRAMETHYL-DIPHENYLACETONE. A SINGLE PRODUCT IN SOLUTION AND IN THE SOLID STATE

Reaction of tetramethyl-diphenylacetone to tetramethyl-diphenylethane in the solid state proceeded smoothly and nearly to completion for powdered samples irradiated at 20 °C with $\lambda \geq 350$ nm. The formation of a single product in the solid state suggests that the phase diagram of this system should be relatively simple. Careful analysis reveals the absence of phenyl acetaldehyde, styrene or isopropyl benzene from radical pair disproportionation reactions which occur in small amounts in solution. It was found that the melting points of the reactant and the product are relatively high and similar. It was also noticed that CO remains trapped within the reacted crystal as gas evolution was observed when samples were dissolved for analysis.

Scheme 10

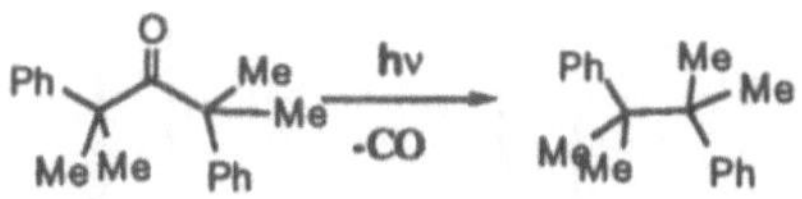

3 (m.p. = 115°C) **4 (m.p. = 114°C)**

Repeated, but not exhaustive recrystallization of the reactant and the product at ambient temperatures and pressures produced a single polymorph in each case. X-Ray analysis of compounds **3** [44] and **4** [45] revealed the same orthorhombic space group (Pbca). The packing motifs of **3** and **4** are fairly similar but not identical. The structural similarity between **3** and **4** is evident from an excellent structural overlap of the reactant and the product which suggest the likelihood of mixed-crystallization (Figure 3).

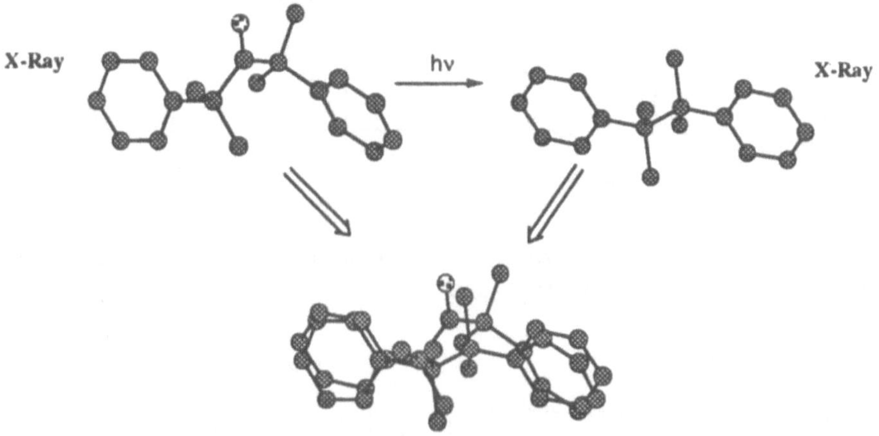

Figure 3. X-Ray molecular structures of compound **3**, its solid state product **4**, and a superposition of the two structures (r.m.s = 0.724).

The solid state ^{13}C NMR spectra of the stable phases of **3** and **4** were obtained along with spectra of samples irradiated for various time intervals. We could not detect clear NMR spectral evidence of the product in the phase of the reactant or the reactant in the product phase but efforts along these lines are still in progress. Interestingly, the spectra of partially reacted samples could be reproduced fairly well by a weighted sum of the spectra corresponding to the pure reactant and product (Figure 4). FT-IR spectra were in qualitative agreement with the ^{13}C NMR results (see Figure 5). Preliminary thermal analysis of the reactant, the product and several partially reacted samples suggested a binary phase system with partial mixed solubility and a eutectic point at about 85 °C (Figure 6). The thermograms of early conversion values were characterized by broadening of the reactant melting endotherm and the appearance of a broad shoulder that displaces to lower temperatures until it reaches the characteristic transition of the eutectic point.

Spectral results and thermal analysis clearly support a solid-to-solid reaction with partial solid state solubility and a eutectic point that is well above the temperature of the photolysis experiment. To investigate the effect of phase transformation on the reaction changes in efficiency were investigated.

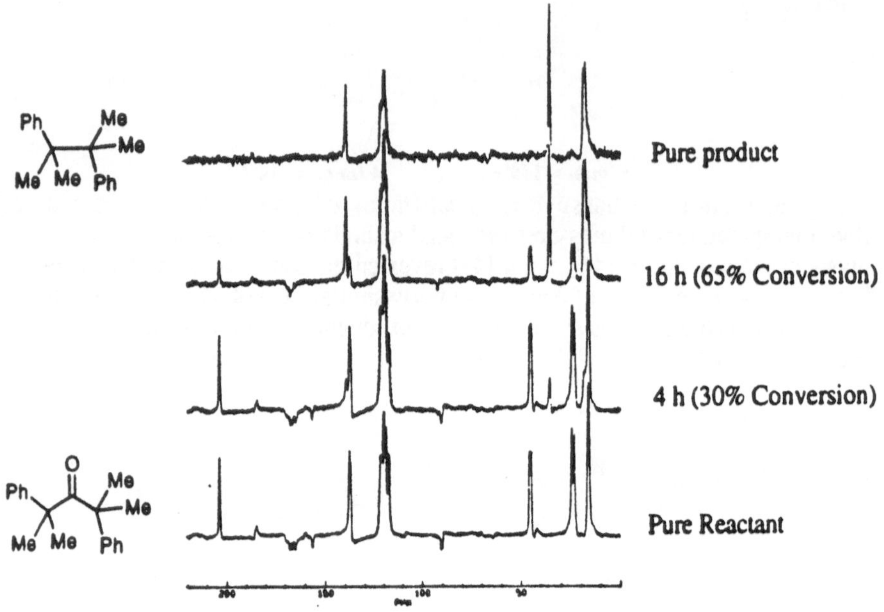

Figure 4. Solid state ^{13}C CPMAS NMR (with total spinning side band suppression, TOSS) of compounds **3** (bottom) and **4** (top) and partially reacted samples (middle).

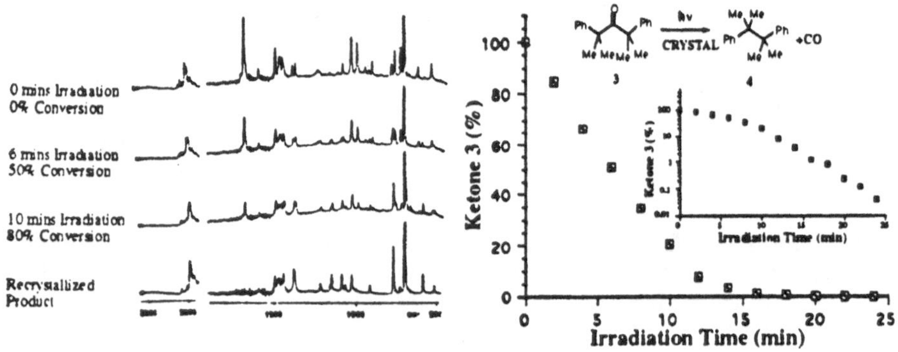

Figure 5. Left: FT-IR spectra of **3** in a KBr matrix as a function of irradiation time. Right: Analysis of the FT-IR data in terms of reactant left vs. irradiation time for compound **3**. A semi-logarithmic plot of the same data is shown in the inset.

In the case of compound **3**, optically transparent KBr matrices were obtained to give between 0.1 and 0.3 absorbance units at the absorption maximum of the CO stretch at 1687 cm^{-1}. This dilution allows for applicability of the Beer-Lambert law and for high optical quality. Pellets were mounted on holders that allowed for reproducible exposure to UV-light as well as reproducible positioning within the FT-IR spectrometer. As shown in Figure 5, irradiations were carried out for 2 minute periods and each was followed by FT-IR measurements. Total irradiation times of up to 25 minute were sufficient to observe the CO band of the starting material almost completely disappear. Formation of the product in the same solid phase as that which occurs in large scale solid-to-solid irradiations was verified from the spectra. Analysis of the absorption data produced plots such as that in Figure 5b. A close inspection of the semi-logarithmic plot in the inset reveals significant curvature indicating that the reaction becomes more efficient as product accumulates.

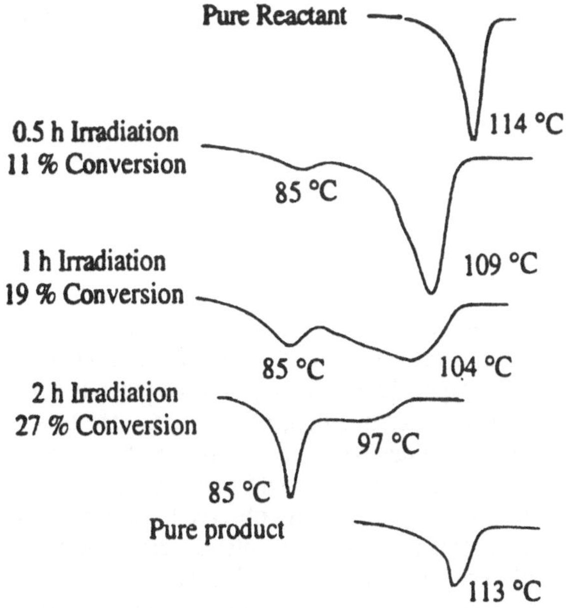

Figure 6. Thermograms of Ketone **3** as a function of irradiation time and conversion. The bottom melting endotherm corresponds to the recrystallized product.

4.3. STEREOSPECIFIC SOLID STATE DECARBONYLATION OF *cis*-2,5-DIPHENYL-2,6-DIHYDROXY-CYCLOHEXANONE: A STEADY STATE PHASE SEPARATION MECHANISM

In contrast to tetramethyl-diphenylacetone **3**, the reaction of 2,6-diphenyl-,2,6-dihydroxycyclohexanone **5** may give two stereoisomeric products and changes in solid state selectivity with increasing conversion can be followed as a measure of reaction control. Decarbonylation of compound **5** in solution produces cyclopentane diols ***cis*-6** and ***trans*-6** in a 50:50 ratio [26]. The reaction occurs selectively in the crystalline state with high efficiency and to very high conversions. A summary of the solid state reaction in Scheme 11 illustrates the melting points of the pure reactant (146°C) and product (104°C) phases along with some mechanistic details of the reaction.

As illustrated in Scheme 11 and Figure 7, the reactant crystallizes with the phenyl groups in the axial position. The equatorial hydroxyl groups are involved in hydrogen bonding with centrosymmetric and translationally related molecules to form an extended linear motif. As with compound **3**, the reaction is proposed to proceed from the triplet excited state of the ketone to give acylalkyl biradical BR-1 which decarbonylates to give dialkyl biradical BR-2. In solution, BR-2 may adopt extended conformations and undergo bond rotation so that both *cis*- and *trans*- isomers are formed with equal probabilities. In the solid state, on the other hand, bond rotation in BR-2 is hindered and closure occurs with high preference towards the observed major product.

Scheme 11

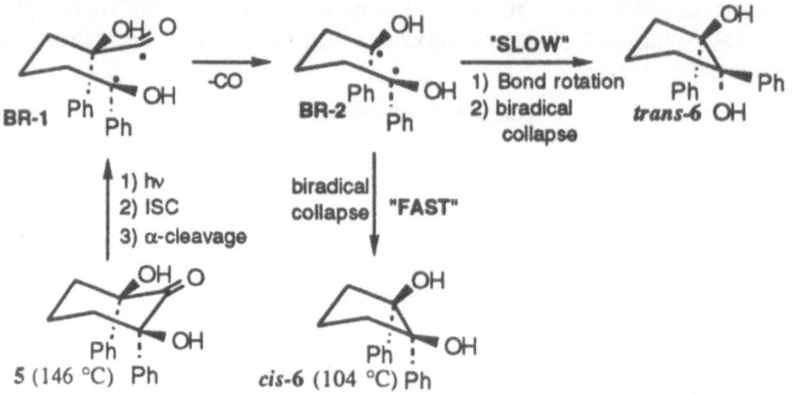

Solid state irradiations were carried out at λ≥350 nm with powdered samples and single crystals. As with compound **3**, carbon monoxide remains trapped within the solid. Experiments carried out with *ca.* 500 mg showed that reaction may be carried out in a solid-to-solid fashion, to completion, and without affecting the relatively high stereoselectivity for product formation (*ca. cis:trans* = 20:1). With experiments in rigid hydroxylic glasses such as sucrose we could demonstrate that rigidity alone is not sufficient to account for the high selectivity observed. X-Ray analysis of the reactant and the recrystallized product revealed completely different packing arrangements. Compound **5** crystallizes in the space group P $\bar{1}$ and compound **cis-6** crystallizes in the space group C2/c with two molecules per asymmetric unit [46] . However, as shown in Figure 7, an excellent structural overlap of molecules of **5** and **6** is consistent with a topochemical reaction and suggests a good solid state solubility of each molecule within the crystal lattice of the other. Compound **cis-6** crystallizes with two very similar molecules per asymmetric unit and both of them give excellent overlap with **5**.

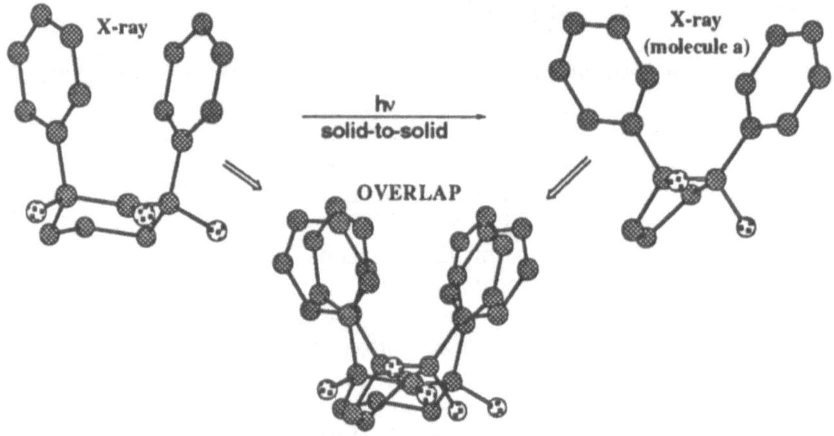

Figure 7. X-Ray structures of compound **5**, and one of the crystallographically distinct structures of its solid state product, ***cis*-6**. A superposition of the two structures gives an r.m.s = 0.874. The molecule of ***cis*-6** not shown gives r.m.s = 0.947.

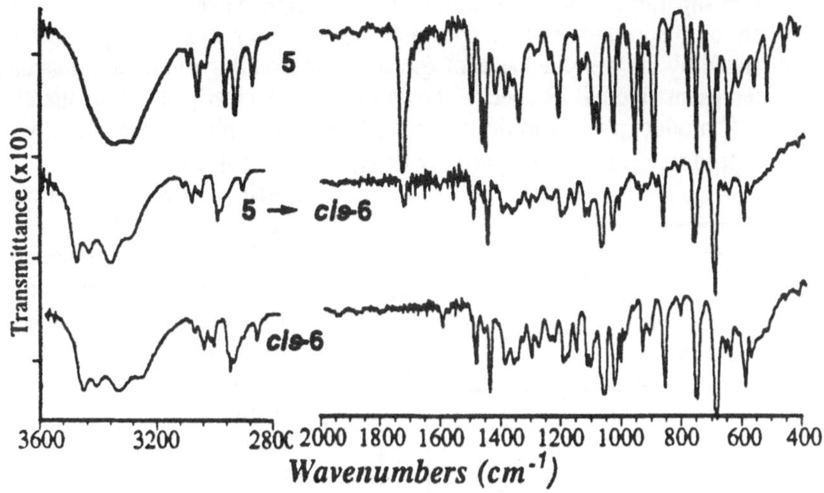

Figure 8. Solid state FT-IR spectra (KBr) of compound **5**, a reacted sample of **5** and a sample of pure **6**. The identity of the as-reacted and recrystallized phases of **6** is suggested by the two bottom spectra.

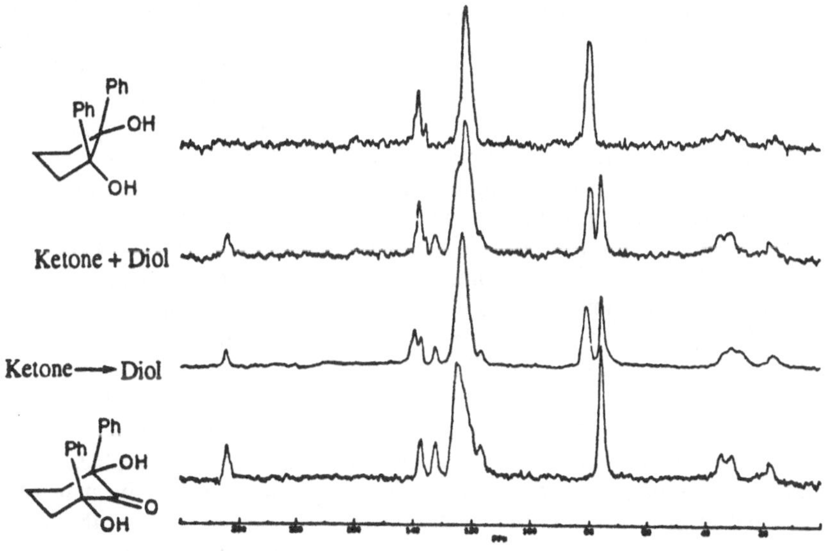

Figure 9. Solid state ^{13}C CPMAS NMR (TOSS) spectra of (from top to bottom) pure photoproduct *cis*-**6**, weighted addition of the spectra of *cis*-**6** and starting material **5**, a partially reacted sample containing **5** and *cis*-**6**, and a sample of pure **5**.

^{13}C CPMAS NMR and FT-IR spectral analysis of the pure reactant, the recrystallized product and several partially irradiated samples suggests that reaction proceeds by phase separation. No polymorphic modifications have been observed for either the reactant or the product, and the spectra of partially photolyzed samples can be efficiently simulated by the weighted sum of spectra corresponding to the pure reactant and product. Representative FT-IR spectra shown in Figure 8 clearly show the identity of the "as-formed" product phase and the recrystallized product phase. The identity of the reactant and product phases in the partially reacted samples is also evident in ^{13}C NMR CPMAS spectra shown in Figure 9.

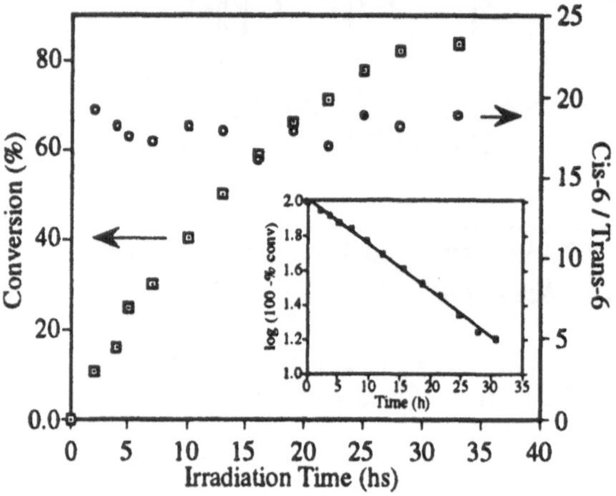

Figure 10. Percent conversion (squares) and cis/trans ratio (circles) from photolysis of **5** as a function of irradiation time. Shown in the inset is a a first order plot for the reaction.

The solid state reactivity of compound **5** is interesting from several points of view. First of all, spectral analysis is consistent with a phase separation reaction where the product crystallizes in a phase that is similar to that obtained upon crystallization from solvents. At the same time, the environmental control of the reaction cavity is maintained throughout the reaction. Mechanistically, this indicates that the rate of formation of the *cis* isomer remains much higher than the rate of formation of the *trans* isomer regardless of how much product is present in the sample. A kinetic analysis in KBr matrices and in semipreparative samples of **5** also resulted in a remarkably clean (apparent) first order reaction with a linear semilogarithmic plot of reactant vs. irradiation time (Figure 10). This behavior indicates that reaction efficiency (which is related to rate of reaction) remains unchanged as the product accumulates in the crystal of the reactant.

DSC Analysis of compounds **5** and **cis-6** were carried out to characterize the pure compounds. They gave sharp peaks at their respective melting points. Partially reacted samples revealed transitions corresponding to the eutectic mixture at 85°C along with the broadened endotherms of the depressed melting point of the reactant. From the smallest to the largest conversion values we could not detect transitions that we could assign to mixed crystalline phases. We speculate that mixed crystalline phases in this system do not accumulate but occur under a steady state. A steady state phase separation mechanism would microscopically increase the concentration of the product in the crystal phase of the reactant until phase separation occurs. If correct, this interpretation may imply that the rate of reaction in microscopic domains already containing some product may be larger than in the pure phase of the reactant.

4.4. 1,2-SHIFTS IN CRYSTALLINE DIAZO COMPOUNDS. A HIGHLY SELECTIVE SOLID STATE REACTION THROUGH METASTABLE PHASES

The reaction in Scheme 2d constitutes one of the most interesting examples of reaction control observed in our laboratory. It involves the transformation of a crystalline diazo compound into a crystalline alkene in high yields and with unprecedented selectivity. When the reaction is carried out in solution, products from migration of hydrogen and phenyl substituents are observed in yields that vary slightly with temperature and solvent polarity [47, 48] . Reactions carried out in the solid state with $\lambda \geq 450$ nm proceed stereoselectively to give the *cis*-stilbene type product, **8Z**, with conversions up to 95% (Scheme 2d). Solid state reactions carried out under conditions where the photoproduct absorbs (e.g., $\lambda \geq 300$ nm) result in lower selectivities that change as a function of time.

Scheme 12

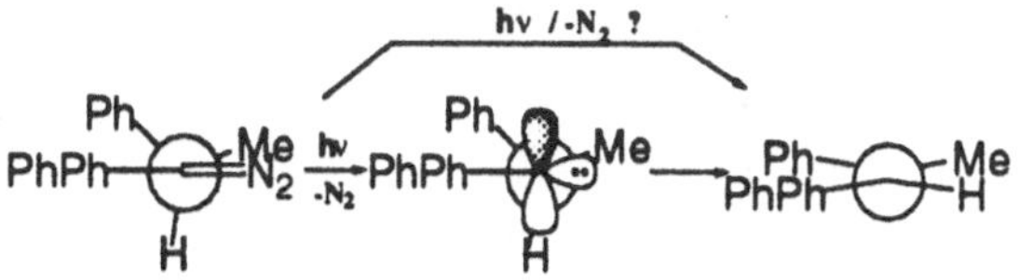

7 **Carbene 7** **8Z**

The mechanism for solid state control appears to be based on the rigidity of the crystal lattice and conformational control that predisposes the reactant for the formation of the observed photoproduct. In the case of carbene 1,2-H shifts, it is expected that the migrating group should be aligned with the empty p-orbital of the p,sp^2-hybridized singlet state carbene. As shown in a Newmann representation of the solid state conformation of diazo **7** (Scheme 12), the solid state arrangement facilitates the migration of the small hydrogen group while least motion of the bulky phenyl group insures the formation of the *cis*-alkene **8Z**.

The reaction may be formulated as shown in Scheme 2d, and as discussed above, with the intermediacy of an arylalkyl carbene. However, it has been recently proposed that 1,2-R shifts may also be formulated as excited state reactions of the diazo compound involving a concerted elimination of nitrogen and migration of one of the α-substituents. It is possible that both mechanisms may contribute to product formation; they would both be subject to least motion control.

As in previous examples, repeated crystallization of compounds 7 and 8Z yielded a single polymorph for each compound [46]. The packing structures of the reactant and the product coincidentally have the same space group but quite different packing structures. The product crystallizes with two molecules per asymmetric unit which differ only in the rotational angle of the biphenyl substituent. As shown in Figure 1, the structural overlap between **7** and one of the two molecular structures of **8Z** is excellent and with the exception of the rotated 4-phenyl group, overlap with the second molecule is also excellent.

Spectroscopic analysis of the solid state reaction of compound **7** differs substantially from those of compounds **3** and **5** where separation of the product to an identifiable stable phase was possible. Both the FT-IR and ^{13}C CPMAS NMR spectra of partially irradiated samples of **7** give spectra with components that differ form those the pure phases of the reactant and the product. For instance, ^{13}C CPMAS NMR spectra corresponding to the recrystallized phases of the reactant (bottom) and the product (top) may be compared with those of a partially irradiated samples. The two molecules per asymmetric unit in the case of **8Z** are evident from the duplicity of some of the lines in the top spectrum. However, it is clear from the spectra that the "as reacted" product is quite different from the one on the top of the figure. The ^{13}C NMR spectra of partially reacted samples suggests the possibility of an accumulated mixed crystalline phase and answers were searched for by DSC.

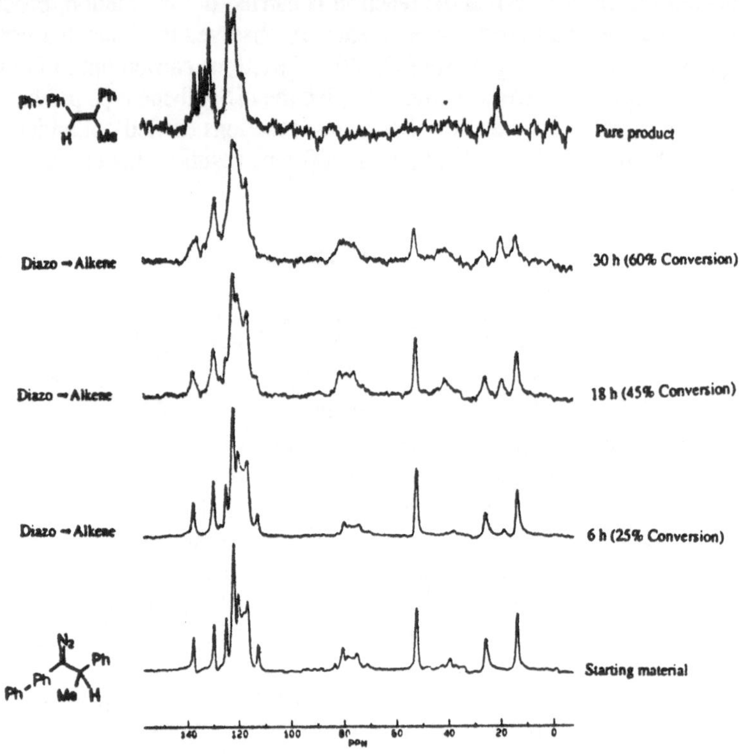

Figure 11. Solid state ^{13}C CPMAS NMR (TOSS) spectra (from top to bottom) of recrystallized photoproduct **8Z**, partially reacted samples after 30, 18 and 6 h of irradiation, and a sample of pure **7**.

Thermal analysis of compounds **7**, **8Z** and partially reacted samples of **7** containing **8Z** are complicated by the fact that compound **7** reacts thermally under conditions of the DSC measurements (Figure 12). Pure samples of compounds **7** and **8Z** melt at 95 and 105 °C, respectively. DSC thermograms of pure diazo **7** give a sharp melting endotherm followed by a highly exothermic reaction involving denitrogenation of the diazo group. The results indicate that a thermal reaction in the pure crystal is much slower than reaction in the melt. However, we found that partially reacted samples facilitate thermal denitrogenation since the reaction exotherm occurs before the melting transition even after relatively modest conversion values, and in spite of a small depression of the melting point. Along this complex behavior, we could not distinguish a eutectic point and liquid phases do not seem to exist below *ca.* 60 °C. Progressive broadening and slow lowering of the melting endotherm without a eutectic transition suggests the involvement of mixed crystalline phases. Studies addressing the possibility of forming solid solutions of compound **7** and **8Z** by mixed crystallization are currently in progress.

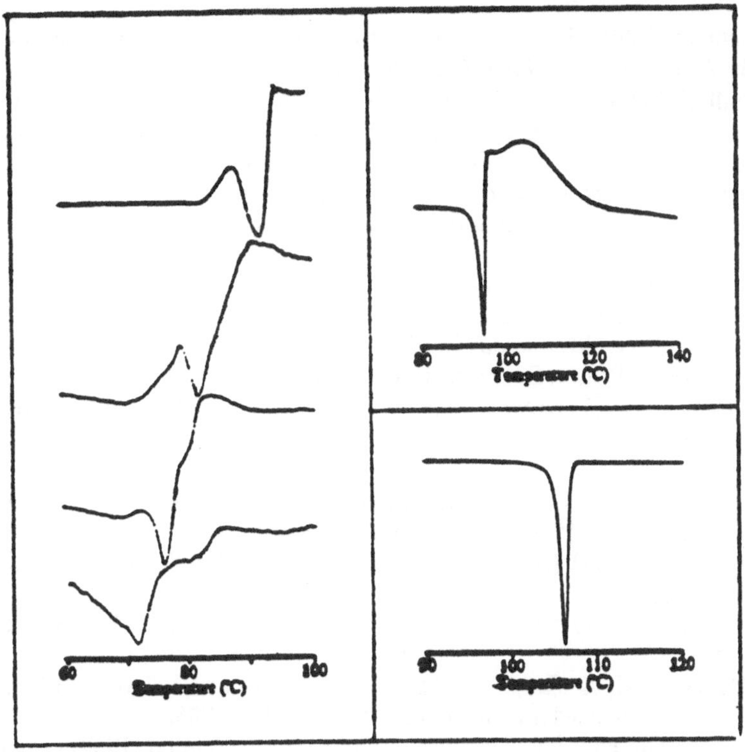

Figure 12. Left: DSC Thermograms of Diazo **7** as a function of irradiation time and conversion. Right: DSC thermograms of pure compounds **7** and **8Z**.

5. Conclusions

Needs of high reaction control and current environmental concerns constitute strong motivations for the development of solvent-free synthetic procedures in crystalline solids. We have shown that solid state reactions may be carried out to completion and with stereoselectivity and stereospecificity characteristic of their pure crystalline phases. While closer examination of several solid state reactions is in progress, it seems clear that the preparative potential of solid state chemistry is excellent.

6. Acknowledgments

We gratefully acknowledge the National Science Foundation, the Petroleum Research Fund, and the University of California for their generous support of this research. The National Science Foundation is also acknowledged for a graduate research fellowship to A.E.C.

7. References

[1] D. O. Cowan and F. M. Wlygul (1986) The Organic Solid State, *Chemical and Enginnering News* **July**, 28.

[2] A. Buchachenko (1989) Molecular Ferromagnetics, *Mol. Cryst. Liq. Cryst.* **176**, 307-

20.
[3] I. Iwamura (1990) High-Spin Organic Molecules and Spin Alignment in Organic Molecular Assemblies, *Adv. Phys. Org. Chem.* **26**, 1.
[4] F. Wudl, F. Closs, P. M. Allemand, S. Cos, G. Hinkelmann, G. Srdanov and C. Fite (1989) Donnors, Acceptors and the FOM Model, *Mol. Cryst. Liq. Cryst.* **176**, 249-258.
[5] F. Toda (1995) Solid State Organic Chemistry: Efficient Reactions, Remarkable Yields, and Stereoselectivity, *Acc. Chem. Res.* **28**, 480-487.
[6] R. Lamartine (1989) Organic solid state chemistry. Recent developments, *Bull. Soc. Chim. Fr.* 237-46.
[7] V. Ramamurthy and K. Venkatesan (1987) Photochemical Reactions of Organic Crystals, *Chem. Rev.* **87**, 433-81.
[8] J. R. Scheffer, M. Garcia-Garibay and O. Nalamasu (1987) *in Organic Photochemistry*, A. Padwa, Ed., Marcel Dekker, Inc., New York.
[9] G. R. Desiraju (1987) *Organic Solid State Chemistry*, Elsevier, Amsterdam.
[10] J. B. Gabarskczyk and D. W. Jones (1991) *Organic Crystal Chemistry*, Oxford, Cambridge.
[11] G. M. J. Schmidt (1976) *Solid State Photochemistry*, Verlag Chmie, New York.
[12] H. W. Kohlshutter (1918) *Anorg. Allg. Chem.* **105**, 121.
[13] M. D. Cohen (1975) The Photochemistry of Organic Solids, *Angew. Chem. Int. Ed. Engl.* **14**, 386-93.
[14] J. R. Scheffer (1987) *in Solid State Organic Chemistry*, G. R. Desiraju, Ed., VCH, Amsterdam.
[15] L. Casswell, M. A. Garcia-Garibay, J. R. Scheffer and J. Trotter (1993) Optical Activity Can Be Generated from Nothing, *J. Chem. Ed.* **70**, 785.
[16] M. Garcia-Garibay, J. R. Scheffer, J. Trotter and F. Wireko (1987) Generation of Optical Activity Through Solid State Reaction of a Racemic Mixture that Crystallizes in a Chiral Space Group, *Tetrahedron Lett.* **28**, 4789.
[17] J. R. Scheffer and M. Garcia-Garibay (1989) *in Photochemistry of Solid Surfaces*, T. Matsuura and M. Anpo, Ed., Elsevier, Amsterdam.
[18] F. Toda (1988) Enantiocontrol of photoreactions in the solid state, *Mol. Cryst. Liq. Cryst.* **161(Pt. B)**, 355-62.
[19] G. Wegner (1977) Solid State Polymerization Reactions, *Pure Appl. Chem.* **49**, 443.
[20] D. Bloor and R. R. Chance (1985) *Polydiacetylenes*, M. Nijhoff, Dordrecht.
[21] N. J. Karch, E. T. Koh, B. L. Whitsel and J. M. McBride (1975) X-ray and electron paramagnetic resonance structural investigation of oxygen discrimination during the collapse of methyl-benzoyloxy radical pairs in crystalline acetyl benzoyl peroxide, *J. Am. Chem. Soc.* **97**, 6729-43.
[22] J. M. McBride and M. R. Gisler (1979) Isotopic and optical studies of the decomposition of crystalline dibenzoyl peroxide, *Mol. Cryst. Liq. Cryst.* **52**, 425-36.
[23] M. A. Garcia-Garibay, A. Gamarnik, L. Pang and W. S. Jenks (1994) Excited State Intramolecular Hydrogen Atom Transfer at Ultra-Low Temperatures. Evidence for Tunneling and Activated Mechanisms in oMethylanthrone, *J. Am. Chem. Soc.* **116**, 12095-6.
[24] M. A. Garcia-Garibay, W. S. Jenks and L. Pang (1996) Heterogeneous Hydrogen Atom Transfer in the Excited State of Methylbenozphenone in EPA glasses at 77K, *J. Photochem. Photobiol. A* **in press**,
[25] T. Choi and M. Garcia-Garibay Rational Control of a Solid State Organic Reaction., manuscript in preparation.
[26] T. Choi, D. Cizmeciyan, S. Khan and M. A. Garcia-Garibay (1995) An Efficient

Solid-to-Solid Reaction via a Steady-State Phase-Separation Mechanism, *J. Am. Chem. Soc.* **118**, 12893-4.
[27] H. S. Shin and M. A. Garcia-Garibay Transforming a Non-Selective Carbene Rearrangement into a Highly Selective Process by Using Crystalline Media, *J. Am. Chem. Soc.* manuscript submitted.
[28] S. Celebi, S. Leyva, D. A. Modarelli and M. S. Platz (1993) 1,2-Hydrogen Migration and Alkene Formation in the Photoexcited States of Alkylphenyldiazomethanes, *J. Am. Chem. Soc.* **115**, 8613-8620.
[29] D. A. Modarelli, S. Morgan and M. S. Platz (1992) Carbene Formation, Hydrogen Migration, and Fluorescence in the Excited States of Dialkyldiazirenes, *J. Am. Chem. Soc.* **114**, 7034-7041.
[30] A. Constable, M. Garcai-Gariibay and K. N. Houk (1996) *Unpublished results*
[31] G. M. J. Schmidt (1964) Topochemistry. Part III. The Crystal Chemistry of Some trans-Cinnamic Aciids, *J. Chem. Soc.* 2014-2021.
[32] A. G. Schultz, A. G. Taverns, R. E. Taylor, F. S. Tham and R. K. Kullnig (1992) Solid State Photochemistry. Remarkable Effects of the Packing of Molecules in Crystals on the Diastereoselectivity of the Intramolecular 2 + 2 Photocycloaddition of a 4-(3'-Butenyl)-2,5-Cyclohexenedione, *J. Am. Chem. Soc.* **114**, 8725-8727.
[33] M. D. Hollingsworth and J. M. McBride (1985) Specific long-range effects on relaxation of local stress during a solid-state reaction, *J. Am. Chem. Soc.* **107**, 1792-3.
[34] J. M. McBride (1983) The role of local stress in solid-state radical reactions, *Acc. Chem. Res.* **16**, 304-12.
[35] J. M. McBride, B. E. Segmuller, M. D. Hollingsworth, D. E. Mills and B. A. Weber (1986) Mechanical stress and reactivity in organic solids, *Science (Washington, D. C.)* **234 (4778)**, 830-5.
[36] P. R. Pokkuluri, J. R. Scheffer and J. Trotter (1989) Surface versus bulk reactivity in solid-state organic chemistry, *Tetrahedron Lett.* **30**, 1601-4.
[37] A. I. Kitaigorodskii (1984) *"Mixed Crystals"*, Springer Verlag, Berlin.
[38] Y. Ohashi (1988) Dynamical Structure Analysis of Crystalline-State Racemization, *Acc. Chem. Res.* **21**, 268-74.
[39] J. M. Thomas (1981) Diffusionless Reactions and Crystal Engineering, *Nature* **289**, 633.
[40] C. R. Theocharis and W. Jones (1987) *in Organic Solid State Chemistry*, G. Desiraju, Ed., Elsevier, Amsterdam.
[41] V. Enkelmann, G. Wegner, K. Novak and K. B. Wagener (1994) Single-Crystal-to-Single-Crystal-Photodimerization of Cinnamic Acid, *J. Am. Chem. Soc.* **115**, 10390.
[42] W. Jones, H. Nakanishi and J. M. Thomas (1980) *J. Chem. Soc. Chem. Commun.* 610.
[43] T. Choi, C. Loo and R. Yakura (1995)
[44] Compound 3: 1,3-Diphenyl-1,1,3,3-tetramethylacetone, coloreless prisms obtained fom ethanol, m.p. = 144 °C, space group Pbca, Z=8, a=12.959(9) Å, b=19.23(1) Å, c=12..535(8) Å, V=3124(4) Å^3, R=0.047 Rw=0.056.
[45] G. Kratt, B. H.-D., H. J. Lindner and C. Ruchardt (1983) Synthese, Struktur and Spannung Symetrischer Tetraalkyldiarylethane, *Chem. Ber.* **116**, 3235.
[46] . Compound 7: 1-(4-phenyl)phenyl-2-phenyl-1-diazopropane , ruby red needles obtained fom pentane, m.p.= 95 °C, space group $P2_1/c$, Z=4, a=12.1088 Å, b=5.8895Å, c=7222.049Å, =96.067°, Rw=0.058; Compound 8Z: Z-i-(4-phenyl)phenyl-2-phenyl-propene, withe needles obtained fom ethanol, m.p.= 105°C, space group $P2_1/c$, Z=8,

a=18.898Å, b=6.255Å, c=26.08 Å, β=94.77°; Rw=0.055.
[47] M. A. Garcia-Garibay, C. Theroff, S. H. Shin and J. Jernelius (1993) Solvent Effects on the Singlet-Triplet Equilibrium and Reactivity of a Ground Triplet State Arylalkyl Carbene, *Tetrahedron Lett.* **52**, 8415.
[48] M. A. Garcia-Garibay (**1993**) The Reactivity of Arylcarbenes with Methanol. Triplet-State Reactivity or Spin-State Equilibrium as a Moving Target?, *J. Am. Chem. Soc.* ***115***, 7011-12.

MAGNETIC, ELECTRICAL AND SPECTROSCOPIC STUDIES OF ALKALIDES AND ELECTRIDES

Experimental Methods and Examples

JAMES L. DYE
Department of Chemistry and Center for Fundamental Materials Research
Michigan State University, East Lansing, Michigan 48824

The properties of alkalides and electrides have been extensively described in a number of review articles, but the methodology has been less completely covered. This presentation describes the special methods used to synthesize and study these air– and temperature–sensitive compounds. The topics include: (1) vacuum-line synthesis and the methods used for crystal growth, handling and structure determination; (2) magnetic susceptibilities; (3) powder and single crystal NMR spectra; (4) EPR, ENDOR and ESEEM studies; (5) DC and AC conductivities; (6) reflectance, absorbance and luminescence spectra; (7) the preparation and study of thin films by solvent evaporation and by high-vacuum codeposition of the metal and complexant; (8) photoelectron emission; and (9) computer methods for visualization of cavities and channels in electrides. The methods are illustrated with selected results from more than 20 years of alkalide-electride research.

1. Introduction

Crystalline alkalides and electrides are prepared by the reaction of one or more alkali metals with a cyclic or bicyclic polyether or polyamine that is kinetically inert to reduction and forms a strong complex with the alkali cation. To date, this has involved crown ethers, **1**, cryptands, **2**, and cyclens, **3**. In principle, the synthesis is straightforward. The alkali

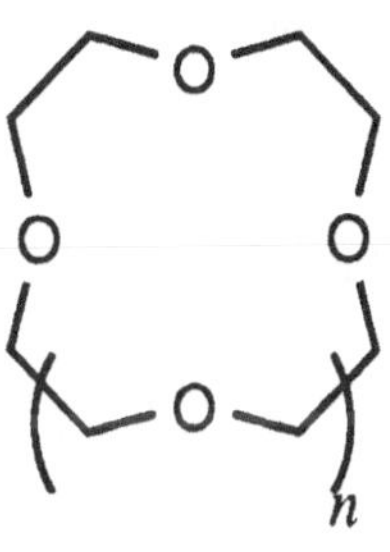

1

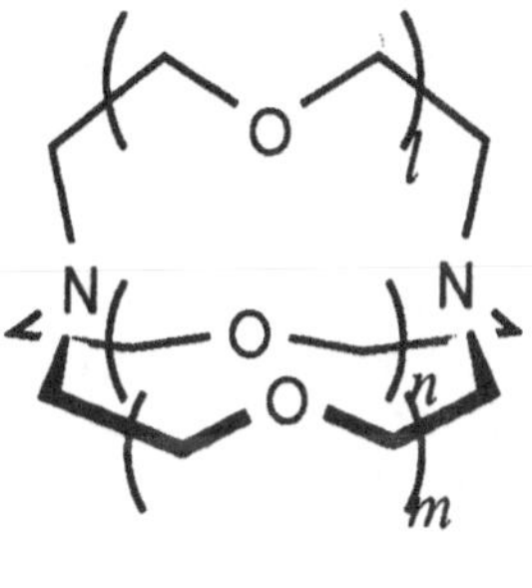

2

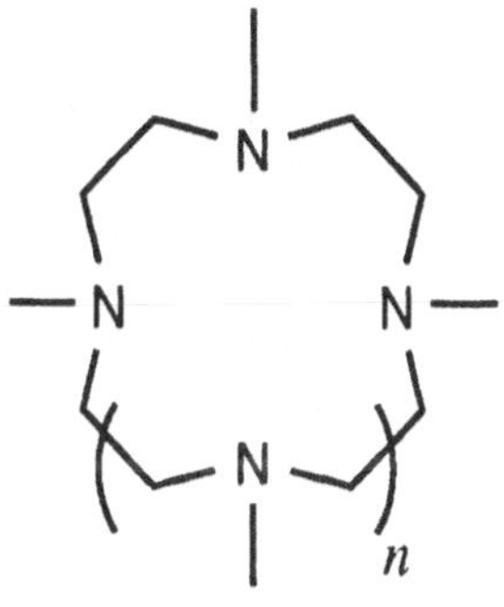

3

L. Echegoyen and A.E. Kaifer (eds.), Physical Supramolecular Chemistry, 313–336.

metal is solubilized by the complexant in an inert solvent such as methylamine ($MeNH_2$) dimethyl ether (Me_2O) or tetrahydrofuran (THF) in the ratio 2:*n* (metal to complexant) for alkalides or 1:*n* for electrides, in which $n = 1$ for cryptands and non-sandwich crown ether compounds and $n = 2$ for crown ether sandwich complexes. The solution then contains the complexed cation M^+C or M^+C_2 (in which C represents the complexant) and either M^- or e^- according to

$$2M(s) + nC\,(soln) \rightarrow M^+C_n + M^- \tag{1}$$

$$M(s) + nC\,(soln) \rightarrow M^+C_n + e^-. \tag{2}$$

The addition of a less polar co–solvent such as trimethylamine (Me_3N), diethyl ether (Et_2O) or ethylamine ($EtNH_2$) followed by slow cooling or slow evaporation of the more polar solvent results in the formation of alkalide or electride crystals. Mixed alkalides such as $Cs^+(C222)Na^-$ can be formed by using two alkali metals. (Note: the abbreviations Cnml for cryptand[n.m.l] and nCm for n-crown-m are used throughout this paper.)

Of course, the methods used to synthesize and study alkalides and electrides are not as simple as the preceding description might imply. No alkalide or electride is thermodynamically stable to irreversible decomposition and no alkali metal solution in the solvents mentioned is truly inert to reduction. To make matters worse, the decomposition of solutions and of crystalline alkalides and electrides is, in many cases, autocatalytic, so that rapid decomposition often follows the initiation of decomposition or oxidation.

Over the years, we have developed techniques for the synthesis and study of materials that are both air-sensitive and subject to thermal decomposition. The properties of alkalides and electrides have been extensively reviewed[1-17] and no attempt will be made here to provide complete descriptions of these studies. Instead, the special methods that we have used to avoid problems of contamination and decomposition are described. It should be noted that most of our work has focused on the behavior of pure crystalline compounds. This requires much more attention to purity of materials and handling techniques than is required to use alkalides and electrides as reducing agents. For example, our production of nanoscale metal and alloy particles by reduction of dissolved transition and post-transition metal salts with M^- or e^- in solution is very forgiving of partial decomposition and is truly routine.[18-20] Similarly, reduction of organic compounds by alkalides and electrides has proven useful in a number of laboratories and the methods used do not require the rigorous synthesis and handling techniques described here. In particular, the extensive synthetic applications of alkalides and electrides by Ohsawa[21-25] and Jedlinski[26-30] are characterized by the use of rather standard organometallic methods.

Because of space limitations, many figures and diagrams of apparatus that would be appropriate here are left out, but specific reference is made to published figures in such cases. Alkalides and electrides are fascinating compounds with many unusual properties. As our work on these materials decreases as a result of my retirement, it is my hope that others will continue to prepare and study them, especially electrides, which provide unique examples of nearly free electrons confined to a complex maze of cavities and channels.

This description of the methods we use in such studies may provide a head-start for continued investigations of alkalides and electrides in other laboratories.

2. Synthesis and Handling

2.1. VACUUM-LINE METHODS

Although Schlenk-line techniques are standard in organometallic synthesis and might well be used successfully in alkalide-electride research, we have preferred to use vacuum-line methods of synthesis and materials purification. The "Wayda-Dye" double manifold, available from several suppliers of glassware, has been fully described in two articles that contain diagrams and photographs of the equipment used.[31,32] We find the use of Cajon® flexible stainless steel tubing, unions and Ultra-Torr® connectors to be particularly appropriate for vacuum-line distillations and crystallizations. The ability to move or shake the glass vessels under vacuum without breaking the line is particularly desirable. We also find that the complete avoidance of greased stopcocks and tapers helps to keep our apparatus clean and avoids the trapping of air and moisture. Initial pump-down of a new line may be slow as trapped air is released from O-ring assemblies, but once accomplished, the vacuum line can be used for months to provide $\sim 10^{-5}$ torr evacuation. We use external tees and traps to avoid contamination of the line and never distill solvents through the vacuum line. Initial pump-down via the low-vacuum part of the double line followed by connection to the high-vacuum part provides rapid and complete evacuation of external apparatus.

2.2. PURIFICATION OF MATERIALS

In an ideal world, all starting materials for synthesis would be pure enough to simply mix the metal, complexant and solvent and proceed with the synthesis. In the real world, however, this is seldom the case. Commercially available complexants commonly are only about 98% pure, solvents may contain stabilizers, oxidation products, etc. Fortunately, alkali metals are available in high purity and, except for the tendency of lithium to form nitrides, pose few problems.

It is necessary to distinguish among purity, reducibility and catalytic activity of the materials and apparatus used in synthesis. For example, in using crown ethers and cryptands, small concentrations of complexants with different ring sizes or of an open chain polyether might not be detrimental, but water, oxygen, hydroxylic groups, carboxylate groups, etc. are "deadly" because of their ease of reduction. The major concern is to remove such impurities and to keep them out.

2.2.1. Solvent Purification

Because ammonia, methylamine and dimethyl ether have high vapor pressures at room temperature, they are stored after purification in 1L. stainless steel cylinders (Matheson, Cat. No. 8HD1000) equipped with a pressure-vacuum gauge (Matheson, Cat. No.63-3104) and Ultra-Torr® connectors. The cylinders are small enough to allow them to be cooled in an isopropanol-dry ice bath for transfer of the solvent. When solvent is needed in a glass synthesis cell, it is easily and safely distilled from the cylinder because the pressure can be monitored.

Reducible impurities are removed from NH_3 and $MeNH_2$ by adding sodium or NaK alloy to a 1-2L borosilicate flask and distilling in the solvent. After standing for several hours at -40 to -80 °C, the solvent is distilled into the storage cylinder. The solvents Me_2O, THF, Me_3N and Et_2O are allowed to react with benzophenone and an excess of NaK before being distilled into the storage tank (Me_2O) or glass storage vessel (THF, Me_3N and Et_2O). Just before use it is advisable to distill the amount needed into an intermediate vessel over NaK as a final precaution. A frit should be used between this vessel and the synthesis cell to avoid spray carryover of NaK.

2.2.2. Complexant Purification

Different complexants may require different purification methods and researchers are advised to check the literature for specific techniques. For example, recrystallization of 18C6 from acetonitrile is an effective initial step. To remove reducible impurities, we sometimes introduce an alkali metal that is not strongly complexed by the complexant along with a suitable solvent for the alkali metal (e.g. $MeNH_2$). For liquid complexants sometimes short-path vacuum distillation from an alkali metal is used. In other cases this can lead to decomposition, so a test run followed by ^1H NMR is advisable. Often, vacuum sublimation or distillation is all that is needed, with the temperature of the condenser set to collect most of the desired material but no so low that more volatile impurities are also condensed. A good indication of satisfactory solvent cleanup is the immediate formation of stable blue solutions with the target alkali metal in a solvent such as Me_2O or THF.

2.2.3. Apparatus Cleaning

Decomposition of alkali metal solutions is catalyzed by some transition metal compounds such as iron oxide or chromium salts. Surface catalysis at acidic sites may also be involved. While all the factors that lead to decomposition are not understood, a cleaning ritual has been developed over the years that usually avoids decomposition problems. The importance of proper materials purification, apparatus cleaning, and temperature control during synthesis is illustrated by the fact that early syntheses were successful only about 50% of the time, with partial or complete decomposition being common. By careful attention to the proper methods, only accidental breakage, improper glass seal-offs or inadvertent admission of air now prevent successful syntheses. Because solution decomposition is a greater problem than decomposition of cold crystalline alkalides and electrides, it is adviseable to complete the precipitation or crystallization steps as quickly as possible.

The key apparatus used to prepare crystalline alkalides and electrides is the "K-cell" shown in ***Figure 1***. One of the sample tubes is left open during the cleaning process to facilitate proper rinsing. Seal-off is made after cleaning and drying. The K-cell is first quickly but thoroughly cleaned with an HF-based cleaner consisting of approximately 35% HNO_3, 5% HF(concentrated), 60% distilled water and a small amount of acid-stable detergent such as Alconox®. The proportions are not critical but, of course, cleaning must be carried out in a well-ventilated hood with adequate face, hand and body protection. Cleaning time should be kept as short as possible to avoid excess attack on the frit by the HF. After complete rinsing with distilled water, *aqua regia* is used to remove traces of detergent and/or surface impurities. Thorough rinsing in the hood is followed by at least five complete rinsing cycles with conductance water in a sink. To be confident that rinsing is complete, we often heat the cell with a small amount of water during the final rinsing [CAUTION: extreme care must be taken to avoid contact with spurts of acid or steam that can come from the open tubes during cleaning].

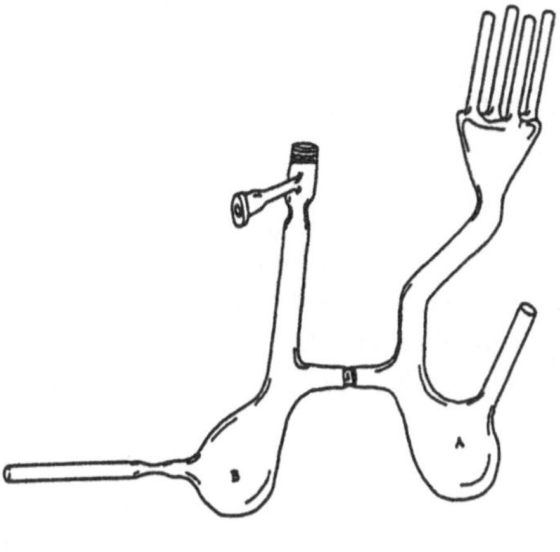

Figure 1. K-Cell used for vacuum-line synthesis.

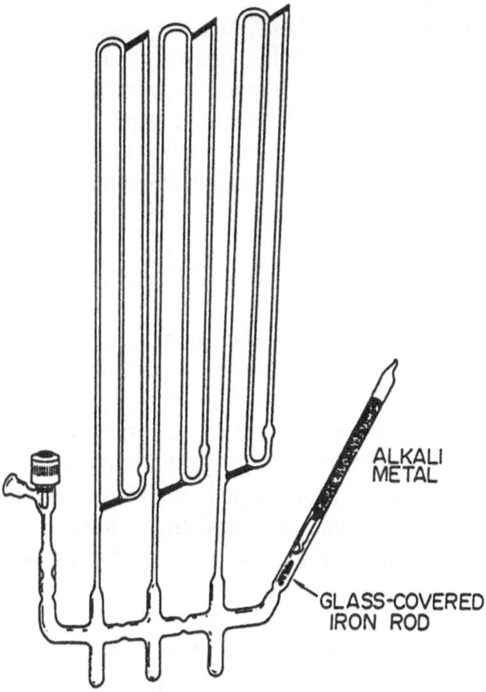

Figure 2. "Trombone" used for alkali metal distribution.

Borosilicate K-cells are satisfactory in most cases in spite of some contamination of non-sodium solutions by sodium from the borosilicate. An exception is the synthesis of lithium-containing alkalides and electrides. The ease of exchange of Li^+ with Na^+ from the borosilicate requires the use of fused silica K-cells.

Although crystalline alkalides and electrides can be stored indefinitely in sealed glass tubes at low temperatures, in many experiments they must contact other materials. In most cases, the solid alkalides and electrides do not react with metals or plastics such as polyethylene or Delrin®. An exception is Teflon® which is blackened significantly in contact with these compounds.

2.3. ALKALI METAL HANDLING

Within a good glove box, lithium, sodium and potassium can be cut and weighed directly without difficulty. Rubidium and cesium are so reactive that they should be exposed only briefly to the glove box atmosphere. A convenient way to deliver known amounts of Na, K, Rb, or Cs into the synthesis apparatus is to use measured lengths of the metal in glass tubes with pre-measured diameters. These metals can be purchased in high purity in break-seal ampoules under argon. We distribute the 5g. samples into known diameter tubing by using the "trombone" apparatus shown in ***Figure 2***. The lengths and diameters of the tubing are chosen to accommodate all of the sample, leaving about 50% open space. The "wells" into which the metal is first melted are of such a size that a 5g. sample will nearly fill all three wells. After evacuation, the break-seal is broken with a magnet-in-glass and the heated metal is allowed to flow into the apparatus, filling the wells. After seal-off into three separate sections, they are inverted and the metal is induced to disperse into the three sections of decreasing diameter. Because the inside diameters have been previously measured with a calibrated microscope, the desired amount of metal for an experiment can be "chased" to the end with a flame, sealed off and stored in the glove-box for use when needed. Measurement of the length with calipers can deliver the desired amount to within about ± 10%. If desired, the filled tube can be weighed and the weight of the empty tube can be determined later.

2.4. CRYSTALLIZATION AND SAMPLE HANDLING

Simple cooling of a solution of alkalide or electride to which a less polar co-solvent has been added will cause the formation of copious amounts of polycrystalline sample. The powder can be isolated by pouring away the solvent and its subsequent removal by distillation and by washing several times with the co-solvent. Removal of solvent by evacuation leaves a polycrystalline powder that can be distributed into the sample fingers for seal-off. Cryptated and sandwiched cations usually lead to no solvent of crystallization and can be prepared solvent-free. However, 1:1 crown ether complexes frequently require solvent for stabilization so that hard pumping will destroy the crystallinity of the alkalide or electride.

Larger single crystals are grown by either slow cooling (1°-2° per hour) or by slow removal of the more volatile polar solvent (usually Me_2O). Typically, the final stages of crystal growth are completed at dry-ice temperatures. Selection of suitable crystals for x-ray diffraction studies is made in a nitrogen-filled glove bag equipped with a microscope and a cold nitrogen stream. The latter is constrained to flow through a copper block which contains an inclined depression and an examination well for the crystals. They are covered with purified octane, pentane or heptane and suitable crystals are moved and attached to a pre-cooled goniometer pin. The sample is then moved to the diffractometer in a stream of cold nitrogen for structure determination.[33]

A word is in order about the use of glove bags. As is well known, they are slowly permeable to gases, including oxygen, and it is difficult to remove traces of water. We keep a dewar of liquid nitrogen in the bag as well as a large metal block partially immersed in liquid nitrogen in a Styrofoam® insulated evaporating dish. This tends to capture moisture and thus decrease the humidity. For particularly sensitive experiments, we have used a double glove bag with nitrogen purging of the outer bag.[34] Whenever possible, a helium (or argon)-filled glove box is used. Cold samples can be introduced into the box within wells in a large copper block cooled to liquid N_2 temperatures. A glove bag over the entrance port is then needed to avoid condensation. In addition to the usual freezer, we use a cold-well within the box that is cooled by circulating cold nitrogen through tubing soldered to the outside of the copper cold-well. The tubing enters and leaves the glove box through a conflat flange welded to the box.

2.5. SUMMARY

The detailed descriptions given above are included because synthesis of pure crystalline alkalides and electrides is critical for further studies. Even with the greatest care, however, surprises can be encountered. For example, the first crystalline electride, $Cs^+(18C6)_2e^-$, was studied for years before it was realized that a slow order-disorder transition occurs at ~230 K and that this transition is irreversible.[35] The disordered material has different NMR, EPR and susceptibility behavior than the ordered crystals even though the transition occurs without appreciable decomposition. Similarly, $Cs^+(15C5)_2e^-$ undergoes such a transition at ~265 K, but in this case, it is slowly reversible.[36] Whenever non-reproducible behavior is observed with an alkalide or electride, one should search for solid-state transitions. Studies by Differential Scanning Calorimetry (DSC) are useful in this regard. DSC is also useful in determining the decomposition temperatures. One often observes endothermic melting followed by substantial exothermic decomposition. With cryptands and crown ethers, decomposition occurs by reduction of the C-O bond to yield an alkoxide and (presumably) an aliphatic radical. When an -O-CH_2-CH_2-O- unit is attacked, ethylene and the di-alkoxide are formed and some butane is formed as well.[37]

3. Measurement Techniques

The simplest way to prepare a powdered alkalide or electride is to evaporate the solvent from a solution that has the proper stoichiometry. This, of course, has obvious drawbacks. Any deviations from the desired stoichiometry are maintained. Impurities and decomposition products are present and the solid may be inhomogeneous because of different rates of crystallization from different species. Simple precipitation of most of the dissolved material by adding a less polar solvent and/or by cooling can eliminate some of these difficulties but may also produce impure materials. Recrystallization in a separate step can help to remove such problems, but the best procedure is to grow crystals of the desired alkalide or electride, check their identity by x-ray diffraction and crush the crystals if a powdered sample is desired. Even with this procedure, crystallite orientation can introduce difficulties,[38] but extensive grinding at liquid N_2 temperatures can minimize such effects.

3.1. MAGNETIC SUSCEPTIBILITY

Between 10 and 100 mg of sample are required for the measurement of the magnetic susceptibility (χ) as a function of temperature and magnetic field in a SQUID susceptometer. Thin-walled Delrin® buckets with lids can be loaded in a glove box or a glove bag but the sample and container must be kept at -50 °C or below during handling. Alternatively, the sample can be sealed in a fused silica tube. We have used a copper block with a hole through its center to keep the sample cold during the pump-down and helium-fill procedure. The block and sample are first cooled with liquid N_2 and inserted into the SQUID loading chamber with the sample in the hole in the cold block. Evacuation and back-filling can be done several times without fear of sample warm-up.

Although sample decomposition is usually a problem, it can be a blessing in the measurement of χ. Alkalides and electrides usually decompose at room temperature or below to yield diamagnetic decomposition products. By allowing the samples to decompose in the measurement container, the correct background for subtraction can be experimentally determined.[39] Since alkalides are intrinsically diamagnetic, the magnitude of any Curie-Weiss contribution to χ can be used to determine the concentration of defect electrons. Similarly, for electrides with only weak inter-electron magnetic coupling, the magnitude of the Curie-Weiss slope can be used to determine whether decomposition had occurred prior to measurement. The varied behavior of the magnetic susceptibility of different electrides is illustrated by Figure 4of Ref.[10] (see also Figure 6 of Ref. [16]). All electrides show antiferromagnetic coupling with effective J-values that range from -3.7K for $Cs^+(15C5)_2e^-$ to -410K for $K^+(C222)e^-$.[17]

3.2. SOLID STATE NMR

Because of the paramagnetism of trapped electrons (even in rather pure alkalides) we have not been able to carry out ^{13}C or ^2H solid-state NMR studies of alkalides and electrides. However, static or magic-angle spinning (MAS) alkali metal NMR spectra can be obtained with either powders or single crystals.[40-47] The large quadrupole coupling constants of ^{39}K, ^{85}Rb and ^{87}Rb make studies of these nuclei difficult, although spin-echo techniques have been used successfully.[48] The most extensive studies have been made with ^{133}Cs, ^{23}Na and ^7Li NMR, all of which yield reasonable line widths and adequate signal-to-noise ratios.

Rotors made of Kel-F® or zirconia can be loaded while cold in a glove bag that is attached to the bore of the superconducting magnet. The probe, to which the glove bag is attached, is kept cold with flowing nitrogen during insertion of the rotor and assembly of the system. Studies of single crystal alkalides and electrides require a crystal rotator within the probe. Since rotation about two independent axes is difficult to achieve, we used a home-built probe that provides rotation of the sample holder around an axis that is perpendicular to the field. The sample is mounted in a cubical box made of the G.E. polymer Rexolite®. After obtaining data for rotation of the box around one axis, the box is flipped 90° for the second data set and again 90° for the third set.[45] If the crystal has a three-fold or higher rotation axis, the symmetry of the NMR pattern can be used to determine the absolute orientation. If not, then the relation of the rotation axes to the crystal axes cannot be determined without additional information. It would be desirable at this stage to be able to determine the crystal orientation by x-ray reflection but we have not been able to accomplish this.

The easiest information to obtain from alkali metal NMR spectra of alkalides and electrides is the identification of the cation and anion and a check on the presence of mixtures of crystals or mixed species in a single crystal. For example, attempts to prepare mixed crystals of composition $Cs^+(18C6)_2(Cs^-)_x(e^-)_{1-x}$ were shown by ^{133}Cs NMR to be mixtures of co-precipitated electride and ceside. (See Figure 3 of Ref. [44]). In another application, polycrystalline samples made from Cs, Na and 18C6 were shown by ^{133}Cs and ^{23}Na NMR to contain Cs^+ and Na^- but not Cs^- and Na^+ (See Figure 3 of Ref. [40]. [Note: At the time we assumed the stoichiometry to be $Cs^+(18C6)Na^-$ but it was later shown to be $Cs^+(18C6)_2Na^-$]. Measurements of this type proved invaluable in the days before we were able to determine crystal structures.

Alkali metal NMR studies of electrides provided early information on the nature of electron-trapping in electrides. The temperature dependence of the cation chemical shift, together with magnetic susceptibility data, yielded the unpaired electron contact density at the alkali metal nucleus.[35,49,50] With both $Cs^+(18C6)_2e^-$ and $Cs^+(15C5)_2e^-$ this proved to be very small, corresponding to less than 0.06% of the value for the cesium atom. This strongly indicated expulsion of most of the electron density from the region occupied by Cs^+ and the sandwich complexant and led to our view that the unpaired electron density has its maximum value in the cavities that serve as anionic sites in corresponding alkalides. Strong support for this view has come from recent theoretical studies of model systems[51,52] and of $Cs^+(15C5)_2e^-$.[53] Although counter-arguments were made by Golden and Tuttle,[54,55] their objections have been challenged.[56] Both experiment and theory now leave little doubt that the so-called "F-center model" of electrides, in which electrons tend to occupy all of the anionic sites, is essentially correct.

3.3. ELECTRON PARAMAGNETIC RESONANCE (EPR)

3.3.1. *Polycrystalline Alkalides and Electrides*

Polycrystalline samples of alkalides and electrides have been extensively studied by EPR spectroscopy.[57-59] Since alkali metal anions in the ground state have paired electrons in *s*-orbitals they are EPR silent. However, alkali metal solutions always contain some solvated electrons in equilibrium with the anions, so that crystalline alkalides always contain some defect trapped electrons. Thus, EPR studies of alkalides provide information about the nature of the electron-trapping sites.

The EPR spectra due to electrons trapped in alkalides fall into two distinct categories. When the cation is well-shielded from the trapped electron so that the electron interacts rather weakly with a number of complexed cations, there is no resolved hyperfine structure and a typical powder pattern is present. When, however, a single cation is exposed to trapped electron density in the anionic site as in $M^+(HMHCY)Na^-$ (M = K, Rb, Cs) and $Rb^+(18C6)Rb^-$ a well-defined hyperfine EPR spectrum is observed.[6,58,60] Examples are shown in Figure 5 of Ref. [6] and Figure 2 of Ref. [61] .

Although the fused silica EPR tubes can be filled with alkalide or electride in a glove box or glove bag, the need to keep them cold at all times must be kept in mind. With the discovery of a disordering transition of $Cs^+(18C6)_2e^-$ at ~230 K[35] EPR studies of this electride (as well as NMR studies) had to be repeated on samples that were always kept below the disordering temperature. To insure the absence of such transitions and of contamination during transfer, it is preferable to have the EPR tube attached as one of the sample "fingers" of the K-cell. In this way, the sample can be transferred easily while cold, and inadvertent contamination by air or moisture can be avoided.

Small samples of insulating electrides such as $Cs^+(18C6)_2e^-$ and $Cs^+(15C5)_2e^-$ can be readily studied by EPR although crystals must be finely ground at liquid N_2 temperatures to avoid orientation effects. The electrides $K^+(C222)e^-$ and $Li^+(C211)e^-$ are such strong microwave absorbers that only the tiniest speck in a sample tube could be studied with our conventional (Bruker) EPR instrument. Larger samples caused loss of the AFC lock. This phenomenon demonstrates the high microwave conductivity of these samples.

3.3.2. *ENDOR and ESEEM Studies*

The amount of information that can be obtained with conventional EPR studies is limited because of the presence of unresolved hyperfine couplings. Electron-Nuclear Double Resonance (ENDOR) and Electron Spin Echo Envelope Modulation (ESEEM) provide much more detailed information. For example, EPR studies of Cs^+(HMHCY)Na^- provide the hyperfine coupling to the nearest cesium, but little else. With ENDOR we were able to discern couplings to protons, to Na^- and to other Cs^+ ions.[60] ESEEM studies showed two different interactions to Cs^+ (besides that which gives the EPR hyperfine coupling).[62] We concluded that an admixture of *p*-character to the *s*-type wave function of the trapped electron was responsible.

3.3.3. *Conclusions*

The EPR, ENDOR and ESEEM studies of alkalides make it clear that the trapped electrons occupy primarily otherwise empty anionic sites. Comparison of the EPR spectra of the isostructural compounds $Cs^+(18C6)_2Na^-$ and $Cs^+(18C6)_2e^-$ show that the electron trapping sites are similar, further evidence for the stoichiometric F-center model of electrides.[59] Because one can study the EPR spectra over the temperature range from under 4K to the decomposition temperature, the effect of any solid-state transitions can be determined.

We have not yet found a way to study the complete orientation-dependence of the EPR spectra of single crystals. Only hints of the behavior have been obtained by rotation about an axis perpendicular to the field.[57] In the future, we plan to use pulsed EPR methods and FT analysis to study single crystals.

3.4. CONDUCTIVITIES

One might think that electrides are good electronic conductors since they *formally* have a half-filled *s*-band. But, except for $Li(NH_3)_4$, all electrides show activated electronic transport. The "localized" electrides, $Cs^+(18C6)_2e^-$ and $Cs^+(15C5)_2e^-$, are insulators with specific conductances below 10^{-10} ohm^{-1} cm^{-1} at -30 °C and below.[63] Examination of the structures shows why *pure* electrides are probably all insulators. The cavities contain one electron each, and the Coulomb repulsion expected for double-occupancy would be so great that these compounds would behave as Mott insulators. Even the conductivity of a "delocalized" electride such as $K^+(C222)e^-$, which is *10 orders of magnitude* more conductive,[63] is probably the result of defects (missing electrons) rather than being intrinsic.

3.4.1. *Two-probe DC Conductivities*

The first alkalide, $Na^+(C222)Na^-$ looks so much like metallic gold that we thought surely it must be a metal. Early measurements with two contacts, however, showed it to have a very high resistance. In fact, careful measurements as a function of temperature, including measurements along one axis of a single crystal,[64] showed it to behave as an intrinsic semiconductor with a band-gap of 2.4 eV. This is the only alkalide whose conductivity is not dominated by defect electrons. All others have preparation-dependent conductivities.

Two-probe methods with inert electrodes are only useful when the resistance is high; otherwise, electrode effects dominate the measurements. The design of the two-probe conductivity cell is shown in ***Figure 3a***.[65] The spring-loaded electrode applies a pressure of about 50 atm on powders in the capillary. Studies of the conductivity vs pressure in this cell showed no effect after reaching a pressure of about 25 atm. By keeping pressure on the sample, expansion effects as a function of temperature are minimized. Originally, controlled temperature nitrogen flow was, used in the rather complex apparatus shown in Figure 23 of Ref.[7], but later we found that suspension of the cell at a variable distance above the liquid nitrogen level in a "soft" dewar worked just as well and is much simpler.[63]

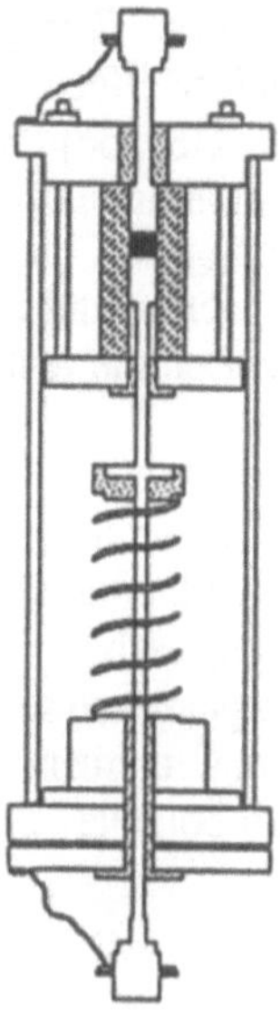

Figure 3a. Two-probe conductivity cell.

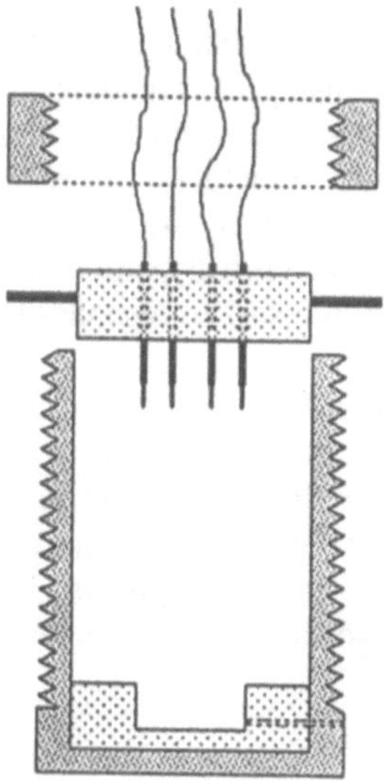

Figure 3b. Four- probe conductivity cell.

When the resistance is small, as with $K^+(C222)e^-$, electrode effects dominate, probably as a result of the Schottky barrier that results when an inert metal such as gold or steel makes contact with a non-metal that has a much lower work function. Most of this electrode effect can be eliminated by coating the electrodes with potassium metal. With a work function that more nearly matches that of the electride, the Schottky barrier is greatly reduced. This technique has been especially effective with four-probe DC measurements and with the AC methods described below.

3.4.2. *Four-probe Methods*
The simplest four-probe technique utilizes pre-pressed pellets of the electride in the cell shown in ***Figure 3b***.[65] The pellet press, surrounded by a glove bag, is cooled with liquid nitrogen and was used to produce cylindrical $K^+(C222)e^-$ pellets of diameter 4.76 mm and height about 2 mm. The four electrodes are spring-loaded "pogo-sticks" and were coated with potassium metal. The electrodes were coated in a glove box and the cold pellets were transferred directly from the press to the pre-cooled cell.

Recently, in work to be published, we have prepared thin stoichiometric films of $K^+(C222)e^-$ by the high-vacuum codeposition method described below in connection with optical spectra. The sapphire substrate was equipped with four gold film electrodes to permit four-probe studies of the conductivity. This eliminated pronounced electrode effects found earlier when the film resistance was measured with two-probe techniques.

3.4.3. *Impedance Spectroscopy*

Studies by DC methods of samples with appreciable conductivity give reliable results only when reversible reactions occur at the electrodes. Polarization that results in the formation of depletion regions near the electrodes can give high resistances and time-dependence. In addition, grain-boundary resistances can dominate a measurement, especially when the conductivity is anisotropic, as with pseudo- 1D or 2D conductors. The presence of polarization, or "blocking" at the electrodes as well as grain-boundary effects can often be detected by using AC methods, known as impedance spectroscopy (IS).[66] We use such studies at frequencies up to 5 MHz. The measurement yields the complex impedance with the real part related to the resistance and the imaginary contribution largely due to capacitance. For an ideal equivalent circuit consisting of a capacitor and a resistor in parallel, the result is a semicircle through the origin. The intercept on the real axis as the frequency is reduced corresponds to the DC resistance.

Few samples show ideal behavior and electrides are no exception. The pronounced effect of the Schottky barrier and its near removal by coating the electrodes with potassium is shown in ***Figure 4.*** [65] Note that the intercept on the real axis decreases by two orders of magnitude for the coated electrodes. In addition a marked voltage dependence seen with gold electrodes virtually disappears. Even in this case, the high frequency intercept is not at the origin, suggesting that the apparent resistance is largely due to grain-boundary effects.

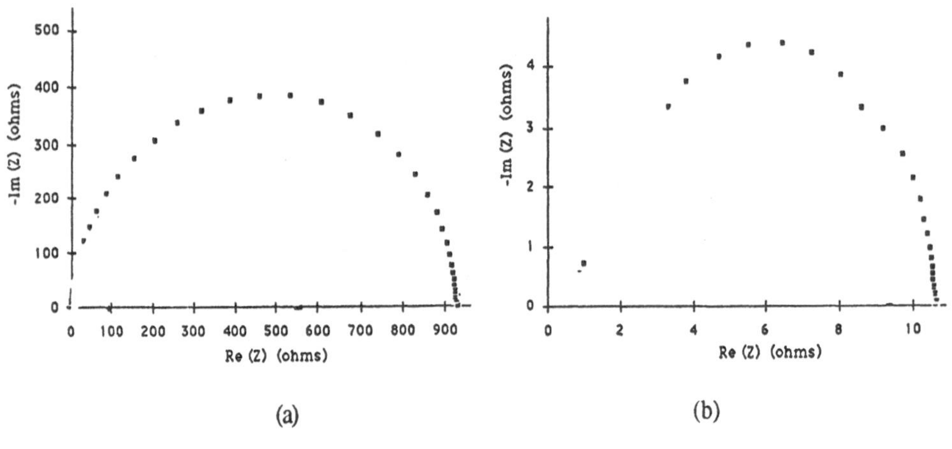

Figure 4. Impedance spectra of $K^+(C222)e^-$ at 90 K,
(a) With gold electrodes; (b) With potassium-coated electrodes.

3.5. OPTICAL STUDIES

Perhaps next to crystal structures in importance to the understanding of alkalides and electrides are optical studies. These include absorption spectra of thin films, fluorescence spectra of powders and single crystals, reflectance spectra from single crystals and photoelectron emission spectra. Both the intrinsic behavior of pure compounds and the marked effects of defect electrons in alkalides are important. In addition, high intensity laser pulses that can essentially depopulate the ground state cause both reversible and irreversible changes that tell us something about the nature of the excited state. The most complete studies have been made with the very stable alkalide, $Na^+(C222)Na^-$. The methods developed for such studies should prove useful in future experiments on other alkalides and on electrides. In keeping with the spirit of this paper, the emphasis will be on the techniques used. The reader is referred to other papers for the results and their interpretation.

3.5.1. *Thin Films by Solvent Evaporation*

Shortly after the synthesis of the first alkalide, a very simple method was developed to obtain qualitative absorption spectra of thin, solvent-free alkalide and electride films.[67] A quartz optical cell was sealed to a side-arm vessel (***Figure 5***) and to a Kontes Teflon® stopcock through which a solvent such as $MeNH_2$ or Me_2O could be admitted. The appropriate sample of alkalide or electride is added to the cell or is formed in situ by attaching the apparatus to a K-cell. Most of the solution is poured out of the optical cell, leaving a liquid film on the cell window. The side-arm vessel is then immersed in liquid nitrogen while shaking the cell to maintain the liquid film on the window while the solvent evaporates. With practice, a reasonably uniform solid film can be obtained that is thin enough to permit an absorption spectrum to be obtained.[68,69]

In early studies, it was necessary to insert the cell into the cell compartment of a spectrophotometer and to keep the cell cold with a stream of cold nitrogen or with a thermostated jacket. We now use a Guided Wave Model 260 Spectrophotometer that brings the light to the sample and back to the instrument via quartz fiber optics. This makes it very easy to keep the sample cold and, indeed, to study the temperature dependence of the absorption spectrum.

There are three major drawbacks to such studies. First, the thickness cannot be controlled or measured so that extinction coefficients and oscillator strengths remain unknown. Second, the films do not have uniform thickness. As a result, the spectral shapes are distorted and peaks are broader than they should be. Third, there is no guarantee that the films are uniform in composition. Sequential precipitation of different compounds could occur. But in spite of these problems, this simple method provided a wealth of information about the nature of alkalides and electrides. Some of the spectra obtained in this way are shown in Figure 2 of Ref. [67].

3.5.2. *Thin Films by Vapor Codeposition of Metal and Complexant*

While distilling cesium metal into one compartment of a glass apparatus that contained 18C6 in an adjoining compartment, we noticed that blue films formed on the walls in the vicinity of the connecting tube. Evidently, some cesium vapor deposited on the walls in the presence of the slightly volatile crown ether. This suggested the possibility of direct synthesis of alkalide and electride films by co-deposition of metal and complexant under vacuum. Preliminary tests[70] with a simple glass apparatus (see Figure 1 of Ref. [70]) confirmed the feasibility of the method and led to the development of apparatus for the preparation of stoichiometric films of known thickness.[71,72]

Since publication of the detailed description of this apparatus,[72] we have made a number of improvements.[73] The fused silica substrates used to collect the films did not permit studies at controlled and known temperatures below about -40 °C because of the poor thermal conductivity of fused silica. We now use sapphire substrates of thickness 1.0 mm that make contact with a cooled copper block via a thin layer of silicone grease. The much greater thermal conductivity of sapphire insures that the substrate temperature remains the same as that of the block.

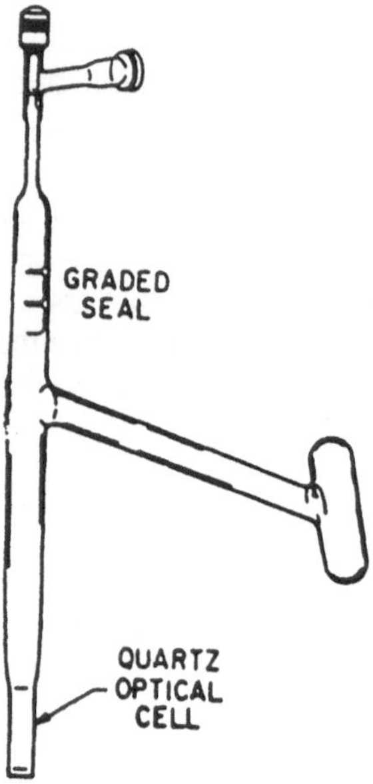

Figure 5 Optical cell for obtaining the absorption spectra of solvent-evaporated films.

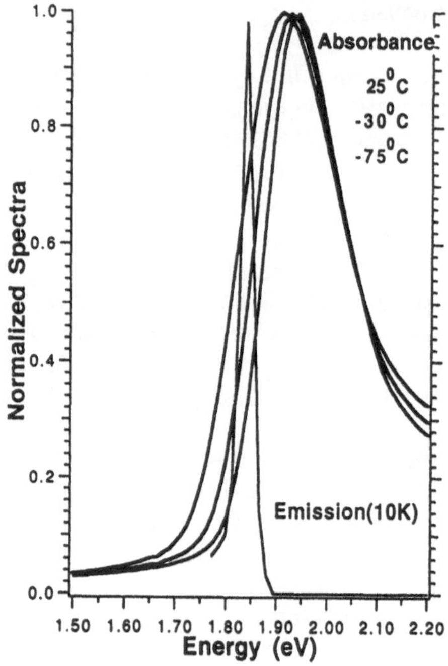

Figure 6. Absorption spectra of co-deposited films of $Na^+(C222)Na^-$ as a function of temperature and comparison with the 10K emission spectrum. The spectra from left to right correspond to decreasing temperature.

Initially, optical spectra were determined by using a complex system consisting of a light source, scanning monochromator, chopper, mirrors, a detector within the vacuum chamber and a lock-in amplifier. All of these components have been replaced by fiber optics and the Guided Wave spectrometer described above. In order to eliminate leaks through and around the fibers, it was necessary to strip the outside covering from the fibers and to epoxy them into tubes that are hard-soldered to a vacuum flange . The added bonus with this system is the ability to scan a complete spectrum (350 - 2000 nm) in about one minute compared with 20 minutes with the previous optical system.

By using computer-controlled ovens, a thermoelectric cooler/heater for the complexant, and quartz oscillator thickness monitors, it is possible to deposit uniform films with controlled stoichiometry and thickness and to determine their spectra. The quantitative nature of the process is demonstrated by the fact that the oscillator strength of a film of $Na^+(C222)Na^-$ is

2.00 ± 0.04 in complete agreement with that expected for a compound with two equivalent electrons in the ground state.[74]

An added advantage of the vapor deposition method is the ability to determine film conductivities by using the four probe method. In studies of $K^+(C222)e^-$ to be published, the technique worked very well and provided us with data that strongly suggest that the conductivity is due to missing electrons (holes). We plan to attempt measurements of the Hall coefficient on such films to determine the sign of the carrier.

Figure 6 shows part of the absorption spectrum of $Na^+(C222)Na^-$ as a function of temperature along with the emission spectrum at ~10K.[74] The temperature dependence of the low energy absorption edge suggests that indirect absorption processes are present in addition to the direct *s* to *p* transition. This view is strengthened by the appearance of a new lower energy emission band at higher temperatures.

3.5.3. *Luminescence Spectra*

By using YAG and dye lasers for excitation, pronounced emission was observed at temperatures ranging from < 10K to room temperature.[75-77] At the time these studies were made the time-resolution was ~0.5 ns so that the earliest processes were masked by the instrument response function. The time and wavelength dependence of the fluorescence spectra at 10-30 K in the case of $Na^+(C222)Na^-$ proved to be complex, with a time-shift of the emission peak to lower energies. On the low energy side of the peak there was actually growth of the emission followed by decay, while at higher energies there was biphasic decay. (See Figures 7 and 11 of Ref. [75]).

Quenching of the fluorescence of $Na^+(C222)Na^-$ by trapped electrons suggested that somehow the excitation was able to migrate to regions that contained defect electrons. To test this idea, a microscopic system was constructed[78,79] that permitted excitation of a 10-20 μm spot on a single crystal and examination of fluorescence at various distances from the excitation site (see Figures 1 and 2 of Ref. [79]). The results showed that the excited state of $Na^+(C222)Na^-$ can migrate with exceptional velocity in a single crystal (some 10^7 cm s^{-1}). This was interpreted as a result of the formation of an exciton-polariton (a "hybrid" between an exciton and a photon).

In addition to direct fluorescence, the exciton-polariton can be trapped at a defect site such as a trapped electron or it can interact with phonons to localize the excitation. The result is an initial high energy rapid fluorescence, followed by fluorescence from the exciton-polariton, and finally, energy-transfer to the lattice yielding lower energy fluorescence. More recent studies have shown that early trapping of the exciton in single crystals occurs with partial retention of the initial polarization.[77] With the advent of femtosecond lasers, it would be profitable to examine the early stages of fluorescence.

3.5.4. *Reflectance Spectra*

It is not possible to obtain the absorption spectra of single crystals of alkalides or electrides because of their opacity. A useful alternative, that we are just beginning to exploit, is the determination of polarized reflectance spectra of single crystals.[74] The method is relatively straightforward, but requires a microscope with Cassagranian optics such as that developed by Welber for diamond anvil pressure studies of crystals.[80]

A schematic diagram of our apparatus is shown in ***Figure*** *7*.[81] Essentially, light polarized with its **E** and **H** vectors parallel to the surface of the crystal is reflected back into the microscope, picked up with fiber optics and delivered to the Guided Wave spectrometer. By reflecting the light from various faces of a single crystal that is sealed into a cooled fused silica optical cell, the reflectance can be obtained and scanned over the wavelength region from 350 to 2000 nm. So far, we have only obtained quantitative reflectances for $Na^+(C222)Na^-$, but the system should be useful in the study of any single crystal alkalide or electride.

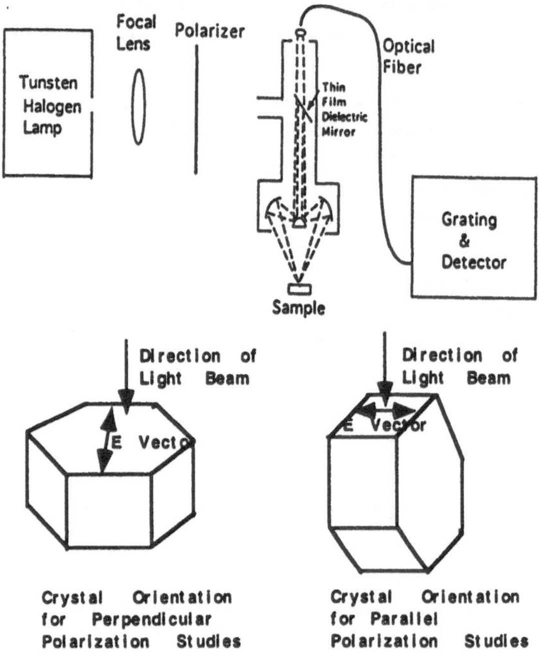

Figure 7. Schematic diagram of the apparatus for determining reflectance spectra of $Na^+(C222)Na^-$. The cone of light striking the sample has a spread of $\pm 6^\circ$.

Because we could study essentially the entire reflectance peak of $Na^+(C222)Na^-$, it was possible to use Kramers-Kronig relationships to calculate the absorption spectrum as well as the complex dielectric function of single crystals. The use of polarized light permitted us to obtain the anisotropy of these properties as well. The correspondence of the calculated absorption spectrum with that of thin vapor-deposited films was excellent and also indicated that the crystallites formed by deposition were preferentially oriented.[74] Methodology of this type should be very useful for the study of the optical properties of any high-quality opaque crystals.

3.5.5. *Photoelectron Emission*

The ionization potential of Na^- in the gas phase is only 0.54 eV. Thus, we anticipate that alkalides and electrides should have weakly-bound electrons. We have explored the binding of electrons in $Na^+(C222)Na^-$ by measuring the photoelectron emission spectrum.[82]

Early work with a number of alkalides and electrides demonstrated the feasibility of such measurements,[83] but reproducibility was poor when samples were introduced by using a glove-bag. Dissolution of the sample in $MeNH_2$ or Me_2O *in situ*, followed by evaporation of the solvent, produced fresh powder that improved reproducibility, but also left defects that were so close to the vacuum level that thermionic emission in the dark was observed that persisted to temperatures as low as -80 °C.[84]

In order to avoid contamination, a new photoemission apparatus was constructed that permitted introduction of freshly crystallized samples without exposure to a glove bag atmosphere.[81] The equipment will be described in detail in a separate publication, and the results with $Na^+(C222)Na^-$ have recently appeared in print.[82] Here we give only a brief description of the methodology and results.

The apparatus contains three sections, each separately evacuable. In the first section, pumped to about 10^{-5} torr, a pre-scored glass sample tube is introduced and broken with the sample transport arm. A spatula arm is used to transfer the sample to the cathode cup on the transport arm, after which the sample is transferred through a gate valve into the second vacuum chamber. Because of the thermal sensitivity of our samples, each arm is vacuum-jacketed so that the portion that contacts the sample can be kept cold with a stream of nitrogen. In the second chamber, which is evacuated to less than 10^{-6} torr with a cryopump, the cathode cup is threaded to a vertical transport "wand" that moves the sample into the third chamber through a gate valve via an opening just large enough to allow the wand to pass. This measurement chamber is pumped with a turbomolecular pump to ~10^{-8} torr and contains a spherical collector, a hemispherical grid and an opening through which light from a xenon lamp and monochromator and/or from a YAG laser can be introduced. The sample can be "wet" with or dissolved in pre-purified $MeNH_2$ or Me_2O in the second chamber, or larger crystals can be used if desired. Fine powders tend to jump out of the cathode cup unless first exposed to liquid solvent. Large crystals that are "surface dampened" with solvent yield reproducible photoemission spectra. If carefully grown, crystals of $Na^+(C222)Na^-$ show no effects due to defect electrons.

The photoemission spectrum of "defect-free" samples of $Na^+(C222)Na^-$ is independent of exposure to 10 ns laser pulses of wavelength 532 nm and energy just below the threshold for irreversible bleaching. From such spectra we obtain an emission threshold of 3.10 ± 0.05 eV, about the same as that of sodium metal. The earlier lack of reproducibility was traced in these later experiments to the presence of defect sites and defect electrons. Excitation with 532 nm laser pulses caused a dramatic increase in the (already present) low energy emission due to trapped electrons, as well as extension of this emission to ~1000 nm. We concluded that the enhanced lower energy photoelectron emission was caused by defect electrons excited into long-lived traps by the light. Decay of the post-exposure response to pulses at the YAG fundamental (1065 nm) yielded lifetimes of these electrons of ~160 s at -60 °C.

It is clear that studies of photoelectron emission could provide a powerful way to determine the energy levels and defect states in other alkalides and electrides. Unfortunately, such measurements depend critically on the nature of the sample surface, so that the apparatus required is complex and difficult to use.

4. Computer Visualization Of Cavities And Channels In Electrides

Theoretical studies[51-53] now make it clear that the electron density in electrides tends to concentrate in the cavities and the channels that connect them. Therefore, it is important to be able to visualize the geometry of such cavities and channels and to relate their structures to properties. For some time we have felt that the magnetic coupling between the electron spins and the conductivity of electrides depended on the size and length of the channels that interconnect the cavities.

We have recently developed a FORTRAN program[85] that permits easy visualization of the cavities and channels in electrides with the aid of 3D isosurface construction by commercial programs.[86,87] The method starts with a set of coordinates of the atoms in a collection of unit cells large enough to contain a number of cavities and channels. These coordinates are obtained from the crystal structure with a commercial program such as BIOGRAF® or BIOSYM®. Each atom is assigned a Van der Waals (VdW) radius and the FORTRAN program creates a grid of points (up to 80 x 80 x 80) that span the structure. Points within the VdW radius of any atom are assigned the value zero, while others are assigned a scaled value (1 to 255) proportional to the distance to the nearest VdW surface.

The grid of points is transferred to a commercial 3D Visualization program such as AVS® or EXPLORER®, which permits construction of isosurfaces at any desired distance from the atoms. The connectivity, shape and size of cavities and channels are readily observed by

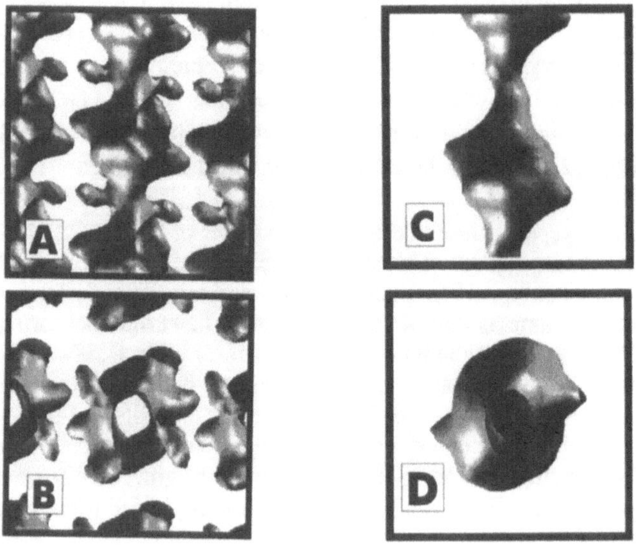

Figure 8. Representations of the cavity-channel geometry of the electride $Cs^+(15C5)_2e^-$ at various distances from the Van der Waals surfaces of the atoms. Views a and b are perpendicular and parallel respectively to the major channel axis and are isosurfaces at 0.3 Å from the VdW surfaces of the atoms. The corresponding views c and d are at 0.6 Å from the surfaces.

looking at isosurfaces at progressively farther distances from the atoms. The results are illustrated in ***Figure 8,*** which shows several views of the cavity-channel isosurfaces for $Cs^+(15C5)_2e^-$ at various distances from the VdW surfaces of the atoms. We can use this method to conclude that each cavity is large enough to contain a sphere of diameter 4.0 Å and that the major channels have a minimum diameter of 1.5 Å and a length of ~ 4.0 Å. So far there have been no theoretical calculations of the spectra, magnetic coupling and conductivity expected for such a "plumbers nightmare" of cavities and channels filled with electrons. *Qualitative* comparisons of the cavity-channel geometries with the effective dimensionalities and coupling constants of electrides suggest strongly that a correlation exists. We hope that theoretical studies will be forthcoming.

5. Summary

The results of more than 20 years of study of alkalides and electrides cannot be treated in depth in a paper of this length. In this presentation, emphasis has been placed on the special methods that we have used to study these reactive and unstable compounds. More detailed descriptions of the results of such syntheses and studies are available in the original papers and in a number of reviews. This paper provides a brief guide to such publications and some descriptions of methods that are scattered among a number of them, with emphasis on the special handling techniques that have proved successful in our research. It is likely that some of the procedures described here will also prove useful in studies of other reactive and thermally unstable materials.

Acknowledgment

This research was supported in part by U.S.-NSF Grant DMR 94-02016 and by the MSU Center for Fundamental Materials Research.

References

1. Dye, J. L.; Andrews, C. W.; Mathews, S. E. (1975) Strategies for the Preparation of Compounds of Alkali Metal Anions, *Journal of Physical Chemistry* **79,** 3065-3070.
2. Dye, J. L. (1977) Anions of the Alkali Metals, *Scientific American* **237,** 92-105.
3. Dye, J. L. (1977) Alkali Metal Anions, An Unusual Oxidation State, *Journal of Chemical Education* **54,** 332-339.
4. Dye, J. L. (1980) Preparation and Analysis of Metal/Solvent Solutions and the Formation of Alkali Metal Anions, *Journal of Physical Chemistry* **84,** 1084-1090.
5. Dye, J. L.; Ellaboudy, A. (1984) Crystalline Alkalides and Electrides, *Chemistry in Britain* **20,** 210-215.
6. Dye, J. L. (1984) Recent Developments in the Synthesis of Alkalides and Electrides, *Journal of Physical Chemistry* **88,** 3842-3846.
7. Dye, J. L. (1984) Electrides, Negatively Charged Metal Ions and Related Phenomena, *Progress in Inorganic Chemistry* **32,** 327-441.
8. Dye, J. L.; DeBacker, M. G. (1987) Physical and Chemical Properties of Alkalides and Electrides, *Annual Review of Physical Chemistry* **38,** 271-301.
9. Dye, J. L. (1987) Electrides, *Scientific American* **257,** 66-75.

10. Dye, J. L. (1989) Valency and Charge Distribution in Alkalide and Electride Salts, in *Valency, The Robert A. Welch Foundation Conference on Chemical Research*; Robert A. Welch Foundation: Houston, Texas pp 65-91.
11. Dye, J. L. (1990) Electrides: Ionic Salts with Electrons as the Anions, *Science* **247**, 663-668.
12. Dye, J. L.; Huang, R. H. (1990) Electride Structures and Properties, *Chemistry in Britain* **26**, 239-244.
13. Dye, J. L. (1991) Electrides and Alkalides: Comparison with Metal Solutions, in *J. Phys. IV Colloque C5, Metals in Solution*; P. Damay and F. Leclerq, Eds.; les editions de physique: Paris pp 259-282.
14. Dye, J. L.; Jackson, J. E.; Cauliez, P. (1992) Synthesis, Properties and Uses of Alkalides and Electrides, in *Organic Synthesis for Materials and Life Sciences*; Z. Yoshida and Y. Ohshiro, Eds.; VCH Publishers: New York pp 243-270.
15. Dye, J. L. (1993) Anionic Electrons in Electrides, *Nature* **365**, 10-11.
16. Dye, J. L. (1993) Alkalides and Electrides, *Chemtracts-Inorganic Chemistry* **5**, 243-270.
17. Wagner, M. J. (1994) Ph.D. Dissertation, Michigan State University, East Lansing, Michigan.
18. Tsai, K. -L.; Dye, J. L. (1991) Nanoscale Metal Particles by Homogeneous Reduction with Alkalides or Electrides, *Journal of the American Chemical Society* **113**, 1605-1652.
19. Dye, J. L.; Tsai, K. -L. (1991) Small Alloy Particles by Co-reduction of Soluble Precursors with Alkalides or Electrides in Aprotic Solvents, *Faraday Discussions of the Royal Society of Chemistry* **92**, 45-55.
20. Tsai, K. -L.; Dye, J. L. (1993) Synthesis, Properties, and Characterization of Nanometer-Size Metal Particles by Homogeneous Reduction with Alkalides and Electrides in Aprotic Solvents, *Chemistry of Materials* **5**, 540-546.
21. Ohsawa, T.; Takagaki, T.; Haneda, A.; Oishi, T. (1981) Dissolving Metal Reduction by Crown Ether---Hydrogenolysis of Alkyl Fluorides, *Tetrahedron Letters* **22**, 2583-2586.
22. Ohsawa, T.; Takagaki, T.; Ikehara, F.; Takahashi, Y.; Oishi, T. (1982) Reductive Removal of Sulfonyl Groups: Cleavage of Sulfonamides and Sulfonates by Alkali Metal Combined With Crown Ether., *Chemical Pharmaceutical Bulletin* **30**, 3178-3186.
23. Ohsawa, T.; Oishi, T. (1984) Dissolving Metal Reduction: K Metal-Crown Ether-Toluene System for Reductive Deflurorination., *Journal of Inclusion Phenomena* **2**, 185-194.
24. Ohsawa, T.; Kobayashi, T.; Mizuguchi, Y.; Saitoh, T.; Oishi, T. (1985) Dissolving Metal Reduction With Crown Ether: Reductive Decyanation, *Tetrahedron Letters* **26**, 6103-6106.
25. Ohsawa, T.; Mitsuda, N.; Nezu, J.; Oishi, T. (1989) Dissolving Metal Reduction With Crown Ether: Reduction Removal of 1-Cyano Groups., *Tetrahedron Letters* **30**, 845.
26. Jedlinski, Z.; Kowalczuk, M.; Kurcok, P.; Grzegorzek, M.; Ermel, J. (1987) A Novel Route to α-Substituted γ-Lactones via Lactone Enolates, *Journal of Organic Chemistry* **52**, 4601-4602.
27. Jedlinski, Z.; Kowalczuk, M.; Misiolek, A. (1988) An Unexpected Outcome of the Reaction Between β-Lactones and Dissolved Potassium., *Journal of the Chemical Society Chemical Communications* 1261-1262.
28. Jedlinski, Z.; Misiolek, A.; Kurcok, P. (1989) Enolate Anions. 2. Reaction Between Potassium Solutions Containing Crown Ethers and β-Lactones, *Journal of Organic Chemistry* **54**, 1500-1501.
29. Jedlinski, Z.; Misiolek, A.; Glowkowski, W. (1990) Alkali Metal Complexes in Organic Synthesis: Novel Route to Ketone Enolates, *Synthesis Letters* 213-214.

30. Jedlinski, Z.; Misiolek, A.; Glowkowski, W.; Janeczek, H.; Wolinska, A. (1990) Reactions of Alkali Metal Anions. XV. Reaction of Ketones with Alkali Metal Anions, *Tetrahedron Letters* **30,** 3547-3558.
31. Wayda, A. L.; Dye, J. L. (1985) A Versatile System for Vacuum-Line Manipulations, *Journal of Chemical Education* **62,** 356-359.
32. Wayda, A. L.; Bianconi, P. A.; Dye, J. L. (1987) Multipurpose Vacuum Line Designs for Use in Synthetic Organometallic Chemistry, in *Experimental Organometallic Chemistry*; A. L. Wayda and M. Y. Darensbourg, Eds.; American Chemical Society: Washington, D.C. pp 116-135.
33. Ward, D. L.; Huang, R. H.; Dye, J. L. (1988) Structures of Alkalides and Electrides. I. Structure of Potassium Cryptand[2.2.2] Electride, $K^+(C_{18}H_{36}N_2O_6)e^-$, *Acta Crystallographica* **C44,** 1374-1376.
34. Wagner, M. J.; Dye, J. L.; Pérez-Cordero, E.; Buigas, R.; Echegoyen, L. (1995) Static Polycrystalline Magnetic Susceptibility and Four-Probe Single Crystal Conductivity Studies of $[Ru(bpy)_3]^0$, *Journal of the American Chemical Society* **117,** 1318-1323.
35. Wagner, M. J.; Huang, R. H.; Dye, J. L. (1993) First Crystalline Electride Revisited: New Magnetic Susceptibility Studies of Cs^+(18-crown-6)$_2$$e^-$, *Journal of Physical Chemistry* **97,** 3982-3984.
36. Dawes, S. B.; Eglin, J. L.; Moeggenborg, K. J.; Kim, J.; Dye, J. L. (1991) Cs^+(15-crown-5)$_2$$e^-$, A Crystalline Antiferromagnetic Electride, *Journal of the American Chemical Society* **113,** 1605-1609.
37. Cauliez, P. M.; Jackson, J. E.; Dye, J. L. (1991) An Unusual Reduction of Ethylene Occurring During the Thermal Decomposition of Alkalides and Electrides, *Tetrahedron Letters* **32,** 5039-5042.
38. Doueff, S.; Tsai, K.; Dye, J. L. (1991) X-ray Powder Diffraction Studies of Alkalides and Electrides that Contain 15-crown-5, *Inorganic Chemistry* **30,** 849-851.
39. Landers, J. S.; Dye, J. L.; Stacy, A.; Sienko, M. J. (1981) Temperature-Dependent Electron Spin Interactions in Lithium[2.1.1] Cryptate Electride Powders and Films, *Journal of Physical Chemistry* **85,** 1096-1099.
40. Ellaboudy, A.; Tinkham, M. L.; VanEck, B.; Dye, J. L.; Smith, P. B. (1984) Magic-Angle Spinning ^{23}Na Nuclear Magnetic Resonance Studies of Crystalline Sodides, *Journal of Physical Chemistry* **88,** 3852-3855.
41. Ellaboudy, A.; Dye, J. L. (1986) Lineshapes in Sodium-23 NMR Spectra of Crystalline Sodides, *Journal of Magnetic Resonance* **66,** 491-502.
42. Tinkham, M. L.; Dye, J. L. (1985) First Observation by ^{39}K NMR of K^- in Solution and in Crystalline Potassides, *Journal of the American Chemical Society* **107,** 6129-6130.
43. Tinkham, M. L.; Ellaboudy, A.; Dye, J. L.; Smith, P. B. (1986) Detection of Rb^- in Crystalline Rubidides by ^{87}Rb Nuclear Magnetic Resonance, *Journal of Physical Chemistry* **90,** 14-16.
44. Dawes, S. B.; Ellaboudy, A. S.; Dye, J. L. (1987) Cesium-133 Solid State Nuclear Magnetic Resonance Spectroscopy of Alkalides and Electrides, *Journal of the American Chemical Society* **109,** 3508-3513.
45. Kim, J.; Dye, J. L. (1990) Single-Crystal ^{23}Na NMR Study of Na^+(cryptand[2.2.2])Na^-, *Journal of Physical Chemistry* **94,** 5399-5402.
46. Wagner, M. J.; McMills, L. E. H.; Ellaboudy, A. S.; Eglin, J. L.; Dye, J. L.; Edwards, P. P.; Pyper, N. C. (1992) NMR Study of Crown Ether Motion in Two Alkalides, *Journal of Physical Chemistry* **96,** 9656-9660.

47. Edwards, P. P.; Ellaboudy, A. S.; Holton, D. M. (1985) NMR Spectrum of the Potassium Anion K^-, *Nature (London)* **317,** 242-244.
48. Kim, J.; Eglin, J. L.; Ellaboudy, A. S.; McMills, L. E. H.; Huang, S.; Dye, J. L. (1996) ^{87}Rb, ^{85}Rb, and ^{39}K NMR Studies of Alkalides, Electrides and Related Compounds, *Journal of Physical Chemistry* **in press,**
49. Dawes, S. B.; Ward, D. L.; Huang, R. H.; Dye, J. L. (1986) First Electride Crystal Structure, *Journal of the American Chemical Society* **108,** 3534-3535.
50. Dawes, S. B. (1986) Ph.D. Dissertation , Michigan State University, East Lansing, Michigan.
51. Rencsok, R.; Kaplan, T. A.; Harrison, J. F. (1990) On the Electronic Structure of Electrides, *Journal of Chemical Physics* **93,** 5875-5882.
52. Rencsok, R.; Kaplan, T. A.; Harrison, J. F. (1993) Electronic Structure of Li(9-crown-3)$_2$: A Molecule with a Rydberg-type Ground State, *Journal of Chemical Physics* **98,** 9758-9764.
53. Singh, D. J.; Krakauer, H.; Haas, C.; Pickett, W. E. (1993) Theoretical Determination that Electrons Act as Anions in the Electride Cs^+(15-crown-5)$_2e^-$, *Nature (London)* **365,** 39-42.
54. Golden, S.; Tuttle, T. R., Jr. (1992) Spectral-Moment Constraints on Electride-Electron Locations in the Crystalline Electride [Cs(18-Crown-6)$_2$], *Physical Review B* **45,** 913-918.
55. Golden, S.; Tuttle, T. R., Jr. (1994) Electride-Electron Locations in the Crystalline Electride [Cs(18-Crown-6)$_2$]: Nonuniqueness of Current Quantum-Mechanical Models, *Physical Review B* **50,** 8059-8062.
56. Kaplan, T. A.; Rencsok, R.; Harrison, J. F. (1994) Cesium Crown-Ether Electrides: Extraordinary or Ordinary Mott Insulators?, *Physical Review B* **50,** 8054-8058.
57. Issa, D.; Ellaboudy, A.; Janakiraman, R.; Dye, J. L. (1984) Magnetic Susceptibilities and EPR Spectra of a Ceside and an Electride, *Journal of Physical Chemistry* **88,** 3847-3851.
58. Shin, D. -H. (1992) Ph.D. Dissertation , Michigan State University, East Lansing, Michigan.
59. Shin, D. -H.; Dye, J. L.; Budil, D. E.; Earle, K. A.; Freed, J. H. (1993) 250 GHz and 9.5 GHz EPR Studies of an Electride and Two Alkalides, *Journal of Physical Chemistry* **97,** 1213-1219.
60. Ellaboudy, A. S.; Bender, C. J.; Kim, J.; Shin, D. -H.; Kuchenmeister, M. E.; Babcock, G. T.; Dye, J. L. (1991) Multinuclear ENDOR Study of Trapped Electrons in the Polycrystalline Sodide Cs^+(HMHCY) Na^-, *Journal of the American Chemical Society* **113,** 2347-2352.
61. Shin, D.-H.; Ellaboudy, A. S.; Dye, J. L.; DeBacker, M. G. (1991) EPR Studies of Electrons Trapped in Sodides, *Journal of Physical Chemistry* **95,** 7085-7089.
62. McCracken, J.; Shin, D. -H.; Dye, J. L. (1992) Pulsed EPR Studies of Polycrystalline Cesium Hexamethyl Hexacyclen Sodide, *Applied Magnetic Resonance* **3,** 305-316.
63. Moeggenborg, K. J.; Papaioannou, J.; Dye, J. L. (1991) Powder Conductivities of Three Electrides, *Chemistry of Materials* **3,** 514-520.
64. Papaioannou, J.; Jaenicke, S.; Dye, J. L. (1987) Electronic Properties of Sodium-C222-Sodide, *Journal of Solid State Chemistry* **67,** 122-130.
65. Moeggenborg, K. J. (1990) Ph.D. Dissertation , Michigan State University, East Lansing, Michigan.
66. MacDonald, J. R. (1987)*Impedance Spectroscopy*; John Wiley & Sons: New York.
67. Dye, J. L.; Yemen, M. R.; DaGue, M. G.; Lehn, J. (1978) Optical Spectra of Alkali Metal Anion and 'Electride' Films, *Journal of Chemical Physics* **68,** 1665-1670.

68. DaGue, M. G.; Landers, J. S.; Lewis, H. L.; Dye, J. L. (1979) Alkali Metal Anions and Trapped Electrons Formed by Evaporating Metal-Ammonia Solutions which Contain Cryptands, *Chemical Physics Letters* **66**, 169-172.
69. Dye, J. L.; DaGue, M. G.; Yemen, M. R.; Landers, J. S.; Lewis, H. L. (1980) Transmission Spectra of Thin Films which Contain Alkali Metal Anions and/or Trapped Electrons, *Journal of Physical Chemistry* **84**, 1096-1103.
70. Le, L. D.; Issa, D.; VanEck, B.; Dye, J. L. (1982) Preparation of Alkalide and Electride Films by Direct Vapor Deposition, *Journal of Physical Chemistry* **86**, 7-10.
71. Jaenicke, S.; Faber, M. K.; Dye, J. L.; Pratt, W. P., Jr. (1987) Optical Reflectivity and Absorption Measurements of Sodium (C222) Sodide, *Journal of Solid State Chemistry* **68**, 239-246.
72. Skowyra, J. B.; Dye, J. L.; Pratt, W. P., Jr. (1989) Apparatus for Codeposition and Layered Deposition of Reactive Metals and Volatile Organic Compounds with Control of Stoichiometry and Film Thickness, *Review of Scientific Instruments* **60**, 2666-2672.
73. Hendrickson, J. E. (1994) Ph.D. Dissertation, Michigan State University, East Lansing, Michigan.
74. Hendrickson, J. E.; Kuo, C. T.; Xie, Q.; Pratt, W. P., Jr.; Dye, J. L. (1996) Reflection and Absorption Spectra of Na^+(C222)Na^-, *Journal of Physical Chemistry* **in press.**
75. Bannwart, R. S.; Solin, S. A.; DeBacker, M. G.; Dye, J. L. (1989) Low Temperature Photoluminescence of Na^+(Cryptand[2.2.2])Na^- Excited with a Picosecond Laser, *Journal of the American Chemical Society* **111**, 5552-5556.
76. Xu, G.; Park, T.-R.; Bannwart, R. S.; Sieradzan, A.; DeBacker, M. G.; Solin, S. A.; Dye, J. L. (1991) Luminescence Studies of Crystalline Sodides, in *Metals in Solution, J. Phys. IV -Colloque C5*; P. Damay and F. Leclerq, Eds.; les editions de physique: Paris pp 283-290.
77. Xu, G. (1992) PhD. Dissertation, Michigan State University, East Lansing, Michigan.
78. Park, T.; Solin, S. A.; Dye, J. L. (1992) Crystal Size Dependent Emission from Mobile Exciton-Polaritons in Sodium Cryptand Sodide, *Solid State Communications* **81**, 59-63.
79. Park, T.; Solin, S. A.; Dye, J. L. (1992) Exciton-Polariton Luminescence from Sodium Cryptand Sodide [Na^+(C222)Na^-], *Physical Review B* **46(II)**, 817-830.
80. Welber, B. (1974) Diamond Anvil Apparatus, *Review of Scientific Instruments* **47**, 183-186.
81. Kuo, C. -T. (1994) Ph.D. Dissertation Dissertation, Michigan State University, East Lansing, Michigan.
82. Kuo, C. -T.; Dye, J. L.; Pratt, W. P., Jr. (1994) Effect of Laser Pulses on the Photoelectron Emission from Na^+(C222)Na^-, *Journal of Physical Chemistry* **98**, 13575-13582.
83. Jaenicke, S.; Dye, J. L. (1984) Photoelectron Emission from Solid Sodides, *Journal of Solid State Chemistry* **54**, 320-329.
84. Huang, R. H.; Dye, J. L. (1990) Low Temperature (-80°C) Thermionic Electron Emission from Alkalides and Electrides, *Chemical Physics Letters* **166**, 133-136.
85. Nagy, T. F.; Overney, G.; Dye, J. L. (1994) Program for Creating a Grid of Distances from Van der Waals Atomic Surfaces in a Crystal Structure to Permit Construction of Isosurfaces of Channels and Cavities; Program TIBORCHAN., *Unpublished program, this laboratory*. Available by anonymous FTP at argus.cem.msu.edu by path /home/slater1/people/ftp/pub/dye/voids.
86. Dye, J. L.; Wagner, M. J.; Overney, G.; Nagy, T. F. ; Tománek, D. (1995) Cavities and Channels in Electrides, *Unpublished work, this laboratory.*

87. Wagner, M. J.; Dye, J. L. (1995) $[Cs^+(15\text{-crown-5})(18\text{-crown-6})e^-]_6 \cdot (18\text{-crown-6})$: Properties of the First Mixed Crown Ether Electride, *Journal of Solid State Chemistry* **117**, 309-317.

SYNTHESIS, CHARACTERIZATION AND APPLICATIONS OF REDOX ACTIVE SELF-ASSEMBLING MONOLAYERS

M.MASKUS, J.TIRADO, J.HUDSON, R.BRETZ and H.D.ABRUÑA*

Department of Chemistry
Baker Laboratory, Cornell University
Ithaca, New York 14853-1301

1. Abstract

Studies of the synthesis, characterization and applications of redox active transition metal complexes capable of forming self-assembled monolayers onto platinum, gold and silver electrodes are presented. On platinum surfaces the complexes are of the type $[M(bpy)_2ClL]^{+1}$ where M = Os, Ru, bpy = 2,2'-bipyridine and L are bis-pyridine derivatives with different spacer groups. We have investigated effects of the deposition potential, pH, ionic strength and concentration on the thermodynamics and kinetics of the self-assembly process. We have also investigated the exchange dynamics of these self-assembling systems via competition studies of complexes with different metal centers. In-situ STM studies suggest that the formation of a monolayer takes place via random decoration of the surface but the adsorbed molecules are very mobile. Ex-situ molecularly resolved STM images have also been obtained. These molecules form compact monolayers with a rectangular lattice with unit cell dimensions of 9.3Å by 12.4Å.

In the case of gold electrode surfaces we have employed pyridine derivatives with a pendant alkylthiol of varying length as well as a novel terpyridine thiol. In the case of the pyridine derivatives, we have prepared complexes of the type $[M(bpy)_2ClL]^{+1}$ (M=Os) and have carried out studies similar to those mentioned above on platinum and silver surfaces as well.

For the case of the terpyridine thiol ligand, we have prepared various complexes of Co, Os and Cr of the type $[M(tpy\text{-}SH)_2]^{+n}$, and $[M(tpy\text{-}SH)L]^{+n}$ (L = terpyridine, tetrapyridyl pyrazine (tppz)). The free-ligand as well as the metal complexes adsorb strongly onto gold electrode surfaces and in the case of the metal complexes, give rise to compact redox-active monolayers. Potential cycling of an electrode modified with a monolayer of $[Co^{II}(tpy\text{-}SH)(Cl)_2]$ generates an adsorbed layer of $[Co^{II}(tpy\text{-}SH)_2]$ via a novel surface cross chelation reaction. Reaction of the surface immobilized $[Os(tpy\text{-}SH)tppz]^{+2}$ with $[Co(tppz)Cl_2]$ gives rise to a structured interface of the type Au-S-tpy-Os-tppz-Co-tppz with both transition metal complexes being

L. Echegoyen and A.E. Kaifer (eds.), Physical Supramolecular Chemistry, 337–353.

redox active. This system lends itself to the preparation of structured interfaces of deliberate design.

We also describe the application of these monolayer systems for the catalytic reduction of nitric oxide.

2. Introduction

Self-assembled monolayers (SAM) onto a variety of substrate materials are currently the object of intense study because their well defined structural character provides molecular surfaces of great potential for scientific studies and technological exploration [1-8]. The use of these modified surfaces ranges from fundamental examinations concerning the structure of the layers and the thermodynamics and kinetics of adsorption to applications e.g. as sensors, molecular or electronic devices. Numerous SAM systems have been investigated including thiols, disulfides and sulfides on gold, copper, platinum and silver surfaces, pyridines on platinum, fatty acids on metal oxides, silanes on glass, oxides and silica surfaces and others. Electroactive centers connected to a self-assembling moiety facilitate the direct observation of the attached species by the use of electrochemical techniques such as cyclic voltammetry. In the case of redox active SAM, there is also great interest in employing them as platforms on which to carry out studies of rates of electron transfer. [9-13]

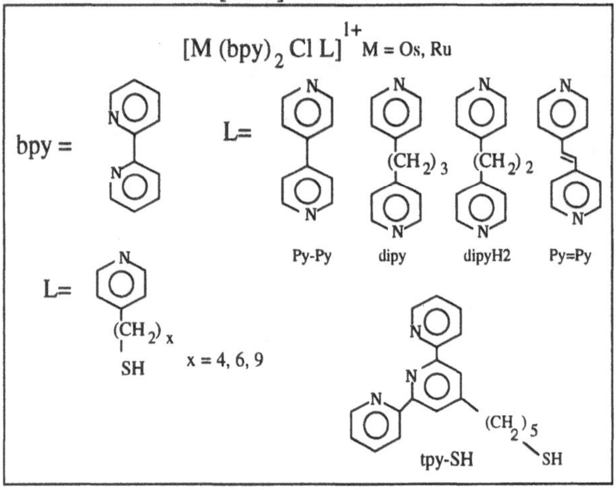

Figure 1. Families of redox-active transition metal complexes capable of adsorbing onto electrode surfaces

We present studies of redox active SAM's based on transition metal complexes. We have studied three different families of these materials:

1. $[M(bpy)_2ClL]^{+1}$ (M = Ru, Os; bpy=2,2'-bipyridine and L is a bis-pyridine ligand (see Figure 1) where one pyridine group is bound to the metal center and the other serves to anchor the complex to platinum and gold surfaces [14-17].

2. $[Os(bpy)_2ClL]^{+1}$ (L = 4-alkylthio substituted pyridine derivatives; see figure 1) [18-19]

3. $[M(tpy\text{-}SH)L]^{+n}$ (M = Co, Os, Cr; tpy-SH = a terpyridine-thiol (see Figure 1) and L = terpyridine, tpy-SH, tetrapyridyl-pyrazine) [20]

For the first two families we present studies of the thermodynamics and kinetics of adsorption and desorption as well as exchange dynamics. In addition we present in- and ex-situ STM data. In the case of the tpy-SH complexes we present the dynamics of monolayer formation as well as a novel surface cross-chelation reaction. We further illustrate the utility of these materials in the preparation of structured interfaces of deliberate design as well as in electrocatalytic applications.

3. Experimental

3.1. REAGENTS AND SYNTHESIS

Reagents and their purification has been previously described. [14-20] The synthesis of the alkylthiol substituted pyridine derivatives was via a thiouronium salt intermediate and has been described previously [18,19]. The synthesis of the terpyridine thiol has been described and was carried out following procedures similar to those of Potts et. al. [21] The synthesis and characterization of the transition metal complexes employed in these studies and which are capable of forming redox-active self-assembling monolayers have been previously described. [14-17]

3.2. INSTRUMENTATION

Cyclic voltammograms were carried out with either an EI-400 bipotentiostat from Ensman Instrumentation and data were collected using a Gateway 2000 PC (486DX) system equipped with an interface card from National Instruments model AT-MIO-16F-5 programmed with LabView, or with an IBM EC-225 Voltammetric analyzer coupled to a Soltec VP-64236 XY recorder. Potentials were measured against a sodium chloride saturated silver/silver chloride (Ag/AgCl) reference electrode without regard for the liquid junction potential. Electrochemical cells of conventional design were employed. Platinum, gold, and silver disk electrodes (sealed in glass) were used. These were pretreated as previously described. [18] STM experiments were carried out with a Digital Instruments Nanoscope III with a locally designed cell. [22] Additional experimental details can be found in our previous work.[14-20]

3.3. PROCEDURES

Procedures for the determination of the adsorption isotherms, kinetics of adsorption and desorption as well as exchange dynamics have been previously described.[14-19]

4. Results and Discussion

4.1. THERMODYNAMICS OF ADSORPTION

Exposure of the appropriate electrode material to a dilute (typically micromolar) solution of the metal complexes described here gives rise to modification of the electrode surface with a film of the metal complex which furthermore retains its redox activity. The thermodynamics of adsorption were studied using cyclic voltammetry to directly measure the surface coverage (Γ in mol/cm^2) of the complexes by integrating the charge under the voltammetric wave and using the relationship:

$$\Gamma = Q/nFA \tag{1}$$

where Q is the charge (Coulombs), n is the number of electrons transferred and A is the electrode area. . (Note: a monolayer of these materials represents approximately $1.3\text{x}10^{-10}$ mol/cm^2). Figure 2 depicts the cyclic voltammetric responses for a platinum electrode modified with about a monolayer of $[Os(bpy)_2dipyCl]^{+1}$ (from a 1.2 μM solution of the complex) as well as a gold electrode modified with a monolayer of $[Co(tpy)(tpy\text{-}SH)]^{+2}$ (from a 1.0 μM solution of the complex), respectively. In both cases the voltammetric wave shape is that anticipated for a surface immobilized redox couple.

Surface coverages were determined both in the deposition solutions as well as in clean supporting electrolyte after rinsing the electrode. While in the deposition solution, the steady state coverages were determined to be 1.1 to 1.5 $\text{x}10^{-10}$ mol/cm^2, surface coverage values typically decreased 10-15% upon rinsing suggesting the presence of material that is weakly associated to the chemisorbed monolayer by hydrophobic interactions or weak intercalation.

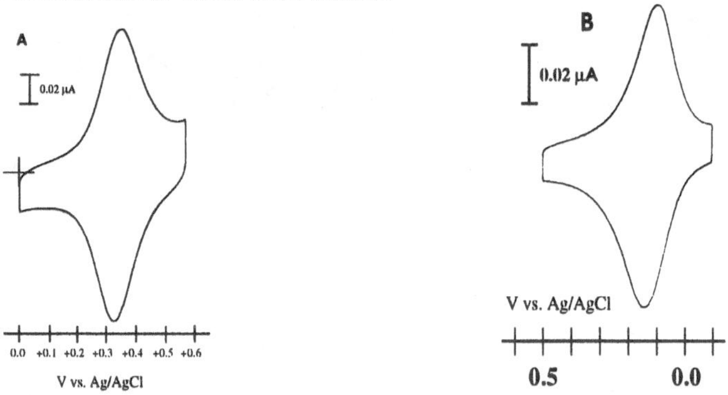

Figure 2. Cyclic voltammogram at 100mV/s in 0.1M $KClO_4$ for (A) a polycrystalline platinum electrode modified with a monolayer of $[Os(bpy)_2dipyCl]^{+1}$ (From reference 15) and for (B) a polycrystalline gold electrode modified with a monolayer of $[Co(tpy\text{-}SH)(tpy)]^{+2}$ (from reference 20).

From plots of surface coverage (Γ) vs solution concentration and fits to different adsorption isotherm models, we were able to determine that although both Langmuir and Frumkin isotherms gave satisfactory fits, the fits to the Frumkin model were consistently better. Given that the Frumkin model (equation 2 where β is the adsorption coefficient, Θ is the fractional coverage and a is the interaction parameter)

$$\beta C^* = (\Theta/1-\Theta) \exp(-2a\Theta) \quad (2)$$

takes into account near-neighbor interactions wehereas the Langmuir model does not, suggests the presence of such near-neighbor interactions. From the fits we obtain that the interaction parameter, a, is typically around 0.2 which implies the presence of small repulsive interactions, likely the result of the net positive charge that these complexes have (+1 in the reduced, and +2 in the oxidized form, respectively). From fits to the isotherms for the complexes in family 1 we obtain that the free energy of adsorption is about 49.5 ±1 kJ/mol consistent with a chemisorptive bond. Moreover, whereas the free energies of adsorption were found to be independent of the potential applied to the electrode during deposition, the coverage values were found to be dependent on

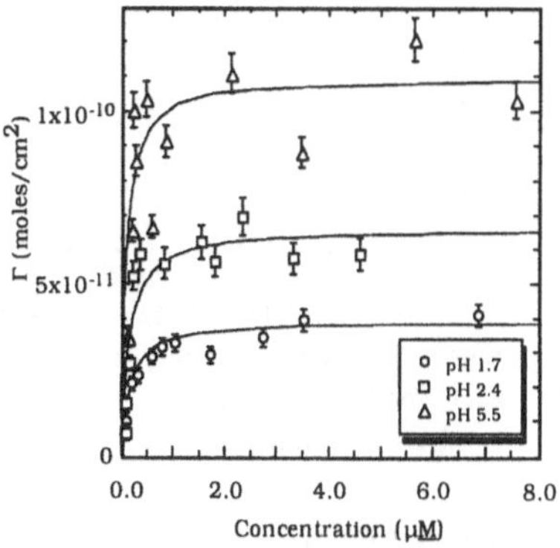

Figure 3. Fit of experimental data (symbols) to the Frumkin isotherm with a = 0.2 for adsorption of $[Os(bpy)_2Cl(py{=}py)]^{+1}$ from solutions of varrying pH to a polycrystalline platinum electrode. (From reference 15).

potential with lower values (by about 10%) obtained at potentials where the complex is present in the oxidized form, carrying a net +2 charge, as opposed to a net +1 charge in the reduced form. This suggests that electrostatic repulsions are likely responsible for the observed effect. However it should also be noted that upon rinsing, there were no differences in the surface coverage values at different potentials suggesting that these electrostatic interactions are relatively weak.

We also investigated the effects of the pH of the deposition solution from 1.2 to 5.5 on the adsorption thermodynamics. There was virtually no effect of pH on the free energy of adsorption. However, as can be seen in Figure 3 the adsorption isotherms were affected in terms of the saturation coverage which decreased with decreasing pH. We ascribe this effect to a simple mass action where at low pH values the pendant pyridine (which is responsible for the adsorption of the complex) is protonated and hence unable to chemisorb.

In the case of the alkylthiol pyridine derivatives (family 2) we find that the amount of material deposited is a strong function of the applied electrode potential. The amount of material deposited onto the working electrode at potentials negative of the point of zero charge (E_{PZC}) for a given metal electrode (Au, Pt or Ag) depended on the potential applied during deposition with increasing amounts obtained for progressively more negative values of the applied potential. However, at potentials positive of the E_{PZC}, the amount deposited was much smaller and virtually independent of the applied potential. A representative data set for adsorption of $[Os(bpy)_2Cl(Py\text{-}(CH_2)_4\text{-}SH)]^{+1}$ is presented in Figure 4 where the above mentioned trend is clearly evident. We ascribed these effects to electrostatic attraction and repulsion.

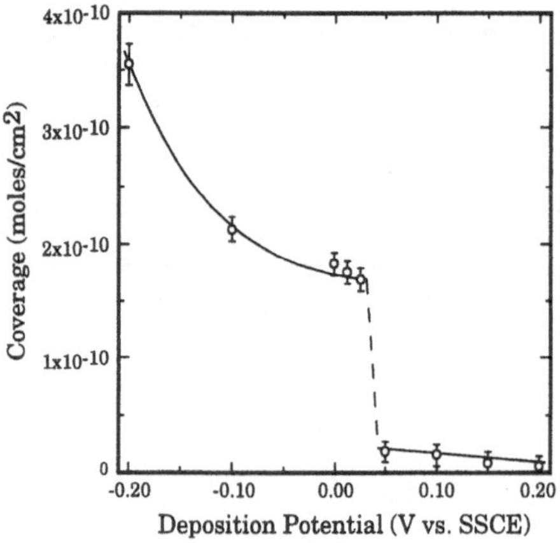

Figure 4. Variation of surface coverage with applied electrode potential during deposition (from an 8.3μM solution of the complex in 0.1M $KClO_4$) for $[Os(bpy)_2\ Cl\ Py\text{-}(CH_2)_4\text{-}SH]^{+1}$ onto a polycrystalline platinum electrode (From reference 18)

4.2. KINETICS OF ADSORPTION AND DESORPTION

We have carried out studies of the kinetics of adsorption and desorption. In general, one can distinguish two limiting cases controlling the adsorption dynamics; diffusion control and kinetic control which are described by equations 3 (with Γ_t and Γ_s representing the coverage values at time t and equilibrium values, respectively and K is a constant) and 4, (with k being the rate constant) respectively.

$$\Gamma_t/\Gamma_s = K\,(C^*/\Gamma_s)\,(Dt)^{1/2} \quad (3)$$

$$\Gamma_t = \Gamma_e\,(1\text{-}\exp(\text{-}kC^*t)) \quad (4)$$

In general we find that for all the complexes studied, the adsorption dynamics are kinetically controlled; that is they obey equation 4. A representative data set for the $[Os(bpy)_2Py{=}Py\ Cl]^{+1}$ complex is presented in Figure 5 along with predictions for both limiting cases. It is clear that the kinetic control model is the more appropriate one. Over the concentration range from 0.07 to 0.32 μM the adsorption rate constant for the above mentioned complex remained virtually invariant with an average value of about 1000 ±100 $M^{-1}s^{-1}$. Qualitatively similar behavior was observed for the other complexes studied.

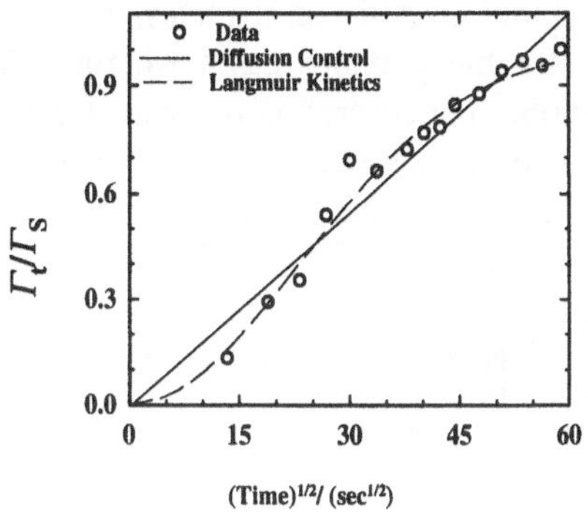

Figure 5. Variation of normalized coverage (Γ_t/Γ_s) vs $t^{1/2}$ for the adsorption of the $[Os(bpy)_2(Py{=}Py)Cl]^{+1}$ complex onto a polycrystalline platinum electrode. (From reference 16)

In terms of desorption, we again observe that a simple exponential expression accurately describes the desorption kinetics. We could express the desorption dynamics in terms of the integrated charge by equation 5

$$Q_{t,d} = (Q_{i,d} - Q_{e,d})\exp(-k_d t) + Q_{e,d} \tag{5}$$

In this equation $Q_{t,d}$, $Q_{e,d}$ and $Q_{i,d}$ are the charge at time *t*, equilibrium charge, and initial charge for the desorbing species, respectively and k_d is the rate constant for desorption. As shown in figure 6 this expression accurately describes the desorption of the $[Os(bpy)_2(dipy)Cl]^{+1}$ complex.

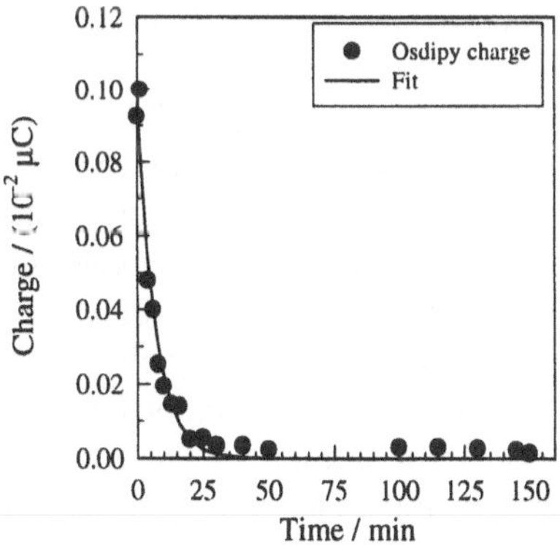

Figure 6. Plot of charge vs. time for the desorption of a monolayer of $[Os(bpy)_2(dipy)Cl]^{+1}$ adsorbed onto a platinum electrode into 0.1 M $KClO_4$ aqueous solution. (From reference 17)

For the case of the alkylthiol pyridine derivatives, we have also investigated the desorption dynamics. As for the previous cases we find that a simple exponential expression accurately describes the desorption process. Figure 7 presents data and the corresponding fits to the exponential expression for $[Os(bpy)_2Cl(Py\text{-}(CH_2)_4\text{-}SH)](PF_6)$ desorbing from a polycrystalline platinum electrode at different temperatures.

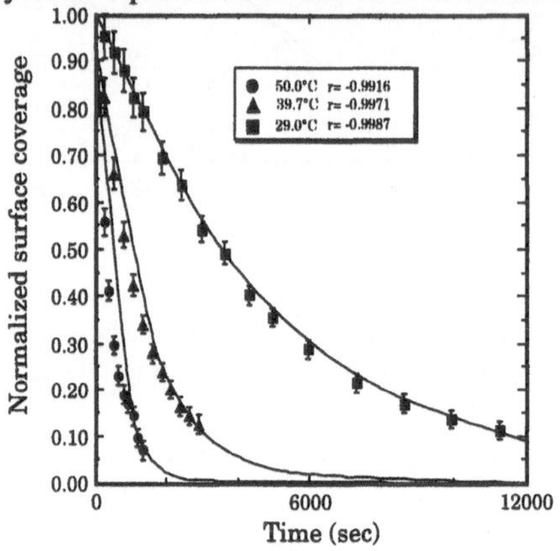

Figure 7. Variation in the normalized surface coverage with time for $[Os(bpy)_2Cl(Py\text{-}(CH_2)_4\text{-}SH)](PF_6)$ desorbed from a polycrystalline platinum electrode continuously cycled between -0.20V and +0.45V in 0.1M $KClO_4$ at 100mV/s at temperatures of 29.0°C, 39.7°C, and 50.0°C. Solid lines are exponential fits to the data. (From reference 19)

Using the Arrhenius equation (6):

$$d \ln k/d\,(1/t) = -E_a/R \qquad (6)$$

where *k* is the rate constant, *T* is the temperature in Kelvin, and *R* is the gas constant, the activation energy (E_a) can be determined from the slope of a plot of ln *k* vs *1/T*. For the data mentioned above for $[Os(bpy)_2Cl(Py\text{-}(CH_2)_4\text{-}SH)](PF_6)$ the value E_a was found to be 87 kJ/mole.

Similar experiments for $[Os(bpy)_2Cl(Py\text{-}(CH_2)_6\text{-}SH)](PF_6)$ and $[Os(bpy)_2Cl(Py\text{-}(CH_2)_9\text{-}SH)](PF_6)$ were carried out and the resulting values of E_a were 105 and 112 kJ/mol respectively a trend which follows Traube's rule, which states that organic weak or non-electrolytes lower the surface tension to a greater degree as the length of the chain increases in a homologous series. In other words, as the chain length increases, the organic material will be less favored to dissolve in aqueous solution, i.e., the solvation energy barrier will increase. The activation energy determined in this study is a measure of the barrier to the desorption of the osmium thiol complex from the electrode surface. Therefore, the trend observed indicates that for these complexes, the barrier to desorption (E_a) increased with increasing carbon chain length. This trend is consistent with the fact that the more hydrophobic a molecule is (the hydrophobicity of the thiol complexes will increase as the hydrocarbon spacer unit is increased in length), the lower its surface tension and the less likely it will move from the hydrophobic environment of the multilayer to the aqueous solution environment.

4.3 EXCHANGE DYNAMICS

By using the closely related complexes $[M(bpy)_2(dipy)Cl]^{+1}$ (M=Ru, Os) we could carry out studies of the exchange dynamics; that is we could study the rate of exchange of one complex by the other. This was possible because whereas these two complexes are virtually identical in size and adsorption strength, their redox responses are well separated (by about 400 mV), thereby providing an ideal system with which to probe exchange dynamics of redox-active self-assembling monolayers. In these studies we could follow the exchange of one complex initially adsorbed (typically at monolayer coverage) by the other under conditions where both complexes are present in solution (competitive) or where only one is present (displacement).

The exchange dynamics were studied by recording cyclic voltammograms at different times during the equilibration (from a homogeneous to a mixed monolayer) process. Figure 8 depicts the formation of a mixed-monolayer from a solution containing a 1:1 ratio of Rudipy/Osdipy. In this case the electrode was initially covered with a monolayer of Osdipy and Rudipy was subsequently injected so as to achieve a 1:1 concentration ratio in solution. It is apparent from the figure that the voltammetric peak due to adsorbed Osdipy decreases while that for the Rudipy complex exhibits a concomitant increase. If the charges associated with these peaks are obtained at different times, the dynamics of exchange can be studied. Figure 8B shows the results obtained from the data in Figure 8A. The anticipated decrease in the Osdipy charge (coverage) as well as the increase in that due to Rudipy are observed until both reach an equilibrium value, determined by their respective solution concentrations. Also plotted in Figure 8B is the total charge which, as can be seen, remains virtually constant during the experiment. This demonstrates that the process being monitored is the surface exchange dynamics at constant surface coverage in which no stacking nor multiple layers are formed. The presence of isopotential points (at +0.45 V in Figure 8A) is also evident. Since isopotential points arise as a result of an equilibrium process between two surface species at constant total coverage, this provides compelling evidence that the exchange process is one that is at equilibrium at all times. This is an important observation since it implies that the time evolution of the changes in surface coverage can be employed to ascertain the kinetics of desorption, displacement or exchange, depending on the specific experiment under consideration.

Based on studies of the dynamics of adsorption mentioned previously we again used simple exponential expressions to describe these adsorption and desorption processes. Equation 5 was used to describe the desorption process whereas equation 7 was used for the adsorption case:

$$Q_{t,a} = Q_{e,a} \, (1\text{-}\exp(-k_a \, t)) \tag{7}$$

As for the previous kinetic studies, we find an excellent fit to the data. In fact, the lines in Figure 8B are based on the use of the above-mentioned equations. Based on these as well as other studies it appears that the exchange dynamics are controlled by the rate of desorption via a dissociative mechanism.

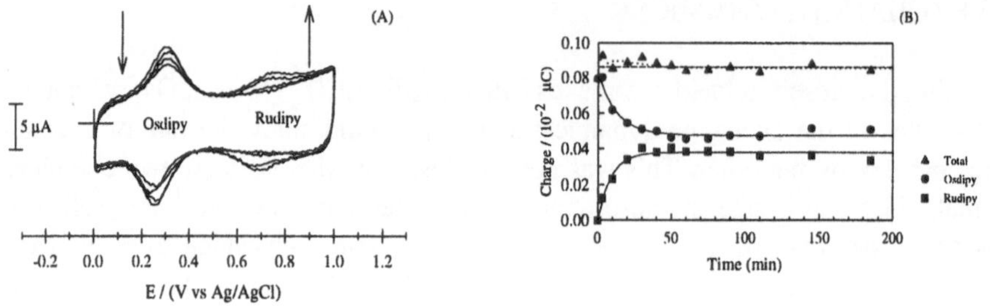

Figure 8. (A) Cyclic voltammograms at different times (i.e.: 0, 3, 10, 20 min) to monitor the exchange dynamics of Osdipy with Rudipy. (B) Charge variation with time for Osdipy (●), Rudipy (■) and Total Charge (Δ). Solid lines are fit to equations 4 and 5 for Rudipy and Osdipy, respectively. The dotted line represents the sum of the two solid lines and the dashed line is the average total charge. (From reference 17)

4.4. STM STUDIES

We have used STM and ECSTM to study the formation and structure of an $[Os(bpy)_2(dipy)Cl]^{+1}$ monolayer on Pt(111). Using cyclic voltammetry, we monitored the coverage of the adsorbing monolayer in real-time while simultaneously employing ECSTM to study the structure of the submonolayer. Our results are consistent with previous studies suggesting that at low coverages the submonolayer spreads uniformly across the sample surface due to electrostatic repulsions between adsorbate molecules and the apparent high mobility of the Osdipy molecules on the sample surface. As the coverage increases, the deposit consists of domains of tightly and loosely packed adsorbate which appear as recessed defects in the ECSTM images. The fully formed monolayer consists of partially solvated Osdipy molecules which are constantly reordering on a time scale too fast to monitor by ECSTM. We also employed ex-situ STM to study the fully formed monolayer crystallized on a Pt(111) substrate. The Osdipy adsorbates appear to crystallize forming tightly packed two dimensional crystals on the sample surface. The crystals pack in a rectangular close packed structure with unit cell dimensions of 9.3 Å by 12.4 Å.

A typical molecularly resolved ex-situ STM image of an $[Os(bpy)_2(dipy)Cl]^{+1}$ monolayer on Pt(111) is presented in Figure 9 where the above mentioned structure is evident. Based on the assumption that each high contrast spot in the image corresponds to a single adsorbed $[Os(bpy)_2(dipy)Cl]^{+1}$ molecule, the observed dimensions of the unit cell were used to calculate a surface coverage of 2.3×10^{-10} mol/cm^2, significantly denser than the electrochemically determined coverage of 1.1×10^{-10} mol/cm^2. It is possible that the observed structure is of a region of high local adsorbate density caused by the adsorbate forming a multilevel structure on the sample surface. However, the structure of the $[Os(bpy)_2(dipy)Cl]^{+1}$ monolayer was modeled using scaled space filling representations of the transition metal head groups (generated using HyperChem, Hypercube Inc.) and this yielded very good fits with experimental suggesting the presence of highly ordered two-dimensional crystals. The model assumes that the head groups preferentially align in two dimensions along the dipole of the osmium chlorine bond and allows space for counterions (perchlorate) to fit between neighboring head groups (not shown). The presence of a high density ex-situ layer is also evidenced by the sparse distribution of adsorbate on the platinum surface which is about 50%.

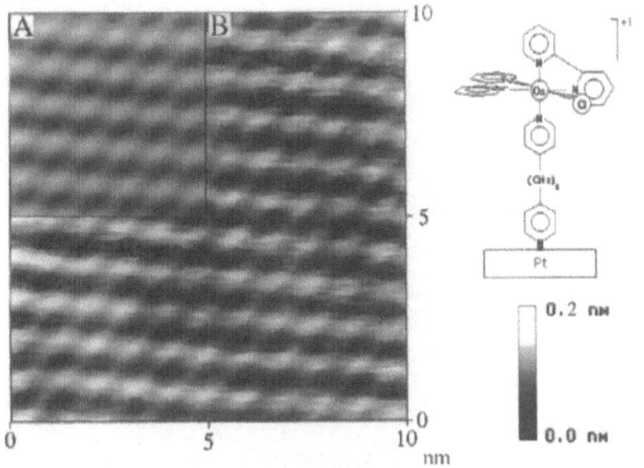

Figure 9. Molecularly resolved ex-situ STM image of an Osdipy monolayer adsorbed onto a Pt(111) substrate (a) Fourier filtered and (b) unfiltered. Tip bias = 100mV, tunneling current = 500pA, and a scan rate = 20 Hz. (From reference 22)

4.5 TERPYRIDINE THIOL

We recently reported on the synthesis of a new thiol modified terpyridine ligand (tpy-SH) as well as of its complexes with Co, Cr and Os. [20] We prepared complexes of the type $[M(tpy\text{-}SH)_2]^{+n}$ and $[M(tpy\text{-}SH)L]^{+n}$ (L = terpyridine, tetrapyridyl pyrazine (tppz)). The free-ligand as well as the metal complexes adsorb strongly onto gold electrode surfaces and in the case of the metal complexes, they retain their redox-active responses. The cyclic voltammetric response exhibits an aging process where it sharpens significantly with time (without loss of coverage) suggesting a slow reorganization of the adsorbed complex on the electrode surface. For example, the voltammetric response for a gold electrode modified with a monolayer of $[Co(tpy\text{-}SH)(tpy)]^{+2}$ initially had a full width at half maximum of about 250 mV. However, if the electrode was allowed to stand in solution for extended time periods, the voltammetric response sharpened considerably, with the full width at half-maximum decreasing to about 160 mV.

One of the main purposes behind the synthesis of this ligand was the preparation of multilayered structured interfaces of deliberate design as shown in scheme I. In an effort to follow the generalized approach depicted in Scheme 1 to consecutively build up electroactive layers onto the gold electrode, an electrode previously modified with a monolayer of tpy-SH was held for 6 h in contact with a concentrated aqueous solution of $CoCl_2$ (0.5 M). We felt that this would give rise to binding of all of the surface immobilized terpyridine sites with a Co^{+2} moiety to give a [Co(tpy-SH)Cl_2] monolayer. The redox response of this material exhibited unusual and unique reactivity which we describe below.

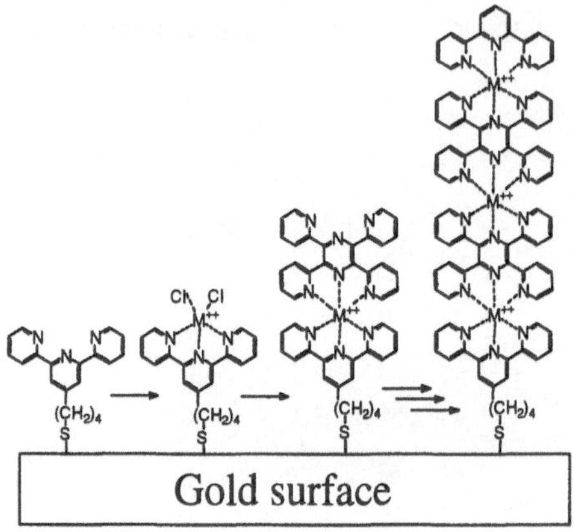

Scheme 1

The cyclic voltammogram (after rinsing) in aqueous 0.1 M $NaClO_4$ of the electrode described above exhibited an irreversible reduction at a peak potential of -0.16 V (inset figure 10). We ascribe this peak to the reduction of the [Co(tpy-SH)Cl_2]-layer. We base this assignment on the fact that the cyclic voltammetric response of a freshly prepared homogenous aqueous solution (0.1 M $NaClO_4$) of [Co(tpy)Cl_2] also exhibits an irreversible peak at -0.14 V which we ascribe to the reduction of [Co(tpy)Cl_2] and its rapid decomposition.

Upon continued cycling of the potential of the modified electrode between +0.40 and -0.40 V, this reduction peak decreases steadily suggesting the loss of $CoCl_2$ moieties from the layer, leaving free terpyridine sites. Concomitant with a decrease in the peak at -0.16 V there is also the emergence of a new reversible redox couple with a formal potential value E°' of +0.13 V (Fig. 10). This process continues until the peak at -0.16 V completely disappears and the reversible couple at +0.13 V (ΔE = 60mV) reaches a maximum. The electrochemical features of this new wave are virtually the same as those of a self-assembled monolayer of $[Co(tpy\text{-}SH)(tpy)]^{+2}$ preformed on gold electrodes.

The presence of well defined isopotential points in the cyclic voltammogram during the course of the above-mentioned process would suggest that there is a transformation between two surface species. [23] We believe that this process corresponds to the transformation of a full monolayer of [Co(tpy-SH)Cl_2] to a $[Co(tpy\text{-}SH)_2]^{+2}$-layer. We propose that after the reduction of the [Co(tpy-SH)Cl_2]-layer, which results in the partial loss of the metal and concomitant generation of uncomplexed ligand tpy-SH, these free terpyridine sites react with nearby [Co(tpy-SH)Cl_2] species to form $[Co(tpy\text{-}SH)_2]^{+2}$. The appearance of very sharply defined isopotential points as described above provides additional evidence in support of such a novel surface cross-chelation reaction process. An analogous case was recently reported by Anson and Lei for [Cu^{+2}-phenanthroline] complexes adsorbed onto graphite electrodes.[24]

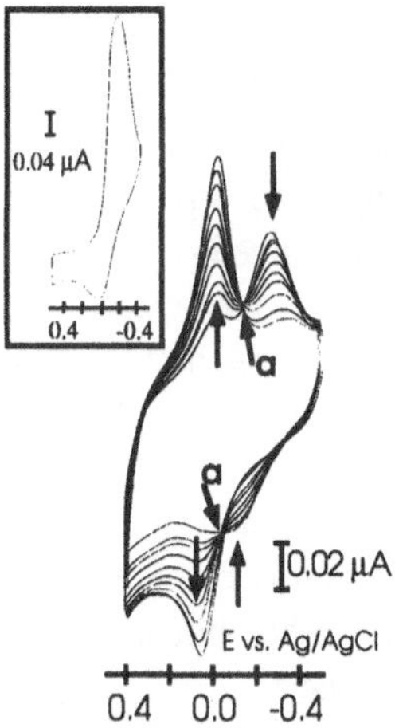

Figure 10. Cyclic voltammograms, recorded at 15 min. intervals at a sweep rate of 0.10 V/s in aqueous 0.1 M $NaClO_4$ of a gold electrode coated with a monolayer of HS-tpy-$CoCl_2$. a: isopotential points. Inset: first cycle. (From reference 20).

Due to the high packing density of the ligand tpy-SH on the gold electrode surface, we were unable to further elaborate on the structure of the adsorbed layer by exposing the [Co(tpy-SH)Cl_2] modified gold surface to solutions of other terdentate ligands including terpyridine (tpy), tetra-2-pyridyl-1,4-pyrazine (tppz) and 2,4,6-tri(2-pyridyl)-1,3,5-triazine (tptz). After immersing the electrode modified with [Co(tpy-SH)Cl_2] into 1 mM solutions of these ligands (for several hours), followed by thorough rinsing, the final cyclic voltammogram in 0.1 M aqueous $NaClO_4$ always showed the response of a $[Co(tpy\text{-}SH)_2]^{+2}$-layer as described above, suggesting that because of steric constraints, the ligands do not have enough space to coordinate to the cobalt centers.

4.6. MULTILAYERED STRUCTURED INTERFACES

There are different approaches that could be followed in order to mitigate such steric constraints. For example, one could employ mixed-monolayers consisting of tpy-SH and an alkyl-thiol of comparable length in order to increase the spacing between adjacent terpyridine sites. However, because of the possibility of island formation, especially among the tpy-SH sites with their π-stacking ability, we chose an alternative method employing a pre-synthesized metal-complex of the type $[M(tpy\text{-}SH)(tppz)]^{+2}(PF_6)_2$. In this case the tpy-SH group would anchor the complex to the gold surface whereas the tppz, with its available terdentate coordination, would allow complexation of a second metal center. In order to ensure the stability of the first layer we chose the very stable osmium analog $[Os(tpy\text{-}SH)(tppz)]^{+2}$. Unfortunately, the Os^{+2}/Os^{+3}-redox couple of such a complex takes place at about +1.10 V and is therefore not accessible on gold electrode surfaces (neither in H_2O (0.1 M $NaClO_4$) nor in CH_3CN (0.1 M TBAP)) due to the interfering gold oxidation reaction at potentials above 1.15 V. However, tpy-SH complexes of osmium also adsorb strongly to gold surfaces. For example, the $[Os(tpy\text{-}SH)_2]^{+2}$ complex, which has a less positive formal potential than $[Os(tpy\text{-}SH)(tppz)]^{+2}$, exhibited an Os^{+2}/Os^{+3}-wave for a modified gold electrode at a potential of +0.93 V in CH_3CN (0.1 M TBAP). This value is very close to that of $[Os(tpy)_2]^{+2}(PF_6)_2$ in solution suggesting that there is no change in the coordination sphere of the osmium complex upon adsorption. In addition, the wave has the shape anticipated for a surface immobilized redox couple. Since a monolayer of $[Os(tpy\text{-}SH)_2]^{+2}$ has no free coordination sites, one would not anticipate further complexation with other metal complexes serving as building blocks. To demonstrate this an electrode modified with $[Os(tpy\text{-}SH)_2]^{+2}$ was held for several hours in a 1.0 mM solution of $[Co(tppz)Cl_2]$. After rinsing with acetonitrile, the cyclic voltammogram of the electrode in CH_3CN (0.1 M TBAP) showed only the response associated with the surface bound $[Os(tpy\text{-}SH)_2]^{+2}$. The fact that no additional redox response is observed, demonstrates that $[Co(tppz)Cl_2]$ neither binds to the $[Os(tpy\text{-}SH)_2]^{+2}$ monolayer nor specifically adsorb to the gold surface.

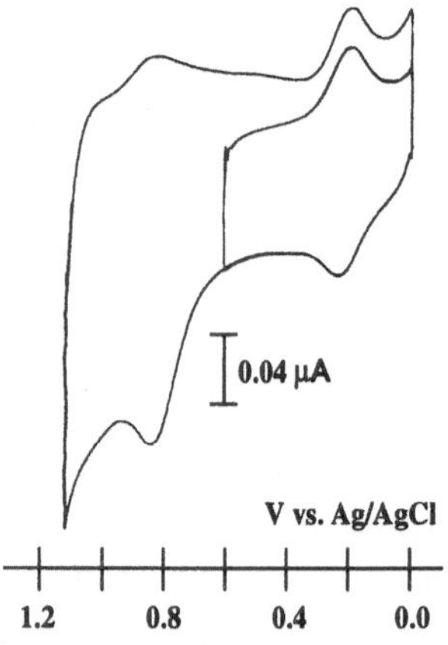

Figure 11. Cyclic voltammogram in CH_3CN (0.1 M TBAP) at a sweep rate of 0.1 V/s for a gold electrode coated with a mixed monolayer consisting of $[Os(tpy\text{-}SH)_2]^{+2}$ and $[Os(tpy\text{-}SH)(tppz)]^{+2}$ and after having been immersed in a 1.0 mM solution of $[Co(tppz)Cl_2]$ overnight. (From reference 20)

A different situation is encountered if one builds up a monolayer from a solution containing both $[Os(tpy\text{-}SH)_2]^{+2}$ and $[Os(tpy\text{-}SH)(tppz)]^{+2}$ complexes. A gold electrode modified with such a solution exhibits a similar electrochemical response (in CH_3CN (0.1 M TBAP)) as one modified with a monolayer of $[Os(tpy\text{-}SH)_2]^{+2}$. Since the Os^{+2}/Os^{+3} couple of $[Os(tpy\text{-}SH)(tppz)]^{+2}$ on gold electrodes cannot be observed (vide-supra), only the response of the symmetrical $[Os(tpy\text{-}SH)_2]^{+2}$ complex at about +0.90 V is observed. However, and in contrast to the previous case, upon exposure of such a modified electrode to a 1.0 mM solution of $[Co(tppz)Cl_2]$ overnight, a new reversible voltammetric wave appeares at +0.26 V (Fig. 11). This surface-confined wave is ascribed to the Co^{+2}/Co^{+3} couple of the $[Co(tppz)_2]^{+2}$ moiety complexed onto the first osmium layer so that the overall structure can be represented as:

Au/-S-tpy-Os-tppz-Co-tppz

The formal potential of the cobalt center is shifted negatively by about 100 mV relative to the value for $[Co(tppz)_2]^{+2}(PF_6)_2$ in solution (+0.39 V in CH_3CN). The same trend can be seen in aqueous 0.1 M $NaClO_4$ where the Co^{+2}/Co^{+3} couple (immobilized on the surface) has a formal potential of +0.21 V, compared to +0.31 V for $[Co(tppz)_2]^{+2}$ in solution. This observation is consistent with earlier studies of multi-metallic complexes of tppz in solution where such effects were ascribed to ligand-meditated metal-metal interactions [25]. The fact that there appear to be significant metal-metal interactions in these surface confined structures opens up the possibility of exploiting such effects in numerous applications.

Differences in the voltammetric response in water vs. acetonitrile for the "bilayer" modified gold electrode were also apparent. Whereas in CH_3CN (0.1M TBAP) the Co^{+2}/Co^{+3} wave is relatively sharp and has a ΔE_p of 80mV, in aqueous $NaClO_4$ the wave broadens significantly and ΔE_p increases to about 120 mV. This suggests that the charge transfer kinetics are slower in water than in acetonitrile as was noted earlier for the $[Co(tpy\text{-}SH)(tpy)]^{+2}$-monolayer.

4.7. ELECTROCATALYTIC APPLICATIONS

Finally, we have prepared gold electrodes modified with a layer of the chromium complex $[Cr(tpy\text{-}SH)_2]^{+3}$ and have explored potential electrocatalytic applications, especially with regards to the reduction of nitric oxide (NO).

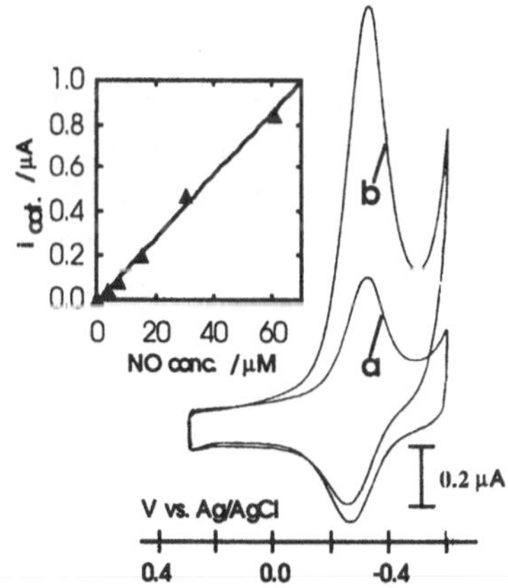

Figure 12. Cyclic voltammograms in aqueous 0.1 M NaClO4 at a sweep rate of 0.4 V/s for a gold electrode coated with a monolayer of [Cr(tpy-SH)2]+3 in presence of a) 0 mM and b) 60 mM nitric oxidde. Inset: Plot of the catalytical current (Icat.) versus nitric oxide concentration for the response at -0.3 V for the above described electrode. (From reference 20)

We have previously demonstrated that electropolymerized films of the analogous complex $[Cr(v\text{-}tpy)_2]^{+3}$ exhibit a strong electrocatalytic effect towards NO [26] and thus we were interested in ascertaining whether such activity was retained by adsorbed layers of $[Cr(tpy\text{-}SH)_2]^{+3}$.

Under an atmosphere of N_2, the immobilized complex exhibits a sharp symmetric and reversible voltammetric wave centered at about -0.29 V (Fig. 12a). This value is close to that obtained for the $[Cr(tpy)_2]^{+3}$ complex in solution ($E^{o\prime}$= -0.38 V), although shifted to more positive values because of the above mentioned reasons. In the presence of NO there is a dramatic increase in the cathodic wave, indicating a very strong electrocatalytic effect (Fig. 12b). The magnitude of the catalytic current is directly proportional to the solution concentration of NO (Fig. 12, inset). We are currently exploring the use of these modified electrodes for in vivo determination of cellular NO evolution.

5. Conclusions

We have presented a brief overview of the synthesis, characterization and application of redox-active self assembling monolayers based on transition metal complexes. These and related materials hold great promise in terms of furthering our understanding of the structure and properties of structured interfaces especially at the microscopic level. Moreover, they represent potential building blocks in the emerging field of nanotechnology and may find numerous applications in these and related fields.

Acknowledgements

This work was supported by National Science Foundation (Grant No. DMR-9107116) and the Office of Naval Research (AASERT Program). We are grateful to the Materials Science Center at Cornell University for the use of their STM facilities. M.M acknowledges support by the Swiss National Science Foundation. J.T. acknowledges support by the Corning Research Foundation. J.E.H. acknowledges support by a Department of Education Fellowship and a NSF training grant. R.L.B. acknowledges support by the U.S. Department of Education through a fellowship administered by the Materials Science Department of Cornell University.

6. Literature Cited

1. Swalen, J.D.; Allara, D.L.; Andrade, J.D.; Chandross, E.A.; Garoff, S.; Isreaelachvili, J.; McCarthy, T.J.; Murray, R.; Pease, R.F.; Rabolt, J.F.; Wynne, K.J.; Yu, H. (1987) *Langmuir* **3**, 932.
2. Ulman, A. (1991) *An Introduction to Ultrathin Organic Films from Langmuir-Blodgett to Self-Assembly*, Academic Press: San Diego, CA.
3. Whitesides, G.M.; Laibinis, P.E. (1990) *Langmuir*, **6**, 87.
4. Gaines, G.L., Jr. (1966) *Insoluable Monolayers*, Interscience:New York.
5. Sagiv, J. (1980) *J. Am. Chem. Soc.*, **102**, 92.
6. Murray, R. W. (1992) *Molecular Design of Electrode Surfaces*, John Wiley & Sons, NY, NY.
7. Bard, A. J. (1994) *Integrated Chemical Systems*, John Wiley & Sons, NY, NY.
8. Bard, A.J., Abruña, H.D., Chidsey, C.E.; Faulkner, L.R.; Feldberg, S.W.; Itaya, K.; Majda, M.; Melroy, M.; Murray, R.W., Porter, M.D., Soriaga, M.P., White, H.S. (1993) *J. Phys. Chem.*, **97**, 7147.
9. Chidsey, C. E. D. (1991) *Science*, **251**, 919.
10. Miller, C., Graeztel, M. (1991) *J. Phys. Chem.*, **95**, 5225.
11. Finklea, H. O., Ravenscroft, M. S., Snider, D. A. (1993) *Langmuir*, **9**, 223.
12. Curtin, L.S., Peck, R.S., Tender, L.M., Murray, R.W., Rowe, G.K., Creager, S.E. (1993) *Anal. Chem.*, **65**, 386.
13. Groat, K.A., Creager, S.E. (1993) *Langmuir* , **9**, 3668.
14. Acevedo, D., Abruña, H. D. (1991) *J. Phys. Chem.*, **95**, 9590.
15. Acevedo, D., Bretz, R.L., Tirado, J.D., Abruña, H.D. (1994) *Langmuir*, **10**, 1300.
16. Tirado, J. D., Acevedo, D., Bretz, R. L., Abruña, H. D. (1994) *Langmuir*, **10**, 1971.
17. Tirado, J. D., Abruña, H. D. *J. Phys. Chem.* (in press).
18. Bretz, R., Abruña, H. D. (1995) *J. Electroanal. Chem.*, **388**, 123.
19. Bretz, R., Abruña, H. D. *J. Electroanal. Chem.* (in press).
20. Maskus, M., Abruña, H. D. *J. Am. Chem. Soc.* (submitted).
21. Potts, K. T., Usifer, D. A., Guadalupe, A. R., Abruña, H. D. (1987) *J. Am. Chem. Soc.*, **109**, 3961.
22. Hudson, J.,Abruña, H. D. *J. Phys. Chem.* (in press).
23. Untereker, D. F.,Bruckenstein, S. (1972) *Anal. Chem.*, **44**, 1009.
24. Y. Lei, Y., Anson, F. C. (1994) *Inorg. Chem.*, **33**, 5003.
25. Arana, C. R., Abruña, H. D. (1993) Inorg. Chem., **32**, 194.
26. Pariente, F.; Maskus, M.; Wu, Q.; Toffanin, A.; Schapleigh, J.; Abruña, H. D. *Anal. Chem.* (submitted).

ELECTRONIC AND MAGNETIC PROPERTIES OF THIN ORGANIZED MOLECULAR SYSTEMS

P. DELHAES
Centre de Recherche Paul Pascal
CNRS, Université de Bordeaux I
Avenue Albert Schweitzer
33600 PESSAC, FRANCE

1. Introduction

In this review, we will examine the particularities of organized molecular media based on charge transfer associations. We will be interested mainly by their electronic and magnetic properties compared to the case of solid state molecular conductors and superconductors. For that purpose, one should define on one hand what the different charge transfer materials are available, and on the other hand, what kind of organized soft media are currently under studies.

Since the pioneer works of Mulliken, Forster [1] and Szent-Gïorgy [2], these molecular systems have given rise to applications in biology, chemistry and physics with in particular the novative concept of induced supramolecular organization involving a cooperative phenomena.

The recent developments of molecular conductors and superconductors have been based on the use of charge transfer salts i.e. charge transfer complexes (CTC) and radical ion salts (RIS) respectively [3]. In

L. Echegoyen and A.E. Kaifer (eds.), Physical Supramolecular Chemistry, 355–378.

such a situation, π-type intermolecular electronic transfers have to be considered; in a more general case different possibilities are existing with, in particular, the interplay between electron transfer and proton transfer interactions (because electron transfers are always labelled C.T. in the litterature, we will use it in the following parts). The second part will be devoted to the general outlook of charge transfer molecular associations in homogeneous media i.e. liquid solutions and solid phases.

We will introduce therefore π donor and π acceptor molecules and their radical-ions involved as molecular bricks for the electronic and magnetic properties. The criteria for a conducting state in these molecular solids are both the necessity of a controlled mixed valence state in these solids and a continuous path of overlapping π type orbitals, at least in a privilegied direction [4]. A special attention will be paid to prototypical π donors as tetrathiafulvalene (TTF) and derivatives and π acceptors as tetracyanoquino-dimethane (TCNQ) or fullerene (C_{60}).

The goal in an organized media is to control the short range interactions in order to obtain either a spontaneous organization as in the case of thermotropic liquid crystals or to induce thin organized films thanks in particular to the Langmuir-Blodgett (L-B) technique [5]. This approach requires a partial ordering and a reduced dimensionality associated to a certain degree of anisotropy of these molecular objects. The selection and tailoring of these special molecules is of prime interest because they are expected to combine an electroactive character inducing a C.T. together with a mesomorphic and/or an amphiphilic one. Indeed, we will show that the key for building up these thin organized films with induced anisotropic properties is the design and synthesis of multifunctional molecules.

Then, the second step is the fabrication of supramolecular assemblies displaying pre-designated architecture. The necessity of a molecular organization not only between monolayers but also inside each layer, before and after the deposit onto a substrate will be presented.

The requirements to obtain a conducting or a magnetic L-B film will be examined ; a few selected examples will be presented showing in particular the specific behaviors which are occuring as interfacial properties. Finally, these endeavors will be placed under the scope of so-called molecular electronics.

2. Charge transfer molecular associations

The functionality of a molecule is manifested through the uptake or the release of electrons, protons or other ions, and changes in dipole moments or spin states. Its internal structure can be also modified through for example a H-bond making-breaking process or a conformational change (as in cis-trans isomers).

All these events are directly linked with the local surroundings of each molecule in an homogeneous, or not, medium as for example the polarity or pH for a given solution.

In this brief review, we just want to point out the following points :

i) intermolecular versus intramolecular charge transfers,

ii) proton transfers and electron transfers,

iii) molecular associations and aggregates.

The two first paragraphs will be devoted to the molecules considered alone and in solution ; then in the third part, the supramolecular organization and the resulting solid state cristallographic structures will be envisaged as a reference.

2.1. INTERMOLECULAR AND INTRAMOLECULAR CHARGE TRANSFERS

We restrict ourselves to the already π electron donors (D) or π electron acceptors (A) which give rise to a stable compound with a definite stoichieometry if an electron transfer between the two protagonists is present. This strong charge transfer following Mulliken's theory [3], exists if the acceptor (A) has a large electronic affinity (E_A) and the donor (D) relatively low ionization potential (I_D). The associated dimer model as presented in figure 1, can be either an almost neutral or a ionic ground state depending upon the relative position of the electronic frontier orbitals. In any case, the signature of this charge transfer association is the presence of a new intense C.T. band in the visible or near IR range which induces often a color change [1][2].

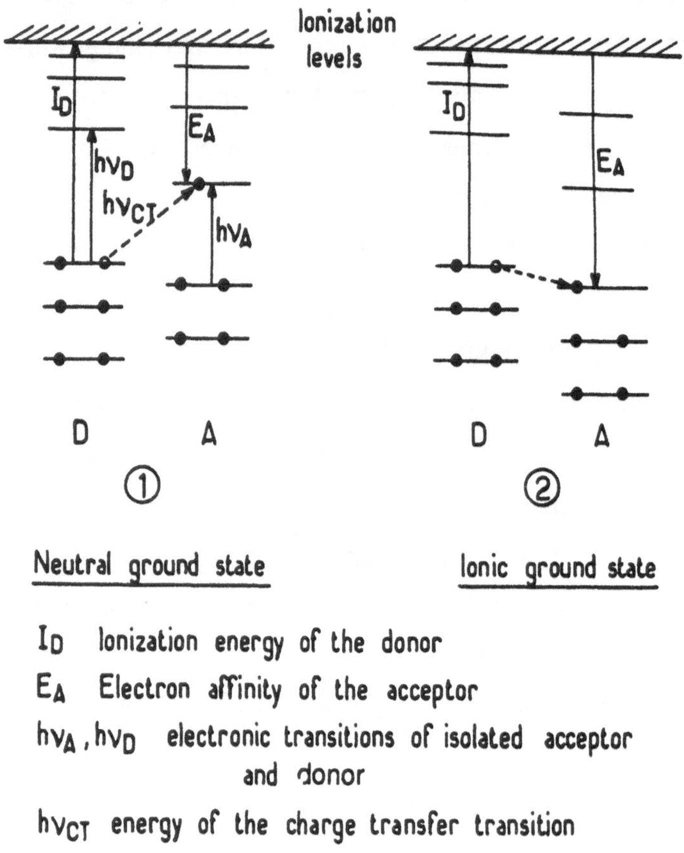

Figure 1. Fundamental electronic transitions in charge transfer complexes (CTC)

The different types of electronic excitations have been already reviewed. For an ionic ground state, they are intermolecular C.T. existing at a lower energy than any intramolecular exciton [3]. In the case of a neutral ground state, one possibility is to induce a photo-excited C.T. in absence of any chemical reaction [6].

An alternative way which has also been developped is to link the donor and acceptor moieties using a covalent bridge which is either σ (D-σ-A) or π (D-π-A) type. In such case, the electron transfer is thought to occur through bond rather than through space and it is called intramolecular C.T. The tailoring of such molecules has been used for non-linear electrical applications and optical properties. Following these tracks, we can quote the use of D-σ-A compounds as organic rectifiers : a molecular diode can be realized inside a strong and ordered monolayer [7]. Non-linear optical properties (second and third order electrical susceptibilities) are also interesting in D-π-A type molecules with in particular the so-called push-pull effect [8].

Nevertheless, we are mainly interested by intermolecular electron transfers which involve supramolecular phenomena starting from the dimer model. We will examine this resulting effect in section 2.3 where we will define the criteria to get π mobile electrons inside an organized phase.

2.2. PROTON AND ELECTRON TRANSFERS

The intrinsic molecular characteristics as the size, symmetry and aromatic character are fundamental to determine a strong donor or acceptor character. As defined on figure 1, the ionization energy as well as the electron affinity can be determined in the gas phase. It has been shown however that the relative donor or acceptor abilities may also be influenced by their molecular environment and intermolecular interactions. The cyclic voltammetry shows fully reversible oxido-reduction waves which are strongly solvent dependent trough its dielectric constant in aprotic solvants. An extrapolation to the zero solvation limit reproduces the gas phase results [9].

Other solution parameters are the acidity or basicity activity and the ionic strength. It has been observed on quinones and other molecules that the redox and dissociation properties are similar because of changing of the charge distribution in the surrounding molecular system [10]. One outstanding example concerns TCNQ [11]. As shown on figure 2, the interplay of electron and proton transfer is demonstrated for redox potentials occuring at different pH [12]. It must be noticed however that it is quite difficult to protonate TTF molecule, excepted with hard acids [13].

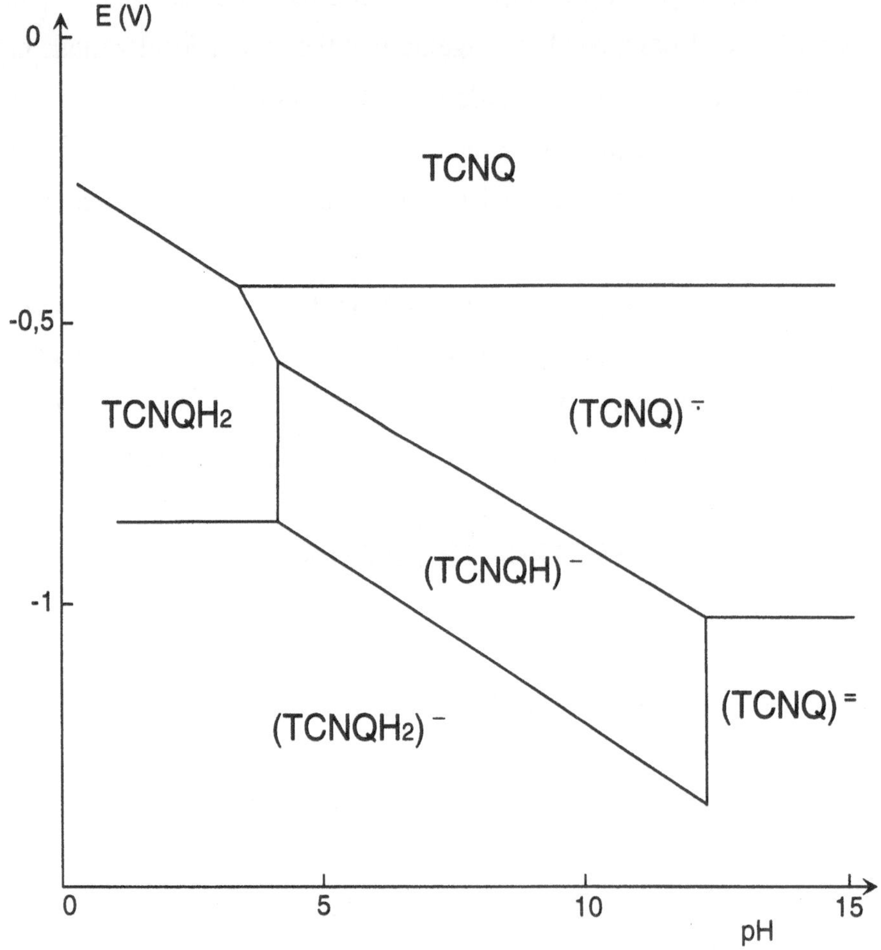

Figure 2. Electrochemical diagramme of TCNQ showing reduction potential in acetoniirile versus pH (from ref. 11)

Several approaches have been realized to combine proton tunneling with electron transfer in H-bonded C.T.C., then to control them by some external pertubation [10] [14]. These characteristics can be associated with the fact that H bondings are effective for molecular recognition.

2.3. MOLECULAR ASSOCIATIONS AND AGGREGATES

The intermolecular electron transfers induce the formation in solution of bimolecular associations which can be either heterodimers (AD) or homodimers, A_2 or D_2 type [1]. Indeed, it has been demonstrated the occurence of completely ionized dimers in presence of reducing or oxidizing agents for TCNQ [15] as well for TTF [16] molecules respectively. The formation of conducting C.T. salts which requires the presence of segregated stacks or layers of mixed valence systems is a consequence of these molecular associations. Only dimers have been clearly evidenced even if after formation of the solid phase more complex aggregates are observed (trimers, tetramers, pentamers...) [4]. It must be noted that in other systems, as phtalocyanines for example, the self-aggregating process is much more developped. In any case, this basic situation due in general to "π–π" interactions will be in favour of a supramolecular organization in soft media (known as J or H aggregates in L-B films).

3. Conducting and magnetic L-B films

The comparison drawn between conducting C.T. crystalline salts and L-B films has evidenced that the Langmuir technique is a new way to arrange these electroactive molecules [5] and modulate therefore their physical properties. In the next part, we will analyse the design of new multifunctional molecular blocks (i.e. electroactive, amphiphilic eventually mesogenic) as presented on tables 1 and 2, then their interfacial properties.

Label and reference	Molecular structure
TCNQ Cn (10 ≤ n ≤ 22) [22]	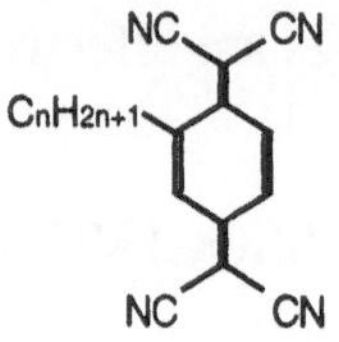
Methanofullerenes :	
[23]	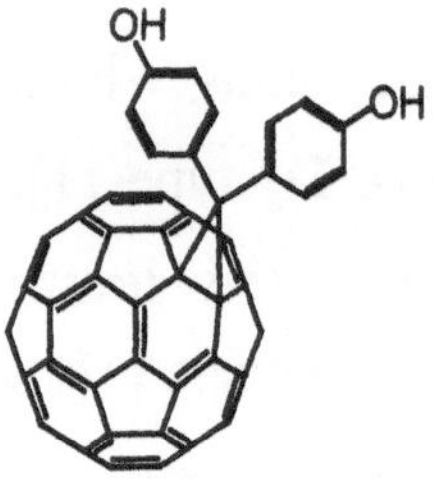
[24]	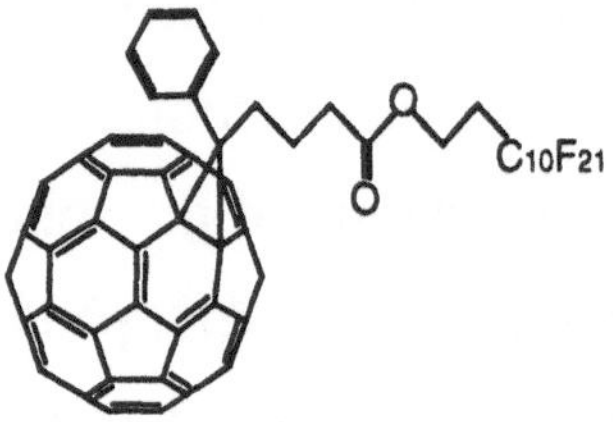
[25]	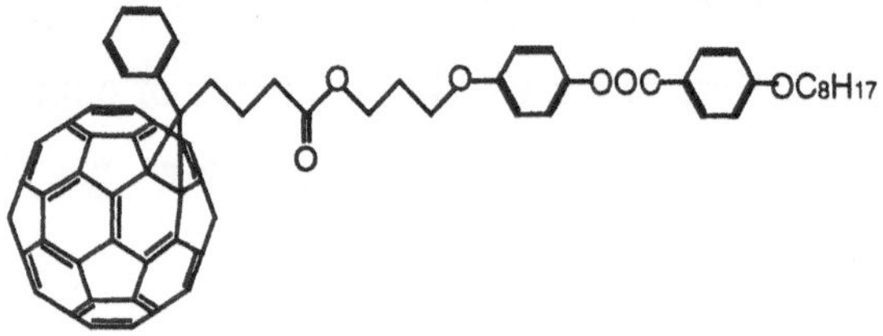

Table 1 : Selected amphiphilic π donors

Label and reference	Molecular structure
Ionic TTF derivative [36]	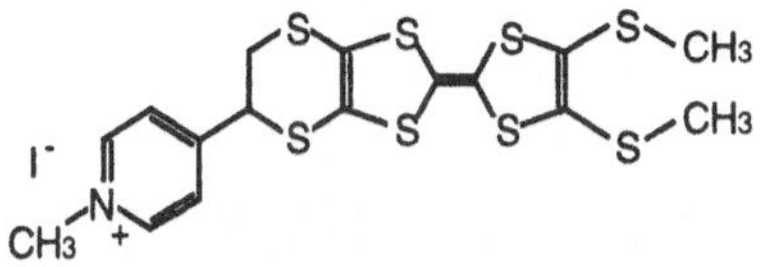
Semi-fluorinated TTF derivatives [27]	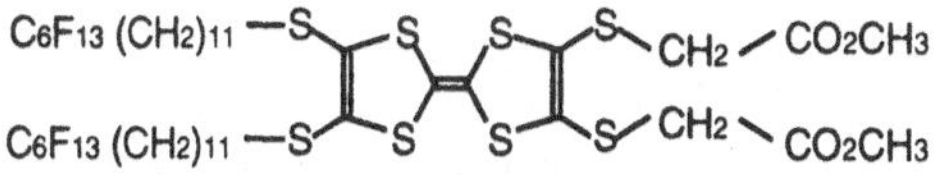 $C_6F_{13}(CH_2)_{11}$—S... CH_2—CO_2CH_3
[24]	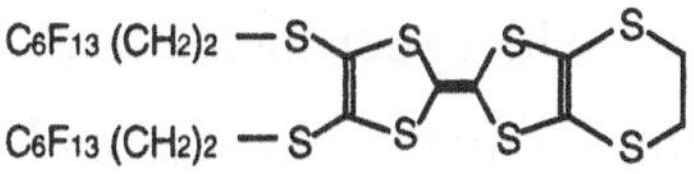 $C_6F_{13}(CH_2)_2$—S...
G-TTF C$_n$ [28]	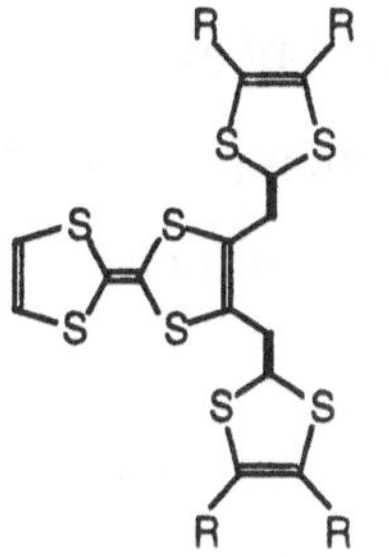 R = COOC$_n$H$_{2n+1}$ (n = 1, 2, 4, 8)
bis amino TTF [29]	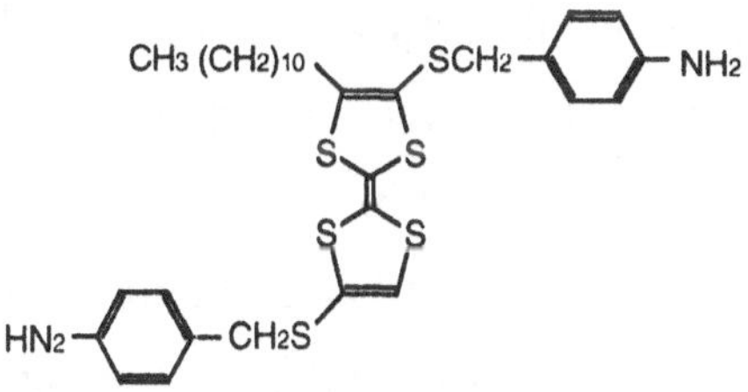 $CH_3(CH_2)_{10}$... SCH_2—NH$_2$

Table 2 : Selected amphiphilic π acceptors based on TTF

Then we will give selected examples on physical properties obtained from L-B films : ESR, optical properties and electrical conductivity. Our starting point for this analysis will be based on recent review papers on this subject [5] [17].

3.1. GENERAL STRATEGY AND DESIGN OF MOLECULAR BLOCKS

The requirements for the formation and deposition of good quality monolayers have been developped for a long time as presented in several monographs. The experiments have moved from classical fatty acids to more complex active molecules and even to different kinds of polymers. With the increasing variety of molecules studied, the influence of experimental constraints but also the sophistication of the characterization methods have to be deepen [18]. Without any technical details relative to the Langmuir-Blodgett technique, some points are relevant to the present purpose :

i) For the film characterisation at the air-water interface, it turns out that the gaseous phase might not exist even for a low density of many amphiphile molecules. As soon as these molecules are interacting together, cluster or islands are formed. Molecular aggregation processes are effective as self assembly formed in a two-dimensional system. This possibility to form small surface micelles in monolayers has been thermodynamically described by Israelachvili [19] but the experimental evidences are rather scarce.

ii) In principle, only the monolayers in a liquid like or solid phase can be used for the fabrication of L-B films. It is necessary to prepare defect-diminished monolayers before transfering at a selected surface pressure. The substrate acting as an external field can largely influence the anchoring effect and the surface induced orientation. One typical example where the cross sectional mismatch between head and tail changes the molecular packing, has been published recently [20]. It appears therefore also necessary to develop the methods of characterising multilayers with in particular a convincing demonstration that an ordered lamellar structure has been achieved.

To obtain a final mixed valence state different ways have been developped :

- One possibility is to work with the polar amphiphilic compounds to look at their chemical and physical stabilities on the water subphase, then oxidize (or reduce) them by a post-treatment after transfer onto a substrate : this is called the heterodoping process.

- At the opposite it is possible to start with an ionic amphiphilic association (CTC or RIS) in a defined mixed valence state which will be supposed to be stable during all the manipulations.

- Another way is the so-called homodoping process initially proposed by Saclay's group [21]. Indeed, a mixture of the neutral electroactive molecule with its ionized form spread at the air-water interface leads to monolayers then multilayers in which the doping is controlled by the ratio between oxidized or reduced molecules and neutral ones. Moreover, recent approaches have shown that this hybrid way is promising through the control of in situ chemical manipulations.

Among the recent works, we have selected two lists of acceptor and donor molecules which are representative of the different classes of compounds. They have been also chosen for their ability to furnish stable pure or mixed monolayers (tables 1 and 2).

A general comment about these acceptor molecules is that excepted the already known TCNQ alkyl chain molecules the new methanofullerenes derivatives [23-25] present specific characteristics. It turns out that it has been very difficult to obtain stable monolayers with these compounds because they form strong 3d aggregates which are necessary to weaken by using different strategies as strong polar groups [23], semi-perfluorated alkyl chains [24], or grafted with a mesogenic group inside a mixed monolayer [25]. Moreover, these methanofullerenes as pristine fullerene are not very good acceptors and only photoinduced electron transfers are expected.

In the case of π donor TTF-derivatives (table 2) several examples have been successfull as an ionic compound [26] or polar molecules functionalized as well as extended TTF compounds [24,27-29] have given rise to stable monolayers on the water surface. It is worth to mention that the presence of perfluoroalkyl chains has improved the monolayer quality before transfer onto a substrate [27][28].

3.2. PHYSICAL PROPERTIES OF L-B FILMS

As already indicated we will describe a few selected examples with some emphasis on the electric, spectroscopic and magnetic (ESR) properties.

- Hetero-doping process on amphiphilic TTF derivatives.

A polar amphiphile molecule is spread on the water surface then transferred onto a substrate. A chemical post-treatment, usually iodine vapours, oxidize the TTF heterocycle creating a full ionic system which is converted into a stable mixed valence state by a controlled annealing treatment [5]. Two kinds of behavior have been observed. On one hand, the mixed valence clusters are stabilized because of a structural re-organization induced by the formation of triiodide chains. In such a situation, a stable high room temperature d.c. conductivity (up to 1 S cm^{-1}) is measured but associated with some structural disorder [30]. On the other hand, with semifluorinated chains grafted on TTF good quality monolayers and high transfer ratio are obtained, however reversible iodination-deiodination cycles are detected [27].

To combine both aspects we have recently worked on giant analogues of TTF (G-$TTFC_n$ see table 2) where a dihydro TTF core bears two conjugated side-arms with various alkyl chains [28]. The interest is due to their increased ability to give electrons together with the reduction of the on-site coulombic interactions because of the increased molecular size. Moreover, their quasi planar shape could induce a molecular stacking as known for discotic molecules.

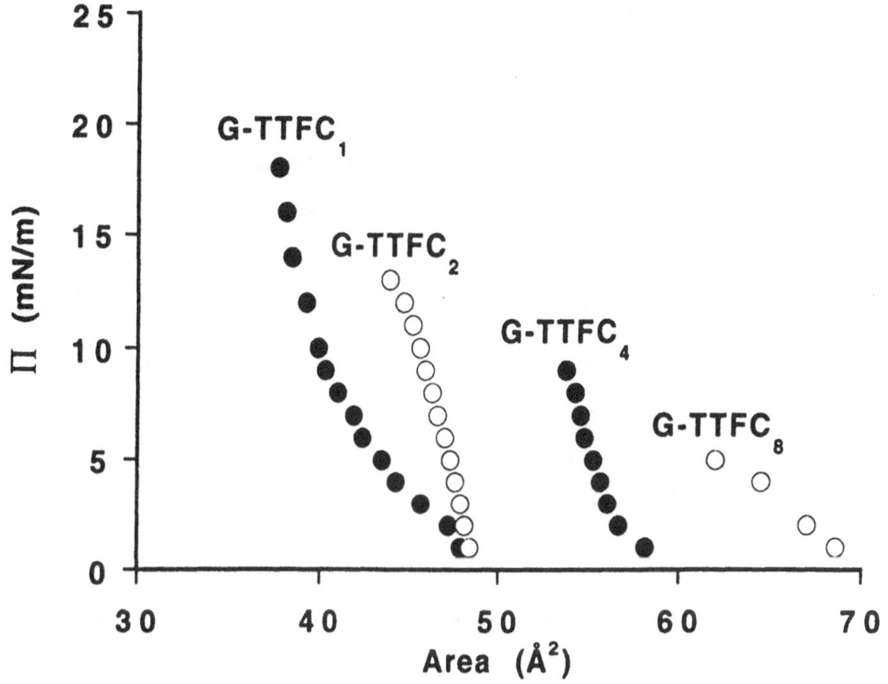

Figure 3. Compression isotherms of G-TTFC_n molecules at 20°C on pure water (pH ≃ 5.5) in a Langmuir trough.

Figure 3 shows the compression isotherms of the different G-TTFC_n compounds. We clearly observed that the increase of alkyl chain length induced a shift of the isotherms towards higher molecular area correlated with a decrease of the collapse pressure. At the opposite of the usual findings the alkyl chains in this series of compounds did not stabilize the monolayers. Indeed they seemed to increase the average distance between the molecules which stand perpendicular to the interface instead of favorizing a core stacking.

After transfer onto a substrate, Y type L-B films are obtained for G-$TTFC_2$ and G-$TTFC_4$ neutral molecules. Their oxidation by iodine vapour furnishes a spectacular colour change from pink to brown; then after a few days, the L-B films turned progressively to a green phase which is a semi-conductor ($\sigma_{RT} \simeq 10^{-3} - 10^{-4}$ Scm^{-1}).

Both the electronic absorption and ESR spectra have shown that the brown phase is completely ionized and ESR silent indicating perhaps the presence of dicationic state. Then the green phase is a mixed valence state which can be destroyed by a further heating. Indeed a iodination-deiodination cycle with presumably three different oxidation states has been evidenced which could be applied for a chemical sensor device.

- Charge transfer complexes as ionic amphiphilic associations.

One other way as already described [5] is to start with mixed valence compound which will be stable in the spreading solution then on the Langmuir trough. Among all the recent attempts one case has conducted to a highly conducting L-B film based on semi-amphiphilic CTC BEDO-$TCNQC_{10}$ (BEDO is bis(ethylenedioxy)tetrathiafulvalene (see figure 4) [31].

It has been shown that a powdered CTC BEDO : $TCNQC_{10}$: H_2O of stoichieometry (10 : 4 : 1) gives a stable monolayer in presence of a fatty acid with a valuable in situ d.c. conductivity $\sigma_{RT} \sim 0.6$ S cm^{-1}. After transfer onto a substrate and deposition of 20 layers d.c. conductivity values up to 10 S cm^{-1} are measured. Around room temperature a metallic behavior is observed in agreement with the Hall effect determination [32]. This result has been confirmed by the evidence of conduction electron spin resonance. The temperature dependence of the ESR signal has shown the presence of a constant Pauli type paramagnetism which is characteristic of the presence of metallic domains (figure 4) [33].

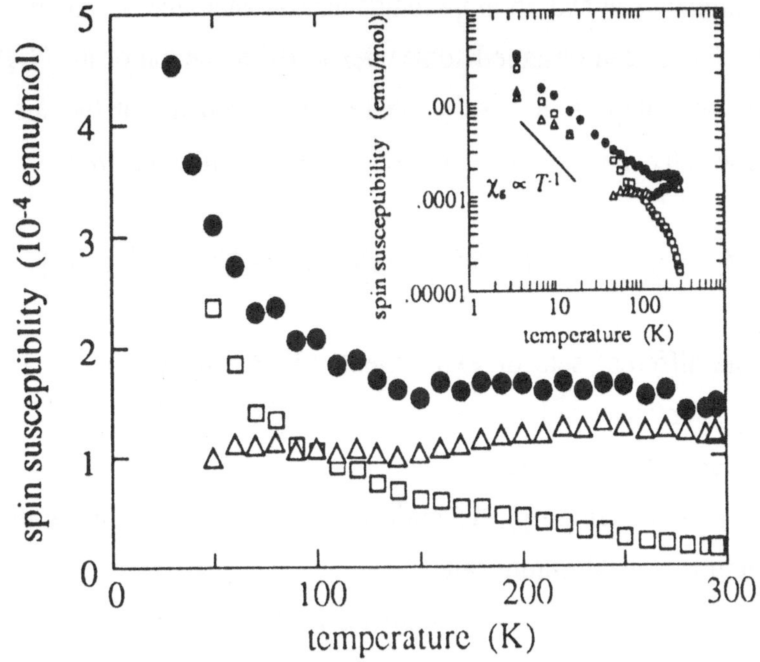

Figure 4. Spin susceptibility temperature dependence of $(BEDO)_{0.4}$ $(TCNQC_{18})_1$ L-B film (•). The decomposition has been done considering the presence of two kinds of species, one following a Curie law (▫) and the other a pauli term (Δ) (from ref. 33).

Finally, it must be noticed that the transfer technique is not the standard one, i.e. by vertical lifting, but the horizontal process. Using this touching technique the aggregates or crystallites with about four surimposed layers are transferred as shown by AFM images on L-B films [34]. Besides, because of the use of a fatty acid in these mixed monolayers a percolation type problem is present as already demonstrated in a previous case [30]. This is nevertheless an interesting approach to pursue for getting a metallic monolayer.

- Homodoping process and in situ chemical manipulations.

As already cited [21], this approach allows us to obtain mixed valence clusters thanks to a controlled ratio between a fully ionic semiamphiphilic salt and an amphiphilic neutral analogue based on TCNQ molecule. It has been observed some unexpected reduction of the C_nTCNQ on stable monolayers, an exchange reaction occuring at the air water interface which can be formally written under the following way:

$$[M^{n+}(C^-)_n] + (TCNQCn^\circ)_m \rightarrow (M^{n+}) :(TCNQCn)_m^{-n} + \{C\}_n \downarrow$$

The different sets of experiments [35-38] have lead to conclude about the spontaneous charge exchange of C_nTCNQ with another anion under certain circumstances:

*M is an amphiphilic cation as a quaternary ammonium, a phosphonium or a sulfonium salt.

*C^- is a water partially soluble inorganic (i.e. I^-, BF_4^-...) or organic anion ($TCNQ^-$, $TCNQF_4^-$) which is recombined in water under an unspecified form.

Two series of experiments have been carried out, in-situ investigations at the gas-water interface and on L-B films after transfer onto a substrate. Different mixtures of all these salts in solution with the neutral octadecyl TCNQ derivative have been prepared (characterized by n : m ratio). We will report in situ manipulation relative to 3-methylthio, 4-5 bis (octadecylthio) dithiolium (or 1-2 dithiolium) TCNQ salt [36]. Thanks to polarization-modulated IR reflection-absorption spectroscopy on surfaces (PMIRRAS) we have carried in situ experiments on the water surface (H_2O or D_2O) as shown on figure 5.

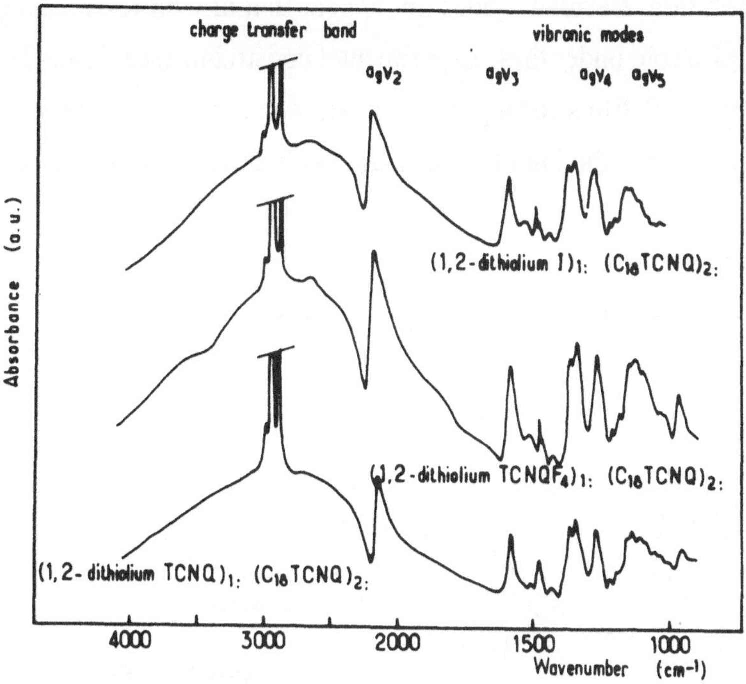

Figure 5. IR absorption spectra of (1-2 dithiolium-R)$_1$:(C18TCNQ)$_2$ LB films (R=I, $TCNQF_4$, TCNQ) made of 50 monolayers (note the change of the abscissa scale at 2000 cm^{-1}).

These spectra show clearly two features:

i) A broad and intense dispersive band located around 3400 cm^{-1} and 2600 cm^{-1} for H_2O and D_2O subphases respectively. This dispersion is mostly due to bound water molecules associated with the ionic polar heads.

ii) A large intermolecular CT band is observed below 3000 cm^{-1} associated with vibronic modes of $(TCNQC_{18})_2^-$ dimers [38].

These findings demonstrate that there is a spontaneous formation of mixed valence clusters at the air-water interface which are not present in the organic solution. Besides it does not appear that any proton exchange on TCNQ play a role under these experimental constraints (see figure 2). After transfer, the L-B films for a given stoichiometry (i.e. 1:2) offer a similar feature whatever is the initial counter-anion C^- as shown on figure 6.

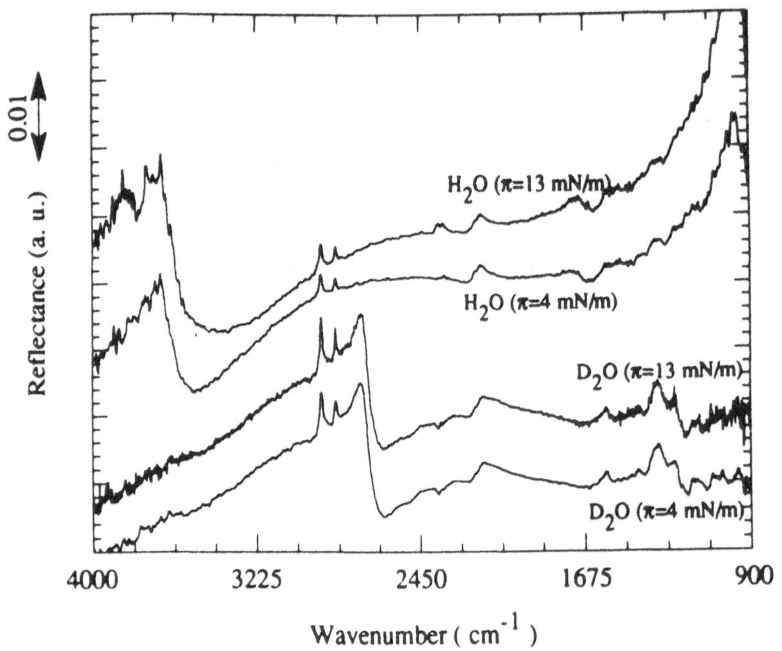

Figure 6. IR surface reflectance spectra obtained on a monolayer of (1-2 dithiolium TCNQ)$_1$:(C_{18}TCNQ)$_2$ floating on H_2O and D_2O subphases at two different surface pressures (pH = 5.5).

As for conducting cristalline TCNQ salts [4] a large CT absorption band around 3000 cm^{-1} is observed, accompanied by a series of vibronic modes due to the coupling between the π-electron and the molecular vibrations [5].

This is the signature for the presence of mixed valence clusters which should give rise to conducting layers. It has been found that the d.c. electrical conductivity is function of the number of deposited layers (N), reaching a constant value for $N \geq 20$ where $\sigma_{RT} \sim 10^{-2}$ S cm^{-1} exhibiting a semi-conducting behavior characteristic of some hopping process. The ESR spectra show two interesting points: the temperature dependence of the spin susceptibility obeys essentially to a Curie law and the linewidth anisotropy exhibits a rotation diagramme versus magnetic field with a minimum value at the so-called magic angle [36]. This behavior is characteristic of spin-spin dipolar interactions in low dimensional systems for localized spins.

The conclusion drawn for these studies is that spins as well as electronic charges are more or less localized in these L-B films because of the absence of a in-plane long range organisation and an effective delocalization of the π electrons [39].

These results open the way to the in-situ control of charge transfers through the understanding of the mechanisms governing the homodoping process. As already pointed out by Mobius and coll. [40] there are strong differences between electron transfers inside a homogeneous solution and a monolayer. In particular the following factors have to be considered :

- geometrical correlations between D and A molecules (orientational effect),
- charge and energy delocalizations and transfers,
- interfacial phenomena as local redox potential and local pH.

These last effects have been already investigated in particular the interfacial perturbations on polarity and pH [41] and the effect on surface potential in presence of charged amphiphilic molecules [42]. A complementary approach is to use in situ cyclic voltammetry on a monolayer [43] or direct chemical oxido-reduction by dissolution of compounds in the water subphase. It appears finally that specific or induced in situ redox processes are occuring at the gas-water interface which have to be controlled to obtain a final mixed valence state.

4. Concluding remarks

In the review we have shown that thin organized molecular films can be realized thanks to the Langmuir-Blodgett technique. The purpose is to control the molecular organization of electroactive and amphiphilic molecules for revealing specific physical properties.

We have evidenced that for obtaining a mixed valence system which will give rise to conducting monolayers it is necessary to define at which step it will be useful to shift from a polar amphiphile to a ionized one through a controlled redox reaction. It appears that several ways are possible but the most promising one seems to be the homodoping technique.

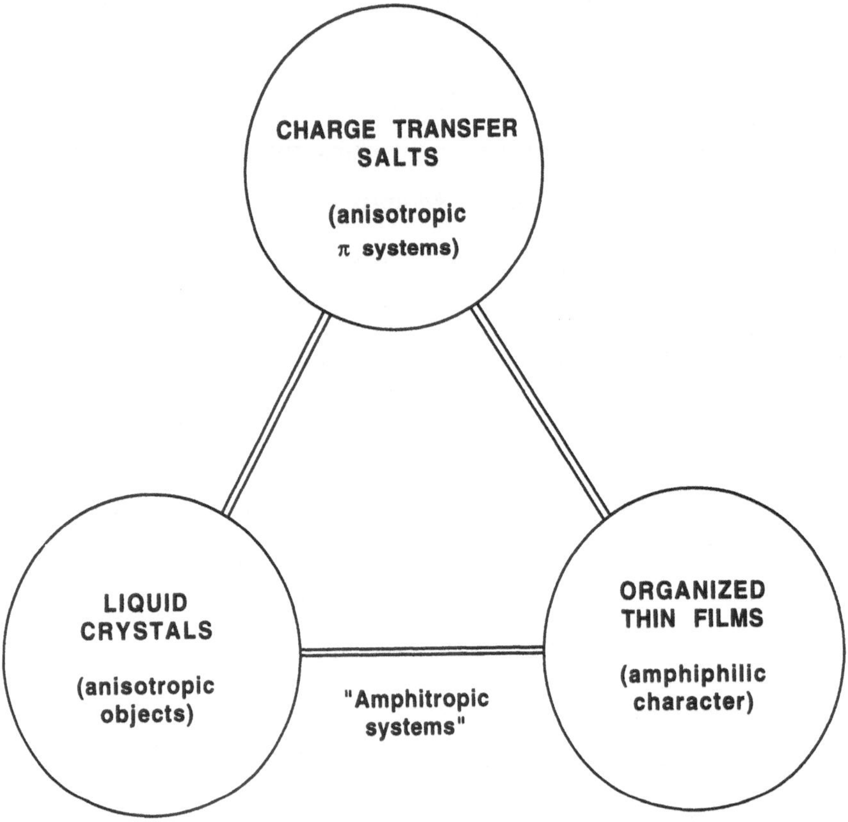

Figure 7. Sketch of the multifunctional molecular characteristics.

Nevertheless in the future it will be necessary to control the in-plane molecular organization. The existence of a better long range organization associated with the presence of aggregates inside the floating layer is a natural way to improve the electronic and magnetic properties of these molecular assemblies. It appears recently that an orientational in plane arrangement can be realized by controlling the flow orientation of aggregates or mesogenic compounds during the transfer process [44].

Indeed, playing with multifunctional molecules, i.e. presenting an electroactive, an amphiphilic and a mesogenic character as presented on figure 7, a multilayered constrained system will present a strong anisotropy of its physical properties [39]. This is a way to realize in the future new sophisticated devices based on these magnetic or electronic characteristics at a supramolecular level.

References

1. Forster, R., (Academic Press, 1969) Organic charge transfer complexes, *Organic Chemistry* **15**.
2. Szent-Gïorgy, A., (Academic Press, 1960) Introduction to a submolecular biology.
3. Soos, Z.G., (1974) *Ann. Rev. Phys. Chem.* **25**, 121-180.
4. Delhaès, P., (1990) Lower dimensional systems and molecular electronics, *Nato Asi series* **248**, 43-66 (Ed. by R.M. Metzger, R.M. Day, and G.C. Papavassiliou).
5. Delhaès, P. and Yartsev, V.M., (John Wiley and Sons, 1993) Spectroscopy of new materials, Ed. by Clark R.J.H. and Hester R.E. 22, chapter 5, 199-289.
6. Malak, P., (Springer-Verlag Berlin-Heidelberg, 1993) Topics in current chemistry 168, 1-43
7. Metzger, R.M. and Panetta, C.A., (1991) *New J. Chem.* **15**, 209-

238.

8. Denning, R.G., (John Wiley and sons 1993) *Spectroscopy of new materials* **22**, 1-60, (Ed. by R.J.H. Clark and R.E. Hester).

9. Wudl, F., Allemand, P.M., Delhaès, P., Soos, Z. and Hinkelmann, K. (1989) *Mol. Cryst. Liq. Cryst.* **171**, 179-192.

10. Akutagawa, T. and Saito, G., (1995) *Bull. Chem. Soc. Jpn* **68**, 1753-1773.

11. Buvet, R., Dupuis, P., Neel, J. and Perichon, J. (1969) *Bull. Soc. Chim. Fr.* **11**, 3991-3946.

12. Inzelt, G. and Chambers, J.Q. (1989) *J. Electroanal. Chem.* **266**, 265-271.

13. Giffard, M., Alonso, P., Garin, J., Gorgues, A., NGuyen, T.P., Richomme, P., Robert, A., Roncali, J. and Uriel, S. (1994) *Advanced Materials* **6**, 298-300.

14. Mitani, T. (1989) *Mol. Cryst. Liq. Cryst.* **171**, 343-355.

15. Boyd, R.H. and Phillips, W.D. (1965), *J. Chem. Phys.* **43**, 2927-2930.

16. Torrance, J.B., Scott, B.A., Welber, B., Kaufman, F.B. and Seiden, P.E. (1979) *Phys. Rev. B* **19**, 730-741.

17. Bryce, M.J. and Petty, M.C. (1995) *Nature* **374**, 771-776.

18. Tregold, R.H. (1995) *J. Mater. Chem.* **5**, 1095-1106.

19. Israelachvili, J. (1994) *Langmuir* **10**, 3774-3781.

20. Garnaes, J., Larsen, N.B., Bjornholm, T., Jorgensen, M., Kjaer, K., Als-Nielsen, J., Jorgensen, J.F. and Zasadzinski, J.A. (1994) *Science* **264**, 1301-1304.

21. Ruaudel-Teixier, A., Vandevyver, M., Roulliay, M., Bourgoin, J.P., Barraud, A., Lequan, M., Lequan, A.M. (1990) *J. Phys. D : Appl. Phys.* **23**, 987-990.

22. Kubota, M., Ozaki, Y., Araki, T., Ohki, S. and Iriyama, K. (1991) *Langmuir* **7**, 774-7

23. Ravaine, S., Le Pecq, F., Mingotaud, C., Delhaès, P., Hummelen,

J.C., Wudl, F. and Patterson, L.K. (1995) *J. Phys. Chem.* **99**, 9551-9557.

24. Ravaine, S., Agricole, B., Mingotaud, C., Cousseau, J. and Delhaès, P. (1995) Chem. *Phys. Letters* **242**, 478-482.

25. Ravaine, S., Vicentini, F., Mauzac, M. and Delhaès, P. (1995) *New. J. Chem.* **29**, 1-3.

26. Goldenberg, L.M., Becker, J.Y., Paz-Tal Levi, O., Khodurkovsky, V.Yu., Bryce, M.R. and Petty, M.C. (1995) *J. Chem. Soc. Chem. Commun.* 475-477.

27. Dupart, E., Agricole, B., Ravaine, S., Mingotaud, C., Fichet, C., Delhaès, P., Ohnuki, H., Munger, G. and Leblanc, R.M. (1994) *Thin Solid Films* **243**, 575-580.

28. Morrisson, V., Mingotaud, C., Agricole, B., Salle, M., Gorgues, A., Garrigou-Lagrange, C. and Delhaès, P. (1995) *J. Mater. Chem.* **5**, 1617-1624.

29. Le Hoerff, T., Lhorcy, D., Floner, D., Moinet, C., Bittner, S., Mingotaud, C., Delhaès, P. and Robert, A. (1995) *J. Mater. Chem.* **5**, 1589-1592.

30. Dourthe, C., Izumi, M., Garrigou-Lagrange, C., Buffeteau, T., Desbat, B. and Delhaès, P. (1992) *J. Phys. Chem.* **96**, 2812-2820.

31. Isotalo, H., Yunome, G., Abe, M., Horiuchi, S., Yamochi, M., Saito, G., Tachibana, H., Nakamura, T. and Matsumoto, M. (1994) *J. Chem. Soc. Chem. Comm.* 573-574.

32. Takenaga, M., Aboulla, A., Kasai, A., Nakamura, A., Nakamura, T., Matsumoto, M., Horiuchi, S., Yamochi, H. and Saito, G. (1994) *Appl. Phys. Lett.* **64**, 2602-2604.

33. Ikegami, K., Kuroda, S., Nakamura, T., Yunome, G., Matsumoto, M., Horiuchi, S., Yamochi, H. and Saito, G. (1994) *Phys. Rev. B* **49**, 10806-10811

34. Nakamura, T., Yunome, G., Azumi, R., Tanaka, M., Yumura, M., Matsumoto, M., Moriuchi, S., Yamochi, H. and Saito, G. (1993)

Synthetic metals 55-57, 3853-3855.

35. Ahuja, R.C., Matsumoto, M. and Möbius, D. (1992) *J. Phys. Chem.* **96**, 1855-1860.

36. Fichet, O., Agricole, B., Amiell, J., Gionis, V. and Delhaès, P. (1993) *C.R. Acad. Sci. Paris* **316** (II), 335-340.

37. Perez, J., Bourgoin, J.P., Barisone, C., Vandevyver, M. and Barraud, A. (1994) *Thin Solid Films* **244**, 1043-1049.

38. Fichet, O., Agricole, B., Kassi, H., Desbat, B., Gionis, V., Leblanc, R.M., Garrigou-Lagrange, C. and Delhaès, P. (1994) *Thin Solid Films* **243**, 592-595.

39. Delhaes, P. (1996) in Nato Asi School on "Localized and itinerant molecular magnetism from molecular assemblies to the devices" (Kluwer ed. to appear).

40. Möbius, D., Ahuja, R.C., Caminati, G., Feng Chi, L., Cordroch, W., Li, Z.M. and Matsumoto, M. (1992) Dynamics and mechanism of photoinduced transfer and related phenomena, Elsevier Sc. Pub., 377-393.

41. Mingotaud, C., Chauvet, J.P., Patterson, L.K. (1994) *Thin Solid Films* **242**, 243-248.

42. Palacin, S. and Barraud, A. (1991) *Colloids and Surfaces* **52**, 123-147.

43. Goldenberg, L.M., Cooke G., Pearson, C., Monkman, A.P., Bryce, M.R. and Petty, M.C. (1994) *Thin Solid Films* **238**, 280-284.

44. Mingotaud, C., Agricole, B., Jego, C. (1995) *J. Phys. Chem.* **99**, 17068-17070.

THERMODYNAMIC AND KINETIC STABILITY OF POLYNUCLEAR COMPLEXES

A.F. WILLIAMS, R.F. CARINA, L. CHARBONNIÈRE, P. G. DESMARTIN† AND C. PIGUET
Department of Inorganic Chemistry
University of Geneva
30 quai Ernest Ansermet
CH 1211 Genève 4, Switzerland

ABSTRACT. Self-assembly of polynuclear metal complexes generally takes place under conditions of thermodynamic control, and the resulting polynuclear complexes are often extremely labile. The investigation of thermodynamic and kinetic stability of the complexes can give information on the relative importance of different structural features. The linking of a ligand to more than one metal centre can greatly reduce its lability and confer kinetic stability on the resulting complex. Two examples are given of the manipulation of thermodynamic and kinetic stability in the synthesis of polynuclear complexes.

1. Introduction

One of the most atttractive aspects of supramolecular chemistry is the development of new and efficient routes to complex chemical structures which are frequently of great aesthetic appeal. These methods frequently use metal ions to bind smaller molecular units together in a stereospecific way *via* a self-assembly reaction. Many metal ions show clear stereochemical preferences for their coordination sphere, and these preferences, allied with suitably structured ligands, allow the stereochemical control of the self-assembly process. The kinetic and thermodynamic factors involved in self-assembly have been less well studied, yet they are just as important as the stereochemical considerations.

$$\text{Components} \rightarrow \text{Self-assembly} \rightarrow \text{Structured Supermolecule}$$

L. Echegoyen and A.E. Kaifer (eds.), Physical Supramolecular Chemistry, 379–392.

While the components contain the structural information in their stereochemical properties, the kinetics and thermodynamics of the interactions involved in self-assembly must allow this information to be expressed efficiently and completely. The reactions discussed in this chapter belong to the category described by Lindsey [1] as strict self-assembly, in which the product of self-assembly is thermodynamically stable. Obviously, since the assembly process involves the formation of several bonds, the kinetics of formation of each bond must be rapid. It is equally true, although perhaps less obvious, that the kinetic barriers for the dissociation of the bonds should be low, so that an error in the self-assembly process may be corrected, and so that the system can range over the potential energy surface in search of the free energy minimum. Since the kinetic barriers must be low in both directions, the free energy change per bond upon self-assembly must equally be fairly low, and for this reason interactions such as hydrogen bonds or coordinate bonds are particularly suitable for self-assembly reactions. The definition of supramolecular chemistry as "the chemistry of the non-covalent bond" [2] has a sound thermodynamic basis.

Since the self-assembly will involve the formation of several bonds between components, the overall free energy change may be quite negative, and the resulting molecule very stable. Nonetheless, the basic lability of each bond makes the system vulnerable to a change in pressure, temperature, or most importantly, the activities of the components, which may result in a shift in the equilibrium position leading either to a "spontaneous self-fragmentation" into the components or to the formation of other species. The development of self-assembly routes to complex supermolecules demands a good understanding of the thermodynamic and kinetic stability of these supramolecular systems.

The predominance of X-ray crystal structure determination as a method of characterisation of supramolecular systems tends to mask the presence of dynamic equilibria. A molecule in a crystal has lost translational and rotational degrees of freedom, and rapid intermolecular rearrangements are frozen out. Although structures determined in the solid state are very frequently maintained in solution, they may be in dynamic equilibrium with partly or wholly dissociated species, and this equilibrium may well be undetectable even by a detailed analysis of the crystal structure.

Our objective in this chapter is to present the thermodynamic and kinetic properties of some simple supramolecular systems assembled around metal ions. We shall endeavour to show the importance of these effects in the self-assembly process, and to show how a careful understanding of them can allow the improvement and the extension of supramolecular synthesis.

2. Double Helical Systems of Copper(I)

Some years ago [3] we reported the synthesis of a simple double helix formed by the assembly of two bis-(2,6-bis(1-methyl)benzimidazo-2-yl)pyridine (**1a**) ligands around two copper(I) ions (Figure 1, Scheme 1)

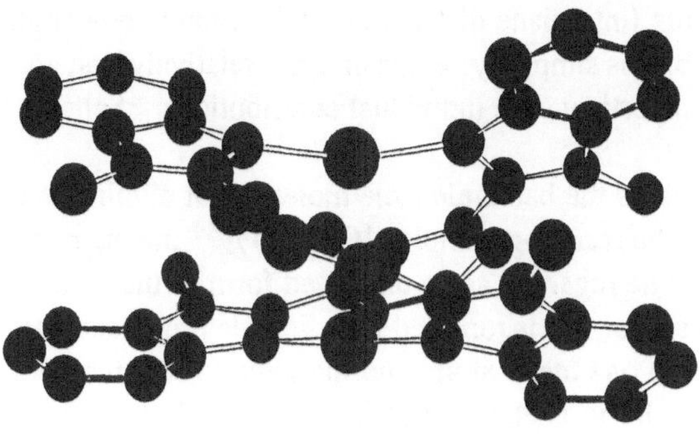

Figure 1. Crystal structure of $[Cu_2(1a)_2]^{2+}$

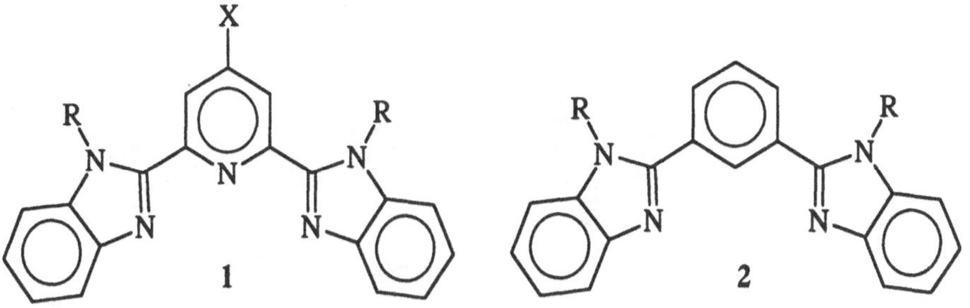

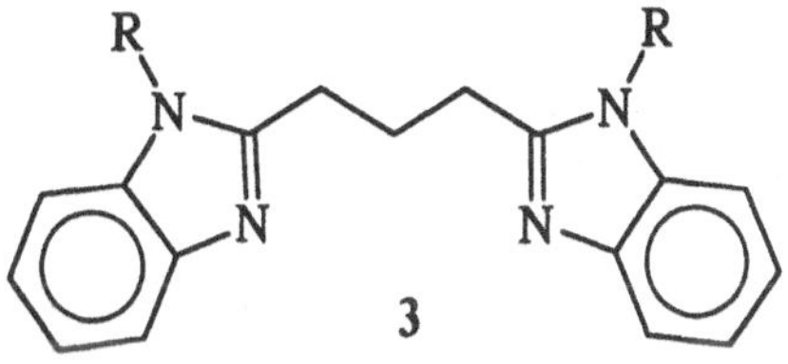

1a R = Me, X = H
1b R = dmb, X = H,
1c R = dmb, X = Ph
1d R = dmb, X = 4-$Me_2NC_6H_4$
2a R = Me
2b R = dmb
2c R =Et
3a R = Me,
4a R = Me
4b R = dmb

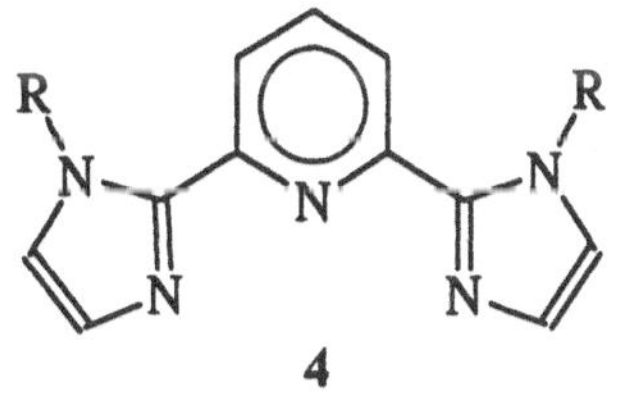

dmb = 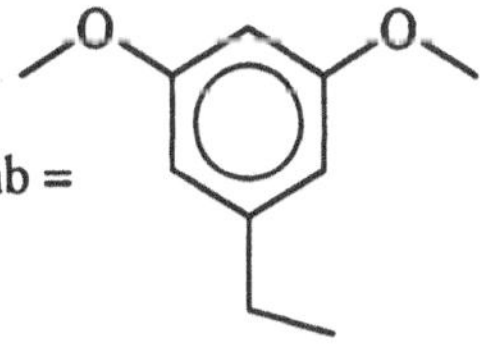

Scheme 1.

Analysis of the structure showed three features which could contribute to the self-assembly: (i) short (average 1.92 Å) Cu-N bonds between the copper ions and the benzimidazole moieties, giving a basically linear coordination of the metal ions; (ii) a much longer bridging interaction with the pyridyl fragments (average Cu-N distance 2.5 Å); (iii) an apparent intramolecular stacking interaction between the superposed benzimidazole units (interplane distance *ca.* 3.4 Å, interplane angles of *ca.* 3°). The attractive feature of this simple system is that it is relatively easy to modify the ligand in such a way as to investigate the individual contributions. Scheme 1 shows the ligands used in this study.

Ligand **2** maintains the benzimidazole moieties but eliminates the possibility of the bridging pyridine: the resulting complex $[Cu_2(\mathbf{2a})_2]^{2+}$ maintains the dinuclear structure (*Figure 2*) but may be regarded as an untwisted form of the parent complex [4]. If the rigid metaphenyl spacer in **2** is replaced by a flexible propylene unit, **3a** is able to act as a bidentate ligand to one metal atom, and the dimeric structure is lost [5]. Finally the ligand **4** keeps the basic chelating possibilities of **1**, but the intramolecular stacking interaction is no longer possible; the structure of $[Cu_2(\mathbf{4a})_2]^{2+}$ is basically the same as $[Cu_2(\mathbf{1a})_2]^{2+}$ although the complex shows intermolecular stacking in the solid state, forming columnar stacks of $[Cu_2(\mathbf{4a})_2]^{2+}$ units of the same chirality [6] (*Figure 2*).

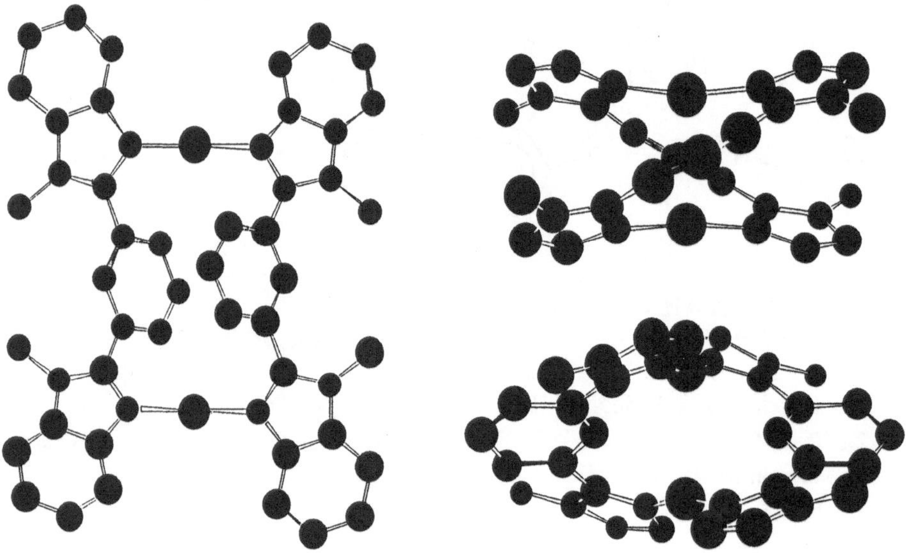

Figure 2. Crystal structures of $[Cu_2(\mathbf{2a})_2]^{2+}$ (left) and $[Cu_2(\mathbf{4a})_2]^{2+}$ (right).

These structural results suggest that the rigid spacer between the benzimidazole units is essential, as is the bridging pyridine, but that the stacking interaction is not indispensable. However, the structural studies are all or nothing observations: either the helical structure is formed or it is not. Studies of thermodynamic stability in solution can give a better idea of the energies involved in the different interactions. Stability constants were measured either by UV-visible spectrophotometry or by potentiometry, and the results are summarised in Table 1.

Table 1. Stability constants and kinetic data for $[Cu_2L_2]^{2+}$

Ligand	$\log_{10}\beta_{298}$ DMF[a]	MeCN[b]	Kinetics of racemisation (DMF) $\Delta G^{\circ\ddagger}_{298}$ kJ/mol	$\Delta H^{\circ\ddagger}$ kJ/mol
1b	14.5(3)	8.6(2)	65.1(4)	80(4)
1c	15.0(1)		68.5(4)	68(4)
1d	16.0(1)		73.9(4)	79(2)
2b		7.7(2)		
4b	12.8(4)	10.0(2)	50.8(4)	126(13)

[a] + 5% MeCN; [b] + 10% DMF

The stability constants in DMF show that substituents at the 4-position of the pyridyl moiety inflence the stabitity of the complex, implying a significant rôle for the bridging interaction of the pyridyl. The complex of ligand **4b**, where the stacking interaction is absent, is slightly less stable than **1b**, as one might expect. If, however we move to a more competitive solvent such as acetonitrile, then the stability constants drop considerably, but the complex of ligand **4b** is now more stable than that of **1b**. Complex $[Cu_2(\mathbf{2b})_2]^{2+}$ is rapidly oxidised in DMF, but in acetonitrile it is sufficiently stable for the stability constant to be measured, and it is found to be less than one log unit smaller than the complex of **1b**.

The use of the 3,5-dimethoxybenzyl substituents at the benzimidazole nitrogens not only increases the solubility of the ligands enough to allow studies in solution, but also gives two diastereotopic benzylic protons in chiral helical complexes. This is useful not only to confirm the helical structure [4], but also, by studying the temperature dependence of the signals, allows us to assess the kinetic stability of the complexes. The free energies of activation for racemisation at 298 K and the calculated enthalpies of activation are given in Table 1. In spite of the thermodynamic stability of these complexes, most are undergoing rapid exchange at or slightly above room temperature. The free energies of activation correlate reasonably well with the stability constants measured in DMF, which suggests that the racemisation is occurring via a dissociative mechanism, and this is confirmed by the observation that the racemisation is faster in the more strongly coordinating solvent acetonitrile, but slower in weakly coordinating nitromethane. The complex of ligand **2b** shows no diastereotopic effect even at low temperatures, implying it to be much more labile. The complex of ligand **4b** is more labile than those of ligands **1b-d**, but the enthalpy of activation is, surprisingly, much higher.

Taken together, the thermodynamic and kinetic results show that even under conditions where the complexation is essentially complete, the complexes are labile and that racemisation occurs on a millisecond timescale. The solvent exerts a strong influence on the thermodynamic and kinetic stability. The disappearance of the helical structure with ligand 2 and the increase in thermodynamic and kinetic stability upon substituting donor groups in the 4-position of the pyridyl moiety show the importance of the pyridyl unit in forming the structure, even though the Cu-bridging-N distances are long. The analysis of the relative importance of the stacking is not simple. In DMF we observe as expected a drop in thermodynamic and kinetic stability on going from1 to **4**, but the complex $[Cu_2(4b)_2]^{2+}$ is actually more stable than $[Cu_2(1b)_2]^{2+}$ in acetonitrile, and the activation energy for racemisation in DMF is also greater. A probable explanation lies in the variation of the Cu-imidazole (or benzimidazole) bond distances. These are appreciably shorter in $[Cu_2(4a)_2]^{2+}$ (1.87 Å) than in $[Cu_2(1a)_2]^{2+}$ (1.92 Å), so that the absence of a stacking interaction is more or less compensated by a stronger Cu-N bond. The Cu-benzimidazole distance in $[Cu_2(2a)_2]^{2+}$ (1.890 Å) is shorter than that in $[Cu_2(1a)_2]^{2+}$ suggesting that the formation of the close stack between benzimidazoles in $[Cu_2(1a)_2]^{2+}$ may lengthen the Cu-N bond; this may equally explain why the difference in stability of the complexes of **1** and **2** in acetonitrile is surprisingly small.

3. Triple helical systems of octahedral metals.

The effect of constraints on metal-ligand bond lengths is clearly shown in the chemistry of the ligands **5** and **6** which possess two well separated bidentate sites, and which adopt a non-planar conformation as a result of repulsion between the hydrogen atoms at the 4 position of the benzimidazole units. With metals favoring octahedral geometry,**5** and **6** form complexes of composition $[M_2L_3]^{2n+}$ with a well-developed triple helical structure (Figure 3), each metal having an octahedral coordination. Such structures have been known for many years, although they were not immediately described as triple helices [7].

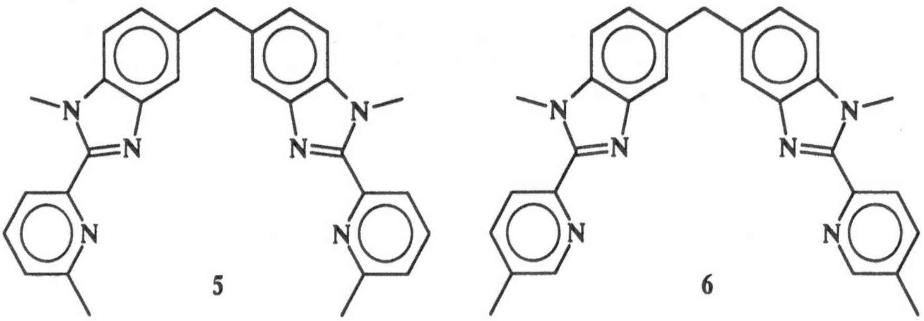

The first reported complex was $[Co_2(5)_3]^{4+}$ [8], which rather surprisingly could not be oxidised to the analogous cobalt(III) complex. Examination of the crystal structure showed that the methyl groups in position 6 of the pyridine units were very close, and suggested a steric interaction, which was confirmed by the considerable difference of the cobalt-pyridine (average 2.29 Å) and the cobalt-benzimidazole (average 2.07 Å) Co-

N distances. The unexpectedly long Co-N(pyridine) distances arise from the repulsions between methyl groups. Cobalt(III) complexes typically show Co-N(pyridine) distances 0.19 Å shorter than cobalt(II) complexes, so that if the Co-N distance is kept unusually long by the steric repulsion, oxidation is strongly disfavored.

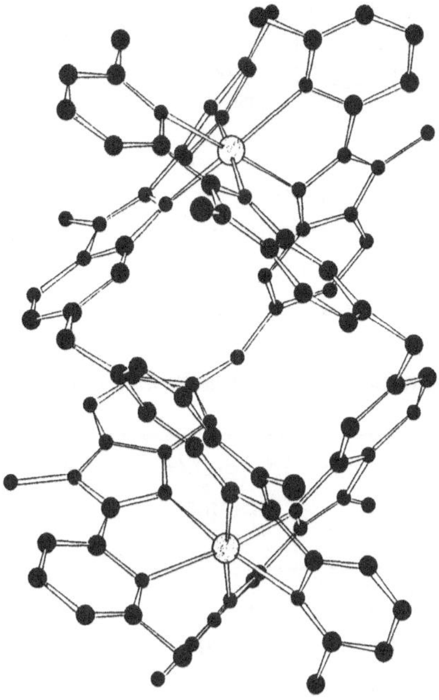

Figure 3. Crystal structure of $[Co_2(5)_3]^{4+}$

Replacement of ligand **5** by ligand **6** (in which the methyl groups were shifted to the 5-position of the pyridine) gave $[Co_2(6)_3]^{4+}$ for which the average cobalt-N(pyridine) distance was reduced to 2.17 Å, thereby confirming the steric effect in the 6-methyl complex [9]. The cobalt(II) complex could now be oxidised cleanly to $[Co_2(6)_3]^{6+}$. The thermodynamic stability of $[Co_2(6)_3]^{4+}$ is so great that complexation is essentially complete at concentrations of 10^{-5} M, and spectroscopy could only give a lower estimate of the stability constant for complex formation which appears to be at least three orders of magnitude greater than that for formation of $[Co_2(5)_3]^{4+}$. The steric effect was also observed in the iron(II) complexes $[Fe_2(5,6)_3]^{4+}$, where the longer bond lengths in the sterically hindered complex of **5** give a yellow high spin complex, while **6** gives a dark red low spin complex [10].

The availability of a kinetically stable cobalt(III) triple helix allowed us to effect the resolution of the two enantiomers [11] and measure the circular dichroism spectrum. Reduction of enantiomerically pure $[Co_2(6)_3]^{6+}$ gave enantiomerically pure $[Co_2(6)_3]^{4+}$ whose racemisation kinetics were studied to estimate the kinetic stability of the triple helix. The kinetics followed the expected first-order rate law, but were much slower than normally expected for the labile cobalt(II) ion, with a half-life at room temperature of 13 hours[12]. This surprising result prompted the investigation of the analogous mononuclear complex $[Co(7)_3]^{2+}$ in order to determine whether the unexpected kinetic stability was associated with the nature of the inner coordination sphere.

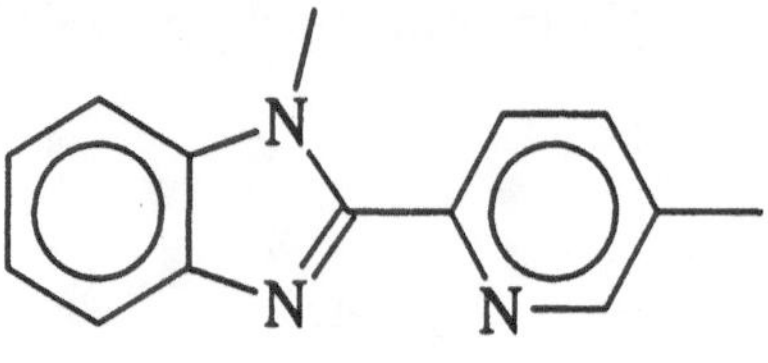

7

The complex $[Co(7)_3]^{2+}$ exists as an equilibrium mixture of meridional and facial isomers, whose interconversion may readily be followed by NMR line broadening [13]. The kinetic data for *fac-mer* conversion of $[Co(7)_3]^{2+}$ and racemisation of $(+)[Co_2(6)_3]^{4+}$ are compared in Table 2.

Table 2. Kinetic data for mononuclear and dinuclear complexes.

	$[Co(7)_3]^{2+}$ fac→mer	$(+)[Co_2(6)_3]^{4+}$ racemisation
k_{298}, s^{-1}	11(1)	$1.4(2)\times10^{-5}$
$\Delta H^{\ddagger}$, kJ/mol	57(3)	99(2)
$\Delta S^{\ddagger}$, J/mol/K	-34(8)	-6(7)

The results for $[Co(7)_3]^{2+}$ are typical for the rapid kinetics of cobalt(II) and establish that the kinetic stability of $[Co_2(6)_3]^{4+}$ is not due to the intrinsic nature of the coordination sphere. Variable pressure measurements of the isomerisation of $[Co(7)_3]^{2+}$ suggest the mechanism to be dissociative, and the much greater enthalpy of activation for the racemisation of the triple helix implies that it involves a greater degree of bond breaking. This could be understood if the mechanism involves the breaking of two metal-ligand linkages as shown in Scheme 2.

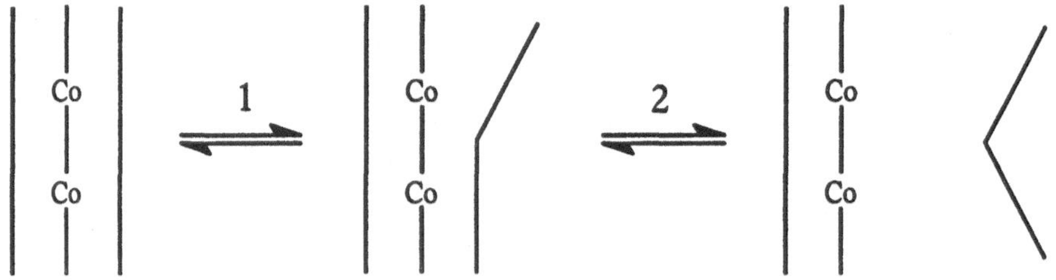

Scheme 2

The reverse of reaction 1 is extremely favorable thermodynamically, and may compete very efficiently with reaction 2, the dissociation of the ligand from the second cobalt ion. Support for this hypothesis comes from the observation that the racemisation is independent of pH around pH 5 but if the pH is lowered to below 2 the complex dissociates within the time of mixing. Under these acid conditions the free coordination site of the ligand will be protonated, the reverse of reaction 1 becomes thermodynamically unfavorable, and rapid dissociation occurs. The mechanism of racemisation of (+)$[Co_2(6)_3]^{4+}$ is still under investigation, but it seems clear that if the formation of the complex is thermodynamically favorable, the necessity to break several metal ligand bonds to break up the structure can confer considerable kinetic stability to these self-assembled species.

4. Thermodynamic control of synthesis.

Since the self-assembly reactions discussed here are under thermodynamic control, it is possible to use thermodynamics to control the synthesis. Two examples of such control are given to conclude this chapter.

The ligand **2a** discussed above undergoes a cyclometallation reaction with palladium (II) acetate to form a trimeric species $[Pd_3(\mathbf{2a}\text{-H})_3(OAc)_3]$ in virtually quantitative yield, presumably via the mononuclear intermediate shown in Scheme 3.

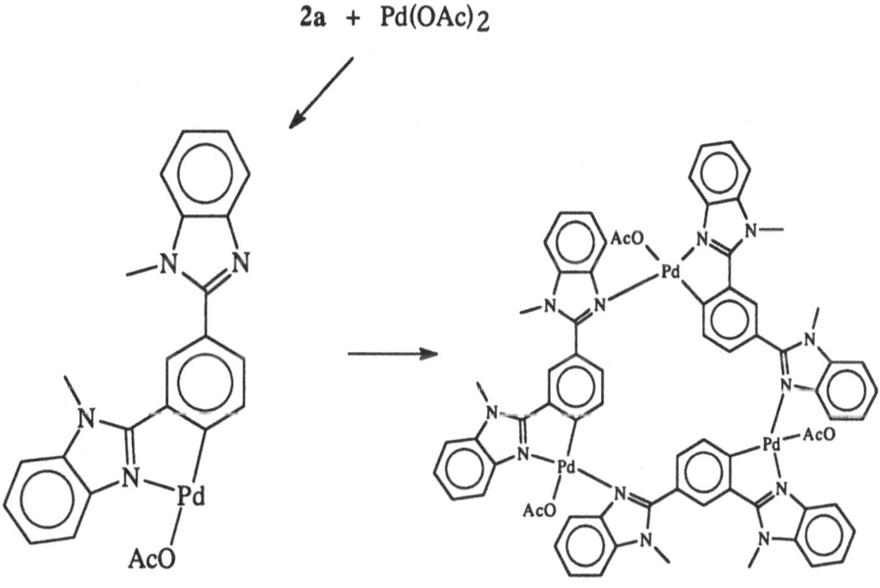

Scheme 3

NMR and mass spectrometry show that the trimeric structure is maintained in solution, but that the monodentate acetate groups undergo very rapid exchange. If the trimer is reacted with three equivalents of palladium acetate, an insoluble yellow product of composition $Pd_2(\mathbf{2a}\text{-2H})(OAc)_2$ is obtained in almost quantitative yield, implying that a second cyclometallation has occurred. The insolubility of this compound in all solvents was an obstacle to its structural characterisation until we realised that the addition of acid would labilise the acetates. In order to have a diastereotopic probe for chirality in the complex [4] we used the ethyl derivative of the ligand, **2c**. Pd_2(2c-2H)$(OAc)_2$ dissolves either in a DMSO solution of HBr to give [Pd_2(2c-2H)$Br_2$$(DMSO)_2$] or in p-toluene-sulfonic acid (TsOH) in DMSO to give [Pd_2(2c-2H)$(OTs)_2$$(DMSO)_2$], which could be characterised by NMR. The symmetry of the NMR spectrum confirmed the formation of the dimetallated fragment Pd_2(2c-2H), **8** whose planar nature was established by the A_2X_3 signal of the ethyl substituents.

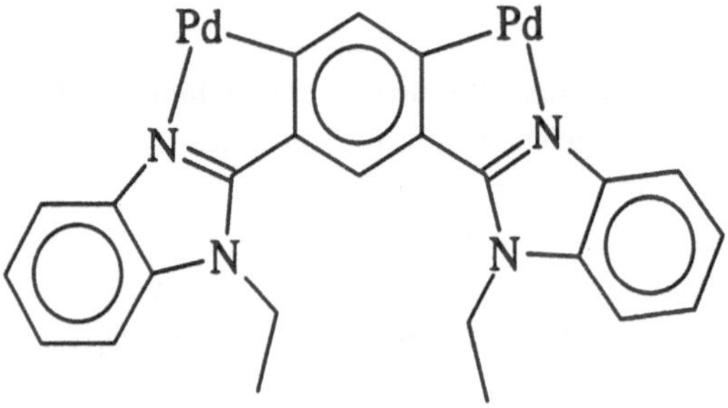

8

If hydroxyacetate anion was added to a solution of [Pd_2(2c-2H)$(OTs)_2$$(DMSO)_2$], the carboxylate groups coordinate to the metal and give [Pd_2(2c-2H)$(HOCH_2COO)_2$] which is sufficiently soluble to be recrystallised. The crystal structure of the compound shows it to be a dimer, [Pd_2(2c-2H)$(HOCH_2COO)_2$]$_2$, in which two fragments **8** are held together by four carboxylate bridges (Figure 4).

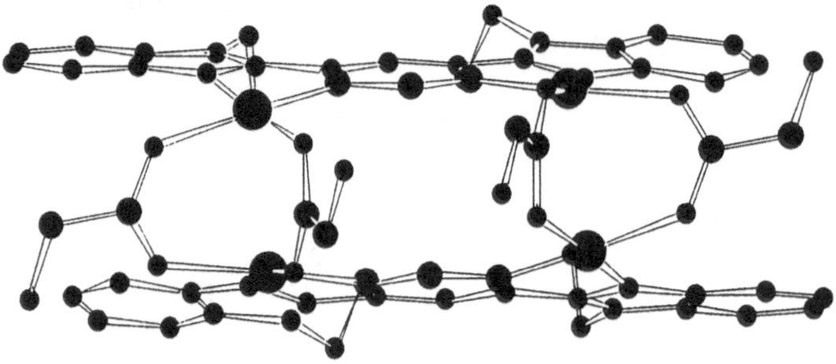

Figure 4. Crystal structure of [Pd_2(2c-2H)$(HOCH_2COO)_2$]$_2$. The molecule has a crystallographic twofold axis perpendicular to the plane of the paper.

The methyl groups of the ethyl substituents of **2c** are bent into the cleft between the two halves of the molecule which are slightly staggered so as to allow the methyl groups to intermesh. In contrast to the very labile monodentate acetate groups in [Pd_3(2a-H)$_3$$(OAc)_3$], the bridging acetate groups show no signs of exchange on the

NMR time scale. This recalls the kinetic stability of ligand **6** bound to two cobalt ions in the previous section. It is interesting to note that the eight metal-carboxylate bonds needed to assemble $[Pd_2(2c\text{-}2H)(HOCH_2COO)_2]_2$ may be made or broken reversibly. In other work we have also found the carboxylate ligand to be readily added or removed by control of the acidity of the solution [15], and this makes it a potentially interesting unit for supramolecular synthesis. Finally, we may note that the only reasonable mechanism for the formation of $[Pd_2(2c\text{-}2H)(OAc)_2]_2$ is through the mononuclear intermediate in Scheme 3. This establishes that the formation of the trimer is indeed reversible.

Clearly there are limits to the number of components that may be assembled in one self-assembly reaction. While it is theoretically possible to do so using ligands containing many different metal binding sites, in practice the synthesis of the ligands becomes increasingly tedious. A possible alternative is to use successive assembly steps, in which the product of the first assembly has a large number of spatially arranged binding sites, and can participate in a second step. This clearly requires that the product of the first assembly should be stable (either kinetically or thermodynamically) during the second step. The last example we discuss shows one means of achieving this. The ligand 5,5'-dicarboxy-2,2'-dipyridyl (**9**) is a readily available bidentate ligand which complexes iron(II) and cobalt(II) to give octahedral complexes. These complexes are labile, but oxidation of cobalt(II) to cobalt(III) affords the kinetically stable complex $[Co(9)_3]^{3-}$ in which the carboxylate ligands project away from the complex and are available to form further links.

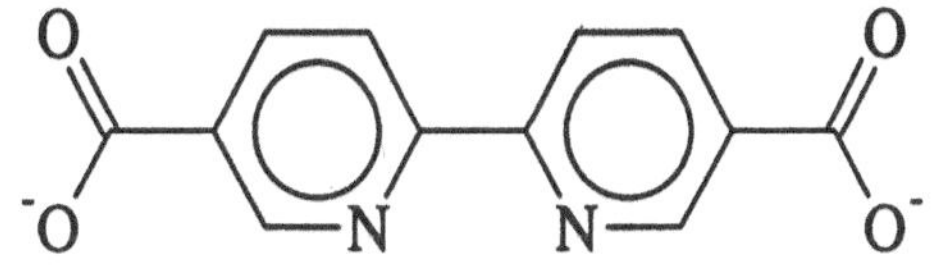

9

Attempts to link these units together with complexation reactions have so far been unsuccessful, but simple protonation of the complex leads to the neutral complex $[Co(9H)_3]$ which possesses three free carboxylates and three carboxylic acid fragments. The hydrogen atoms bound to the carboxylate are localized on one face of the complex (the upper face in Figure 5) and are directed away from the centre. The carboxylate groups, on the lower face of the complex, are equally directed away from the cobalt atom, and are available to form hydrogen bonds with carboxylic acid groups of neighbouring complexes as shown in Figure 6.

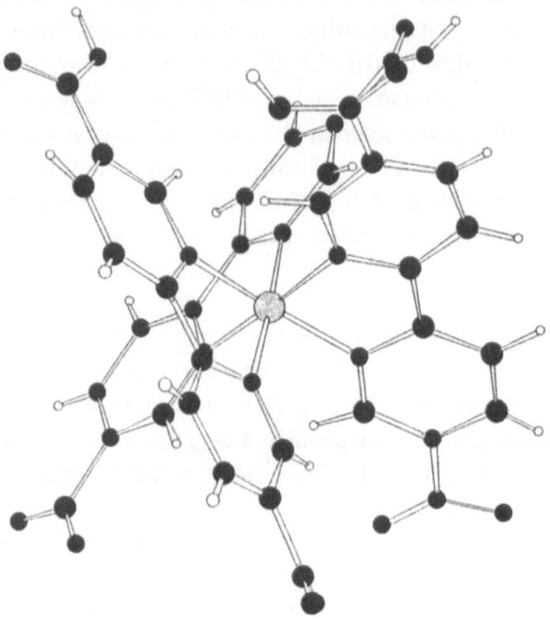

Figure 5. Crystal structure of $[Co(9H)_3]$ showing the three protons located on one face of the complex at the top of the figure.

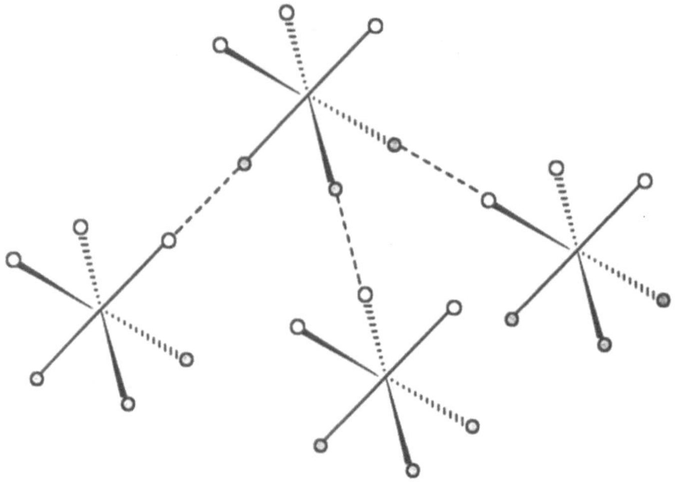

Figure 6. The development of the rhombohedral hydrogen bonded network between $[Co(9H)_3]$ units. Free carboxylates (shaded circles) hydrogen bond to carboxylic acids (open circles) related by a translation along the rhombohedral axes.

This is indeed the basis of the structure found for $[Co(9H)_3].10H_2O$ [16]. The unit cell is pseudo-rhombohedral. The $[Co(9H)_3]$ units are hydrogen bonded to other complexes to give a rhombohedral network as expected from the spatial distribution of the hydrogen bonding donors and acceptors. Since the individual units are related to each other by a translation along one cell axis, they must have the same chirality. The resulting lattice is thus homochiral, but a similar lattice of opposite chirality interpenetrates the first to give racemic crystals. In spite of this interpenetration, the crystal lattice has quite low density, and the cavities are filled up with water molecules, which are however fairly easily removed by drying. The series of reactions (synthesis of the complex in basic solution, in which the hydrogen bonding interaction is "switched off", the introduction of kinetic stability by oxidation, followed by "turning on" the hydrogen bonding by addition of protons) forms a self-assembling reaction scheme.

5. Conclusions

Much attention has been paid in supramolecular chemistry to the design of the components, or, in Lehn's terminology, "programming the structural information" into the components [2]. In this chapter we have paid more attention to the energetics of the self-assembly process itself. The examples we have chosen were intended to show how the study of the energetics is a valuable source of information on the relative importance of the different structural features of the components for the self-assembly process, as illustrated by the copper systems. Even though the individual bonds formed may be labile, the results on the cobalt(II) triple helix and the palladium dimer show that the product can have considerable kinetic stability when its disintegration requires the simultaneous rupture of several bonds. Control of the self-assembly is possible by modifying the thermodynamic conditions, most importantly activities. A thorough understanding of mechanism has proved vital for synthesis in molecular chemistry, and it seems reasonable to assume that it will also contribute greatly to supramolecular synthesis.

Acknowledgements.

We thank Dr Gérald Bernardinelli, Laboratory of X-ray Crystallography, University of Geneva, for the resolution of all crystal structures reported here, and Mr Bernard Bocquet for technical assistance. This research was supported by the Swiss National Science Foundation.

References

1. Lindsey, J.S. *New J. Chem.* 1991, 15, 153-180.
2. Lehn, J.-M., *Supramolecular Chemistry*, 1995, VCH Weinheim, Germany.
3. Piguet, C.; Bernardinelli, G.; Williams, A.F. *Inorg. Chem.* 1989, 28, 2920-2925.
4. Rüttiman , S.; Piguet C.; Bernardinelli, G.; Bocquet, B.; Williams, A.F. *J. Amer. Chem. Soc.* 1992, 114, 4230-4237.
5. Bernardinelli, G.; Kübel Pollak, A.; Williams, A.F. *Chimia* 1992, 46, 155-158.
6. Carina, R.F.; Bernardinelli, G.; Williams, A.F. *Angew. Chem. Int. Ed.* 1993, 32, 1463-1465.
7. Scarrow, R.C.; White, D.L.; Raymond, K.N. *J. Amer. Chem. Soc.* 1985, 107, 6540-6546; Serr, B.R.; Andersen, K.A.; Elliot, C.M.; Andersen, O.P. *Inorg. Chem.* 1988, 27, 4499-4504.
8. Williams, A.F.; Piguet, C.; Bernardinelli, G. *Angew. Chem. Int. Ed.* 1991, 30, 1490-1492.
9. Piguet, C.; Bernardinelli, G.; Bocquet, B.; Schaad, O.; Williams, A.F. *Inorg. Chem.* 1994, 33, 4112-4121.
10. Charbonnière, L. J., PhD. Thesis, University of Geneva, 1996.
11. Charbonnière, L. J.; Bernardinelli, G.; Piguet, C.; Sargeson, A.M.; Williams, A.F. *J. Chem. Soc. Chem. Commun.* 1994, 1419-1420.
12. Charbonnière, L. J.; Williams, A.F.; Gilet, F.; Bernauer, K. *J. Chem. Soc. Chem. Commun.* 1996, 39-40.
13. Charbonnière, L. J.; Williams, A.F.; Frey, U.; Lye, P.; Merbach, A.E., to be published.
14. Rüttimann, S.; Bernardinelli, G.; Williams, A.F. *Angew. Chem. Int. Ed.* 1993, 32, 392-394.
15. Carina R.F., PhD. Thesis, University of Geneva, 1994.
16. Desmartin, P.G.; Bernardinelli, G.; Williams, A.F. *New J. Chem.* 1995, 19, 1109-1112.

DYNAMICS OF SUPRAMOLECULAR ASSEMBLIES IN SOLUTION AND SOLID STATE

ZHIGANG CHEN, LOUIS MERCIER, JAMES J. TUNNEY, CHRISTIAN DETELLIER

Ottawa-Carleton Chemistry Institute
University of Ottawa, Ottawa (Ont)
Canada K1N 6N5

Abstract

Dynamics of supramolecular assemblies are defined and described both in solution and in solid-state by high-resolution NMR. The dynamics of macrocyclic complexes of cationic species in solution can be described in terms of *inter*supramolecular ligand exchange and of *intra*supramolecular ligand facial exchange. These processes were investigated for several systems, including crown ethers and calixarenes. More particularly, macrocyclic complexes of lanthanides were investigated by ^{139}La, ^{1}H, ^{13}C, and ^{17}O NMR. In some systems, such as in the case of 15C5-La(III), the two faces of the coordinated macrocycle become inequivalent on the ^{1}H NMR timescale. The exchange between the "inner" (close to La(III)) and the "outer" (close to the solvent) protons could then be studied, showing that the exchange is controlled by dissociation/recombination of the ligand. In the case of 18C6, the complexation process was much slower, and the two faces of the coordinated macrocycle could not be differentiated. In both cases, the exchange mechanism and the kinetic data are in agreement with structures in solution very similar to those in the solid state.

Several organo-mineral nanocomposites were prepared, with covalent attachment of organic units to the interlamellar surfaces of naturally occurring layered minerals. More particularly, oxyethylene units were grafted on a layered polysilicic acid, magadiite, and on kaolinite. The mobility of the carbon chains was characterized by dipolar dephased CPMAS ^{13}C NMR, which permitted the clear differentiation between grafted and intercalated organic molecules.

L. Echegoyen and A.E. Kaifer (eds.), Physical Supramolecular Chemistry, 393–411.

These experiments provide the tools to follow the chemical linkage of small organic guests to the extended, layered, structure of their large, polymeric, inorganic hosts.

1. Introduction

While a lot of work is devoted to the synthesis of new molecular receptors leading to the formation of sophisticated supramolecular assemblies, some questions are less frequently asked, such as the nature of the pathways followed by the molecular entities to associate and form organized structures.

In solution, the formation of complexes of cations with macrocyclic ligands can follow several mechanistic pathways [1]. This has been shown in the case of crown ethers [1,2], and, more recently, in the cases of aza-crown ethers containing functional groups as sidearms [3,4] and of calixarenes [5], for example. In the solid state, while the intercalation of organic molecules in layered materials, and more particularly in smectite clays, is extensively studied [6], while also the organic reactivity of these intercalated molecules is receiving a lot of attention [7], the covalent grafting of these intercalated guests to the inorganic host to form organo-inorganic nanocomposites is much less described and studied [8-10].

Two examples are given here, in solution and in the solid state, of studies on the nature of the mechanisms involved in the formation of such organized structures.

2. Lanthanum Nitrate - Crown Ether Complexes in Acetonitrile Solution

Lanthanoid cations are increasingly used as probes in biochemical systems [11] and their characteristics of complexation by specific ligands in solution are of interest [12]. However, the data uon the kinetics and mechanisms of their complexation by macrocyclic ligands in solution are rather scarce [13] and certainly much less understood than their alkali metal cations counterparts [1].

Metal cation NMR in solution, particularly of quadrupolar nuclei, is a very sensitive probe of the metal cation direct electronic environment [14]. Among the lanthanide series, the quadrupolar ^{139}La nucleus ($I = 7/2$; $Q = 0.21 \times 10^{-28}$ m^2) is the most attractive nucleus, with a receptivity compared to ^{13}C of 342. In this work, its NMR spectroscopy was used in combination with ^1H,

^{13}C and ^{17}O nuclei to follow, describe and understand the mechanisms of complexation of the La(III) cation by macrocyclic ligands in solution.

2.1. STOICHIOMETRY

The knowledge of the stoichiometry of a complex is the prerequisite for any study of its mechanism of formation in solution. Even if, generally, lanthanoid cations form 1:1 complexes with crown ethers as evidenced by a large number of crystal structures [15,16], a variety of structures can be found, with different stoichiometries [17], or resulting from the formation of outer-sphere complexes [18], or from dimerization [15,19].

In this study, the hexahydrated nitrate salt of La(III) was dissolved in anhydrous acetonitrile, and complexed by various crown ethers. The nature of the first coordination sphere of the La (III) cation in acetonitrile solutions of $La(NO_3)_3.6H_2O$ has been previously investigated by ^{139}La and ^{17}O NMR [20]. It was shown that all three nitrate anions are coordinated to La(III) [20, 21]. There is a competition between water and acetonitrile molecules for the occupancy of the La(III) first coordination sphere, with two species, anhydrous and monohydrated $La(NO_3)_3$, in rapid chemical exchange on the ^{17}O NMR timescale. Consequently, several questions are associated with the complexation of La(III) by a crown ether in these acetonitrile solutions: is water totally expelled from the first coordination sphere of La(III), are the three nitrate anions still coordinated after the complexation, and what is the number of crown ether ligands?

Figure 1 shows a series of ^{139}La NMR spectra (42.37 MHz) for a series of $\rho = [18C6]_0/[La]_0$ values. The ^{139}La NMR spectrum of a 0.015 M solution of $La(NO_3)_3.6H_2O$ in acetonitrile is characterized by a chemical shift of 9.6 ppm (referenced to 0.10 M $La(NO_3)_3$ 20% D_2O aqueous solution) and a linewidth of 2.5 kHz (Figure 1). This ^{139}La NMR signal results from an equilibrium involving $\{La(NO_3)_3\ (AN)_x\}$ and $\{La(NO_3)_3\ (H_2O)\ (AN)_y\}$ [20]. During the titration of La(III) by 18C6, another ^{139}La NMR signal, corresponding to the 18C6 complexed species, appears at lower frequency, with a chemical shift of -139 ppm and a linewidth of 4.4 kHz. In the region $0 < \rho < 1$, the resonances characterizing the solvated and the crown ether complexed La(III) species coexist. A temperature variation between 300 K and 335 K does not affect the lineshape of the ^{139}La signals, indicating a very slow exchange between the two sites. The spectra were decomposed in two Lorentzian lines, using a procedure for an imperfectly phase-corrected spectrum, as described previously [22]. Figure 2 shows a direct linear relationship between the population of uncomplexed La(III) as a function of ρ: it demonstrates the formation of a 1:1 La(III)-18C6 complex, with an equilibrium constant of formation larger than 10^4 [1]. Similar results can be obtained from ^1H and ^{13}C NMR studies of the ligand.

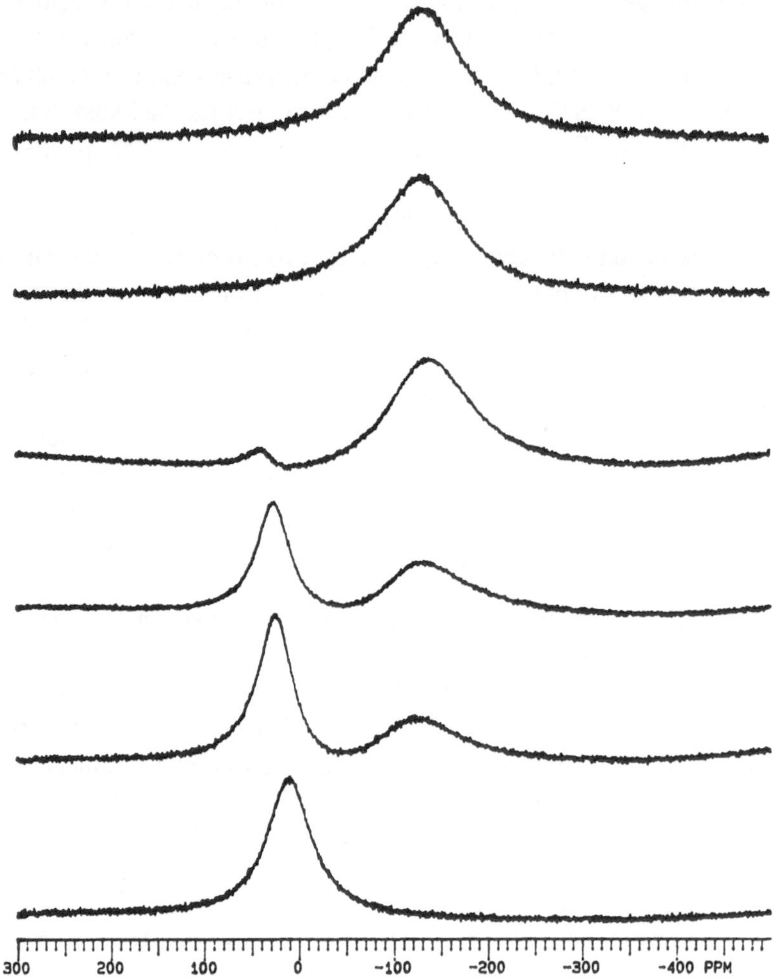

FIGURE 1: ^{139}La NMR spectra of $La(NO_3)_3.6H_2O$ solutions in acetonitrile in the presence of 18C6 for various ρ = $[18C6]_0/[La]_0$ values. $[La(III)]_0$=0.015 M. T =300K. Top to bottom, ρ =1.20, 1.00, 0.90, 0.70, 0.50, 0.

These data are in agreement with a log K_f of 4.4 previously reported [23] for a similar system. Stoichiometries of 4:3 and 1:1 have been reported for the 18C6-La(III) complex in the solid state, the 4:3 complex dissociating in acetonitrile solution [23]. The large differences of chemical shifts and linewidths between solvated and complexed La(III) sites suggest that 18C6 occupies the first coordination sphere of La(III) in the complexed species.

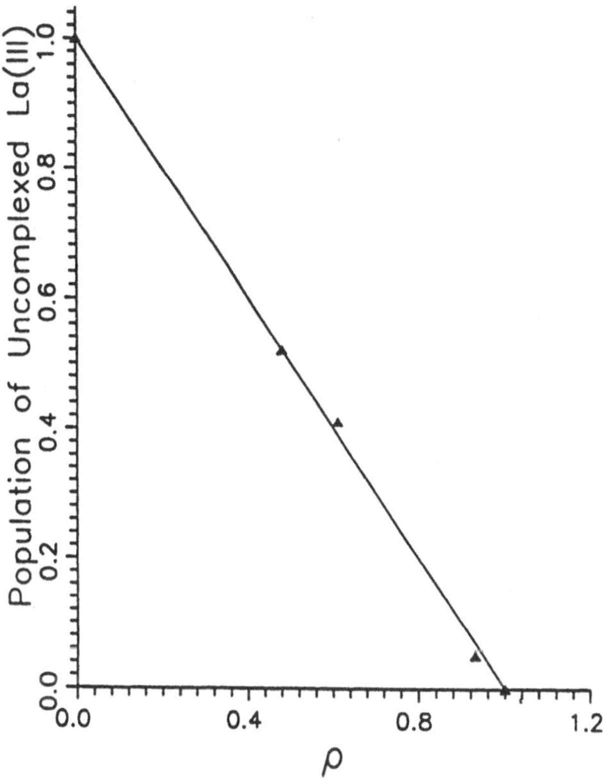

FIGURE 2: Normalized population of uncomplexed La(III) as a function of ρ. T = 300K. $[La(III)]_0$ = 0.015M. The data points were obtained from a non-linear regression analysis decomposing the experimental ^{139}La NMR spectra shown in Figure 1.

The chemical shifts and linewidths of the solvated La(III) species vary with ρ, in agreement with our previous study on the nature of La(III) species in acetonitrile solutions [20].

A similar study could be done in the case of benzo-15-crown-5 (B15C5) and of 15C5, showing also the formation of a 1:1 complex with La(III). In the case of B15C5, the complex is characterized by a ^{139}La chemical shift of -44 ppm, and a linewidth of 1.9 kHz. ^{1}H and ^{13}C NMR data confirm the formation of a 1:1 complex with log K_f $>10^4$. Figure 3 shows a series of ^{13}C NMR spectra of B15C5 in the presence of increasing quantities of La(III). The exchange of the B15C5 ligand between its solvated and coordinated sites is moderately fast on the ^{13}C NMR timescale, permitting a kinetic and mechanistic study of the exchange.

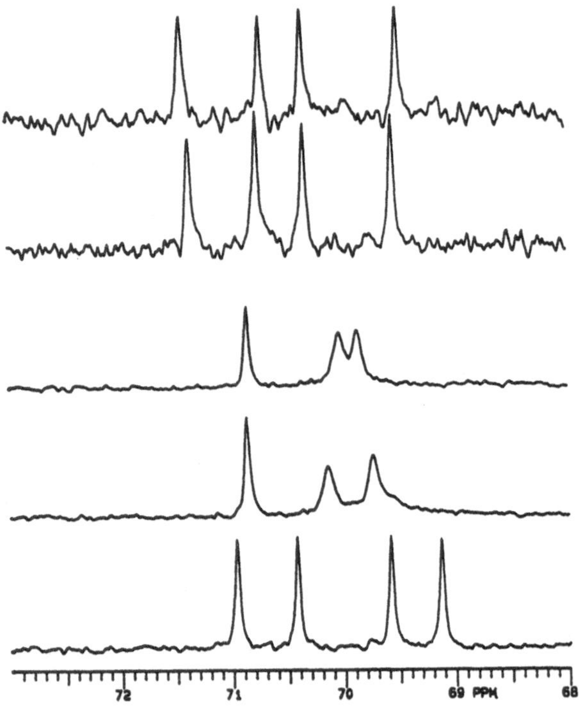

FIGURE 3: ^{13}C NMR spectra of B15C5 solutions in acetonitrile in the presence of $La(NO_3)_3.6H_2O$ for various values of R $=[La(III)]_0/[B15C5]_0$. $[B15C5]_0$ =0.015M. T = 300K. Top to bottom: R =1.20, 1.00, 0.40, 0.20, 0.

Since water molecules are directly coordinated to La(III) in acetonitrile [20], the ^{1}H and ^{17}O NMR spectra of water were also recorded with various ratios of hexahydrate lanthanum nitrate and B15C5 in acetonitrile. Figure 4 shows the variation of the water ^{1}H chemical shifts as a function of R = $[La(NO_3)_3.6H_2O]_0/[B15C5]_0$. For R <1.0, in which case all the lanthanum nitrate is complexed by B15C5, the water chemical shift is constant, corresponding to the expected value of water in acetonitrile (2.13 ppm) [21], as indicated on Figure 4. When R >1.0, lanthanum nitrate is in excess in the solution, water molecules coordinate [20], and a strong chemical shift variation is observed. These data indicate that water is completely expelled from the first coordination sphere of B15C5 complexed La(III). A very similar behaviour has been reported in the case of the formation of La(III) complexes of 1,10-phenanthroline [21] and 2,2'-bipyridine [24] in acetonitrile.

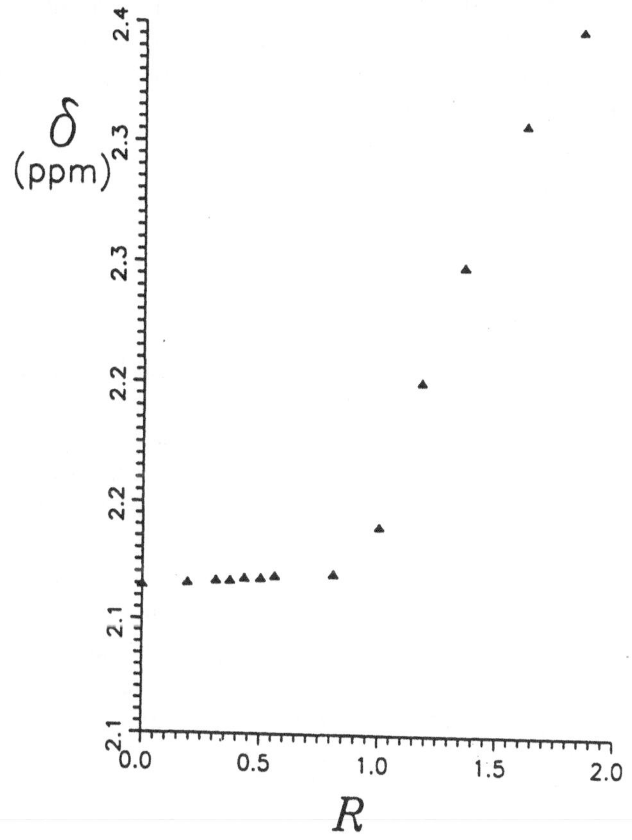

FIGURE 4: Water ^{1}H NMR chemical shifts as a function of R $=[La(NO_3)_3.6H_2O]_0/[B15C5]_0$.

One can conclude that the coordination number of La(III) in these crown ether complexes in acetonitrile solution is 11 or 12, corresponding to the chelation of three nitrate anions and the coordination of B15C5 or 15C5 and of 18C6, respectively. The structure of these complexes should be very similar to the one in the solid state, as it was found for the 1,10-phenanthroline ligand [21]. The crystal structure of $La(NO_3)_3.18C6$ has been reported [25]. The lanthanum cation is 12-coordinated, with its coordination sphere comprising the six crown ether oxygens and two oxygens from each of the three bidentate nitrates. Two nitrates are on one side of the crown ether ring, and one nitrate is on the other side.

2.2. KINETICS AND MECHANISMS OF COMPLEXATION.

Since only 1:1 complexes of La(III) with crown ethers were detected in the systems used in the present work, the exchange of La(III) between its uncomplexed and complexed sites involves only two sites, and can be described by the simple Equation 1,

$$[La(III)]_s \underset{k_b}{\overset{k_a}{\rightleftarrows}} [La(III)]_c \qquad (1)$$

where $[La(III)]_s$ and $[La(III)]_c$ represent respectively the La(III) cation in the solvated site and in the site complexed by a crown ether. k_a ($=\tau_a^{-1}$) and k_b ($=\tau_b^{-1}$) are the reciprocal lifetimes of the La(III) cation in these two sites. Figure 5 shows ^{139}La NMR spectra of lanthanum nitrate in the presence of B15C5 in acetonitrile. Two sites are observed, which can be shown, in contrast with the 18C6 case shown on Figure 1, to be in the moderately slow exchange regime.

The pseudo-first order rate constants of Equation 1 determined from the ^{139}La NMR lineshape analysis should then be related to a mechanism of exchange [1]. Equation 2 shows the case of a dissociation / recombination mechanism, in which the exchange is due to a full dissociation of the complex, with a departure of the ligand from the first coordination sphere, followed by a recomplexation of the cation by a different ligand.

$$\{La(NO_3)_3,C\} \overset{k_{-1}}{\rightleftarrows} \{La(NO_3)_3\,(AN)_x\} + C \qquad (2)$$

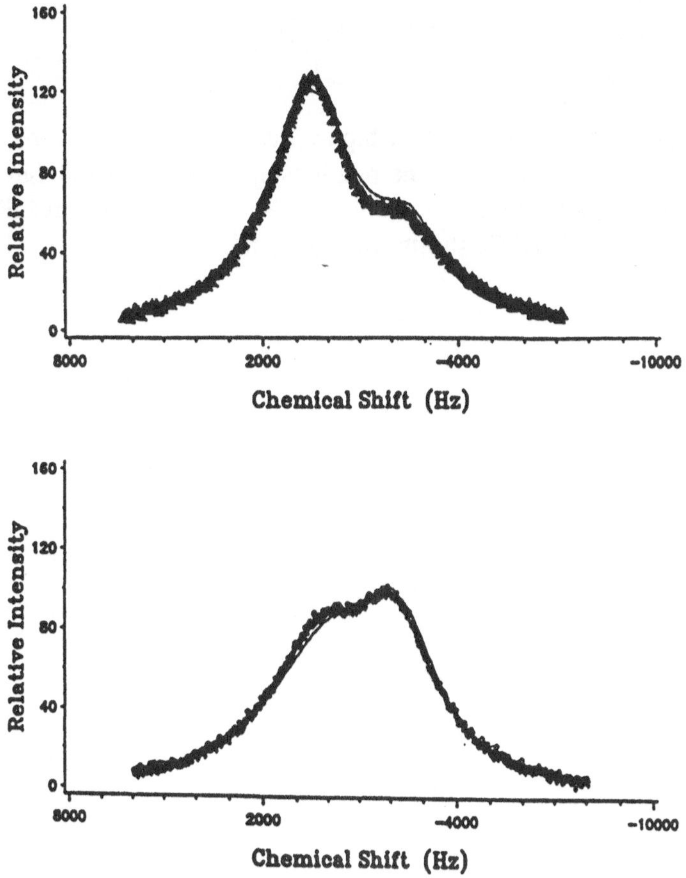

FIGURE 5: ^{139}La NMR spectra of $La(NO_3)_3 .6H_2O$ in the presence of B15C5 in acetonitrile. The points are experimental and the curves are calculated from the equations describing the chemical exchange of two uncoupled sites [26,27]. Top: $\rho = [B15C5]_0/[La]_0 = 0.18$; bottom: $\rho = 0.40$. $[La(III)]_0 = 0.0050$ M. T =300K.

It can be easily shown [1] that, in the case of a large equilibrium constant of complexation, the pseudo-first order rate constants of Equation 1 can be related to k_{-1} by Equation 3, with ρ as previously defined if one observes the metal nucleus (^{139}La here), or to Equation 4, with R as previously defined, if one observes a nucleus from the ligand (^{13}C for example).

$$k_a + k_b = k_{-1} \ (1-\rho)^{-1} \qquad (3)$$

$$k_a' + k_b' = k_{-1} (1-R)^{-1} \tag{4}$$

Figure 6 shows the relationship between $k_a + k_b$ and $(1-\rho)^{-1}$. In the limits of the experimental error, the relationship is linear, extrapolating to the origin, and giving a value of $k_{-1} = (2.4 \pm 0.2) \times 10^3\ s^{-1}$. Similar results can be obtained from the ^{13}C NMR spectra of the ligand.

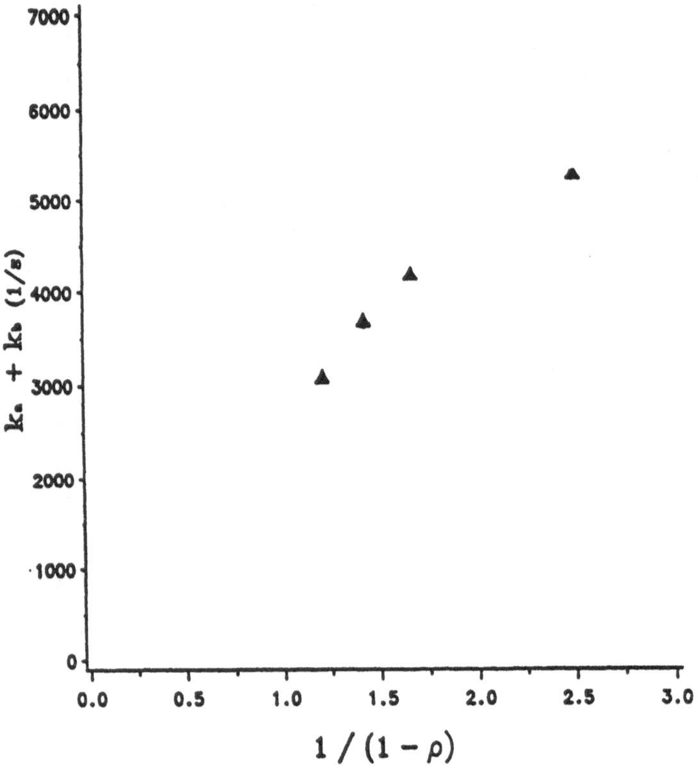

FIGURE 6: $(k_a + k_b)$ as a function of $(1-\rho)^{-1}$ (see Equation 3). The data were obtained from the analysis of the ^{139}La NMR spectra as shown on Figure 5. $[La(III)]_0 = 0.0050$ M and T= 300 K.

The extrapolation of Equation 3 to the origin, in the limits of the experimental errors, coupled with a similar observation in ^{13}C NMR on Equation 4, rules out any major influence from an other type of exchange mechanism, such as a cation associative interchange [1], or a ligand associative interchange [1,4].

2.3. LIGAND FACIAL EXCHANGE

While the ^{1}H NMR spectrum of the complex $\{La(III)(NO_3)_3.18C6\}$ gives only one single line at 3.84 ppm for the ether protons, magnetically equivalent, the situation is very different in the cases of 15C5 and of B15C5. Figure 7 shows the ^{1}H NMR spectra of 15C5 and of the complex $\{La(III)(NO_3)_3.15C5\}$ in acetonitrile. The signal of the complexed 15C5 is no longer a single line, but a broadened AA'BB' system. At lower temperature, the two multiplets are well resolved, and show the typical AA'BB' pattern, indicating a chemical exchange between them. The ^{13}C spectra do show only one single line for complexed 15C5 carbons.

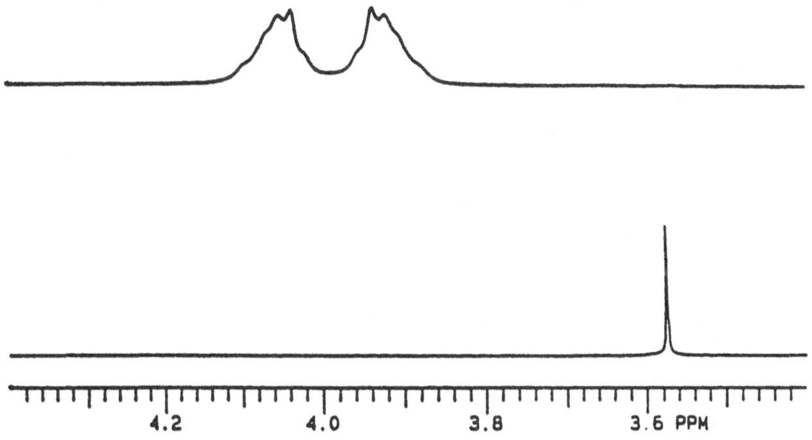

FIGURE 7: ^{1}H NMR spectra of 15C5 solutions in acetonitrile. $[15C5]_0$ =0.0032 M. T =300K. Top: R =$[La(NO_3)_3.6H_2O]_0$ / $[15C5]_0$ =1.00; bottom: R = 0.

The ^1H NMR pattern shows the non equivalence of its CH_2 protons when 15C5 is coordinated to La(III), making the two faces of the crown ether non identical. In contrast to the 18C6 case, this is in agreement with a structure where the La(III) cation is out of the plane of the crown ether with the three chelating nitrate groups on the other side. The exchange equilibrium shown by the ^1H NMR spectra of Figure 7 is schematized on Scheme 1.

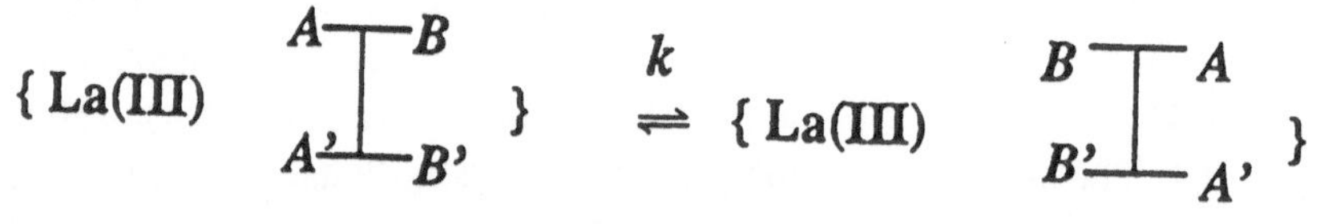

SCHEME 1

A full lineshape analysis of the AA'BB' exchanging system gives , at 300K, a rate k of 35 $\pm$ 5 s^{-1}. This value is very close to the value of k_{-1} of 50 $\pm$ 5 s^{-1} which was independently determined at the same temperature by ^{13}C NMR for the dissociation process given by Equation 2 in the case of 15C5. This shows that both processes are controlled by the dissociation / recombination process given in equation 2, and that the two spectroscopically independent kinetic phenomena observed on Figures 3 and 7 are one and the same process driven by dissociation and recombination of the complex.

3. Organo-Mineral Nanocomposites

Magadiite is a rare mineral, found in dry beds of alkaline lakes [28], but can be conveniently synthesized in the laboratory by straigthforward hydrothermal techniques [29]. It is a layered silicate with the chemical formula $Na_2H_2Si_{14}O_{30}$.x H_2O [30]. Hydrated sodium cations are located between the silicate sheets. Acid treatment results in the formation of a layered silicic acid bearing abundant hydroxyl groups on its interlamellar surfaces, forming the basis of the interlamellar reactivity of these minerals. Various pillaring processes have been reported for these layered silicates, including, for example, reactions with tetraethylorthosilicate [30] and silane coupling agents [31].

Recently, we have reported the covalent functionalization of magadiite interlayers by thermal condensation of ethylene glycol to silanol groups [32]. The interlamellar ethylene glycol grafting was demonstrated by a combination of characterization methods, including X-ray powder diffraction, TGA, and ^{29}Si CP-MAS NMR. Figure 8 shows the TGA profiles of two different samples of magadiite treated by ethylene glycol. In one case, ethylene glycol is intercalated (EG/MAG), forming a host-guest complex, and in the other profile, ethylene glycol has partially reacted with magadiite, forming a material with two types of ethylene glycol molecules, intercalated (EG/MAG) and grafted (EG-MAG) in which case host and guest have chemically reacted [32].

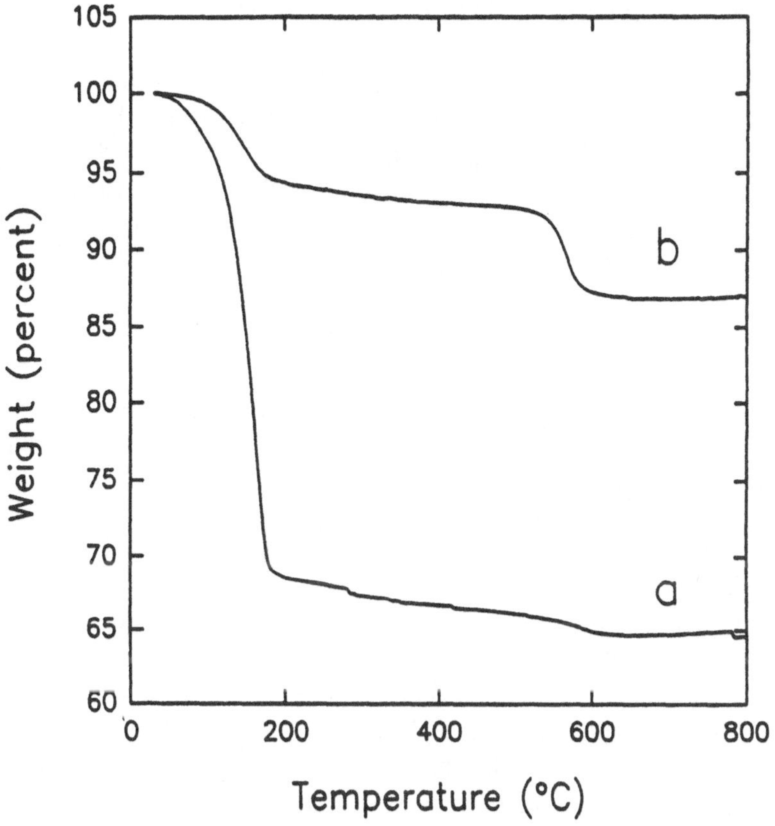

FIGURE 8: TGA profiles for (a) ethylene glycol intercalated into H-magadiite (EG/MAG) and (b) ethylene glycol grafted into H-magadiite (EG-MAG)

Dipolar dephased CPMAS ^{13}C NMR [33] provides a very convenient and simple spectroscopic tool to clearly differentiate between grafted and intercalated organic molecules. If, in the liquid state, ^{13}C-^{1}H dipolar interactions are negligeable due to fast motional averaging, in the solid state, however, the rigid conformations adopted by C-H bonds can result in a significant dipolar relaxation mechanism for ^{13}C. The dipolar dephasing pulse sequence is similar to that of CPMAS, with the addition of a time, called dephasing time, which can be varied, between the cross-polarization and the acquisition steps. During that time, ^{13}C nuclei will lose some, or all, of their polarization. So, dipolar dephasing studies measure the time for a polarized carbon to lose its magnetization after the proton locking field is terminated [33], and molecular groups more rigidly held would be expected to have greater signal attenuation than more mobile ones. The ratio of I_{DD}/I_0, where I_{DD} and I_0 are the CPMAS ^{13}C signal intensities obtained respectively with and without dipolar dephasing conditions, can be used as a semiquantitative measure of the dynamic state of a molecule or molecular group [34]. Equation 5, where t is the dipolar dephasing time and T_2 is the transverse relaxation time constant, has been proposed to describe the signal decay for carbons strongly coupled to protons, such as in a methylenic - CH_2- group [33].

$$I_{DD} = I_0 \exp(-t^2/(2T_2^2)) \qquad (5)$$

In cases where the two types of ethylene glycol were present in magadiite, EG-MAG and EG/MAG, with a ^{13}C spectrum too broad to permit a clear differentiation of the signals from each type of ethylene glycol, a dipolar dephasing experiment (with a dephasing time of 40 μs), permitted to separate the two components from the observed ^{13}C CPMAS signal [32].

Several other diols were also grafted on magadiite. The variation of the dephasing time allowed a test of Equation 5, and afforded values of T_2, which could demonstrate different modes of binding for short and longer chain diols.

Kaolinite is a 1:1 layered aluminosilicate whose aluminate side is made of a gibbsite-type aluminol surface, hydrogen-bonded to the silicate side of an adjacent layer. It provides the interesting case of a layered mineral with a asymmetric interlamellar environment. We have recently reported the grafting of ethylene glycol units to the interlamellar surface of kaolinite via Al-O-C bonds [10,35]. Two different intercalates of ethylene glycol in kaolinite were contrasted. In one case, the ethylene glycol molecules are intercalated, held in the interlamellar spaces by intermolecular forces, and characterized by a d_{001} spacing of 10.8 Å ("Kao-EG 10.8Å"). In the other case, the ethylene glycol units are chemically bound to the aluminol surface, and characterized by a d_{001} spacing of 9.4 Å ("Kao-EG 9.4Å"). The dipolar dephasing experiment confirms that the EG units in Kao-EG 9.4Å are more rigidly held in the interlamellar space than in Kao-EG 10.8Å, since after a 40 μs dephasing time, I_{DD}/I_0 was only 0.28 in the first case and 0.61 in the latter (Table 1).

TABLE 1. ^{13}C DD/CPMAS results for Kao-EG 10.8Å and Kao-EG 9.4Å. The spinning rate was 4kHz, and the dephasing time was 40 ms.

Product	δ ^{13}C (ppm)	$\nu_{1/2}$ (Hz)	I_{DD}/I_0
Kao-EG 10.8Å	65	55	0.61
Kao-EG 9.4Å	65	120	0.24

4. Conclusions

The dynamics of macrocyclic complexes of cationic species in solution could be described in terms of *inter*supramolecular ligand exchange and of *intra*supramolecular ligand facial exchange. In the case of the 15C5-La(III) complex, the two faces of the coordinated macrocycle are unequivalent on the ^1H NMR timescale. The exchange between the "inner" (close to La(III)) and the "outer" (close to the solvent) protons could then be studied, showing that the exchange is controlled by dissociation/recombination of the ligand.

Dipolar dephasing experiments have provided an useful tool to follow the chemical linkage of small organic guests to the extended, layered, structure of their large, polymeric, inorganic hosts. This was shown in the case of the grafting of ethylene glycol units to the interlamellar surfaces of two minerals, magadiite and kaolinite.

Acknowledgments

The Natural Sciences and Engineering Council of Canada (NSERCC) is gratefully acknowledged for continuous support. LM thanks also NSERCC for a postgraduate scholarship. We thank Dr G. Facey (Univ. of Ottawa) for his help in recording NMR spectra.

5. References

1. Detellier, C. (1996) Complexation Mechanisms, in G. Gokel (ed.), *Comprehensive Supramolecular Chemistry* Vol.1, in press.
2. Brière, K.M., Detellier, C. (1992) Metal interchange of crown ether - alkali metal cation complexes in solution. ^{7}Li nuclear magnetic resonance study of the exchange kinetics of lithium 15-crown-5 and lithium monobenzo-15-crown -5 in nitromethane, *J. Phys. Chem.*, **96**, 2185-2189.
3. Li, Y., Gokel, G., Hernández, J. and Echegoyen, L. (1994) Cation-exchange kinetics of sodium complexes with amide- and ester-substituted crown ethers in homogeneous solution, *J. Amer. Chem. Soc.*, **116**, 3087-3096.
4. Li, Y., Echegoyen, L. (1994) Unusual mechanistic pathways and dynamic processes during cation exchange between substituted crown ether - sodium complexes in homogeneous solution, *J. Amer. Chem. Soc.*, **116**, 6832-6840.
5. Blixt, J., Detellier, C. (1995) Kinetics and mechanism of the sodium cation complexation by 5,11,17,23-tetra-*p*-*tert*-butyl-25,26,27,28-tetramethoxy-calix[4]arene in solution, *J. Amer. Chem. Soc.*, **117**, 8536-8540.
6. Barrer, R.M. (1989) Shape-selective sorbents based on clay minerals: a review, *Clays Clay Min.*, **37**, 385-395.
7. Cornélis, A. and Laszlo, P. (1994) Molding clays into efficient catalysts, *Synlett.*, 155-161.
8. Choudary, B.M., Subba Rao, Y.V. and Prasad, B.P. (1991) New triphase catalysts from montmorillonite, *Clays Clay Min.*, **39**, 329-332.
9. Mercier, L. and Detellier, C. (1995) Preparation, characterization, and applications as heavy metal sorbents of covalently grafted thiol functionalities on the interlamellar surface of montmorillonite, *Environ. Sci. Technol.*, **29**, 1318-1323.
10. Tunney, J.J., Detellier, C. (1993) Interlamellar covalent grafting of organic units on kaolinite, *Chem. Mater.*, **5**, 747-748.
11. Evans, C.H. (1990) *Biochemistry of the Lanthanides*, Plenum Press, New York.

12. Bünzli, J.C.G.(1987) Complexes with synthetic ionophores, in K.A.Jr. Gschneider and L. Eyring (eds.), *Handbook on the Chemistry and Physics of Rare Earths*, Elsevier Sci. Publ., New York, Chapter 60, pp 321-394.
13. Izatt, R.M., Pawlak, K., Bradshaw, J.S. and Bruening, R.L. (1991) Thermodynamic and kinetic data for macrocyclic interaction with cations and anions, *Chem. Rev.*, **91**, 1721-2085.
14. Detellier, C. (1983) Alkali metals, in P. Laszlo (ed.), *NMR of Newly Accessible Nuclei*, Vol.2, Acad. Press, New York, Chapter 5, pp 105-151.
15. Rogers, R.D., Rollins, A.N., Etzenhouser, R.D., Voss, E.J. and Bauer, C.B. (1993) Structural investigation into the steric control of polyether complexation in the lanthanide series: macrocyclic 18-crown-6 versus acyclic pentaethylene glycol, *Inorg. Chem.* **32**, 3451-3462.
16. Lu, T., Gan, X., Tan, M., Li, C. and Yu, K. (1993) Studies on crown ether complexes-XXVIII. Synthesis, characterization and structure of the complexes of heavier lanthanide nitrates (Gd-Lu) with dibenzo-24-crown-8, *Polyhedron*, **12**, 1641-1646.
17. Nicolò, F., Plancherel, D., Bünzli, J.C.G. and Chapuis, G. (1987) Synthesis, crystal and molecular structure of the 3:2 complex between europium nitrate and the A-isomer of dicyclohexyl-18-crown-6: conformational study of the ligand, *Helv. Chim. Acta* **70**, 1798-1806.
18. Rogers, R.D., Rollins, A.N., Etzenhouser, R.D., and Henry, R.F. (1990) *f*-Element/crown ether complexes. 27. The synthesis and crystal structure of $[Ce(NO_3)_3(OH_2)(12\text{-crown-}4)]$.12-crown-4, *J. Inclusion. Phen. Molec. Recogn. Chem.* **8**, 375-382.
19. Rebizant, J., Spirlet, M.R., Barthélemy, P.P., and Desreux, J.F. (1987) Solid state and solution structures of the lanthanide complexes with cryptand (2.2.1): crystallographic and NMR studies of a dimeric praseodymium (2.2.1) cryptate containing two μ-hydroxo bridges, *J. Inclusion Phen.* **5**, 505-513.
20. Chen, Z., Detellier, C. (1994) A ^{139}La and ^{17}O nuclear magnetic resonance study of the nature of La(III) first coordination sphere in acetonitrile solutions of hydrated lanthanum nitrate, *Can. J. Chem.*, **72**, 1797-1802.

21. Fréchette, M., Butler, I.R., Hynes, R., and Detellier, C. (1992) Structures in solution and in the solid state of the complexes of lanthanum (III) with 1,10-phenanthroline. X-ray crystallographic and ^{1}H, ^{13}C, ^{17}O, and ^{139}La solution NMR studies, *Inorg. Chem.*, **31**, 1650-1656.
22. Chen, Z., Dettman, H., Detellier, C. (1989) Characterization of the 18-crown-6 lanthanum(III) complex in methanol using ^{139}La nuclear magnetic resonance, *Polyhedron*, **8**, 2029-2033.
23. Bünzli, J.C.G., Wessner, D. (1981) Complexes of lanthanoid nitrates with 15-crown-5 and 18-crown-6 ethers, *Helv. Chim. Acta*, **64**, 582-598.
24. Fréchette, M. (1993) Characterization of the 2,2'-bipyridine chelates of lanthanum(III) in acetonitrile. A ^{1}H, ^{13}C, ^{17}O and ^{139}La NMR study, *Can. J. Chem.*, **71**, 377-383.
25. Backer-Dirks, J.D.J., Cooke, J.E., Galas, A.M.R., Ghotra, J.S., Gray, C.J., Hart, F.A., Hursthouse, M.B. (1980) Complexes of lanthanide ions with the crown ether 1,4,7,10,13,16- hexaoxacyclo - octadecane, *J. Chem. Soc. Dalton Trans.*, 2191-2198.
26. Sandström, J. (1982) *Dynamic NMR spectroscopy*, Academic Press, New York.
27. Detellier, C. (1991) Reaction kinetics and exchange, in A.I. Popov and K. Hallenga (eds.), *Modern NMR techniques and their application in chemistry, Practical Spectroscopy Series, Vol. 11*, Marcel Dekker, New York, pp 521-566.
28. Eugster, H.P. (1967) Hydrous sodium silicates from lake Magadi, Kenya: precursors of bedded chert, *Science*, **157**, 1177-1180.
29. Lagaly ,G., Beneke, K. and Weiss, A. (1975) Magadiite and H-magadiite: I. Sodium magadiite and some of its derivatives, *Amer. Mineral.*, **60**, 642-649.
30. Dailey, J.S., Pinnavaia, T.J. (1992) Silica-pillared derivatives of H^{+}-magadiite, a crystalline hydrated silica, *Chem. Mater.*, **4**, 855-863.
31. Ruiz-Hitzky, E., Rojo, J.M. (1980) Intracrystalline grafting on layer silicic acids, *Nature*, **287**, 28-30.
32. Mercier, L., Facey, G.A. and Detellier, C. (1994) Organo-layered silicates. Interlamellar intercalation and grafting of ethylene glycol in magadiite, *J. Chem. Soc., Chem. Comm.*, 2111-2112.

33. Alemany, L.B., Grant, D.M., Alger, T.D. and Pugmire, R.J. (1983) Cross polarization and magic angle sample spinning NMR spectra of model organic compounds. 3. Effect of the ^{13}C -^{1}H dipolar interaction on cross polarization and carbon-proton dephasing, *J. Amer. Chem. Soc.*, **105**, 6697-6704.
34. Ripmeester, J.A., Burlinson, N.E. (1985) Chiral discrimination and solid-state ^{13}C NMR Application to tri-*o*-thymotide clathrates, *J. Amer. Chem. Soc.*, **107**, 3713-3714.
35. Tunney, J.J., Detellier, C. (1994) Preparation and characterization of two distinct ethylene glycol derivatives of kaolinite, *Clays Clay Min.*, **42**, 552-560.

THE DESIGN OF METAL ION SELECTIVITY INTO LIGANDS AND SOME PRACTICAL APPLICATIONS

Xian Xin Zhang, Andrei V. Bordunov, Reed M. Izatt, and Jerald S. Bradshaw
Department of Chemistry and Biochemistry
Brigham Young University
Provo, UT 84602, USA

Abstract. New macrocyclic compounds showing improved metal ion selectivity have been designed and synthesized based on information obtained from characterization and quantitation of the host-guest interactions. Replacement of three ethyleneoxy units of cryptand [2.2.2] by 2-methylenepropyleneoxy or methyleneoxy units results in new cryptands that selectively bind Na^+ over both K^+ and Li^+. A new lariat ether bearing two pendant 5-chloro-8-hydroxyquinoline arms shows a "ligating macrocyclization" effect. This effect results in formation of a pseudo second macroring upon complexation with K^+ and Ba^{2+} and extremely high selectivity for these cations over other alkali and alkaline earth cations. Temperature has an effect on metal ion selectivity by crown ethers. An increase in the selectivity factor of Pb^{2+} over K^+ by 18C6 is observed when temperature is increased from 298.15 K to 398.15 K. Metal ion separations of commercial interest have been achieved by IBC Advanced Technologies, Inc. using principles developed in our laboratory. These separations include removal of impurity cations from hydrometallurgical streams, recovery and purification of precious metals from aqueous streams, removal of heavy metals from environmental streams, and recovery of radioactive nuclides from nuclear waste streams.

1. Introduction

The establishment and development of macrocycle chemistry have provided new approaches and a large variety of new reagents for studies of selectivities of metal ions,[1-4] anions,[1,5] and neutral molecules.[1,2,6,7] Many macrocyclic compounds have been synthesized and their complexation and selectivity properties have been studied.[1-4,6,8-11] A large number of these compounds have been used for analytical and separation purposes.[12-16]

In order to obtain high selectivity for a particular guest molecule, a macrocyclic host needs to have unique selective features that allow it to interact strongly with the guest. Characterization and quantitation of host-guest interactions are two important aspects in the design of macrocyclic compounds for metal ion selectivity. Calorimetric, NMR, UV-visible spectroscopic, and X-ray crystallographic methods are powerful tools for studies of host-guest supramolecular interactions. A large number of thermodynamic data for host-

L. Echegoyen and A.E. Kaifer (eds.), Physical Supramolecular Chemistry, 413–431.

guest interactions have been determined by a calorimetric titration technique in our laboratory. Equilibrium constants (*K*) provide a direct evaluation of metal ion selectivity by the ligands. Enthalpy, entropy, and heat capacity changes provide not only explanations for the magnitudes of log *K* values but also information about structural changes during complexation. NMR and UV-visible spectral studies provide evidence of detailed structural relationships between the host and guest molecules in solution. X-ray crystallography shows the conformations of crystalline host-guest complexes. These studies usually present a clear profile of the origin of selectivity of the host molecules for cations and provide information of value in the design of selectivity into macrocyclic ligands.

In this paper, the design of metal ion selectivity into macrocyclic ligands is illustrated by several examples. Crown ethers, cryptands, lariat ethers, and a new type of lariat ethers, "quasi cryptands", are discussed and a comparison of their selectivity properties for metal ions is presented. The effect of temperature on metal ion selectivity is also discussed. Several applications of highly selective ligands for metal ion separations are presented.

2. Crown Ethers

18-Crown-6 (18C6) forms stable complexes with many metal ions.[1,2] The size-match relationship was an early concept used to describe macrocycle selectivity for metal ions.[1-4,17-22] Among related metal ions, the one with a size that most closely matches the crown cavity forms the most stable complex with the crown ether. Thus, of the alkali metal ions 18C6 forms the most stable complex with K^+ and it shows a K^+/Na^+ selectivity of 1.70 as measured by the difference of the log *K* values ($\Delta\log K$ = 1.70, Table 1) in methanol (MeOH). The size-match selectivity does not hold in some cases where other factors play significant roles in complexation.[19,23,24] One example is found in the aza-crown ethers bearing pendant arms which are called lariat ethers. The size-match relationship between the crown cavity and the cation is usually not observed due to the coordination of flexible side arms of the lariat ethers with the cation trapped in the parent macroring.[25-27] The lariat ethers will be discussed in detail below.

An increase in molecular rigidity of macrocycles usually enhances cation selectivity. As compared with 18C6, the more rigid dicyclohexano-substituted 18C6 ($Cy_2$18C6) shows an increased selectivity factor for K^+ over Na^+. *Cis-syn-cis*-$Cy_2$18C6 selects K^+ over Na^+ by a factor of 1.93 as measured by the $\Delta\log K$ value. As can been seen from Table 1, entropy losses for Na^+ and K^+ interactions with *cis-anti-cis*-$Cy_2$18C6 are smaller than those for their interactions with 18C6, indicating that the Cy_218C6 experiences a smaller conformational change by complexation with the cations, probably due to its rigid structure.

TABLE 1. Log *K*, Δ*H* (kJ/mol), and Δ*S* (J/mol•K) values for the interactions of 18-crown-6 type macrocycles with Na^+ and K^+ in methanol solution at 25°C

Ligand	Cation	log *K*	Δ*H*	Δ*S*	Δlog *K* [a]	Ref.
18C6	Na^+	4.36	-35.1	-34.4		19,20
	K^+	6.06	-56.1	-72.2	1.70	19,20
cis-syn-cis-Cy_218C6	Na^+	4.08				21,22
	K^+	6.01			1.93	21,22
cis-anti-cis-Cy_218C6	Na^+	3.68	-23.4	-8.0		21,22
	K^+	5.38	-43.9	-43.9	1.70	21,22

[a] Δlog *K* = log *K* (K^+) - log *K* (Na^+).

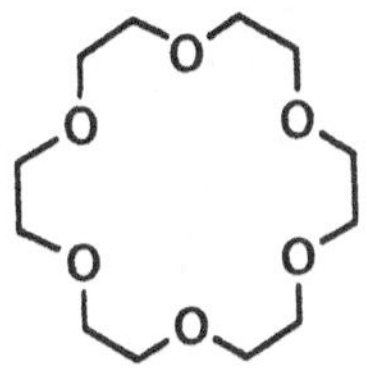

18C6

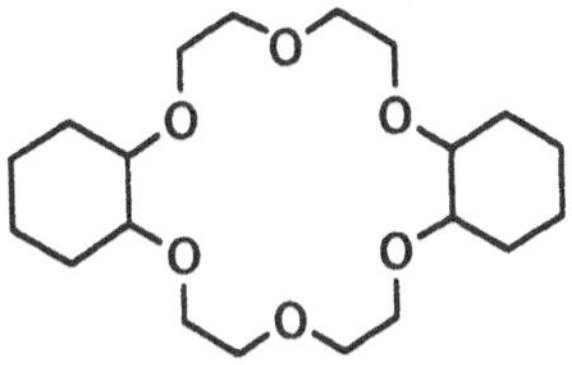

Cy_218C6

Attachment of two cyclohexane groups not only increases the molecular rigidity but also the hydrophobicity so that the ligands can be used for cation concentration and separation by solvent extraction. Another way to increase molecular rigidity and hydrophobicity is to attach two phenyl groups to 18C6 to form dibenzo-18-crown-6 (B_218C6). Although improved selectivity for K^+ over Na^+ by B_218C6 was not observed in homogeneous solvent systems due to the electronic effect,[8] B_218C6 shows excellent solvent extraction properties. Solvent extraction of alkali and alkaline-earth cations with Cy_218C6 and B_218C6 from aqueous solutions into organic phases has been studied extensively.[27-29] In addition, some multivalent metal ions can be effectively extracted by Cy_218C6 and B_218C6 in the presence of K^+.[30-35] These metal ions are extracted as the anionic forms, such as $InBr_4^-$, InI_4^-, $AuCl_4^-$, ReO_4^-, TcO_4^-, and TaF_6^- through formation of ion-pair complexes with the K^+-Cy_218C6 or K^+-B_218C6 cations.

The crystal structure of the K^+-18C6 complex shows that the K^+ is surrounded by a nearly planar hexagon of 18C6 oxygen atoms.[36] An interaction between K^+ and the disordered anion SCN^- indicates that the planar crown ether cannot completely encapsulate the cation. A crystal structure of the complex having a 1:2 K^+:B_218C6 ratio has been resolved.[37] In spite of the presence of two crown ethers, the K^+ is coordinated by only one of them and a water molecule coordinates with K^+ at the axial position (Figure 1). The second B_218C6 is in a "free" state (no interaction with K^+). This observation suggests that the effectiveness of cation binding can be increased by providing ligating donor(s) from a macrocyclic compound itself to bind the cation in the axial direction. The idea led to studies of macrobicyclic cryptands and armed lariat ethers.

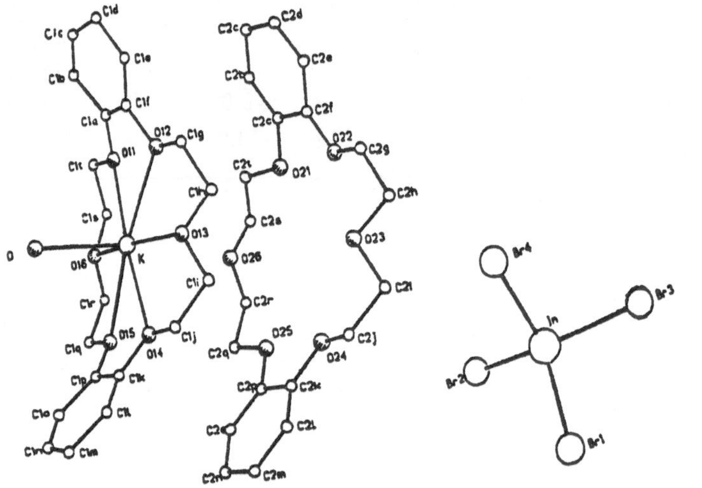

Figure 1. Crystal structure of the K^+-$(B_218C6)_2$ •InB_4^- •H_2O complex

3. Cryptands

Cryptand [2.2.2] can be regarded as a macrocycle having two 18-membered rings. Compared to 18C6, [2.2.2] selects K^+ over Na^+ by a higher factor[38] ($\Delta\log K = 2.48$ in an 80% MeOH solution, see Table 2). The increase in K^+/Na^+ selectivity by [2.2.2] over 18C6 originates from both enthalpy and entropy effects. Compared with 18C6 complexes, K^+-[2.2.2] interaction shows a larger enthalpy gain ($\Delta\Delta H$ = -65.8 - (-45.3) = -20.5 kJ/mol) and a lower entropy loss ($\Delta\Delta S$ = -57.6 - (-61.9) = 4.3 J/mol•K) than does Na^+-[2.2.2] interaction ($\Delta\Delta H$ = -40.5 - (-23.3) = -17.2 kJ/mol, $\Delta\Delta S$ = -20.2 - (-19.8) = -0.4 J/mol•K). These data indicate that the recognition ability of macrobicyclic [2.2.2] for K^+ over Na^+ is higher than that of the planar 18C6.

TABLE 2. Log K, ΔH (kJ/mol), and ΔS (J/mol·K) values for the interactions of cryptands and 18-crown-6 with alkali metal ions in methanol (MeOH) solutions at 25°C

Ligand	Cation	log K	ΔH	ΔS	Δlog K [a]	Solvent	Ref.
[2.2.2]	Na^+	6.04	-40.5	-20.2		80% MeOH	38
	K^+	8.52	-65.8	-57.6	2.48	80% MeOH	38
18C6	Na^+	3.05	-23.3	-19.8		80% MeOH	38
	K^+	4.70	-45.3	-61.9	1.65	80% MeOH	38
[2.2.1]	Li^+	4.69	-10.3	54.7	-	MeOH	40
	Na^+	9.71	-49.8	18.1		MeOH	40
	K^+	8.40	-61.1	-45.0	1.31	MeOH	40
[2.1.1]	Li^+	7.90	-33.9	36.9	-	MeOH	40
	Na^+	6.64	-33.1	15.4		MeOH	40
	K^+	2.36	-23.2	-32.9	4.28	MeOH	40
MP_3[2.2.2]	Li^+	b			-	80% MeOH	41
	Na^+	6.50	-71.7	-116.0		80% MeOH	38
	K^+	3.60	-43.5	-77.0	2.90	80% MeOH	38
M_2[2.2.2]	Li^+	< 2			-	80% MeOH	42
	Na^+	5.5	-36.3	-16.5		80% MeOH	42
	K^+	2.42	-28.0	-47.6	3.1	80% MeOH	42

[a] Δlog K = log K (K^+) - log K (Na^+) for [2.2.2] and 18C6 or Δlog K = log K (Na^+) - log K (K^+) for all others.
[b] No measurable heat other than heat of dilution indicates small ΔH and/or log K values.

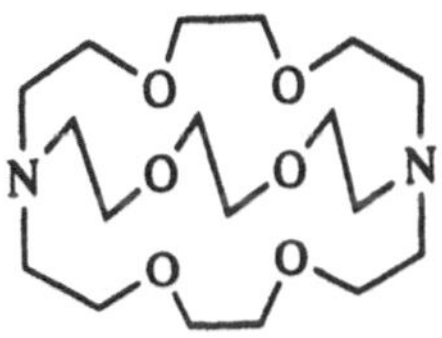

[2.2.2]

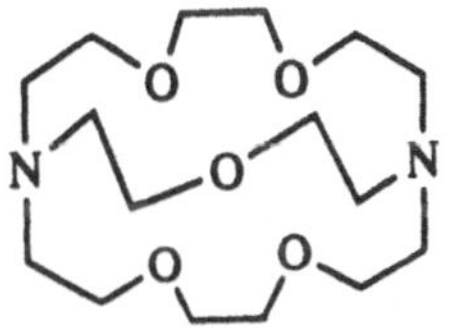

[2.2.1]

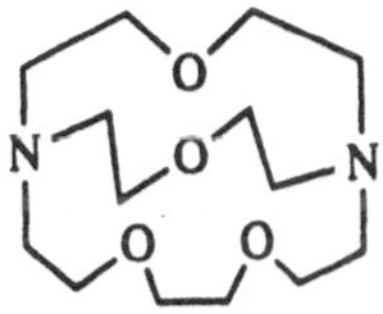

[2.1.1]

For selective binding of Na^+, small sized cryptands have been considered. Both [2.2.1] and [2.1.1] selectively bind Na^+ over K^+ (Table 2). However, the low symmetry of these cryptands results in some disadvantages.[39] [2.1.1] shows a high selectivity for Na^+ over K^+ (Δlog K = 4.28) but Li^+ is also highly selected. The presence of the [1.1] macroring portion is a reason for the high Li^+-binding constant. [2.2.1] selectively binds Na^+ over both K^+ and Li^+ but the Na^+/K^+ selectivity factor is not very high (Δlog K = 1.31). We recently designed and synthesized a new cryptand of high symmetry, MP_3[2.2.2], by introducing three 2-methylenepropylene groups into the cryptand [2.2.2].[43] The replacement of ethyleneoxy units of [2.2.2] by 2-methylenepropyleneoxy units results in a new cryptand that selectively binds Na^+ over K^+ with a high selectivity factor ($\Delta \log K = 2.90$) and also excludes Li^+ (see Table 2). The 2-methylenepropylene substituent has the following properties for high Na^+ selectivity. 1) Formation of six-membered chelate rings through the propyleneoxy group stabilizes a small cation over a large one;[11] 2) three double bonds increase the cryptand rigidity; and 3) the planar structure of the 2-methylenepropylene group results in a proper conformation of the ligand for Na^+ binding.[38]

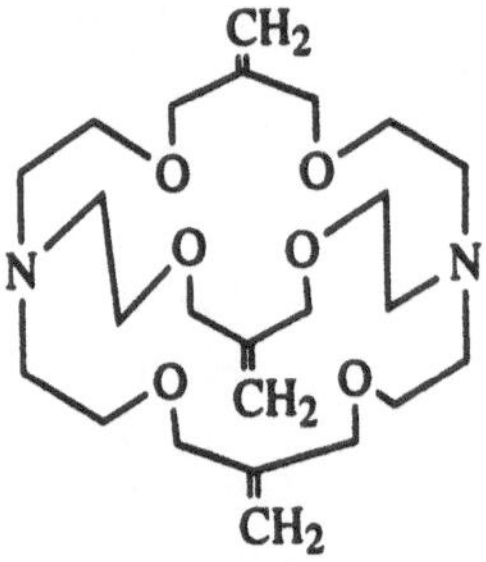

$MP_3[2.2.2]$

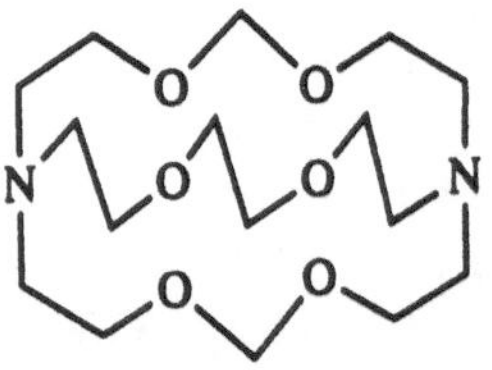

$M_2[2.2.2]$

Decreasing cavity size and keeping the number of oxygen atoms unchanged for the cryptand [2.2.2] are other ways to selectively bind Na^+ over K^+. Cryptand $M_2[2.2.2]$ shows a high selectivity for Na^+ over K^+ ($\Delta \log K = 3.1$, Table 2) due to a significant decrease in cavity size by replacing two ethylene units of [2.2.2] with two methylene groups.[42] Both ΔH (more negative for Na^+-$M_2[2.2.2]$ than for K^+-$M_2[2.2.2]$ complexation) and ΔS (less negative for Na^+-$M_2[2.2.2]$ than for K^+-$M_2[2.2.2]$ complexation) values make contributions to Na^+ selectivity. On the other hand, only the enthalpy change makes a contribution to Na^+ selectivity over K^+ by $MP_3[2.2.2]$. The more unfavorable entropy change for Na^+-$MP_3[2.2.2]$ interaction indicates that the $MP_3[2.2.2]$ experiences a large conformational change to accommodate the small Na^+.

4. Lariat Ethers

Although cryptands form three-dimensional complexes of high stability with cations, complexation and decomplexation rates are usually lower than those for flat crown ethers.[2,44,45] Attachment of pendant ligating arm(s) to a macrocycle results in a new class of ligands: lariat ethers. The lariat ethers retain the high dynamic properties of monocyclic crown ethers[46-48] and meanwhile provide three-dimensional coordination with the cations. Involvement of the side arms in cation binding has been verified by thermodynamic, NMR, IR, and X-ray crystallographic data.

TABLE 3. Log K, ΔH (kJ/mol), and ΔS (J/mol·K) values for the interactions of lariat ethers with Na^+ and K^+ in methanol (MeOH) solution at 25°C

Ligand	Cation	log K	ΔH	ΔS	Δlog K [a]	Ref.
1	Na^+	1.92	-18.1	-24.2		49
	K^+	2.31	-28.9	-52.7	0.39	49
2	Na^+	4.61	-29.0	-9.39		49
	K^+	> 5.5	-47.4	> -53.3	> 0.89	49
3	Na^+	4.5	-28.6	-10.1		49
	K^+	> 5.5	-48.2	> -56.0	> 1.0	49
4	Na^+	4.33				50
	K^+	6.07			1.74	50
5	Na^+	4.75				50
	K^+	5.46			0.71	50

[a] Δlog K = log K (K^+) - log K (Na^+).

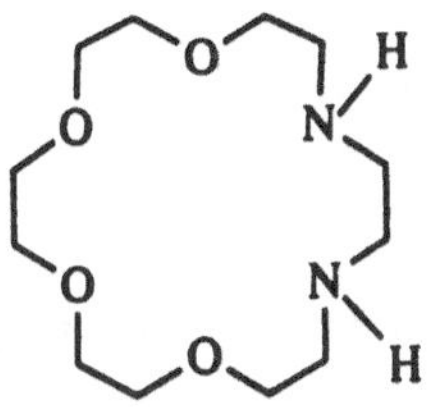

1

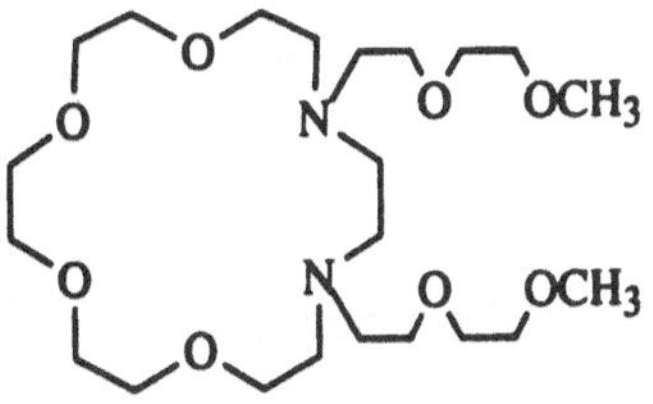

2

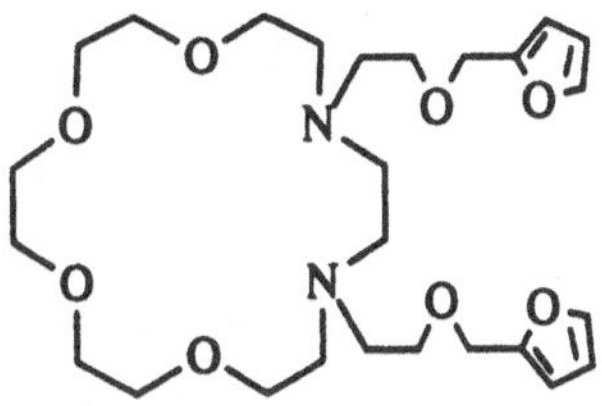

3

The log K (MeOH) values for complexation of Na^+ and K^+ with 1,4-diaza-18-crown-6 (**1**) are 1.92 and 2.31, respectively (Table 3).[49] However, attachment of side arms to **1** to form **2** and **3** results in increases in log K values of over two orders of magnitude. As can be seen from Table 3, the large increase in K^+-binding constants is due to increasingly favorable enthalpy changes. This large increase in -ΔH values for K^+ interaction with **2** and **3** as compared with **1** (18.5 kJ/mol and 19.3 kJ/mol, respectively) indicate that the side arms are involved in cation binding. On the other hand, both enthalpy and entropy changes contribute to the increased Na^+-binding constants with **2** and **3**. The favorable change in ΔH values is attributed to the binding of the side arms with Na^+ while the favorable change in ΔS values suggests that additional solvent molecules are replaced by the side arms. Therefore, the thermodynamic data provide clear evidence for side-arm involvement in cation binding.

Side-arm involvement in lariat ether complexes has been confirmed by Gokel et al. with ^{13}C-NMR relaxation time and lanthanide shift reagent studies,[50] by Kataky et al.

with ^{13}C NMR and IR spectral studies,[51] and by Zhu et al. with a ^{1}H-NMR spectral study.[52] In the solid state, the involvement of the side arm(s) in cation binding has been observed by X-ray crystallographic studies for a number of the lariat ether complexes. The results have been summarized by Fronczek and Gandour.[53]

Compared with cryptands, however, lariat ethers show lower cation selectivity and binding abilities.[1,2,50] Among a series of 18-membered-ring lariat ethers, for example, **4** shows the strongest interaction with K^+.[50,54] However, the log K (MeOH) value for K^+-**4** complexation (6.07, Table 3) is not higher than that for K^+-18C6 interaction (6.06, Table 1). The selectivity factor of **4** for K^+ over Na^+ (Δlog $K = 1.74$) is almost the same as that of 18C6 (see Tables 1 and 3). The double-armed lariat ether **5** has the same number of oxygen atoms as [2.2.2] and 18C6, but it shows both lower K^+-binding constant and K^+/Na^+ selectivity than either 18C6 or [2.2.2] (Tables 1-3).

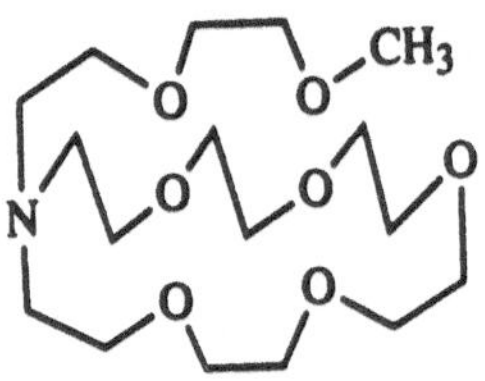

4

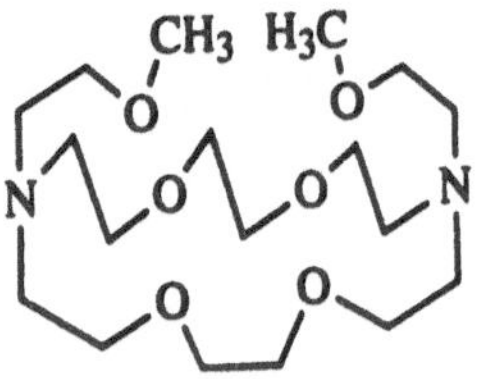

5

Many lariat ethers display lower cation-binding ability than their parent all-oxygen-containing crown ethers. For example, some single- and double-armed aza-18-crown-6 ethers exhibit smaller binding-constant values than 18-crown-6 for alkali and alkaline-earth metal ions.[50,54,55] This is caused by introducing into the macroring nitrogen atom(s) that have lower affinity than the oxygen atom for the cations. The reason for the high log K value for K^+-**4** interaction is that **4** has seven oxygen atoms, more than those of an 18C6 molecule. The seven oxygen atoms (five in the crown ring and two in the side arm) compensate for the weak interaction of the macroring nitrogen with the cation. **5** shows low cation-binding constants and selectivity since it has only six oxygen atoms. The two oxygen donors on the side arms can not compensate for the weak interaction of the two nitrogen atoms in the macroring with the cation. Both **4** and **5** cannot rival cryptand [2.2.2] owing to the lack of the second macroring. The lower selectivity of lariat ethers as compared with cryptands is attributed to high flexibility of the ligands. The flexible lariat ethers can easily modify their conformation to accommodate different sizes of guest molecules resulting in low selectivity.

5. Ligating-Macrocyclization Lariat Ethers

It is desirable to retain the highly dynamic properties of the lariat ethers. One way to increase the cation-binding ability and the selectivity factor of a macrocyclic ligand is to design special pendant arm(s) into the macrocycle. These pendant arms may interact with each other or one pendant arm may interact with the crown ring through an intramolecular interaction so that a pseudo second macroring could be formed during complexation with a cation. The resulting complex would resemble a cryptate and be expected to have the high stability of cryptate complexes and the highly dynamic property of lariat ethers. We recently reported a new lariat ether **6**, a diaza-18-crown-6 derivative bearing two 5-chloro-8-hydroxyquinoline (CHQ) arms, and studied its complexation properties.[56] The CHQ group in **6** has two important roles in cation binding and selectivity. First, CHQ can form a stable five-membered chelate ring with a metal ion thereby increasing the complex stability. Second, it is possible for two CHQs to stack together through π-π interaction, so that a second macroring can be formed.

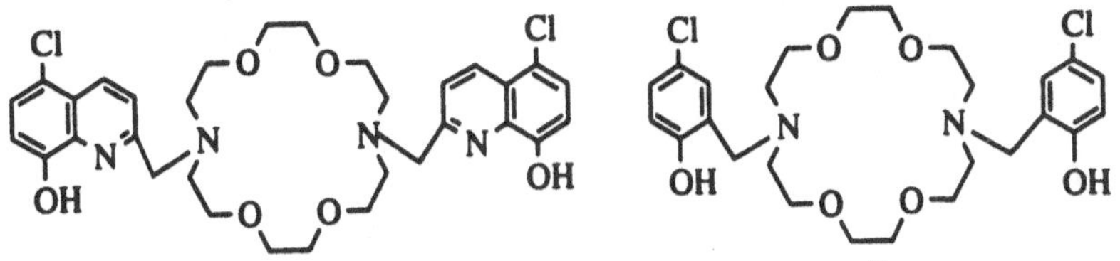

6 **7**

Thermodynamic data in Table 4 indicate that the double-armed lariat ether **6** exhibits unique complexing properties as compared with its phenol-substituted analog **7**. More stable complexes with alkali and alkaline-earth metal ions and much higher selectivities for Ba^{2+} over all other cations studied and for K^+ over Na^+ and Cs^+ are shown by **6**. Weaker interactions in the case of **7** are a result of smaller enthalpy contributions. Log *K* values for the formation of K^+ and Ba^{2+} complexes with **6** are larger than those for K^+ and Ba^{2+} complexes with all other lariat ethers.[1,2,50,54] The log *K* (MeOH) value for the **6**-Ba^{2+} complex (12.2) is the same magnitude as that of the cryptand [2.2.2]-Ba^{2+} complex (12.9 [57,58]). Selectivity factors as measured by Δlog *K* for Ba^{2+} over other alkaline-earth cations and for K^+ over Na^+ are > 7.5 and 2.87, respectively, which are the highest factors ever reported for lariat ethers.[1,2,50,54] In addition, the selectivity of **6** for Ba^{2+} is larger than that of any cryptand studied to date.[1,2,59]

The two CHQ arms play an important role in the formation of highly stable complexes and in the high selectivities for K^+ and Ba^{2+}. Burton and Davis have shown that the phenol ring of CHQ is electron-rich and the pyridine ring electron-deficient.[60] Therefore, π-π interaction could occur between the two CHQ rings if they approach each other in such a way that the pyridine part of the one CHQ overlaps the phenol part of the other CHQ. A possible way to bring the two CHQ rings together is through coordination with a cation that provides the correct template effect. K^+ and Ba^{2+} provide this effect to bring the two CHQ arms together so that a pseudo second macroring is formed. This effect results in a cryptate-like structure and, therefore, highly stable complexes.

TABLE 4. Log *K*, Δ*H* (kJ/mol), and Δ*S* (kJ/mol•K) values for interactions of quinoline- and phenol-substituted diaza-18-crown-6 ligands with cations in methanol at 25°C. Data are from ref. 56.

	6			**7**		
Cation	log *K*	Δ*H*	Δ*S*	log *K*	Δ*H*	Δ*S*
Na^+	3.74	-26.4	-17.1	2.85	-16.0	1.01
K^+	6.61	-58.1	-68.4	2.76	-24.1	-28.2
Cs^+	2.70	-36.9	-72.1	a		
Mg^{2+}	a			a		
Ca^{2+}	4.71	-25.2	5.7	4.48	-3.3	74.8
Sr^{2+}	4.67	-24.6	7.0	< 2		
Ba^{2+}	12.2	-76.1	-21.8	3.52	-32.2	-40.9

[a] No measurable heat other than heat of dilution indicating that Δ*H* and/or log *K* is small.

^{1}H NMR and UV-visible spectra provide the evidence for π-π overlapping between the two CHQs. A comparison of ^{1}H NMR spectra for free **6**, **6**-K^+, and **6**-Ba^{2+} complexes is illustrated in Figure 2. Upon complexation with K^+ and Ba^{2+}, large upfield shifts of CHQ aromatic proton signals can be observed for lariat ether **6**. This is caused by overlapping of the two aromatic rings, which results in a magnetic shielding effect.[61] On the other hand, Na^+ induced much smaller chemical shifts of CHQ protons.[56] It has been observed that π-π overlapping can result in a decrease in UV absorption intensity of aromatic rings.[62] UV-visible spectra of **6** and its complexes in MeOH solution indicate that the absorption intensity of the CHQ of free **6** at 249 nm and 319 nm decreases in its complexes with K^+ and Ba^{2+} (Figure 3a) suggesting an overlapping between the two CHQ rings. On the other hand, **7** exhibits a different UV behavior from that of **6** due to absence of π-π interaction between the two 4-chlorophenol rings. Increased UV absorption intensity upon complexation of **7** with K^+ and Ba^{2+} (Figure 3b) is a result of coordination of the phenols with the cations.[63]

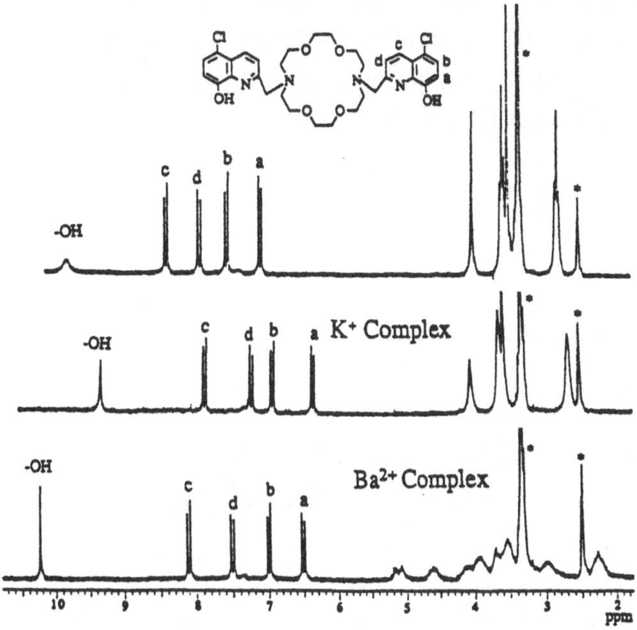

Figure 2. ^{1}H NMR spectra of free **6** and its K^+ and Ba^{2+} complexes in DMSO. The peaks labeled by * are attributed to the solvent.

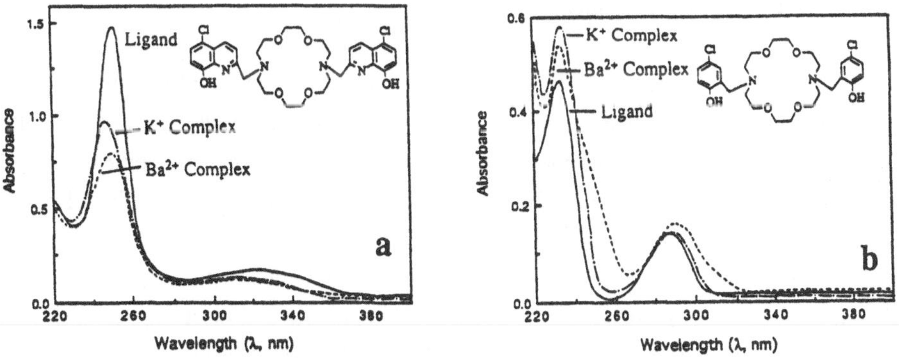

Figure 3. UV spectra of free and complexed **6** (a) and **7** (b) in MeOH. [**6**] = 2.04 x 10^{-5} M and [**7**] = 3.09 x 10^{-5} M.

Crystal structures of the complexes of **6**-Ba^{2+} show that the cation is coordinated by all oxygen and nitrogen atoms of the ligand and a water molecule (Figure 4a). The coordination of the Ba^{2+} by both CHQ arms brings the two aromatic bidentate groups close together so that π-π interactions are possible. The two CHQ rings are nearly parallel. The dihedral angle between the least-square planes of the aromatic groups is 13° and the distance between the centers of the CHQ rings is 3.6 Å. The electron-rich phenol rings overlap the electron-deficient pyridine rings. Therefore, the crystal structure clearly shows a π stacking between the two CHQ rings.

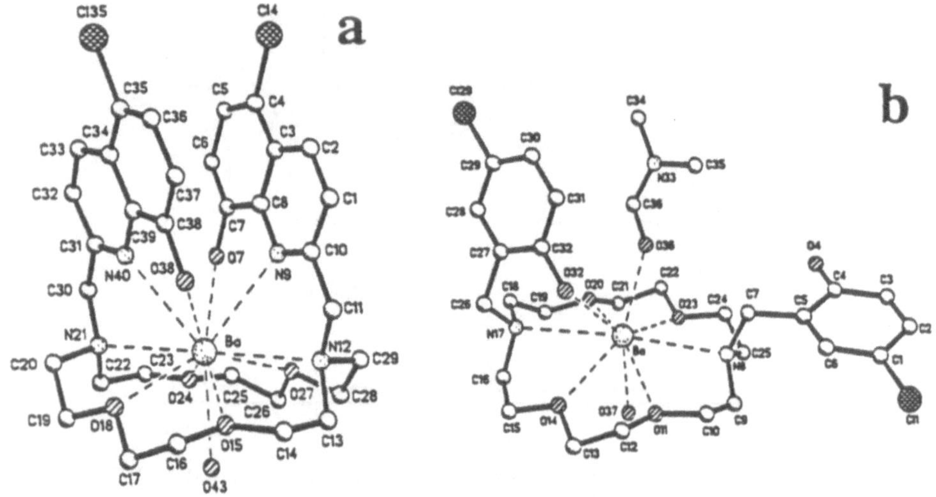

Figure 4. Crystal structures of the **6**-Ba^{2+} (a) and **7**-Ba^{2+} (b) complexes. The hydrogen atoms, the two anions (Br^-), and in the case of **6**-Ba^{2+} the solvent molecules not involved in the coordination are omitted for clarity.

The solid state structure of the **7**-Ba^{2+} complex differs significantly from that of the **6**-Ba^{2+} complex. The two phenol rings are far from each other (Figure 4b) and one of the phenol groups does not interact with the Ba^{2+}. Instead, two solvent molecules (water and DMF) coordinate with the cation. The lack of interaction between the two phenol rings is consistent with a low log *K* value for the formation of the Ba^{2+} complex (Table 4) due to absence of a second macroring. In solutions, interaction between the two phenol groups of **7** in the complex is also unlikely because of geometric and electronic features. Each aromatic arm has only one donor atom and both phenol arms are electron-rich.

In summary, the two CHQ substituents on diaza-18-crown-6 play an important role in cation binding and selectivity by **6**. Not only do they provide donor atoms to coordinate with the cation, but they also stack together upon complexation to form a second macroring. The resulting cryptand-like structure allows very strong complexation with the cation that plays a key role in forming the additional macroring. Coordination with a cation whose size matches the pseudo three-dimensional cavity of the lariat ether induces a conformation that causes the two CHQ rings to stack together to form the second macroring. That is why we refer to this macrocycle as a "ligating-macrocyclization lariat ether". Neither a larger nor a smaller cation could form the complex having a pseudo second macroring since the complex would not have the two ligating side arms together due to an unfavorable size relationship between the cation and the ligand. Therefore, ligating-macrocyclization lariat ethers not only form highly stable complexes with the size-matched cations but also show high selectivities for those cations.

The two pendant arms of a ligating-macrocyclization lariat ether have no appreciable interaction with each other in the uncomplexed state. Upon complexation with a size-matched cation, however, the two arms interact to form a pseudo second macroring resulting in a cryptate-like complex. Hence, the ligating-macrocyclization lariat ethers could also be called "quasi cryptands".

6. Temperature Effect

A change in operating temperature results in a change in metal ion[64,65] and anion[66] separations by macrocycle-based chromatographic systems. In homogeneous solvent systems, few thermodynamic studies have been reported on macrocycle-cation interactions as a function of temperature (283.15 K - 318.15 K).[1,67,68] We recently determined thermodynamic quantities for 18C6 interaction with several metal ions in aqueous solution from 298.15 K to 398.15 K.[69] The results show that temperature has an appreciable effect on cation selectivity by 18C6. Several changes in cation selectivity from 298.15 K to 398.15 K are listed in Table 5. At 298.15 K, 18C6 selects Pb^{2+} over K^+ by a $\Delta\log K$ factor of 2.19. This factor is increased to $\Delta\log K = 2.41$ at 398.15 K. Although the increase in the selectivity factor is not large, the $\log K$ value for K^+-18C6 interaction at 398.15 K (0.86) is small compared with the Pb^{2+}-18C6 interaction (3.27). Thus, the binding sites of 18C6 at this temperature are largely occupied by Pb^{2+} and a small amount of K^+ is captured by the ligand. The selectivity factor of 18C6 for Ba^{2+} over Sr^{2+} decreases with increasing temperature (from $\Delta\log K = 1.15$ at 298.15 K to $\Delta\log K = 0.61$ at 298.15 K). Therefore, it is apparent that changing the temperature does have a marked effect on cation selectivities.

The changes in cation selectivity with temperature is a result of the different rates of equilibrium constant change with temperature. According to the van't Hoff equation,[70] since $(\partial \ln K/\partial T)_p = \Delta H°/RT^2$, the greater the value of $|\Delta H°|$, the greater the change in the log K value with temperature. Because the ΔH values for 18C6 interaction with different cations are different (see Table 5), the rate of the log K change with temperature is different for different cations resulting in a change in cation selectivity with temperature. The van't Hoff equation points out that log K values decrease for exothermic reactions ($\Delta H°$ negative) and increase for endothermic reactions ($\Delta H°$ positive) with increasing temperature. Therefore, a knowledge of $\Delta H°$ values will allow one to predict how the selectivity of a given ligand for metal ions changes with temperature. The larger the difference in the $\Delta H°$ values, the greater the change in metal ion selectivities with temperature.

TABLE 5. Log K, ΔH (kJ/mol), ΔS (J/mol•K), and ΔC_p (J/mol•K) values for the interactions of metal ions with 18C6 in aqueous solution at 298.15 and 398.15 K. Data are from ref. 69.

	298.15 K					398.15 K				
Cation	log K	ΔH	ΔS	ΔC_p	Δlog K [a]	log K	ΔH	ΔS	ΔC_p	Δlog K [a]
K^+	2.08	-24.0	-40	-18		0.86	-33.5	-68	-143	
Pb^{2+}	4.27	-21.6	9.2	20	2.19	3.27	-26.9	-5	-128	2.41
Sr^{2+}	2.72	-15.1	1.25	-19		1.96	-21.2	-16	-100	
Ba^{2+}	3.87	-31.7	-33	126	1.15	2.57	-31.3	-29	-104	0.61

[a] Δlog K = log K (Pb^{2+}) - log K (K^+) or Δlog K = log K (Ba^{2+}) - log K (Sr^{2+}).

It is seen in Table 5 that not only log K but also ΔH and ΔS values change with increasing temperature. This is a result of the change of water structure. It has been concluded that the change of water structure as temperature increases causes the thermodynamic quantities for a chemical reaction in aqueous solution to change dramatically.[71] For the 18C6-cation interactions in Table 5, the decrease in log K values with increasing temperature results from the decreased ΔS values. The negative ΔS values at elevated temperature probably indicate that the water structure becomes more organized with the formation of the complexes. A highly organized shell of water molecules is formed around the complex due to the ion-dipole, dipole-dipole, and hydrophobic interactions.[69,72] Because the increase in temperature decreases the dielectric constant of water and the amount of hydrogen bonding,[72] the effect of the cation charge on the water dipoles becomes more important at high temperature. As a result, significant changes in ΔH and ΔS values for the complexation reactions can be observed with increasing temperature.

A better understanding of the solvent-solute interactions can be obtained from ΔC_p values.[69] Data in Table 5 show that the ΔC_p values for 18C6 complexation decrease significantly with increasing temperature. The positive ΔC_p values for Pb^{2+} and Ba^{2+} at 298.15 K indicate the net release of water molecules from the desolvation of the cations and the ligand upon complexation. On the other hand, the negative ΔC_p values for K^+ and Sr^{2+} at both 298.15 K and 398.15 K and for Pb^{2+} and Ba^{2+} at 398.15 K suggest strong solvation of the complexes under the indicated conditions. This conclusion is consistent with the postulation of highly organized water systems as demonstrated by negative ΔS values at the higher temperature.

7. Practical Applications

The need to develop highly selective reagents capable of discriminating among similar chemical species has been recognized.[73] Such reagents should be capable of concentrating trace amounts of solutes even in the presence of large excesses of matrix solutes and of removing solutes from large quantities of solvent. Based on initial separation studies at BYU by two of us (RMI and JSB) and our students, IBC Advanced Technologies, Inc. (American Fork, Utah) has prepared cation-selective ligands, attached them to appropriate solid supports, and used the resulting materials to accomplish numerous difficult separations of commercial interest. The principles involved in the separations, descriptions of the processes, and examples of actual separations have been described.[74-80] Some commercial applications include removal of radioactive Sr^{2+} and Ra^{2+} from nuclear waste streams, removal of bismuth from copper smelter streams, removal of Pb^{2+} and Hg^{2+} from aqueous streams, and removal of Pd^{2+} and Rh(III) from platinum metal refinery streams.

8. Summary

Metal ion selectivities can be improved by incorporation of different structural elements into macrocyclic compounds. Such structural elements include number and size of the macrorings, different chelate rings and pendant arms, and an increase in molecular rigidity. Temperature has an effect on metal ion selectivity due to the different magnitudes of ΔH values for macrocycle-metal ion interactions. Crown ethers can form stable complexes with metal ions, but usually show lower metal ion selectivity than do cryptands that provide three-dimensional metal ion coordination. Modification of the cryptand bridge structures can increase cation selectivity. Lariat ethers show good dynamic properties but form less stable complexes with metal ions and lower selectivities for metal ions than do the cryptands. A new lariat ether bearing two pendant arms that can interact with each other to form a pseudo second macroring by complexation exhibits both strong interaction with metal ions and high selectivity for K^+ and Ba^{2+} over other similar metal ions.

Acknowledgment

The authors thank the Department of Energy, Office of Basic Energy Sciences for financial support through Grant No. DE-FG02-86ER-13463. We warmly thank the following co-workers for carrying out part of the research described here: N. Kent Dalley, John L. Oscarson, Xiaolan Kou, Ronald L. Bruening, Krzysztof E. Krakowiak, Peiming Wang, Haoyun An, Cheng Y. Zhu, and Tingmin Wang.

References

1. (a) Izatt, R.M., Pawlak, K., Bradshaw, J.S., and Bruening, R.L. (1991) *Chem. Rev.* **91**, 1721-2085. (b) Izatt, R.M., Pawlak, K., Bradshaw, J.S., and Bruening, R.L. (1995) *Chem. Rev.* **95**, 2529-2586.
2. Izatt. R.M., Bradshaw, J.S., Nielsen, S.A., Lamb, J.D., and Christensen, J.J. Sen, D. (1985) *Chem. Rev.* **85**, 271-339.
3. Dietrich, B., Viout, P., and Lehn, J.-M. (1993) *Macrocyclic Chemistry*, VCH Publisher, New York.
4. Patai, S. and Rappoport, Z. (Eds) (1989) *Crown Ethers and Analogs*, John Wiley & Sons, New York.
5. Dietrich, B. (1993) *Pure Appl. Chem.* **65**, 1457-1464.
6. Izatt, R.M., Bradshaw, J.S., Pawlak, K., Bruening, R.L., and Tarbet, B.J. (1992) *Chem. Rev.* **92**, 1261-1354.
7. (a) Schneider, H.J. (1991) *Angew. Chem. Int. Ed. Engl.* **30**, 1417-1436. (b) Franke, J. and Vögtle, F. (1986) *Topics in Current Chemistry*, Vol.132, Springer-Verlag; Berlin, p135.
8. Gokel, G.W. (1991) *Crown Ethers and Cryptands*, U.K.: Royal Society of Chemistry, Cambridge.
9. Bradshaw, J.S., Krakowiak, K.E., and Izatt, R.M. (1993) *Aza-crown Macrocycles*, John Wiley & Sons, New York.
10. Weber, E. (Ed.) (1993) *Supramolecular Chemistry I - Directed Synthesis and Molecular Recognition*, Springer-Verlag; Berlin.
11. Hancock, R. D. and Martell, A. E. (1989) *Chem. Rev.* **89**, 1875-1914.
12. Hiraoka, M. (1982) *Crown Compounds: Their Characteristics and Applications*, Elsevier, Amsterdam.
13. Takagi, M. and Nakamura, H. (1986) *J. Coord. Chem.* **15**, 53-82.

14. Tsukube, H. (1993) *Talanta* **40**, 1313-1324.

15. Tsukube, H. (1987) *J. Coord. Chem.* **16**, 101-129.

16. Pimple, M. (1995) *J. Radioanal. Nucl. Chem.* **194**, 311-318.

17. Pedersen C.J. and Frensdorff, H.K. (1972) *Angew. Chem. Int. Ed. Engl.* **11**, 16-25.

18. Lehn, J.-M. and Sauvage, J.P. (1975) *J. Am. Chem. Soc.* **97**, 6700-6707.

19. Lamb, J.D., Izatt, R.M., Swain, S.W., and Christensen, J.J. (1980) *J. Am. Chem. Soc.* **102**, 475-479.

20. Haymore, B.L., Lamb, J.D., Izatt, R.M., and Christensen, J.J. (1982) *Inorg. Chem.* **21**, 1598-1602.

21. (a) Izatt, R.M., Rytting, J.H., Nelson, D.P., Haymore, B.L., and Christensen, J.J. (1969) *Science* **164**, 443-444. (b) Christensen, J.J., Eatough, D.J., and Izatt, R.M. (1974) *Chem. Rev.* **74**, 351-384.

22. Frensdorf, H.K. (1971) *J. Am. Chem. Soc.* **93**, 600-606.

23. Gokel, G.W., Goli, D.M., Minganti, C., and Echegoyen, L. (1983) *J. Am. Chem. Soc.* **105**, 6786-6788.

24. Gokel, G.W. and Trafton, J.E. (1990) In Inoue, Y. and Gokel, G.W. (eds.), *Cation Binding by Macrocycles,* Marcel Dekker, New York, Chapter 6.

25. Schultz, R.A., Dishong, D.M., and Gokel, G.W. (1982) *J. Am. Chem. Soc.* **104**, 625-626.

26. Gandour, R.D., Fronczek, F.R., Gatto, V.J., Minganti, C., Schultz, R.A., White, B.D., Arnold, K.A., Mazzocchi, D., Miller, S.R., and Gokel, G.W. (1986) *J. Am. Chem. Soc.* **108**, 4078-4088.

27. Katayama, Y., Nita, K., Ueda, M., Nakamura, H., and Takagi, M. (1985) *Anal. Chim. Acta* **173**, 193-209.

28. Takeda, Y. (1984) In Boschke, F.L. (ed.), *Topics in Current Chemistry*, Vol.121, Springer-Verlag, Berlin, p1.

29. Olsher, U., Hankins, M.G., Kim, Y.D., and Bartsch, R.A. (1993) *J. Am. Chem. Soc.* **115**, 3370-3371.

30. (a)Zhang, X.X., Zhou, Z.X., Ma, S.J., and Shu, C. (1993) *Sol. Extr. Ion Exch.* **11**, 585-601. (b) Zhou, Z.X. and Zhang, X.X. (1988) *Huaxue Xuebao* (*Acta Chim. Sinica*), 46, 496-499.

31. Koshima, H. and Onishi, H. (1990) *Anal. Chim. Acta* **232**, 287-292.

32. Jalhoom, M.G. (1986) *Radiochim. Acta* **39**, 195-197.

33. Vibhute, R.G. and Khopkar, S.M. (1989) *Anal. Chim. Acta* **222**, 215-219.

34. Caletka, R., Hausbeck, R., and Krivan, V. (1986) *Talanta* **33**, 219-224.

35. Namdeo, R.P. and Khopkar, S.M. (1995) *India J. Chem. Sect.A* **34A**, 840-842.

36. Seiler, P., Dobler, M., and Dunitz, J.D. (1974) *Acta Crystallogr.* **B30**, 2744-2745.

37. Ye, L., Fan, Y.G., Zhou, Z.X., and Zhang, X.X. (1987) *Wuji Huaxue* (*J. Inorg. Chem.*) **3**, 93-99.

38. Krakowiak, K.E., Zhang, X.X., Bradshaw, J.S., Zhu, C.Y., and Izatt, R.M. (1995) *J. Incl. Phenom. Mol. Recognit. Chem.* in press.

39. Zhang, X.X., Izatt, R.M. Krakowiak, K.E., and Bradshaw, J.S. (1995) *Inorg. Chim. Acta* submitted.

40. Buschmann, H.J. (1986) *Inorg. Chim. Acta* **125**, 31-35.

41. Zhang, X.X., Izatt, R.M., and Bradshaw, J.S. (1995) Unpublished results.

42. Bradshaw, J.S., An, H.Y., Krakowiak, K.E., Wang, T.-M., Zhu, C.Y., and Izatt, R.M. (1992) *J. Org. Chem.* **57**, 6112-6118.
43. Krakowiak, K.E., Bradshaw, J.S., and Izatt, R.M. (1990) *J. Heterocyclic. Chem.* **27**, 1011-1014.
44. Liesegang, G.W., Farrow, M.M., Vazquez, F.A., Purdie, N., and Eyring, E.M. (1977) *J. Am. Chem. Soc.* **99**, 3240-3243.
45. Lehn, J.-M. (1973) *Struct. Bonding* **16**, 1-69.
46. Echegoyen, L., Gokel, G.W., Kim, M.S., Eyring, E. M., and Petrucci, S. (1987) *J. Phys. Chem.* **91**, 3854-3862.
47. Echegoyen, L., Kaifer, A., Durst, H., Schultz, R.A., Dishong, D.M., Goli, D.M., and Gokel, G.W. (1984) *J. Am. Chem. Soc.* **106**, 5100-5103.
48. Eyring, E.M., Petrucci, S. (1990) In Inoue, Y. and Gokel, G.W. (Eds.) *Cation Binding by Macrocycles*, Marcel Dekker, New York, Chapter 4.
49. Izatt, R.M., Zhang, X.X., An, H.Y., Zhu, C.Y., and Bradshaw, J.S. (1994) *Inorg. Chem.* **33**, 1007-1010.
50. (a) Gokel, G. W. and Trafton, J. E. (1990) In Inoue, Y. and Gokel, G.W. (Eds.) *Cation Binding by Macrocycles*, Marcel Dekker, New York, Chapter 6. (b) Gokel, G.W. (1992) *Chem. Soc. Rev.* **21**, 39-47.
51. Kataky, R., Parker, D., Teasdale, A., Hutchinson, J.P., and Buschmann, H.-J. (1992) *J. Chem. Soc. Perkin Trans.* 2 1347-1351.
52. Zhu, C.Y., Izatt, R.M., Wang, T.-M., Huszthy, P., and Bradshaw, J.S. (1993) *Pure Appl. Chem.* **65**, 1485-1492.
53. Fronczek, F.R. and Gandour, R.D. (1990) In Inoue, Y. and Gokel, G.W. (Eds.) *Cation Binding by Macrocycles*, Marcel Dekker, New York, Chapter 7.
54. Arnold, K.A., Hernandez, J.C., Li,C., Mallen, J.V., Nakano, A., Schall, O.F., Trafton, J.E., Tsesarskaja, M., White, B.D., and Gokel, G.W. (1995) *Supramol. Chem.* **5**, 45-60.
55. Gatto, V.J., Arnold, K.A., Viscasiello, A.M., Miller, S.R., Morgan, C.R., and Gokel, G.W. (1986) *J. Org. Chem.* **51**, 5373-5384.
56. Zhang, X.X., Bordunov, A.V., Bradshaw, J.S., Dalley, N.K., Kou, X., and Izatt, R.M. (1995) *J. Am. Chem. Soc.* **117**, 11507-11511.
57. Chantooni, M. K., Jr., Kolthoff, I. M. *J. Solution Chem.* **1985**, *14*, 1-12.
58. Arnaud-Neu, F., Yahya, R., Schwing-Weill, M. J. *J. Chim. Phys. Phys.-Chim. Biol.* **1986**, *83*, 403-408.
59. (a) Lehn, J-M. and Sauvage, J-P. *J. Am. Chem. Soc.* **1975**, *97*, 6700-6707.
(b) Kauffmann, E., Lehn, J-M., and Sauvage, J-P. *Helv. Chim. Acta* **1976**, *59*, 1099-1111.
60. Burton, R.E. and Davis, W.J. (1964) *J. Chem. Soc.* 1766-1771.
61. Sanders, J.K.M. and Hunter, B.K. (1987) *Modern NMR Spectroscopy*, Oxford University Press, Oxford.
62. Colquhoun, H.M., Stoddart, J.F., Williams, D.J., Wolstenholme, J.B., and Zarzycki, R. (1981) *Angew. Chem. Int. Ed. Engl.* **20**, 1051-1053.
63. Kimura, E., Kimura, Y., Yatsunami, T., Shionoya, M., and Koike, T. (1987) *J. Am. Chem. Soc.* **109**, 6212-6213.
64. Fortier, N.E. and Fritz, J.S. (1987) *Talanta* **34**, 415-418.

65. Iwachido, T., Naito, N., Samukawa, F., Ishimaru, K., and Toei, K. (1986) *Bull. Chem. Soc. Jpn.* **59**, 1475-1480.
66. Smith, R.G., Drake, P.A., and Lamb, J.D. (1991) *J. Chromatogr.* **546**, 139-149.
67. Izatt, R.D., Nelson, D.P., Rytting, J.H., Haymore, B.L., and Christensen, J.J. (1971) *J. Am. Chem. Soc.* **93**, 1619-1623.
68. Vasil'ev, V.P., Orlova, T.D., and Goncharova, N.Y. (1992) *Russ. J. Inorg. Chem.* **37**, 1080-1082.
69. Wang, P., Izatt, R.M., Gillespie, S.E., Oscarson, J.L., Zhang, X.X., Wang, C., and Lamb, J.D. (1995) *J. Chem. Soc. Faraday Trans.* **91**, 4207-4213.
70. Levine, I.N. (1988) *Physical Chemistry*, 3rd ed. McGraw-Hill, New York, pp177, 308.
71. Chen, X., Oscarson, J.L., and Izatt, R.M. (1994) *Chem. Rev.* **94**, 465-517.
72. Izatt, R.M., Oscarson, J.L., Gillespie, S.E., Chen, X., Wang, P., and Watt, G.D. (1995) *Pure Appl. Chem.* **67**, 543-549.
73. King, G.J. (Ed.) (1987) *Separation and Purification*, National Academy of Science, U.S..
74. Izatt, R.M., Bruening, R.L., Bruening, M.L., Tarbet, B.J., Krakowiak, K.E., Bradshaw, J.S., and Christensen, J.J. (1988) *Anal. Chem.* **60**, 1825-1826.
75. Bradshaw, J.S., Bruening, R.L., Krakowiak, K.E., Tarbet, B.J., Bruening, M.L., Izatt, R.M., and Christensen, J.J. (1988) *J. Chem. Soc., Chem. Commun.* 812-814.
76. Izatt, R.M., Bradshaw, J.S., Bruening, R.L., Tarbet, B.J., and Krakowiak, K.E. (1993) In Lakshmanan, Bautista, R.G., Somasundaran, P. (Eds.) *Emerging Separation Technologies for Metals and Fuels*, TMS, Warrendale, PA, pp67-75.
77. Izatt, N.E., Bruening, R.L., Anthian, L., Griffin, L.D., Tarbet, B.J., Izatt, R.M., and Bradshaw, J.S. (1994) In Sohn H.Y. (Ed.) *Proceedings: Metallurgical Processes for Early 21st Century*, TMS, Warrendale, PA.
78. Izatt, R.M., Bradshaw, J.S., Bruening, R.L., and Bruening, M.L. (1994) *Am. Lab.* **26**, no.18, 28c-28m.
79. Bruening, R.L., Dale, J.B., Holbrook, W.R., Izatt, N.E., Rytting, M.H., Tarbet, B.J., Bradshaw, J.S., and Izatt, R.M. (1994) In Harris, B. and Krause, E. (Eds.) *Impurity Control and Disposal in Hydrometallurgical Processes: Proceedings of the 24th Annual CIM Hydrometallurgical Meeting*, CIM Publications, Montreal, pp213-221.
80. Izatt, R.M., Bradshaw, J.S., Bruening, R.L.,Tarbet, B.J., and Bruening, M.L. (1995) *Pure Appl. Chem.* **67**, 1069-1074.

ANION SENSING BASED ON THE METAL-LIGAND INTERACTION

L. FABBRIZZI, M. LICCHELLI, P. PALLAVICINI,
L. PARODI, A. POGGI AND A. TAGLIETTI
Dipartimento di Chimica Generale
Università di Pavia
27100 Pavia, Italy

ABSTRACT. Some chosen examples illustrate that coordinative interactions between anions and metal centres hosted by a poly-aza framework can be conveniently used for recognition and sensing purposes. A pair of Cu^{II} ions positioned within an octa-amine cage recognizes the bite length of ambidentate polyatomic anions (e.g. N_3^-, NCO^-, HCO_3^-). When the two Cu^{II} centres are positioned within an hexa-amine ring having two pendant picolyl arms, adequate room is provided for the binding of the imidazolate ion and for the recognition of any molecule bearing an imidazole residue, including histidine in presence of any other aminoacid. Finally, the Zn^{II} complex of an anthracenyl substituted tripodal tetra-amine recognizes aromatic carboxylates: when the anion exhibits either donor or acceptor tendencies, in a photophysical sense, binding is signalled through the quenching of the fluorescence of the appended anthracene fragment.

1. Introduction

Given a substrate, a primary element inspires the design of the corresponding receptor: *shape*. In particular, most synthetic efforts are centred on the preparation of a system offering a concave room that matches the geometrical features of the molecule or ion to be recognized (including *size*). In an historical perspective, the first objects of the molecular recognition were *s* block metal ions, typically spherical substrates. The proposed receptors, *crowns* [1] and *cryptands* [2], offer a cavity of radial symmetry, so that selectivity towards the metals could be modulated by enlarging-shrinking the ring or the cage. Coordinative interactions involving the ethereal oxygen atoms provide the energy gain responsible for binding.

Recognition of anions can be (and has been) approached in a similar way, as far as geometrical aspects (shape and size) are concerned. A substantial difference pertains to the nature of the receptor-substrate interaction. As anions are subject to electrostatic interactions, the receptor should bear a number of positively charged groups positioned along its framework. In this context, monoatomic anions can be incorporated and recognized by polyammonium cages of type **1**, which offer a spherical concavity [3].

L. Echegoyen and A.E. Kaifer (eds.), Physical Supramolecular Chemistry, 433–448.

Again, selectivity can be modulated by varying cavity size, i.e. the length of the aliphatic chains linking the ammonium groups. Ellipsoidal cages of *bistren* type, **2**, were used to encapsulate and recognize linear triatomic anions[4]. As the six secondary nitrogen atoms of the octa-amine cage have to be protonated, anion recognition studies must be carried out in a rather acidic solution.

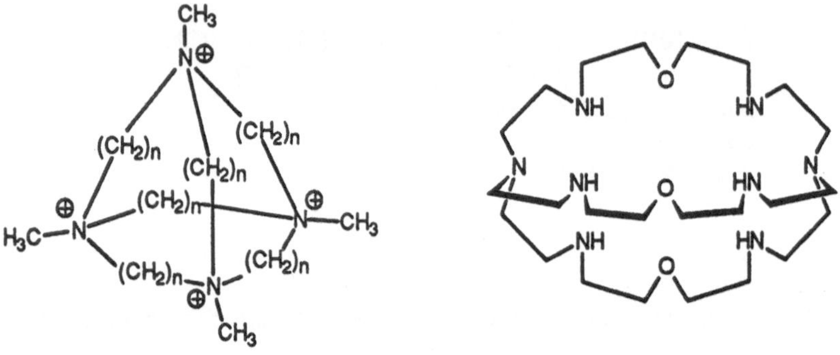

1 **2**

Figure 1. Anion recognition by cages. The receptor-substrate interaction is electrostatic in nature. In the cage **1**, four positive charges (alkylammonium groups) are positioned at the comers of a tetrahedron. The number of $-CH_2-$ groups of the chains linking the alkylammonium groups defines the size of the spherical cavity and determines selectivity towards anions (n=6: Br^-; n=8: I^-) [3]. The *bistren* cage **2**, when protonated at the six secondary amine groups, offers an ellipsoidal cavity and incorporates linear polyatomic anions (e.g. N_3^-) [4].

We considered that another type of interaction could be conveniently utilized for anion binding and recognition: the metal-ligand interaction. Compared to electrostatic interactions, coordinative bonds exhibit two main advantages: (i) usually, they are distinctly stronger than purely Coulombic interactions (especially when the metal centre profits from ligand field stabilization energy effects); (ii) they are directional, which may impose further geometric constraints and increase selectivity (whereas electrostatic interactions are typically a-directional). The receptor framework should contain donor atoms capable of coordinating one or more metal centres. In this sense, polyamine cages of type **2** seem appropriate systems, in view of the well known affinity of the sp^3 nitrogen atoms towards transition metal ions. However, the number of donor atoms positioned in the receptor framework should not saturate the coordinative tendencies of the metal(s): in fact, at least one binding site should be left available for the incoming anion. As a final and quite obvious requirement, metal ions displaying the greatest coordinative affinity towards the donor atom of the envisaged anion should be chosen.

In the following Sections, we will consider some examples of metal containing receptors for anions. They include a dicopper(II) *bistren* cryptate displaying a sharp peak selectivity towards polyatomic ambidentate anions; a dicopper(II) poly-aza macrocycle suitable for binding the imidazolate anion and able to recognize any molecule bearing an imidazole residue; and a zinc(II) polyamine complex capable of signalling the recognition of aromatic carboxylates through the quenching of the fluorescence of an appended anthracene subunit. A common element for the metal containing receptors to be discussed in the following is that, in any case, the anion goes to occupy by its donor atom an axial position of a trigonal bipyramid coordination polyhedron. Mainly for steric reasons this geometrical arragement appears the most convenient for metal binding and recognitive purposes.

2. Anion bite recognized by a dicopper(II) cryptate.

A *bistren* framework can be used to generate an anion receptor operating through metal-ligand interactions [5]. In particular, each one of the two tetramine compartments of the octa-amine cage 3, L, in Figure 2 can host a Cu^{II} ion, to give the dimetallic cryptate 4. The space between the metal centres is large enough to accomodate a polyatomic anion. In particular, an ambidentate anion can position its donor atoms to complete the trigonal bipyramidal coordination polyhedron of each Cu^{II} ion. The two-step process is illustrated in Figure 2.

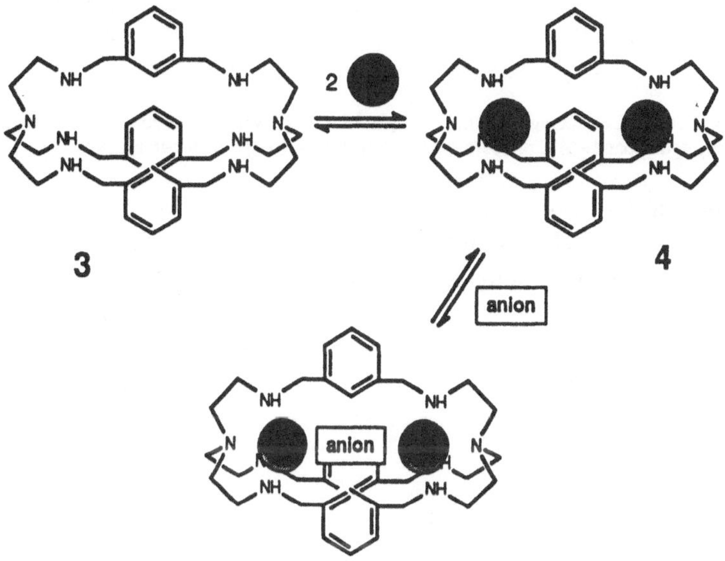

Figure 2. The generation of a receptor for polyatomic anions based on the metal-ligand interaction. The ambidentate anion is expected to position its donor atoms in an axial site of the trigonal bipyramidal coordination polyhedron of each Cu^{II} ion.

pH dependent equilibrium studies have shown that at pH=4, in a solution containing 1 eqv. of L and 2 eqv. of Cu^{2+}, the dimetallic complex species $[Cu^{II}_2L]^{4+}$ begins to form. At pH=5 $[Cu^{II}_2L(OH)]^{3+}$ appears, which at pH≥7 is the only species present at the equilibrium. The stereochemistry of such a complex can be only guessed. It is possible that the OH^- ion bridges the two metal centres, but the coordination by further water molecules cannot be excluded. Thus, a solution of $[Cu^{II}_2L(OH)]^{3+}$, adjusted at pH=8 with a 0.1 M triflic acid/morpholine buffer, was titrated with a standard solution of N_3^-. On anion addition, the pale blue solution took a bright green colour, while an absorption band centred at 410 nm formed. The family of spectra taken in the course of the titration is displayed in Figure 3.

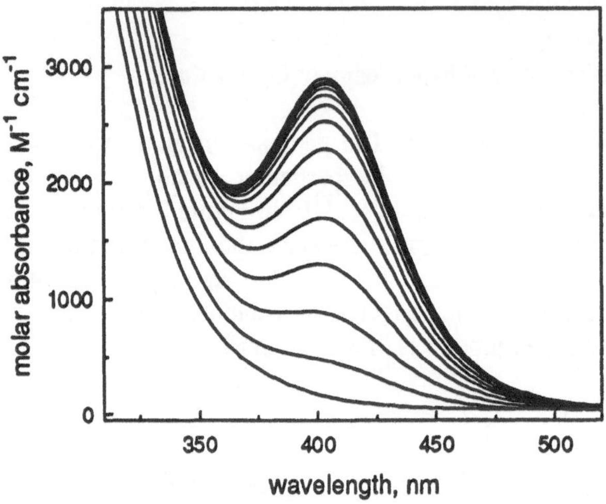

Figure 3. Spectra obtained during the titration of $[Cu^{II}_2L(OH)]^{3+}$ with N_3^- in an aqueous solution adjusted at pH=8. The absorption band at 410 nm corresponds to the $[Cu^{II}_2L(N_3)]^{3+}$ inclusion complex.

The plot of the absorbance at 400 nm vs. the number of eqv. of N_3^-, shown in Figure 4, clearly indicates the formation of a 1:1 cryptate/anion adduct. Non-linear least-squares analysis of titration data gave a logK value of 4.78 ± 0.05.

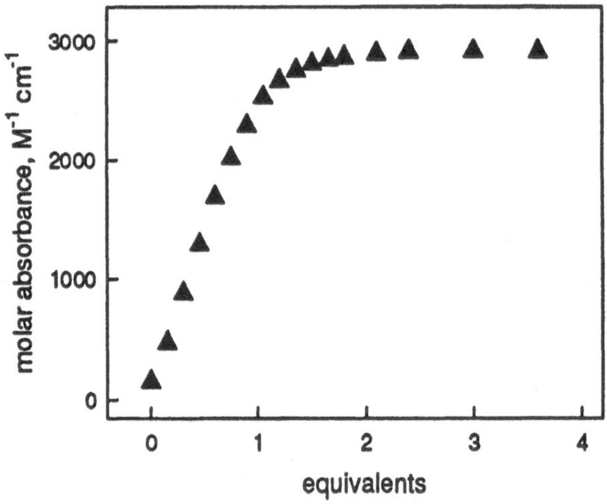

Figure 4. Spectrophotometric titration plot for the formation of the $[Cu^{II}_2L(N_3)]^{3+}$ inclusion complex. Data taken from the spectra in Figure 3.

It should be noted that the process taking place in solution is described by the following exchange equilibrium:

$$[Cu^{II}_2L(OH)]^{3+} + N_3^- = [Cu^{II}_2L(N_3)]^{3+} + OH^- \qquad (1)$$

The constant of equilibrium (1), K_{exch}, is related to the spectrophotometrically measured constant K through the relationship: $K = K_{exch}/[OH^-]$. Thus, K is a conditional constant and its value should decrease with increasing pH. In fact, spectrophotometric titration experiments carried out at pH=9 gave a logK value of 3.7.

The drastic modification of the *d-d* spectrum of the dicopper(II) cryptate on addition of N_3^- would suggest the occurence of a significant change in the coordination of the two Cu^{II} ions. Quite interestingly, on slow evaporation of the solution of the spectrophotometric titration, crystals of the $[Cu^{II}_2L(N_3)](CF_3SO_3)(ClO_4)_2$ salt, suitable for X-ray analysis, were obtained. Crystal structure determination showed that the N_3^- anion is co-linearly bound to the two metal centres and that each Cu^{II} ion experiences a regular trigonal bipyramidal stereochemical arrangement [6].

Spectrophotometric titrations were carried out with further polyatomic anions. In any case, anion addition produced a distinct colour change (to blue-green or green) and a well detectable modification of the absorption spectrum in the UV-vis region, which made it possible the determination of the following logK values (±0.05): NCO^-: 4.60; NCS^-: 2.95; NO_3^-: 2.70; HCO_3^-: 4.58; CH_3COO^-: 2.97; $HCOO^-$: 3.32; SO_4^{2-}: 3.26. The low value observed with the thiocyanate ion may be surprising, if one considers that NCS^- displays a much greater affinity than N_3^- and NCO^- towards Cu^{II} in a five-coordinate arrangement. Moreover, one would not expect the doubly negatively charged SO_4^{2-} to exhibit a lower logK value than the singly charged HCO_3^-. All this evidence indicates that another factor exists besides the energy of the coordinative bonds or the intensity of the electrostatic interactions. We suggest that this factor has a geometric nature and is related to the ease with which the anion places its donor atoms in the fifth coordination site of each Cu^{II} ion. In this sense, we considered for each anion the distance between the proximate donor atoms, which will be indicated as the *bite* length. In the case of the linear triatomic anions the bite length coincides with the distance between the two extreme atoms. As an example, bite lengths for the HCO_3^- and NCO^- anions are indicated in Figure 5.

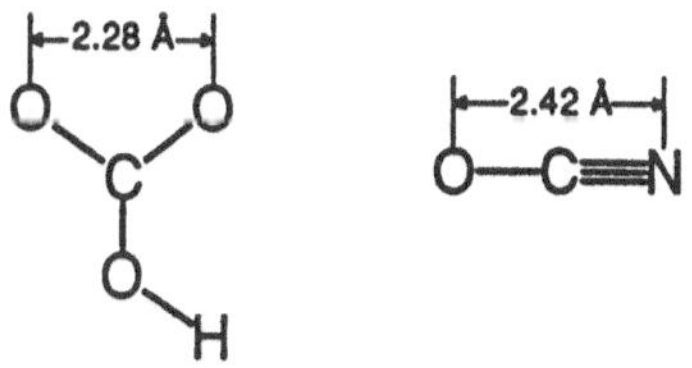

Figure 5. Bite length of anions bridging the two Cu^{II} centres in the cage. Distances were calculated through a semi-empirical method. These values were preferred to those taken from crystallographic studies, which can be altered by the electrostatic interactions with counterions.

Quite interestingly, a plot of logK vs bite length for all the investigated anions (as displayed in Figure 6) indicated a sharp peak selectivity, centred at about 2.3 Å.

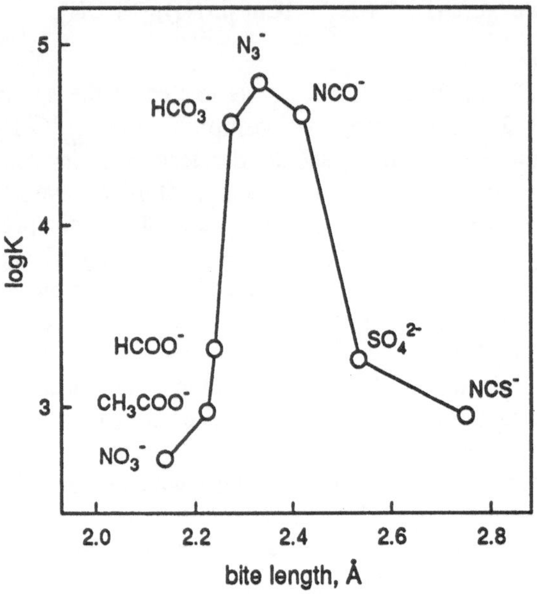

Figure 6. Selectivity in the anion incorporation by the dicopper(II) cryptate **4**. The receptor recognizes bite length. This diagram is reminiscent of the classical logK vs. ionic radius plots observed for the recognition of alkali metals by *crowns* and *cryptands*.

The diagram demonstrates that the linear N_3^- and NCO^- ions and the triangular HCO_3^- anion show the highest affinity as their bite fits well the distance between the two Cu^{II} 'fifth coordination sites' within the cryptate. On the other hand, the linear NCS^- ion has too long a bite and its inclusion should induce an endothermic rearrangement of the cryptate and of its carbon backbone. In a similar way, the triangular NO_3^- has too short a bite and should make the receptor cavity to contract, which causes an endothermic steric rearrangement again and makes log$K$ to decrease. Thus, it is the *co-linear fitting* that determines selectivity. This purely steric effect is dominating and overcomes effects related to coordinative tendencies and electrical charge. As a consequence, the usually strongly coordinating anion SCN^- gives a less stable adduct than other linear triatomic anions, due to its poor fitting. Moreover, the too large SO_4^{2-} anion is bound weakly, in spite of its double negative charge. In conclusion, *the receptor does not recognize the donor tendencies or shape, but simply recognizes the bite length of the anion.*

It also should be noted that, whereas *bistren* cryptands of type **3** are able to incorporate a number of *3d* metals, the role of copper(II) seems essential in view of its unique tendencies to give stable five-coordinate complexes. As a further advantage, due to the drastic modifications of the energy of Cu^{II} d levels following axial coordination, anion inclusion causes distinct colour and spectral changes.

3. Two Cu^{II} ions, pre-positioned in a *bisdien* macrocycle, recognize the imidazolate ion

The practical message is that any ambidentate anion whose donor atoms are about 2.3 Å distant from each other can be recognized by the dicopper(II) octamine cage described in the previous Section. A very special anion fulfilling such a requirement is imidazolate, 5.

Imidazole, ImH, $pK_A = 14.5$, is a weaker acid than H_2O and does not deprotonate in aqueous solution. However, in presence of two pre-positioned Cu^{II} ions, the ImH molecule simultaneously deprotonate and bridges the two metal centres, according to the the equilibrium:

$$[Cu^{II}_2]^{4+} + ImH = [Cu^{II}(Im^-)Cu^{II}]^{3+} \qquad (2)$$

Following the process (2), electrons extensively delocalize along the Cu^{II}-N-N-Cu^{II} system. It is this favourable enthalpic contribution that overcomes the endothermic imidazole deprotonation process. A basic requirement for reaction (2) to proceed is that the two Cu^{II} centres occupy fixed positions, in a rigid coordinative framework, at the correct distance.

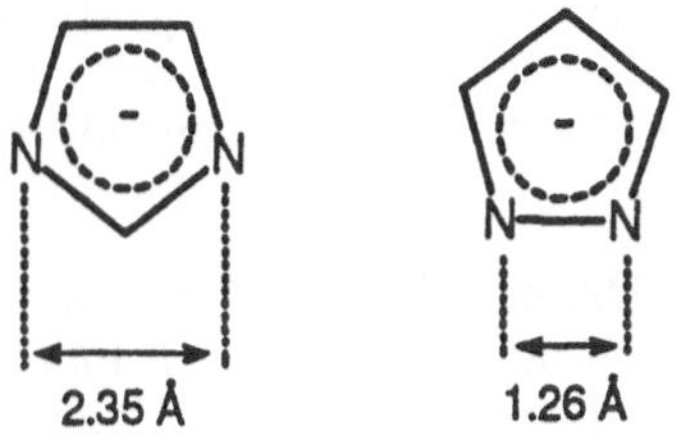

Figure 7. Bite lengths for imidazolate and pyrazolate anions.

In this context, we noted that the N-N distance in the imidazolate anion, as calculated from a semi-empirical method, is 2.35 Å (see Fig. 7) and we considered that the previously discussed dimetallic *bistren* cage, $[Cu^{II}L(OH)]^{3+}$, could be conveniently used for recognizing imidazole (through simultaneous deprotonation and encapsulation). Indeed, on addition of imidazole, the pale blue aqueous solution of $[Cu^{II}L(OH)]^{3+}$ took an intense blue colour, whereas the absorption band of the Cu^{II} poly-aza chromophore, centred at *ca.* 640 nm, shifted at 690 nm and increased its intensity. A rather intense absorption band at around 700 nm is typically observed on formation of Cu^{II}-imidazolate-Cu^{II} bridges. It should also be noted that the mixed-metal Cu^{II}-Zn^{II} complex of cage 3 hosting an imidazolate anion has been recently isolated as a crystalline solid and characterized through X-ray analysis [7]. The $[Cu^{II}Zn^{II}L(Im)]^{3+}$ complex species has been proposed as a SOD model and tested. All the above evidence would indicate $[Cu^{II}L(OH)]^{3+}$ as the ideal candidate for recognizing imidazole and, more interestingly, any molecule bearing an imidazole residue. However, a serious problem exists: incorporation of imidazole by the dicopper(II) cryptate is a thermodynamically favoured, but rather slow process. As an example, on addition of 1 eqv of imidazole to a solution of $[Cu^{II}L(OH)]^{3+}$ adjusted to pH=8, the band centred at 690 nm formed instantaneously, but took several hours to reach its limiting value. The corresponding plot of the absorbance at 690 nm vs. time is shown in Figure 8.

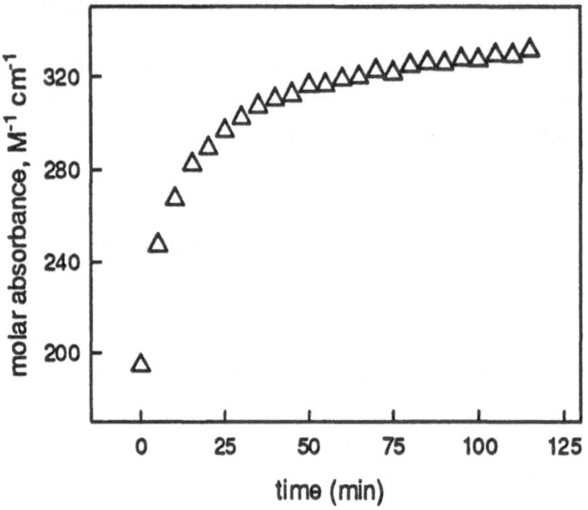

Figure 8. Variation of the intensity of the absorption band of $[Cu^{II}L(OH)]^{3+}$ after the addition of 1 eqv. of imidazole, in a solution buffered at pH=8. The $[Cu^{II}L(OH)]^{3+}$ + ImH = $[Cu^{II}L(Im)]^{3+}$ + H_2O equilibrium goes to completion in hours, probably due to the steric repulsive effects experienced by the imidazole molecule when entering the cage.

On the other hand, on addition of 1 eqv. N_3^- or of any other inorganic anion to the same dicopper(II) cage, the absorption band which forms reaches instantaneously its limiting value. This indicates that incorporation of imidazole by the dicopper(II) cage is not a kinetically simple process, probably due to steric effects related to the closed nature of the receptor. These steric effects do not operate during the incorporation of the small inorganic anions discussed in the previous Section.

The results described above prompted us to design a receptor still containing two pre-positioned Cu^{II} ions, but in an open coordinative domain. The synthetic approach is outlined in Figure 9 [8].

An octa-aza framework was chosen as a receptor in this case, too. A *bisdien* ring was chosen to position the two Cu^{II} ions. In particular, the two *dien* compartments were separated by *p*-xylyl spacers. Coordination of each metal centre (still aiming at a trigonal bipyramidal stereochemistry) was completed by a pendant 2-picolyl arm. The dicopper(II) receptor maintains the qualities of the dicopper(II) *bistren* cage (1. fixedness of the two metal centres, at the proper distance; 2. axial sites available for ambidentate anion coordination), plus one: an open arrangement for easy access of the substrate and fast recognition.

The coordinating behaviour of the octadentate ligand **5** in aqueous solution was investigated through pH titration experiments. In particular, least-squares treatment of titration data showed that in an aqueous solution containing 1 eqv of **5** and 2 eqvs of Cu^{II}, the dimetallic species begins to form at pH=6.5. At pH=9, 100% of the copper(II) is present as a dinuclear complex. An aqueous solution containing 1 eqv of 5 and 2 eqvs of Cu^{II} was adjusted at pH=9 with the morpholine/HNO_3 buffer and was titrated with a standard solution of imidazole. On titration, the original pale blue solution took a progressively more intense blue colour, whereas the absorption band at 690 nm formed and progressively increased its intensity ($\varepsilon = 320\ mol^{-1}\ dm^3\ cm^{-1}$).

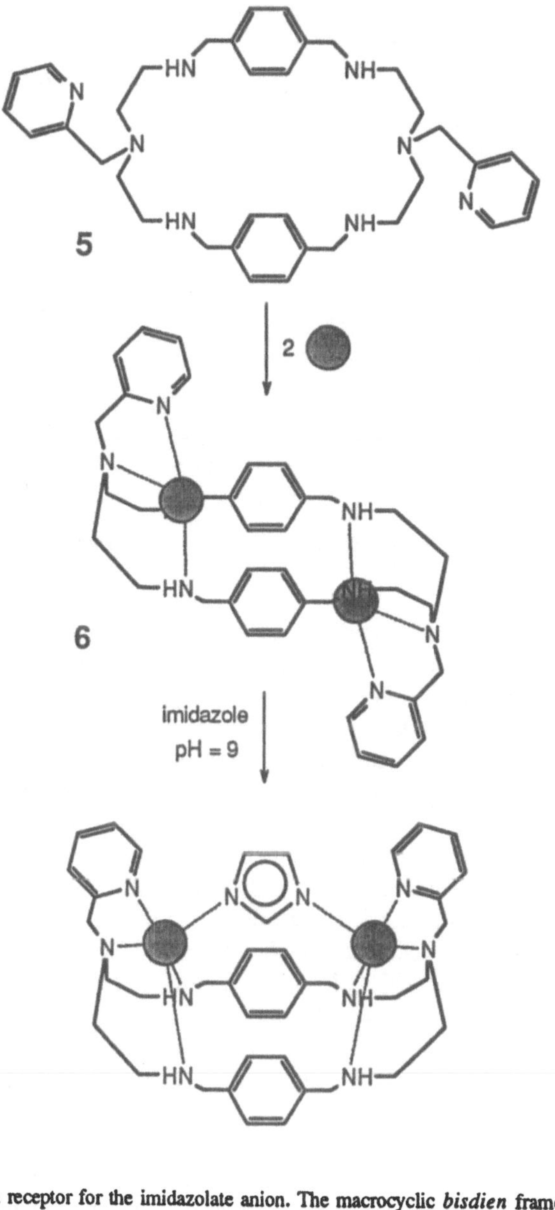

Figure 9. Design of a receptor for the imidazolate anion. The macrocyclic *bisdien* framework keeps the two Cu^{II} ions at the right distance; each appended picolyl arm completes the metal five-coordination; the open nature of the system allows an easy and fast binding of the incoming imidazole residue.

The profile of the spectrophotometric titration (see Figure 10) corresponds to the the formation of a 1:1 receptor/imidazolate adduct. Addition of even a large excess of the titrating solution does not cause any increase of the absorbance or alteration of the spectrum, indicating that further imidazole molecules are not able to remove the 2-picolyl pendant arms from their coordination sites and to modify the 1:1 stoicheiometry. On least-squares treatment of the titration data, a binding constant of 4.7±0.1 log units was calculated for the dimetallic complex/imidazolate adduct.

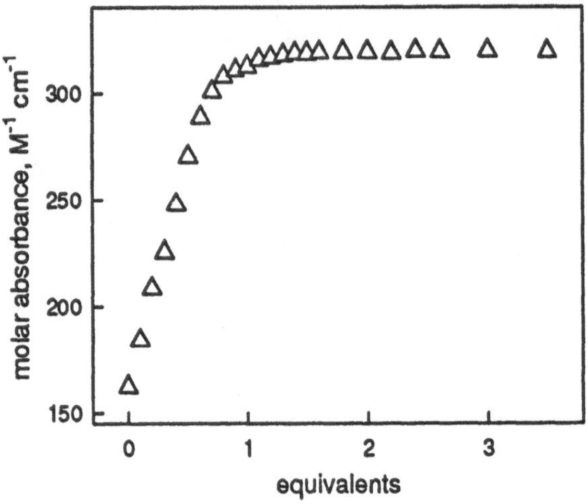

Figure 10. Profile of the spectrophotometric titration of the dimetallic receptor **6** with imidazole, in an aqueous solution buffered at pH=9.

It is worth noting that addition of pyrazole to a solution of **6** buffered at pH=9 did not induce any spectral change. Failure in coordination can be ascribed to the too short bite of the pyrazolate ion (see Fig. 7), which prevents bridging of the two Cu^{II} ions. Furthermore, we wanted to verify whether the dicopper(II) system **6** could act as a receptor for polyatomic inorganic anions, possibly displaying a bite selectivity as that shown by the dicopper(II) cage **4**. Anion binding constants were determined through spectrophotometric titration experiments in aqueous solutions buffered at pH=9. Their values are plotted vs. anion bite length in the diagram of Figure 11.

No selective binding behaviour exists and, in any case, no relationship is observed between logK and anion bite. Lack of selectivity can be ascribed to the more flexible nature of the macrocyclic receptor which allows the accomodation of anions of different size (e.g. NCO^- and NCS^-, logK = 3.2 and 2.9, respectively) without serious distortions of the poly-aza framework, as compared to the much more rigid macro-tricyclic dicopper(II) receptor **4**. Thus, the open and relatively flexible nature of **6** brings the advantage of fast binding of large anions as imidazolate, but involves a loss of selectivity.

However, receptor **6** may play a major role in the recognition of biologically relevant substrates. In fact, the selective affinity of the dimetallic receptor **6** is not restricted to imidazole, but can be extended to any molecule containing an imidazole residue. In particular, the development of the band at 690 nm and similar titration profiles were observed during titration experiments carried out in the same conditions with histamine (logK=4.3±0.1) and with L-histidine (logK=5.5±0.1). The greater binding constant of L-histidine compared to histamine is not surprising, if one considers that at pH=9 the aminoacid bears the negative charge of the carboxylate group (pK_{A1}=6.10), whereas its amine group is only partially protonated (pK_{A2}=9.18).

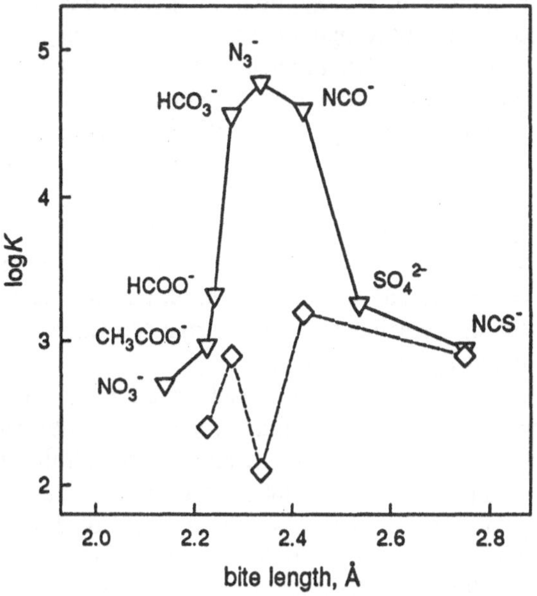

Figure 11. The 'open' dicopper(II) receptor **6** (diamonds) does not show any selectivity in binding inorganic polyatomic anions. Compare to the sharp peak selectivity of the dicopper(II) cryptate receptor **4** (triangles).

On the other hand, most histamine molecules (pK_A=9.83) exist as the ammonium form at pH=9. Thus, the more favourable binding of L-histidine compared to histamine (and to plain imidazole) seems to be ascribed to an additional favourable contribution of electrostatic nature. The strong binding to the dimetallic receptor **6** offers the unique opportunity to recognize L-histidine in presence of other natural aminoacids, as shown by competition experiments. In particular, addition of 1 eqv of one of the following representative aminoacids: L-glycine, L-proline, L-cysteine, L-valine, L-arginine, L-serine, L-tryptophan, to a blue solution containing 1 eqv of **6** and 1 eqv of L-histidine caused a decrease of the intensity of the band at 690 nm lower than 5%. This behaviour is consistent with the lower binding tendencies towards the dimetallic receptor **6** of the carboxylate group, i.e. the anion group of aminoacids capable of coordination. For instance, the acetate ion exhibits a rather low logK value (2.4±0.1). A binding constant of a similar value cannot ensure a serious competition with the imidazolate residue at pH=9. Thus, the open dicopper(II) receptor **6** recognises L-histidine in aqueous solution in presence of any other aminoacid.

4. The recognition of aromatic carboxylates by a Zn^{II} containing receptor is signalled through fluorescence quenching

The two metal containing systems **4** and **6** described in the previous Sections are good *receptors* for anions having a bite around 2.3 Å and for molecules bearing an imidazole residue, respectively. The question is whether **4** and **6** are also good *sensors*. Sensing is something more than recognition: it requires, in addition, signalling [9]. In particular, the system used for recognition should also be able to signal the occurrence of the receptor-substrate interaction through the sharp variation of a given property. In this sense, binding of N_3^- by **4** and of imidazole by **6** induces a colour change and the development of a distinct absorption band, which are well detectable down to a 10^{-4} M concentration. Other properties to be considered for signalling purposes are: the shift of an NMR line; the variation of an electrode potential; the modification of the intensity of a fluorescent emission. Fluorescence is a very advantageous property since: (i) it can be detected at an extremely low concentration ($\leq 10^{-7}$ M); (ii) there exist some well defined mechanisms for its control: electron transfer (ET) and energy transfer. In this Section, we will consider the design of a fluorescent sensor for anions, still based on the metal-ligand interaction.

The design of a sensor can be approached according to two different ways. In a traditionally designed sensor, the signalling property originates from the binding site itself and is drastically altered following the interaction with the substrate. Systems of this type are one-component sensors. Notice that **4** and **6** belong to this category, as the anion binding modifies the energy of the *d* levels of the Cu^{II} ions and causes detectable changes of colour and absorption spectrum. In second generation sensors, recognition and signalling have been physically separated. In fact, they consist of two covalently linked components: one component is devoted to substrate binding; the other one to signalling. Very importantly, a mechanism should exist for transferring the information about the occurred recognition from the binding subunit to the signalling component.

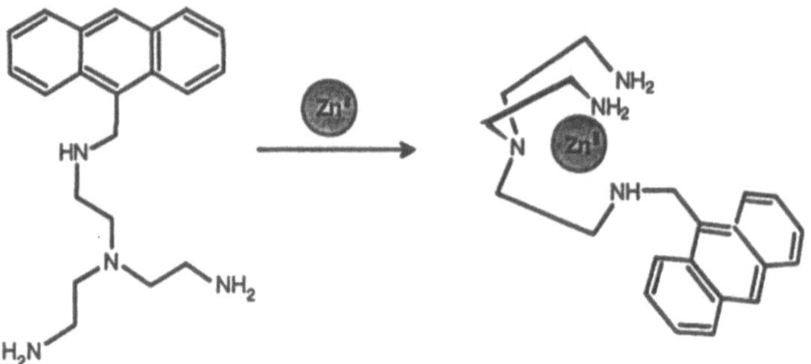

7

Figure 12. Design of a fluorescent sensor for anions. The anthracenyl substituted tripodal tetra-amine **7** coordinates Zn^{II} according to a trigonal bipyramidal stereochemistry: an axial site is available for anion binding. The anthracene fragment is expected to signal the occurrence of the receptor-substrate interaction through a modification of its fluorescent emission.

Figure 12 illustrates a two-component fluorescent sensor for carboxylate anions, which is obtained through the reaction of the *N*-anthracenyl substituted tripodal tetra-amine 7 with the Zn^{II} ion [10]. The Zn^{II} tetra-amine (*tren*) fragment represents the anion binding subunit. In particular, the metal leaves an axial site of the trigonal bipyramidal polyhedron available for anion coordination. The fluorescent subunit - the most classical light-emitting fragment of organic chemistry - is linked to the receptor component through a $-CH_2-$ group.

Spectrophotometric titration experiments showed that the $[Zn^{II}(7)]^{2+}$ system forms a stable 1:1 adduct with the benzoate ion in an ethanolic solution, as shown by significant modifications in the UV absorption spectrum. On the other hand, when the titration was carried out within the spectrofluorimetric cuvette, no change was observed in the characteristically structured emission band of the anthracene fragment, even after the addition of several eqvs of benzoate. Thus, the anthracene subunit witnesses in silence the anion binding to the Zn^{II} centre.

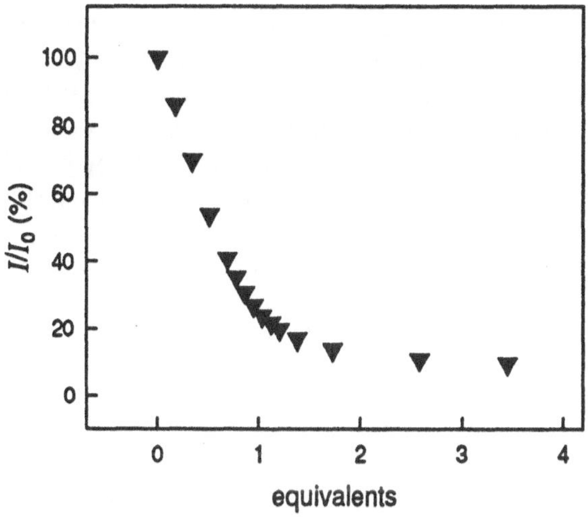

Figure 13. Profile of the spectrofluorimetric titration of $[Zn^{II}(7)]^+$ with the 4-*N,N*-dimethylaminebenzoate ion, in a methanolic solution. Fluorescence intensity, I_F, was taken at 415 nm.

However, on titration with 4-*N,N*-dimethylaminebenzoate, X^-, the intensity of anthracene emission band, I_F, progressively decreased until full quenching. The I_F vs. X^- eqvs plot, shown in Figure 13, indicated the formation of a 1:1 adduct, $[Zn^{II}(7)X]^+$ and non-linear least squares treatment of titration data gave a logK value of 5.45±0.03 for the anion complexation equilibrium: $[Zn^{II}(7)]^+ + X^- = [Zn^{II}(7)X]^+$. This time, the anthracene fragment receives the information on the occurrence of the recognition process and diligently exerts its signalling function. But how is the information transmitted? It is suggested that fluorescence quenching is due to an intramolecular ET process within the $[Zn^{II}(7)X]^+$ adduct, in particular from the Zn^{II} bound 4-*N,N*-dimethylaminobenzoate subunit, DMA, to the photo-excited anthracene fragment, *An. *N,N*-dimethylaniline is a well known electron donor and the free energy change $\Delta G°_{ET}$ associated to the DMA-to-*An ET process, calculated through the thermodynamic cycle in Figure 14, is distinctly negative (-0.4 eV) and accounts for its feasibility.

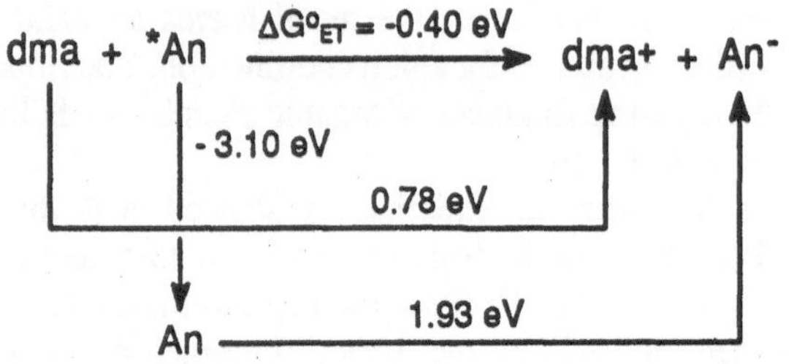

Figure 14. Thermodynamic cycle for the determination of the free energy change, $\Delta G°_{ET}$, associated to the intramolecular electron transfer from the Zn^{II} bound 4-*N,N*-dimethylaminobenzoate to the photoexcited proximate anthracene fragment. Relevant redox potential values refer to the individual components. E^{0-0} has ben obtained from the energy of the emission band of the $[Zn^{II}(7)]^{2+}$ system.

On the other hand, the genuine ET nature of *An quenching in the $[Zn^{II}(7)X]^{+}$ adduct has been confirmed by spectrofluorimetric studies at 77 K. It is known that immobilisation of the solvent molecules in a frozen matrix raises the energy of the ion pair, thus strongly disfavouring the ET process and restoring fluorescence [11]. It was observed that an ethanolic solution containing $[Zn^{II}(7)]^{2+}$ and an excess of 4-*N,N*-dimethylaminobenzoate, not emitting at room temperature, when frozen in liquid nitrogen, restores the fluorescence of the anthracene subunit, demonstrating the ET nature of the intramolecular quenching at room temperature. The mechanism of such a process is pictorially illustrated in Figure 15.

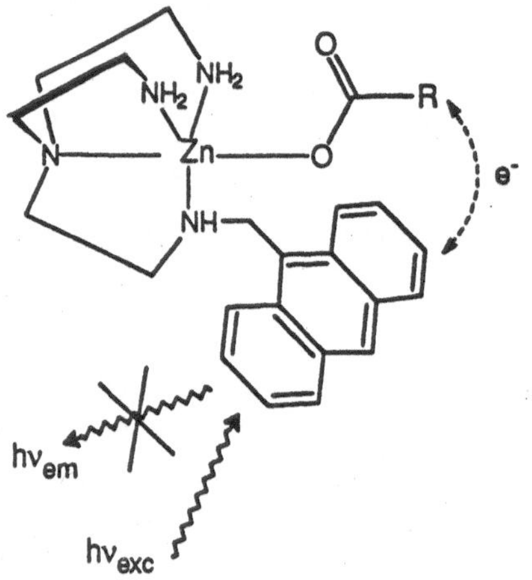

Figure 15. The electron transfer process responsible for fluorescence quenching of the anthracene subunit and for signalling the recognition of the carboxylate anion. R = 4-*N,N*-dimethylaminophenyl (Donor, D), 4-nitrophenyl (Acceptor, A), 9-anthracenyl (D or A), 1-ferrocenyl (D).

It should also be noted that quenching of the excited anthracene fragment of the $[Zn^{II}(7)]^{2+}$ receptor should not be restricted to 4-*N,N*-dimethylaminobenzoate, but should be feasible for any benzoate anion bearing either a *donor* or an *acceptor* group, provided that the thermodynamic requirements for the transfer of an electron *to* or *from* the photo-excited fluorophore are fulfilled. In particular, titration of $[Zn^{II}(7)]^{2+}$ with 4-nitrobenzoate ion in a methanolic solution induces the decrease and quenching of the fluorescence, according to a profile analogous to that displayed in Fig. 13. Nitrobenzene and its derivatives are classical acceptors. Thus, fluorescence quenching has to be ascribed to an intramolecular ET process from *An to the metal bound 4-nitrobenzoate anion. The negative value of the $\Delta G°_{ET}$ quantity (-1.0 eV) accounts for its thermodynamic feasibility. Quite interestingly, decrease and almost complete quenching of the fluorescence is also observed when $[Zn^{II}(7)]^{2+}$ is titrated with 9-anthracenoate anion. Notice that 9-anthracenoate itself is fluorescent. It is suggested that fluorescence quenching originates from an ET process between the two anthracene subunits of the supramolecular adduct: $*An + An = An^- + An^+$. Both the anthracenoate-to-anthracene and anthracene-to-anthracenoate electron transfer processes exhibit a slightly negative $\Delta G°_{ET}$ value (-0.2 eV). In any case, the observed revival of the fluorescence emission in an ethanolic solution frozen at 77 K demonstrated also for the present system the authentic nature of the electron transfer mechanism, responsible for the quenching of the photo-excited aromatic subunit.

The $[Zn^{II}(7)]^{2+}$ system is able to signal the binding of other carboxylate anions than benzoates, provided that they have definite donor or acceptor tendencies in a photophysical sense. For instance, fluorescence quenching is observed also on titration of $[Zn^{II}(7)]^{2+}$ with 1-ferrocenecarboxylate. Ferrocene and its derivatives are quite strong donors. In particular, the ferrocene to *An electron transfer is thermodynamically favoured: $\Delta G°_{ET} = -0.4$ eV. In this case too, the electron transfer nature of the quenching process has been confirmed by the fluorescence revival observed at 77 K.

5. Conclusions

It has been shown that metal-ligand interactions can be conveniently used for anion binding and recognition. In all the examples reported, the metal responsible for anion binding was hosted in a tetra-aza coordinative environment favourable to a trigonal bipyramid coordinative arrangement. In particular, one of the axial positions of the coordination polyhedron was left open for the binding of the envisaged anion. In the copper(II) containing receptors 3 and 5, the presence of two metal centres required ambidentate coordination and provided a stringent element of selectivity related to the anion bite length. The fluorescent sensor 7 contained only one metal and left open room for the approach of anions of varying size and shape. In this case, the selectivity is only determined by the energy of the metal/anion coordinative interaction, a not too selective parameter. However, a novel element of selectivity is introduced by the signal transduction mechanism: in particular, only the interactions with anions displaying distinctive electron donor or electron acceptor tendencies could be signalled and sensed. The Zn^{II} containing system 7 works well with carboxylate anions. Following a similar approach, more sophisticated and powerful systems can be designed for the recognition and fluorescent sensing of anions of varying nature and higher complexity.

6. References

1. C. J. Pedersen, *J. Am. Chem. Soc.*, 1967, *89*, 7017
2. B. Dietrich, J.-M. Lehn, J.-P. Sauvage, *Tetrahedron Letters*, 1969, *34*, 2885
3. F. P. Schimdtchen, *Angew. Chem., Int. Ed. Engl.*, 1977, *16*, 720
4. B. Dietrich, J. Guilhem, J.-M. Lehn, C. Pascard, E. Sonveaux, *Helv. Chim. Acta*, 1984, *67*, 91
5. L. Fabbrizzi, P. Pallavicini, A. Perotti, L. Parodi, and A. Taglietti *Inorg. Chim. Acta*, 1995, *238*, 5.
6. A. M. Manotti Lanfredi, F. Ugozzoli, L. Fabbrizzi, P. Pallavicini, L. Parodi, and A. Taglietti, manuscript in preparation
7. J.-L. Pierre, P.Chautemps, D. M. Refaif, C. G. Béguin, A. El-Marzouki, G. Serratrice, P. Rey, J. Laugier, *J. Chem. Soc., Chem. Comm.*, 1994, 1117.
8. L. Fabbrizzi, P. Pallavicini, L. Parodi, A. Perotti, A.Taglietti, *J. Chem. Soc., Chem. Comm.*, in press.
9. L. Fabbrizzi, A. Poggi, *Chem. Soc. Rev.*, 1995, *24*, 197.
10. G. De Santis, L. Fabbrizzi, M. Licchelli, A. Poggi, A. Taglietti, *Angew. Chem.*, in press
11. J.-P. Sauvage, J.-P. Collin, J.-C. Chambron, S. Guillerez, C. Coudret, V. Balzani, F. Barigelletti, L. De Cola, L. Flamigni, *Chem. Rev.*, 1994, *94*, 993.

NOBLE GAS ATOMS INSIDE FULLERENES

Martin Saunders, R. James Cross, Hugo A. Jiménez-Vázquez, Rinat Shimshi, Anthony Khong

Yale Chemistry Department - New Haven - CT 06520

Introduction

Host-guest chemistry is an important category of supramolecular chemistry and was the subject of many presentations at this workshop. The majority of host-guest systems which have been studied involve multiple, weak attractive interactions between the host and guest. The nature of these interactions and their specificity is a central issue in the study of these systems. Many of the host molecules are flexible in the sense that there are many conformers in equilibrium. The free energies and hence the populations of these conformers can be important. There may be conformers which are present in minor amount in the absence of the guest but which interact strongly with the guest. The host-guest complex may therefore have the host largely in these conformers. In order to investigate these host-guest systems experimentally, the equilibrium constant for association must normally be sufficiently large to result in substantial concentrations of the host-guest complex at attainable concentrations.

Fullerenes are cage compounds made of carbon with cavities large enough to contain most atoms.[1,2] Fullerenes containing atoms can be considered as host-guest systems, but differ from most other systems in several significant ways. The fullerenes are very rigid closed cages. In each case, there is only a single minimum-energy conformer. Its structure may be obtained experimentally or it can be predicted well using quantum mechanics or molecular mechanics calculations. The openings leading to the cavities inside are five and six-membered carbocyclic rings. Even the smallest atom (helium) requires very high energy to penetrate one of these rings. Therefore, if fullerenes with atoms inside are prepared, they are indefinately stable. For this reason the equilibrium constant for binding does not prevent one from detecting these substances and examining their physical properties in detail. The consequences of the non-bonded interactions of the atom inside within the cage can be investigated even if the equilibrium constant is unfavorable. If the atoms inside are noble gas atoms, the interactions are non-specific and

L. Echegoyen and A.E. Kaifer (eds.), Physical Supramolecular Chemistry, 449–457.

can be properly discussed as van der Waals attractive or repulsive interactions. In all cases, the noble gas atoms are constrained to remain close to the carbon atoms of the cage. While the term complex (borrowed from coordination chemistry) may be appropriate for the majority of host-guest systems where the equilibrium between guest in and guest out is fast, substances with atoms inside fullerenes can be better described as compounds. Compounds are defined (Handbook of Chemistry and Physics) as being: "substances containing more than one consitiuent element and having properties, on the whole, different from those which their constituents had as elementary substances. The composition of a given pure compound is perfectly definate, and is always the same no matter how that compound may have been formed."

Three questions which one might consider specifically concerning the noble gas fullerene compounds are:

1. How can these substances be prepared?
2. How can the noble gas inside be detected and its amount measured?
3. What properties of these substances will be affected by the host-guest interaction and how?

Answers to questions 1 and 2 will be discussed here and some preliminary information about 3 will be presented.

Helium Found in Commercial Fullerene

We began our consideration of this problem by considering that fullerenes containing helium might have already been prepared. The Krätschmer-Huffman procedure for making fullerenes[3] employs an arc between graphite electrodes in about 1/5 of an atmosphere of helium. The graphite is known to dissociate to carbon atoms in the hot center of the arc. These atoms associate and eventually form the fullerenes in cooler parts of the arc. If just before the cages finally closes, a helium atom happens to be inside, it will be trapped and will be inside the fullerene which is isolated from the soot which accumulates on the walls of the chamber.

Since the chances of a helium atom being trapped seemed small, we did not expect that very much would be incorporated. We considered and rejected a number of potential ways of detecting the helium since it did not seem likely that they would be sensitive enough to detect the helium-containing molecules in a huge excess of empty fullerene. If there were a means of getting the helium out, it could then be detected with high sensitivity by mass spectroscopy. We began a collaboration with Professor Robert Poreda in the Geology Department of the University of Rochester, who studies noble gases trapped in rocks. Professor Poreda degasses samples of the rock in an ultra-high vacuum chamber. The rock is then heated and the noble gases which come out are measured using a mass spectrometer designed to be particularly sensitive for them.

When Poreda studied a commercial sample of fullerene which we sent him, he found more than enough helium to readily measure.[4] About one in 880,000 molecules contained ^{4}He. We were surprised to learn that he could measure the amount of ^{3}He as well. It turned out to correspond to one atom in 10^{12} molecules of fullerene. The helium was released from the fullerene over a temperature range from 550 to 850° C. The rough rate data obtained by Poreda yielded an activation energy between 70 and 80 kcal/mol. The mechanism of pushing the helium through a ring, as mentioned above, had an estimated barrier of around 200 kcal/mol. We needed to think of a process with a much lower barrier. We considered the possiblity of breaking a carbon-carbon bond of the fullerene as a first step. This would open what we have called a "window" in the fullerene cage. The helium might then come out through this window. The cost of breaking the bond seemed likely to be in the vicinity of the experimental activation energy.

If this were the process for the escape of the helium, we considered that such window opening might occur in ordinary, empty fullerene as well. A noble gas atom might enter through this window. If, as seemed likely, the bond could close to regenerate the intact fullerene cage; then this would be a mechanism for incorporating noble gas atoms. We tested this idea by heating fullerene with ^{3}He (3 atmospheres at 600° C). Poreda found that the ^{3}He content increased by a factor of 5,000,000. This observation was quickly followed by experiments where fullerene was heated with neon and then with krypton. The noble gas was incorporated in these cases as well.[4]

High Pressure - High Temperature Incorporation of Noble Gases

All of these noble gases were incorporated at the level of one or two in a million molecules of fullerene. In order to investigate properties of the noble gas labeled fullerene, we needed a much higher level of incorporation. That obvious variable which we could alter was the pressure under which the labeling reaction was carried out. Through collaboration with Stanley Mroczkowski of the Yale Applied Physics Department, we were able to develop procedures for heating fullerenes with these gases at up to 3000 atmospheres. This resulted in increasing the fraction of incorporation of noble gas to the level of several per thousand. This was enough for detection using conventional mass spectroscopy of the intact fullerenes.[5]

With samples made this way, we have recently been able to study the rate of release of the noble gases using our own mass spectrometer.[6] We have discovered that when the fullerene is purified by sublimation in vacuum and never exposed to air, that it becomes much more stable to heating and also that the rate at which the noble gases come out decreases a great deal. We feel that this implies that the processes of incorporation and release, previously studied with less pure material, must involve some form of catalysis by the impurities. We have also found that the high temperature high pressure process which we use does not get us even close to equilibrium. This was shown by taking labeled material and subjecting it the labeling procedure again. The amount which is found inside virtually doubles and this increase continues through four succesive treatments. Clearly, a catalyst which could get us close to equilibrium in one run would be extremely valuable to us. We have tried several reasonable candidate substances but without success so far. Theoretical calculations based on force-field approaches also predict that the equilibrium amounts of incorporation should be far greater than we have yet achieved experimentally.[7]

^{3}He NMR Spectroscopy

One gas which we have incorporated was of particular interest. ^{3}He has a nucleus with spin=1/2 and a high gyromagnetic ratio. This makes it a superb NMR nucleus. We established a collaboration with Frank Anet at UCLA to look for the helium NMR spectrum for ^{3}He inside the fullerenes. With the material labeled at high pressure he was able to obtain spectra using a probe converted from the proton frequency.[8] The chemical shift of helium inside C_{60} was found 6.3 ppm upfield of the shift of free helium dissolved in the solution. This was a much greater shift upfield than had been predicted. Since we did not expect helium to chemically interact with the pi orbitals of the carbons, we concluded that the shift was telling us that the magnetic field inside C_{60} was less than the field outside. The pi orbitals of the fullerene were diamagnetically shielding the helium. We concluded that this change must be due to what is commonly described as an aromatic ring current in the pi orbitals. There had previously been discussion of the aromatic character of the fullerenes with the prediction that they are non-aromatic. The helium shift is highly relevant experimental information about this. We found that the helium shift inside C_{70} was 28.8 ppm upfield of dissolved helium. This was in accord with the prediction that C_{70} should be more aromatic than C_{60}.

Computer programs which use *ab initio* methods to theoretically calculate chemical shifts were applied to these cases and produced results which are reasonably close to those obtained.[9] It was obviously interesting to look at the higher fullerenes in the same way. The first question which we had concerned whether the high temperature, high pressure labeling method could be applied to these substances. We established collaborations with groups at Rice under the supervision of Billups and at the ETH under the supervision of Diederich in order to study these higher fullerenes.[10] In order to increase the sensitivity of our ^{3}He NMR spectroscopy, we purchased a probe from Nalorac specifically designed for ^{3}He. All the data which we have obtained indicates that the labeling goes at least as well with the higher fullerenes. Since the volume inside is greater, one might expect more to go in at equilibrium. We have no direct information about this, but we did obtain strong ^{3}He spectra. To our surprise, the helium shifts for C_{76}, and isomers of C_{78} and C_{84} were found to move downfield from those of C_{70}. The second surprise was that we saw peaks corresponding to five isomers of C_{78} and nine isomer of C_{84}. Only three isomers of each had previously been reported. The reason that helium spectroscopy is such a powerful technique for detecting these isomers is that each one yields a single, sharp peak in the helium NMR spectrum, well resolved from all the others. In contrast, ^{13}C NMR spectra of these isomers consist of many weak peaks which overlap with the peaks from the other isomers. It is only when the isomers can be fairly well purified and obtained in quantity that the CMR spectrum becomes a useful tool. Recently, Haddon has published a review of all the magnetic properties of the fullerenes including the helium NMR shifts.[11]

If helium NMR is so useful in detecting the isomers of the higher fullerenes, it was worth considering as a method for following chemical transformations of the fullerenes. Fullerenes as a class are quite reactive. A large number of reagents yield products, normally by adding to adjacent pairs of carbons.[12] We could readily obtain products which were labeled with sufficient ^3He for helium NMR by starting with ^3He labeled fullerene. We quickly found that all the products of such reactions show helium peaks which are substantially shifted from those of the parent fullerenes.[13] A common feature of fullerene chemistry is that such addition reactions do not inactivate the molecules for further attack by the same reagents. Thus, it is very easy to obtain bis, tris etc. addition. For addition in the most common fashion, which yields 1,2 addition across junctions between pairs of six-membered rings (6,6 adducts), there are eight isomeric bis-adducts, 44 tris-adducts, etc. Obtaining such complex mixtures of these isomers leads to difficulties in characterizing the products. Separating them by chromatography followed by taking CMR spectra for each one, yields information but is very time-consuming and difficult. In the cases which we have studied so far, all of these isomers appear to yield distinct, sharp peaks in the helium spectra which can be seen directly in the reaction mixtures without separation. We have preliminary information leading to hope that the position of the helium shift for a bis-isomer might be sufficient for specifically identifying its structure.[14]

"Hot Atom" Incorporation of Tritium Inside Fullerenes

In addition to the high temperature - high pressure labeling method described above, we have found another way of getting atoms inside fullerenes. We started by making the known lithium salt of C_{60}. On exposure to slow neutrons, ^6Li cleaves to an alpha particle and a tritium nucleus. The tritium, which is produced with a large amount of kinetic energy, loses this energy through collisions and eventually picks up an electron. It finally gets to an energy range where it is capable of incorporating into molecules. Tritium labeling through this process is known as "hot atom" chemistry. In the case of fullerenes, we expected that the tritium atoms would penetrate the cage and become trapped inside. When our lithium salt was irradiated in a reactor at Brookhaven, material was produced which had the properties of fullerene (followed it on a chromatograph column) but showed substantial tritium radioactivity.[15]

We expected that the tritium was inside, but we could not exclude the possibility that it was attached to the outside of the fullerene cage. In contrast with the noble gases, tritium can form strong bonds with carbon. We were able to demonstrate that the tritium was actually inside, by taking a sample which was standing in our lab for fifteen months and sending it to Poreda for measurement of the ^3He inside the fullerene. Tritium decays with a half-life of 12.3 years and the sole chemical product is ^3He. This beta decay occurs with a very small amount of energy release. Since momentum is conserved, most of this energy goes to the kinetic energy of the electron. The recoil energy of the helium was expected to be very small and we expected that this helium would not escape from its fullerene cage. Poreda found that the ^3He content was approximately that expected from the decay of the known amount of tritium over the time of storage. We concluded that the tritium is inside. These experiments still do not tell us whether the tritium is simply a free atom or is it bound to one of the carbons from the inside of the cage. Further experiments are in progress which we hope will give us more information on this question.

Having radioactively labeled fullerene may be useful in following it conveniently. Actually, our ^3He labeled fullerene is potentially a very much more sensitive tracer. Poreda's analytical procedure yields a blank for ^3He of only 10,000 atoms. If we took one miligram of the ^3He labeled fullerene which we have now, and dissolved it in the Yale 50 meter swimming pool and mixed it thoroughly, it would be possible for Poreda to accurately measure the ^3He obtained from only one mililiter of the water. Tritium radioactivity would be a much more convenient (although less sensitive) means of analysis for the fullerenes. Depending on whether the chemical properties of our tritium labeled fullerene are affected by the tritium, it might also be used for analyzing the reaction products of fullerenes.

Possible Applications of Noble Gas - Fullerene Compounds

The utility of ^3He NMR spectroscopy to follow fullerene chemistry is apparent from the results already obtained and we have been collaborating with a number of groups in applying it.[16,17] One special application which has not yet been achieved is to use it to monitor the escape of helium through a chemically opened gap in the cage. This could be observed by the decrease of the signal for ^{3}He in the fullerene, but doing it in a sealed NMR tube should also allow us to also see the helium which has escaped. Even a transitory opening of a large enough gap could be detected in this way.

While helium NMR requires about 10^{15} spins contributing to a single peak for detection; mass spectroscopy of the released ^{3}He, as mentioned above, is about eleven orders of magnitude more sensitive. It could thus serve as a highly specific, sensitive tracer for such substances as fullerene derivatives which are being tested for biological activity. For example, a drug candidate might be made from helium-labeled fullerene so that it could be later followed in different organs etc. If metabolism of the fullerenes occurs in such a manner so as to open the cage, this could be readily seen through release of the helium. If this occured in an animal, it would breath out the helium. It would be possible to measure it over time without sacrificing the animal.

Investigation of many interesting physical properties of these compounds is presently not possible because we are limited to materials with only a few noble gas atoms in a thousand molecules. Increasing the pressure under which the labeling process is carried out is one way to improve this situation, but this would require very expensive high pressure vessels and would only achieve labeling in proportion to the pressure. As mentioned, catalysts for the labeling should be able to help a great deal, if we can find them. A further possibility is the separation of "full" molecules from "empties". A partial separation of this kind has already been achieved by chromatography.[18]

References

1. a. E. Osawa, Kagaku (Kyoto) 25, 854-863 (1970).
   b. Z. Yoshida and E. Osawa, Aromaticity (Kagakudojin, Kyoto, 1971), p 174.

2. H.W. Kroto, J.R. Heath, S.C. O'Brien, R.F. Curl, R.E. Smalley, Nature, 318, 162 (1985)

3. W. Krätschmer, L.D. Lamb, K. Fostiropoulos, D.R. Huffman,

Nature, 347, 354 (1990)

4. M. Saunders, H.A. Jiménez-Vázquez, R.J. Cross, R.J. Poreda, Science 259, 1428, (1993).

5. M. Saunders, H.A. Jiménez-Vázquez, R.J. Cross, S. Mroczkowski, M.L. Gross, D.E. Giblin, R.J. Poreda, J. Am. Chem. Soc. 116, 2193 (1994).

6. R. Shimshi, A. Khong, H.A. Jiménez-Vázquez, R.J. Cross, and M. Saunders, in press, Tettr. Lett.

7. H.A. Jiménez-Vázquez and R.J. Cross, in press, J. Chem. Phys.

8. M. Saunders, H.A. Jiménez-Vázquez, R.J. Cross, S. Mroczkowski, D.I. Freedberg and F.A.L. Anet, Nature 367, 256 (1994).

9. M. Bühl, W. Thiel, H. Jiao, P. von R. Schleyer, M. Saunders and F.A.L. Anet, JACS, 116, 6005 (1994).

10. M. Saunders, H.A. Jiménez-Vázquez, R.J. Cross, W.E. Billups, C. Gesenberg, A. Gonzalez and W. Luo, R.C. Haddon, F. Diederich and A. Herrmann, JACS, 117, 9305 (1995).

11. R.C. Haddon, Nature 378, 249 (1995).

12. F. Diederich and C. Thilgen, Science 271, 317, (1996).

13. M. Saunders, H.A. Jiménez-Vázquez, B.W. Bangerter, R.J. Cross, S. Mroczkowski, D.I. Freedberg and F.A.L. Anet, JACS, 116, 3621 (1994).

14. unpublished results with D.I. Schuster and S.R. Wilson

15. H.A. Jiménez-Vázquez, R.J. Cross, M. Saunders and R.J. Poreda, Chemical Physics Letters, 229, 111, (1994).

16. M. Saunders, H.A. Jiménez-Vázquez, R.J. Cross, W.E. Billups, C. Gesenberg and D.J. McCord, Tetr. Lett., 1994 p.3869

17. A.B. Smith III, R. Strongin, L. Brard, W.J. Romanow, M. Saunders, H.A. Jiménez-Vázquez, R.J. Cross, JACS, 116, 10831 (1994).

18. M. Saunders, A. Khong, R. Shimshi, H.A. Jiménez-Vázquez and R.J. Cross, Chemical Physics Letters, 248, 127, (1996).

AUTHOR INDEX

SUBJECT INDEX

D

E

N

O

P

Q

R

S

T

U

V

W

X

Z